Advances in
ORGANOMETALLIC CHEMISTRY

VOLUME 33

Advances in Organometallic Chemistry

EDITED BY

F. G. A. STONE

DEPARTMENT OF CHEMISTRY
BAYLOR UNIVERSITY
WACO, TEXAS

ROBERT WEST

DEPARTMENT OF CHEMISTRY
UNIVERSITY OF WISCONSIN
MADISON, WISCONSIN

VOLUME 33

ACADEMIC PRESS, INC.
Harcourt Brace Jovanovich, Publishers
San Diego   New York   Boston
London   Sydney   Tokyo   Toronto

Academic Press, Inc.
San Diego, California 92101

United Kingdom Edition published by
ACADEMIC PRESS LIMITED
24-28 Oval Road, London NW1 7DX

Library of Congress Catalog Card Number: 64-16030

ISBN   0-12-031133-X   (alk. paper)

PRINTED IN THE UNITED STATES OF AMERICA
91 92 93 94   9 8 7 6 5 4 3 2 1

Contents

Boron Atoms in Transition Metal Clusters

CATHERINE E. HOUSECROFT

Organometallic Ions and Ion Pairs

JAY K. KOCHI and T. MICHAEL BOCKMAN

Structure and Bonding in the Parent Hydrides and Multiply Bonded Silicon and Germanium Compounds: from MH_n to $R_2M{=}M'R_2$ and $RM{\equiv}M'R$

ROGER S. GREV

Organotin Heterocycles

KIERAN C. MOLLOY

Halogenoalkyl Complexes of Transition Metals

HOLGER B. FRIEDRICH and JOHN R. MOSS

Bulky or Supracyclopentadienyl Derivatives in Organometallic Chemistry

CHRISTOPH JANIAK and HERBERT SCHUMANN

ADVANCES IN ORGANOMETALLIC CHEMISTRY, VOL. 33

Boron Atoms in Transition Metal Clusters

CATHERINE E. HOUSECROFT

University Chemical Laboratory
Cambridge CB2 1EW, England

I

INTRODUCTION

A plethora of cluster compounds exists in which interactions between transition metal (M) and main group (E) atoms are exhibited. For most main group elements, the number of metal atoms exceeds the number of main group atoms in the $M_x E_y$ cluster core. On the other hand, metalloborane cluster chemistry continues to be dominated by boron-rich systems ($1,2$). Significantly, though, the number of $M_x B_y$ clusters in which $x > y$ has grown significantly during the last few years ($3-5$). The aim of this article is to survey the methods of synthesis, the solution and solid-state structural properties, and the reactivity of transition metal cluster compounds containing small borane fragments or single (naked) boron atoms. The relevance of such species both to isoelectronic organometallic analogs and to solid-state metal borides has been made in two recent reviews ($6,7$). Because one of the most useful tools for routine characterization of boron-containing transition metal clusters is multinuclear NMR spectroscopy ($4,8,9$), an appendix of spectral details for compounds discussed in the text is provided at the end of the article, along with an indication as to whether

1

crystallographic data are available (Appendix I). Detailed structural information, where available, is provided in Appendix II.

II

CLUSTERS BASED ON HOMONUCLEAR DIMETAL UNITS

Each of the dimetallic units $(Cp^*Nb)_2$, $(Cp^*Ta)_2$, $Mn_2(CO)_6$, and $Fe_2(CO)_6$ (where Cp^* is $\eta^5\text{-}C_5Me_5$) supports one or two B_2H_6 moieties in the metalloboranes $Cp^*_2Nb_2(B_2H_6)_2$ (*10*), $Cp^*_2Ta_2(B_2H_6)_2$ (*11*), $Cp^*_2Ta_2(\mu\text{-}X)_2(B_2H_6)$ (X = Cl, Br) (*11*), $HMn_2(CO)_6(B_2H_6)Mn(CO)_4$ (*12,13*), and $Fe_2(CO)_6B_2H_6$ (*14–16*). One significant feature of the coordination mode of the B_2H_6 group to a dimetallic fragment is the tendency toward an ethanelike structure in contrast to the doubly hydrogen-bridged structure present in free diborane. $Cp^*_2Ta_2(\mu\text{-}X)_2(B_2H_6)$ (X = Cl, Br) is formed in the reaction of $Cp^*_2Ta_2(\mu\text{-}X)_4$ with $LiBH_4$ and therefore represents an example of transition metal-mediated homologation of monoborane. Further reaction leads to $Cp^*_2Ta_2(B_2H_6)_2$ and, presumably, parallels the formation of $Cp^*_2Nb_2(B_2H_6)_2$ (*10*). The structure of $Cp^*_2Ta_2(\mu\text{-}Br)_2(B_2H_6)$ (**1**) has been determined (*11*) and illustrates an asymmetrical

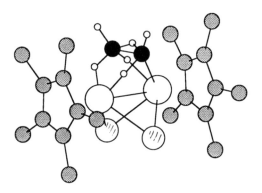

(**1**) $Cp^*_2Ta_2(\mu\text{-}Br)_2(B_2H_6)$

bonding mode for the B_2H_6 ligand. NMR spectroscopic data for (**1**) are consistent with the presence of two isomers. The major isomer is proposed to have a structure equivalent to that in the solid state, and the minor one exhibits a symmetrically bound B_2H_6 ligand with two terminal B—H and four bridging Ta—H—B interactions. The reported high-field ^{1}H-NMR resonances corresponding to B—H—B and Ta—H—B protons in the mixture of isomers are at δ − 4.1, − 7.1, and − 7.6.

The ferraborane $Fe_2(CO)_6B_2H_6$ (2) may be prepared by the copyrolysis of B_5H_9 and $Fe(CO)_5$ (15) or by the reaction of $Fe(CO)_5$ with $BH_3 \cdot THF$ (THF is tetrahydrofuran) and $Li[BHEt_3]$ followed by protonation (16). [The species generated in solution when $Fe(CO)_5$ is treated with $Li[BHEt_3]$ have recently been studied in detail by the use of infrared spectroscopy (17).] An improved route starting from $Fe(CO)_4SMe_2$ and $BH_3 \cdot SMe_2$ and giving a 25% yield of 2 has now been reported (18). At room temperature, cluster 2 is a yellow-brown liquid and is extremely air sensitive. The room temperature NMR spectroscopic properties of 2 suggest that the diborane unit possesses an ethanelike structure and bridges the diiron unit via four Fe—H—B interactions (15); the electronic structure of a symmetrical species has been compared to that of $Fe_2(CO)_6S_2$ (14). However, ^{1}H-NMR data at $-100°C$ confirm that the apparent C_{2v} structure is actually a consequence of a fluxional process and that the static structure of 2 (Fig. 1a) exhibits the same asymmetrical B_2H_6 unit as observed in 1 (16). Mössbauer spectroscopy has confirmed the presence of two inequivalent iron sites in 2. Compound 2 readily deprotonates via loss of an Fe—H—B proton, and the two different *endo*-hydrogen environments in 2 lead to the possibility of two isomers for the conjugate base. Solution spectroscopic data do not unambiguously provide an answer to the isomer question, but the Mössbauer spectrum of $PPN[Fe_2(CO)_6B_2H_5]$ is consistent with the presence of both the isomers shown in Fig. 1b (16).

The Fe_2B_2 cores of both 2 and its anion are hydrogen rich and as such are suitable precursors for cluster expansion reactions. Such reactivity has been exploited in the formation of $HFe_4(CO)_{12}BH_2$ ($\sim 10\%$ yield) by reaction of 2 with $Fe_2(CO)_9$ and of $[HFe_4(CO)_{12}BH]^-$ ($\sim 10\%$ yield) by reaction of $[Fe_2(CO)_6B_2H_5]^-$ with $Fe_2(CO)_9$. The anion reacts more rapidly than does the neutral ferraborane; both reactions are easily monitored by ^{11}B-NMR spectroscopy (16) (see also Section IV).

The red compound $HMn_2(CO)_6(B_2H_6)Mn(CO)_4$ exhibits an interesting bridging mode for the diborane ligand. The latter uses all six of its hydro-

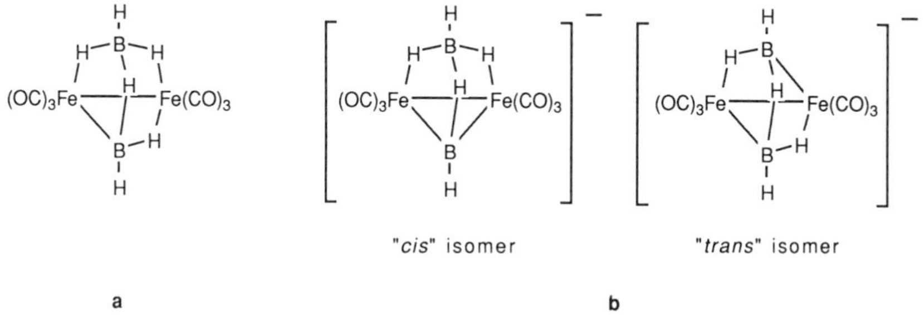

"*cis*" isomer   "*trans*" isomer

a   b

FIG. 1. Proposed structures of (a) $Fe_2(CO)_6B_2H_6$ (2) and (b) the two isomers of $[Fe_2(CO)_6B_2H_5]^-$.

gen atoms in Mn—H—B interactions and thereby links a dimanganese-hexacarbonyl fragment and a manganese tetracarbonyl unit ($12,13$). The structure has been confirmed crystallographically (13); in the ^{1}H-NMR spectrum, a resonance at $\delta - 19$ is consistent with the presence of an Mn—H—Mn hydride, but no signals for the boron-associated protons have been reported.

A dicobalt unit features in each of the two open clusters $Cp_2Co_2S_2B_2H_2$ (3) (19) and $Cp_2Co_2(\mu\text{-PPh}_2)B_2H_5$ (4) where Cp is $\eta^5\text{-}C_5H_5$ (20). Purple 3 is one product of the reaction of cobalt vapor, B_5H_9, and C_5H_6 with either H_2S or SCO. The formula $Cp_2Co_2S_2B_2H_2$ is consistent with a *nido* cage based on a pentagonal bipyramid, and two isomers (Fig. 2a) are possible each of which possesses a single boron environment as required by the ^{11}B-NMR spectral data for 3. An X-ray diffraction study confirms a cage in which an equatorial site of the bipyramid is vacated; this places the two cobalt atoms in apical sites and thus allows 3 to be described in terms of a triple-decker sandwich compound, the central section of which is an open

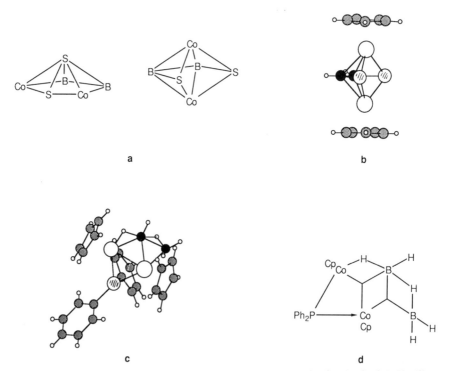

a

b

c

d

FIG. 2. Dicobaltaboranes: (a) possible $Co_2S_2B_2$ core geometries for $Cp_2Co_2S_2B_2H_2$ (3); (b) crystallographically determined structure of 3; (c) crystallographically determined structure of $Cp_2Co_2(\mu\text{-PPh}_2)B_2H_5$ (4); (d) valence bond representation of 4.

rather than the more usual closed ring (Fig. 2b) (*19*). A series of cobalta-boranes (see also Sections III and IV) has been derived from the reaction of $BH_3 \cdot THF$ with $CpCo(PPh_3)_2$. The Lewis acidity of BH_3 allows the latter to abstract triphenylphosphine from $CpCo(PPh_3)_2$, thus creating an unsaturated cobalt center. Cluster assembly follows to give the air-stable, dark red **4** as one product (Fig. 2c). The B_2H_5 ligand is bound in a manner described as being analogous to that of a $\sigma - \pi$ vinyl group, but **4** can also be considered to be an analog of *arachno*-B_4H_{10}, with the valence bond representation shown in Fig. 2d thought to be an important contributor to the cobalt–borane bonding scheme (*20*).

III

CLUSTERS BASED ON HOMONUCLEAR TRIMETAL UNITS

The M_3B core is the most commonly featured basic unit among known metal-rich metalloborane clusters. Before considering this relatively large group of compounds, it is worth mentioning a series of adducts of type $(RO)X_2B \cdot L$ in which RO is a trimetal carbonyl cluster or carbonyl cluster anion and X is H or halogen. Members of the series include $Co_3(CO)_9(\mu_3\text{-}COBH_2 \cdot NR_3)$ (R = Me or Et) (*21*), $Co_3(CO)_9(\mu_3\text{-}COBX_2 \cdot NEt_3)$ (X = F, Cl, Br, or I) (*22–24*), and $HRu_3(CO)_{10}(\mu\text{-}COBH_2 \cdot NMe_3)$ (*25*). The complexes $Co_3(CO)_9(\mu_3\text{-}COBH_2 \cdot NR_3)$ and $Co_3(CO)_9(\mu_3\text{-}COBX_2 \cdot NEt_3)$ (X = Cl, Br, or I) are prepared by the reaction of $Co_2(CO)_8$ and $R_3N \cdot BH_3$ or $Et_3N \cdot BX_3$, respectively. Fluorination of $Co_3(CO)_9(\mu_3\text{-}COBBr_2 \cdot NEt_3)$ using $AgBF_4$ leads to $Co_3(CO)_9(\mu_3\text{-}COBF_2 \cdot NEt_3)$ (*22*).

$Co_3(CO)_9(\mu_3\text{-}COBH_2 \cdot NEt_3)$ (*21*), $Co_3(CO)_9(\mu_3\text{-}COBCl_2 \cdot NEt_3)$ (*23*), and $Co_3(CO)_9(\mu_3\text{-}COBBr_2 \cdot NEt_3)$ (*24*) have been structurally characterized, and, in keeping with the Lewis acidity of the borane group, each shows an elongated C—O bond (1.28, 1.33, and 1.31 Å, respectively) for the boron-attached carbonyl group. A related complex is $HRu_3(CO)_{10}(\mu\text{-}COBH_2 \cdot NMe_3)$ (**5**). It is formed in the reaction of $Ru_3(CO)_{12-x}(NCMe)_x$ (x = 1 or 2) with $BH_3 \cdot NMe_3$ (or $BH_3 \cdot THF$ in the presence of NMe_3) (*25*) (see also Section VI). Here the bridging carbonyl group is 1.262 Å in length. The geometry of the $HRu_3(CO)_{10}(\mu\text{-}CO)$ fragment in **5** may be compared to that in the free anion $[HRu_3(CO)_{10}(\mu\text{-}CO)]^-$. Few differences are observed, but one change is significant, namely, the internal dihedral angle between the Ru_3 and $Ru_2(\mu\text{-}CO)$ planes decreases from 102.9° to 96.7° in response to the withdrawal by the borane ligand of electron density from the μ-CO π^* orbital, which is also an important Ru—$(\mu\text{-}CO)$—Ru bridgehead bonding molecular orbital (*25*).

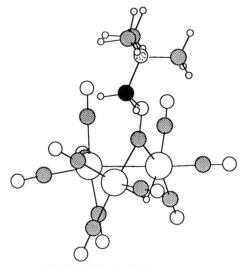

(5) HRu$_3$(CO)$_{10}$(μ-COBH$_2$·NMe$_3$)

Assembly of a tetrahedral Fe$_3$B core is achieved in the reaction of Na[Fe(CO)$_4$C(O)Me] with BH$_3$THF followed by protonation; red-orange HFe$_3$(CO)$_9$BH$_4$ (6) is formed in 5% yield and has been crystallographically

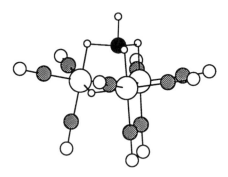

(6) HFe$_3$(CO)$_9$BH$_4$

characterized (26,27). The condensation of Fe(CO)$_3$ and BH$_3$ fragments takes place to give 6 in approximately 50% yield when Fe(CO)$_3$(η^2-cis-cyclooctene) is combined with BH$_3$·L (L = SMe$_2$ or THF). This reaction has been studied in detail, and a mechanism for the formation of 6 has been proposed; competitive pathways give 2 as well as HFe$_4$(CO)$_{12}$BH$_2$ (see Section IV) (18). The yellow ruthenaborane (7), which is an analog of 6, is prepared in 10% yield from $Ru_3(CO)_{12}$ with $BH_3 \cdot THF$ and $Li[BHEt_3]$ followed by treatment with acid (28).

In cluster 6, the $Fe_3(CO)_9$ platform appears to support a μ_3-borohydride ligand, but NMR spectroscopic data suggest that coordination to the transition metal triangle destroys the bonding characteristics usually associated with a ligated $[BH_4]^-$ anion. In borohydride complexes in general, $M-H-B$ and $B-H_{terminal}$ hydrogen atoms interconvert in solution (29). In 6, the terminal (exo) hydrogen atom remains uninvolved in a fluxional process that renders the metal-associated (endo) hydrogen atoms equivalent in solution. Compound 7 has not been characterized by X-ray diffraction, but similarities between the solution NMR spectral properties of 6 and 7 at elevated temperatures imply similarities in structure (28). Interestingly, however, unlike the case for the triferraborane, freezing out the fluxional process for the triruthenaborane results in the observation of two isomeric forms which are equally populated in CD_2Cl_2 solution at room temperature (28). The two isomers of 7 are shown in Fig. 3; the first is isostructural with 6, and the second differs only in the placement of one endo-hydrogen atom. Thus, for 6, the three $Fe-H-B$, one $Fe-H-Fe$, and two $Fe-Fe$ interactions must be favored over any other combination, whereas in 7 the energy difference between three $Ru-H-B$, one $Ru-H-Ru$, and two $Ru-Ru$ interactions and one $Ru-B$, one $Ru-Ru$, two $Ru-H-B$, and two $Ru-H-Ru$ interactions is, presumably, small.

The electronic structure of 6 has been probed by using the Fenske–Hall quantum chemical method, and the factors that influence the distribution of endo-hydrogen atoms have been examined (30). The preference for $M-H-B$ versus $M-H-M$ interactions may also be rationalized in

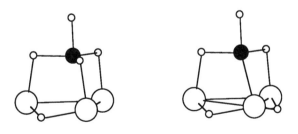

$\bigcirc$ = $Ru(CO)_3$

FIG. 3.   Proposed structures of the two isomers of $Ru_3(CO)_9BH_5$ (7).

terms of differences in the electronegativites of metal and boron atoms, and this argument has been extended to cover clusters M_xE_y in which M is Fe, Ru, or Os, x is 3, E is B, C, or N, and y is 1 or M is Fe or Ru, x is 4, E is B, C, or N, and y is 1 as well as other related compounds (31).

Both metalloboranes 6 and 7 are readily deprotonated to give $[M_3(CO)_9BH_4]^-$ (M = Fe or Ru) in which a low barrier to *endo*-hydrogen mobility is observed (27,32). The NMR spectral signatures of $[Fe_3(CO)_9BH_4]^-$ and $[Ru_3(CO)_9BH_4]^-$ are similar with a single resonance in the ^{11}B-NMR spectrum ($\delta + 6.2$ for M = Fe and $\delta + 22.5$ for M = Ru) and, in addition to a signal for the terminal proton, a broad high-field signal in the ^{1}H-NMR spectrum ($\delta - 13.1$ for M = Fe and $\delta - 12.1$ for M = Ru) that persists down to $-90°$ in both cases. Each ^{1}H-NMR spectrum is consistent with facile exchange of M—H—M and M—H—B *endo*-hydrogen atoms; the *exo*-hydrogen atom remains uninvolved in the process, and when M is Fe, $J_{Hexo-Hendo}$ is 20 Hz. Mössbauer spectral data for $[Fe_3(CO)_9BH_4]^-$ confirm two different iron sites, and this indicates a structural formulation of $[HFe_3(CO)_9BH_3]^-$, that is, the structure shown in Fig. 4. Thus, deprotonation of 6 occurs by loss of an Fe—H—B proton rather than destroying the Fe—H—Fe interaction (27).

Members of a series of clusters each with an Fe_3B core and structurally related to 6 have been spectroscopically characterized. $HFe_3(CO)_9BH_3Me$ and $HFe_3(CO)_9BH_3Et$ are both formed in the reaction of $Fe(CO)_5$ with $BH_3 \cdot THF$ and $Li[BHEt_3]$ (27). Both are orange, air-sensitive compounds and, in solution, exhibit a fluxional behavior reminiscent of 6 but with a higher activation barrier (4,27). Replacement of two *endo*-hydrogen atoms in 6 by an additional CO ligand gives $HFe_3(CO)_{10}BH_2$; this slightly air-sensitive, brown ferraborane is prepared in a similar manner to 6 but in the presence of $Fe(CO)_5$ and at 70°C (33). $HFe_3(CO)_{10}BH_2$ has an osmium analog, although in this case migration of CO to the boron atom has formally occurred (Fig. 5); $H_3Os_3(CO)_9BCO$ (8) is prepared as a yellow, air-stable, sublimable (60°C in vacuum) solid by the hydroboration of $H_2Os_3(CO)_{10}$ in the presence of $BH_3 \cdot NEt_3$ (34). The aminoborane plays an essential role in the mechanism of metalloborane formation. If NEt_3 is

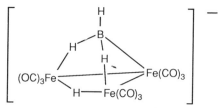

FIG. 4. Proposed structure of $[HFe_3(CO)_9BH_3]^-$.

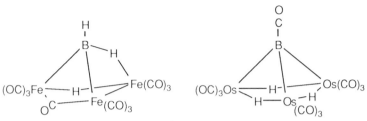

FIG. 5. Comparison of the structures of the isoelectronic (valence electrons only) molecules $HFe_3(CO)_{10}BH_2$ (proposed) and $H_3Os_3(CO)_9BCO$ (8) (determined).

replaced by THF, the trimeric $\{H_3Os_3(CO)_9CO\}_3B_3O_3$ (in which there are no Os—B bonds) forms in place of **8** (*34,35*). In practice, there appears to be no tendency for $HFe_3(CO)_{10}BH_2$ to isomerize to $H_3Fe_3(CO)_9BCO$ nor for $H_3Os_3(CO)_9BCO$ to convert to $HOs_3(CO)_{10}BH_2$. Indeed, interestingly, no osmaboranes with a vertex BH group have yet been reported.

Deprotonation of $HFe_3(CO)_{10}BH_2$ occurs readily and by loss of the Fe—H—Fe proton (*33*); this is in contrast to the case described above for $HFe_3(CO)_9BH_4$. The anion $[Fe_3(CO)_{10}BH_2]^-$ may be prepared directly from $[HFe_3(CO)_9BH_3]^-$ in approximately 80% yield by passing a stream of CO gas (1 atm) for 40 hours through a toluene solution of $[PPN][HFe_3(CO)_9BH_3]$ {where PPN is $[(PPh_3)_2N]^+$ cation} at 45°C (*36*). Other reactions of $[HFe_3(CO)_9BH_3]^-$ with Lewis bases have been studied and illustrate an interesting competition between the boron and iron atoms for the attention of the Lewis base. The nature of the base is clearly important; for example, amines should, and do, prefer to attack the borane moiety. Thus, the reaction of $[HFe_3(CO)_9BH_3]^-$ with NEt_3 decaps the cluster, yielding $Et_3N \cdot BH_3$ (*36*). With $PPhMe_2$, two reaction paths are possible, and the ultimate products depend on the concentration of the phosphine. At low $[PPhMe_2]$, the product is $[Fe_3(CO)_9(PPhMe_2)BH_2]^-$, but note that loss of H_2 has occurred; the reaction does *not* follow a simple phosphine-for-carbonyl substitution pathway. As the concentration of $PPhMe_2$ is increased, cluster degradation competes with cluster substitution, and the products are either (1) $[HFe_3(CO)_9(PPhMe_2)_2]^-$ and $Me_2PhP \cdot BH_3$ or (2) $[Fe_2(CO)_6(PPhMe_2)BH_4]^-$ and $Fe(CO)_3(PPhMe_2)_2$. The kinetics of the system have been investigated, and results are consistent with an associative mechanism, the rate being first order in both cluster anion and $PPhMe_2$ (*36*).

Cluster expansion reactions for which the building blocks are $[HFe_3(CO)_9BH_3]^-$, $[HRu_3(CO)_9BH_3]^-$, $HRu_3(CO)_9BH_4$, or $H_2Ru_3(CO)_9BH_3$ generally lead to tetrametal clusters, although one notable exception is the photolysis of a mixture of the isomers $HRu_3(CO)_9BH_4$ and

"Fe(CO)₃"

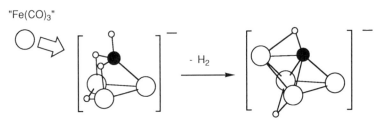

FIG. 6. Schematic representation of the transformation of [HFe₃(CO)₉BH₃]⁻ to [HFe₄(CO)₁₂BH]⁻; the source of the "Fe(CO)₃" fragment is Fe₂(CO)₉.

$H_2Ru_3(CO)_9BH_3$ (see Section VI). The first report of a rational transition metal-rich metalloborane cluster expansion involved the transformation of $[HFe_3(CO)_9BH_3]^-$ to $[HFe_4(CO)_{12}BH]^-$ (Fig. 6). This method uses $Fe_2(CO)_9$ as a source of an additional iron tricarbonyl fragment. The reaction is quantitative, the driving force being the elimination of H_2 gas (37,38). $[HRu_3(CO)_9BH_3]^-$ reacts with $Fe(CO)_5$ rather than $Fe_2(CO)_9$ to give $[HFe_3Ru(CO)_{12}BH]^-$ (see Section V) (32). The photolysis of a mixture of the two isomers of $Ru_3(CO)_9BH_5$ with $Fe(CO)_5$ leads to $HRu_3Fe(CO)_{12}BH_2$ or with $Co_2(CO)_8$ to $HRu_3Co(CO)_{12}BH$ (32). The relationship of the tetrahedral M_3E core to that of a butterfly M_4E cluster (Fig. 6) and the evolution of an initially terminal E—H unit into a bridging E—H—Fe interaction were initially considered in a theoretical sense (39). It is gratifying to see that the bonding analysis put forward for $HFe_4(CO)_{12}CH$ in terms of {$HFe_3(CO)_9CH$} and {$Fe(CO)_3$} fragments (39) (and, thus, by drawing an isoelectronic inference, an analysis for $[HFe_4(CO)_{12}BH]^-$ in terms of {$HFe_3(CO)_9BH$}⁻ and {$Fe(CO)_3$} units) may, in fact, be realized synthetically (37).

As mentioned above, triosmaborane clusters of the type Os_3B and exhibiting a terminal BH group have not, as yet, been reported. However, Shore and co-workers have detailed the chemistry of a group of Os_3B clusters in which the boron atom carries terminal substituents other than a hydrogen atom. The precursor for much of this work is $H_3Os_3(CO)_9BCO$ (**8**), the synthesis of which was described above (34). The reactivity of cluster **8** is summarized in Fig. 7. Substitution of the boron-bound CO ligand by PMe_3 occurs quantitatively in preference to substitution at a metal center; evidence for the formation of a B—P interaction appears in the ¹¹B and ³¹P NMR spectra (J_{BP} 118 Hz) (34). Like **8** (34), $H_3Os_3(CO)_9BPMe_3$ has been structurally characterized (40); the bonding in each cluster has been analyzed by UV-photoelectron spectroscopy and by Fenske–Hall molecular orbital calculations. It has been concluded that the boron atom is capable of acting as a pseudometal atom, and there is

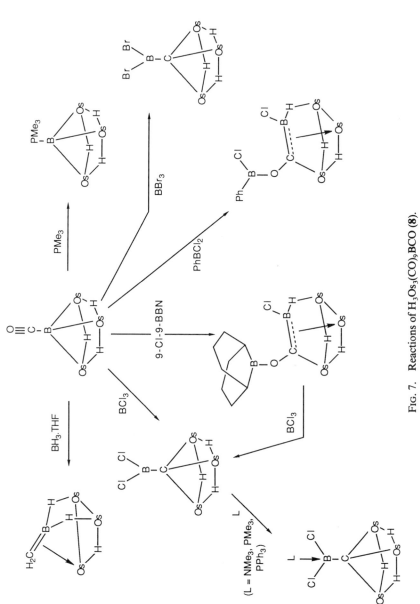

Fig. 7. Reactions of $H_3Os_3(CO)_9BCO$ (**8**).

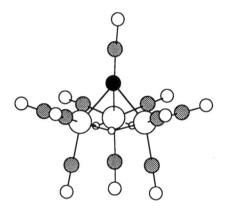

(8) $H_3Os_3(CO)_9BCO$

evidence that a synergistic effect akin to that observed in transition metal carbonyl or phosphine interactions operates in the BL (L = CO or PMe_3) unit (41).

Although reported over a period of several years, patterns in the reactivity of **8** with electrophiles have been brought together and elaborated on in a recent paper by Shore and co-workers (42). Cluster **8** reacts with boron trihalides to give $H_3Os_3(CO)_9CBX_2$ (X = Cl or Br). Note that during this transformation the boron cap is replaced by a carbon cap; a mechanism involving initial attack by BX_3 at a metal carbonyl oxygen atom followed by intramolecular boron–carbon interchange has been proposed. Labeling studies using $^{10}BCl_3$ have confirmed that there is no incorporation of the ^{10}B isotope into the product (42,43).

The structure of $H_3Os_3(CO)_9CBCl_2$ (**9**) has been confirmed crystallographically. The molecule has approximate C_s symmetry, and the $C-BCl_2$ vector deviates by 15° from a line drawn perpendicular to the Os_3 triangle. The B–C bond distance of 1.47 Å implies some π character. This bond length is compared to others in Table I. Power and co-workers have reported ranges of 1.42–1.45 and 1.58–1.62 Å for boron–carbon double and single bonds, respectively (44). Attack by nucleophiles on **9** has been investigated, and adducts $H_3Os_3(CO)_9CBCl_2L$ have been spectroscopically characterized for L equal to NMe_3, PMe_3, and PPh_3. The trimethylphosphine and trimethylamine adducts are stable only below −10°C; above this temperature $H_3Os_3(CO)_9CBCl_2NMe_3$ dissociates to the salt $[Me_3NH][H_2Os_3(CO)_9CBCl_2]$. The triphenylphosphine adduct is stable up to 30°C, at which temperature the phosphine ligand is lost (42).

Reduction of the boron-bound CO group in **8** is achieved by using

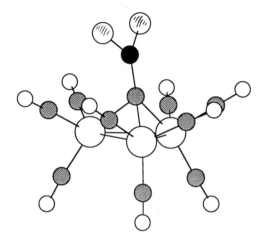

(**9**) $H_3Os_3(CO)_9CBCl_2$

$BH_3 \cdot THF$. The product, $H_3Os_3(CO)_9BCH_2$ (**10**), has been structurally characterized and is a boron analog of the vinylidene cluster $H_2Os_3(CO)_9CCH_2$. The boron atom of the apical $\{B{=}CH_2\}$ group interacts with all three osmium atoms, but the B—C vector subtends an angle of 60° with the plane containing the three osmium atoms (*45*). The *endo*-hydrogen positions (two Os—H—B and one Os—H—Os) have been inferred from the relative Os—Os bond distances and NMR spectral data. The orientation of the $\{B{=}CH_2\}$ group in **10** may be termed "per-

TABLE I

B—C BOND DISTANCES IN TRANSITION METAL CLUSTERS
CONTAINING BORON ATOMS^a

Compound	d_{BC} (Å)	Ref.
$H_3Os_3(CO)_9C(OBC_8H_{14})BCl$ (**11**)	1.46(2)	*46*
$H_3Os_3(CO)_9BCO$ (**8**)	1.469(15)	*34*
$H_3Os_3(CO)_9CBCl_2$ (**9**)	1.47(2)	*43*
$H_3Os_3(CO)_9BCH_2$ (**10**)	1.498(19)	*45*
$HFe_4(CO)_{12}CBH_2$ (**19**)	1.574(6)	*78*
$Fe_3(CO)_9B(H)C(H)C(Me)$ (**13**)	1.596(15)	*53*
	1.597(15)	
$HRu_4(CO)_{12}B(H)C(Ph)CHPh$ (**22**)	1.612(17)	*80*

^{a} Typical distances corresponding to B—C single and double bonds lie in the ranges 1.58–1.62 and 1.42–1.45 Å, respectively (*44*).

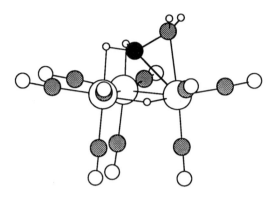

(10) $H_3Os_3(CO)_9BCH_2$

pendicular" in view of the relationship between the B—C vector and one Os—Os edge. A "parallel" mode for a {B—C} unit is observed in $H_3Os_3(CO)_9C(OBPhCl)BCl$ and in $H_3Os_3(CO)_9C(OBC_8H_{14})BCl$ (**11**), the structure of the latter being crystallographically confirmed (*42,46*). These compounds are considered as alkyne rather than vinylidene analogs of organometallic clusters and are formed by the reaction of **8** with $PhBCl_2$ and *B*-chloro-9-borabicyclo[3.3.1]nonane (i.e., *B*-Cl-9-BBN), respectively. $H_3Os_3(CO)_9C(OBC_8H_{14})BCl$ (**11**) is a model (*46*) for an intermediate proposed to form during the reaction of **8** with BCl_3 (*43*), and the detailed structure is clearly of interest. The B—C bond length in **11** is given in Table I and is consistent with there being significant double bond character (*46*).

The chemistry of triruthenium-based clusters is not yet as well developed as that of the triosmium systems, and, in addition to $Ru_3(CO)_9BH_5$ and its conjugate base, only $HRu_3(CO)_9B_2H_5$ and its anion have been reported. In a review by Fehlner (*3*), it was noted that, in view of the full characterization of $H_3Os_3(CO)_9BCH_2$, an early report (*47*) of "$Ru_3(CO)_9B_2H_6$" should be treated with caution since the mass spectral data recorded for this compound could also correspond to a formulation of "$H_3Ru_3(CO)_9BCH_2$." Twelve years after the initial report, the formulation of $Ru_3(CO)_9B_2H_6$ (**12**) (more informatively written as $HRu_3(CO)_9B_2H_5$) was confirmed, and the structure shown in Fig. 8 has been proposed on the basis of ^{11}B- and ^1H-NMR spectroscopic data (*48*). Since only one ^{11}B-NMR resonance is observed, the {B_2} fragment must presumably adopt a "parallel" orientation with respect to the triruthenium framework. In addition, NMR spectroscopic data illustrate that the {B_2H_6} unit is not

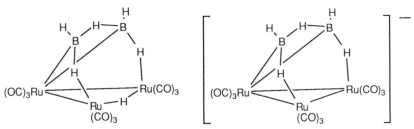

FIG. 8.   Proposed structures of HRu$_3$(CO)$_9$B$_2$H$_5$ (**12**) and its conjugate base.

intact, but rather that one hydrogen atom has adopted an Ru—H—Ru bridging mode (Fig. 8).

Thus, cluster **12** may be considered to be a model compound for a trimetal-supported alkyne [i.e., as described for H$_3$Os$_3$(CO)$_9$C (OBPhCl)BCl and H$_3$Os$_3$(CO)$_9$C(OBC$_8$H$_{14}$)BCl] or may be looked upon (particularly in the light of the *endo*-hydrogen positions) as an analog of pentaborane(9) in which three BH units have been replaced by isolobal Ru(CO)$_3$ fragments. Following the latter description, **12** mimics a member of the series of *nido*-ferraboranes delineated by Fehlner in 1980 (*49*). On treatment with base, **12** loses the Ru—H—Ru proton, thereby leaving the borane ligand (or borane cluster fragment depending on the point of view of the reader) relatively unperturbed (Fig. 8) (*48*). Interestingly, though, the conditions of deprotonation are critical; if the concentration of neutral **12** is high with respect to base added, simple deprotonation competes with cluster aggregation (*50*). Unlike its borane analog [B$_5$H$_8$]$^-$, the anion [Ru$_3$(CO)$_9$B$_2$H$_5$]$^-$ is static at 298 K on the 400-MHz NMR time scale (*48*).

The effects of sequentially substituting Ru(CO)$_3$ for BH fragments in [B$_5$H$_8$]$^-$ cannot as yet be assessed since other members of the series are not yet known, but comparisons with iron analogs are possible (Fig. 9). Fe(CO)$_3$B$_4$H$_8$ contains an apical Fe(CO)$_3$ group; treatment of [Fe(CO)$_3$B$_4$H$_7$]$^-$ with DCl gives the basally deuterated neutral ferraborane, but no spectroscopic data are available to provide information about the fluxional behavior of the anion (*51*). In [Fe$_2$(CO)$_6$B$_3$H$_6$]$^-$, in which the basal plane of the square pyramidal cluster core is perturbed by the entry of one metal atom, the three basally bridging hydrogen atoms are static (*52*). Clearly, the introduction of ruthenium rather than iron will influence bond energetics (*31*) and thus alter activation barriers to any fluxional processes, but, at least, the static nature of [Ru$_3$(CO)$_9$B$_2$H$_5$]$^-$ is not inconsistent with the trends set by [B$_5$H$_8$]$^-$ and [Fe$_2$(CO)$_6$B$_3$H$_6$]$^-$.

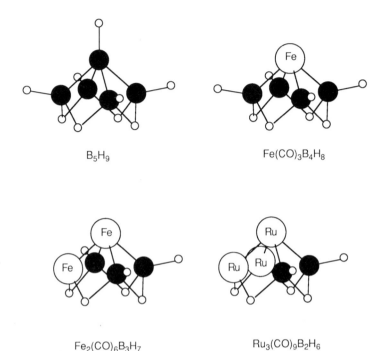

B_5H_9

$Fe(CO)_3B_4H_8$

$Fe_2(CO)_6B_3H_7$

$Ru_3(CO)_9B_2H_6$

FIG. 9. Isolobally related compounds B_5H_9, $Fe(CO)_3B_4H_8$, $Fe_2(CO)_6B_3H_7$, and $Ru_3(CO)_9B_2H_6$. Note that for cases in which one hydride bridges a metal–metal edge, the formula may be usefully rewritten, namely, $HRu_3(CO)_9B_2H_5$.

A novel system containing a coordinated borirene ligand has been produced in the reaction of $[Fe(py)_6][Fe_4(CO)_{13}]$ (where py is pyridine) with $BBrH_2 \cdot SMe_2$ (53). After protonation of the reaction mixture, $Fe_3(CO)_9B(H)C(H)C(Me)$ (13) may be isolated in approximately 5% yield. Structural characterization of 13 confirms a closed, distorted octahedral cage that may be considered as a metal-rich carbaborane cluster or as a borirene ligand supported on a triangular Fe_3 framework. The B—C bond distances given in Table I indicate single bond character. Significantly, 13 is a chiral molecule and therefore has potential for asymmetrical reactivity (53).

Clusters based on homonuclear trimetal units are dominated by those containing transition metals from the iron triad. Clusters constructed on a Co_3 framework are the only additional examples in this section. The first cluster to be reported was $Co_3(CO)_9BNEt_3$ (22). This arises from the reaction of $[Co(CO)_4]^-$ with BBr_3 in the presence of NEt_3 and is clearly related to the triosmium systems of Shore et al. discussed above (Fig. 7)

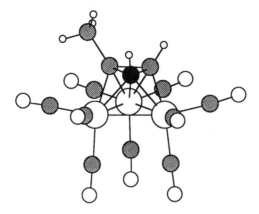

(13) Fe$_3$(CO)$_9$B(H)C(H)C(Me)

and in particular to H$_3$Os$_3$(CO)$_9$BPMe$_3$ (*34,40*). Condensation of mononuclear cobalt and boron fragments to give a cluster with a Co$_3$B core also occurs in the reaction of CpCo(PPh$_3$)$_2$ with 2 equivalents of BH$_3 \cdot$ THF (*54*). For the production of a particular cobaltaborane, reaction conditions are critical (see also Sections II and IV) (*3*). The product here, formed in 7% yield, is brown Cp$_3$Co$_3$(BPh)(PPh) **(14)**. The core of this molecule is a trigonal bipyramid with an equatorial Co$_3$ triangle. A mechanism has been proposed for the formation of **14** via (1) the elimination of Ph$_3$P$\cdot$BH$_3$ and (2) migration of one phenyl group from P to B (*54*).

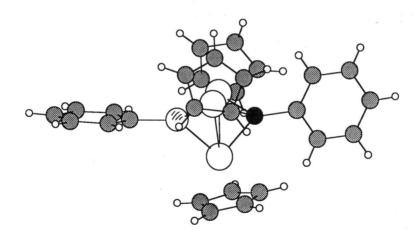

(14) Cp$_3$Co$_3$(BPh)(PPh)

The two tricobalt species $Cp_3Co_3B_3H_5$ (55,56) and $Cp_3Co_3(\mu_3\text{-CO})$ B_3H_3 (57,58) are structurally related to each other and to the ferracarbaborane 13. All have distorted octahedral geometries containing mutually staggered M_3 and E_3 triangles. $Cp_3Co_3B_3H_5$ is an air-stable solid, formed in low yield from the reaction of $[B_5H_8]^-$ with $CoCl_2$ and cyclopentadienyl anion. The two endo-hydrogen atoms in $Cp_3Co_3B_3H_5$ are proposed to lie in Co—H—Co bridging sites, disordered in the solid state over the three equivalent edges (56). The interaction of cobalt vapor with cyclopentadiene and pentaborane(9) leads to $Cp_3Co_3(\mu_3\text{-CO})B_3H_3$ but only in 2.7% yield. It has been suggested that the source of the cluster-bound carbonyl ligand is either metal oxide impurities or the THF solvent used for product extraction. Not surprisingly, the yield of $Cp_3Co_3(\mu_3\text{-CO})$ B_3H_3 may be increased by running the reaction in the presence of carbon monoxide (58).

IV

CLUSTERS BASED ON HOMONUCLEAR TETRAMETAL UNITS

As with the trimetal-based metalloboranes, those with homonuclear tetrametal frameworks are restricted to the iron and cobalt triads. However, the restriction is even greater for M_4 than for M_3 systems as no examples of tetraosmium boron-containing clusters have yet been documented. The series of isolobally related nido clusters originating from B_5H_9 (Fig. 9) can be extended to include a square pyramidal cluster "$H_4Fe_4(CO)_{12}BH$" with two isomers possible having the BH fragment in either an apical or a basal site (6,49). Significantly, although this cluster has not been observed in practice, its dehydrogenated analog has (59,60). The formal loss of two hydrogen atoms reduces the cluster electron count by two and thus causes skeletal rearrangement to the arachno cluster $HFe_4(CO)_{12}BH_2$ (15), in which there is no terminal BH unit, only three endo-hydrogen atoms.

The crystallographically determined structure of 15 confirms the presence of a butterfly arrangement of iron atoms within which the boron atom resides in contact with all four metal atoms [Fe—B = 2.044(6), 2.047(6), 1.966(6), and 1.974(6) Å] and 0.31 Å above the Fe_{wing}—Fe_{wing} vector; the internal dihedral angle of the Fe_4 butterfly is 114.0°. These parameters are compared in Table II with those of related clusters. Compound 15 was first isolated as a product from the reaction of $Fe_2(CO)_6B_2H_6$ with $Fe_2(CO)_9$ (60), and evidence for increased iron–boron interaction is observed in a dramatic change in ^{11}B-NMR spectral shift (8) from $\delta - 24.2$ to $+116.0$.

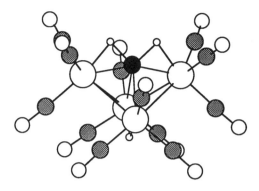

(15) $HFe_4(CO)_{12}BH_2$

[Note that the ^{11}B-NMR chemical shift for **15** was originally reported at
32.1 MHz as δ +106 (*60*)]. By following the formation of **15** by ^{11}B-NMR
spectroscopy, it has been shown that the optimum yield is approximately
10% (*16*). The reaction of $[Fe_2(CO)_6B_2H_5]^-$ with $Fe_2(CO)_9$ proceeds in a
similar manner to give the conjugate base of **15** (see below), but of interest
is that, in this case, the major product is actually the lower nuclearity
cluster $[Fe_3(CO)_{10}BH_2]^-$ (*16*). Metalloborane **15** may also be prepared by
treating a suspension of $Fe_2(CO)_9$ in a hexane solution of $Fe(CO)_5$ with
$BH_3 \cdot THF$ and $Li[BHEt_3]$ followed by acidification (*38*). In Section III, we
noted that the formation of **15** is a competitive pathway in the reaction of
$Fe(CO)_3(\eta^2\text{-}cis\text{-}cyclooctene)$ with $BH_3 \cdot L$ (L = SMe_2 or THF) to give **6**.
The course of the reaction can be forced in favor of **15** if the ratio
$[Fe(CO)_3(\eta^2\text{-}cis\text{-}cyclooctene)]$ to $[BH_3 \cdot L]$ is 4:3, although the yield of the
tetraferraborane is still only 8% (*18*).

Deprotonation of **15** occurs readily with NEt_3 or in a methanolic
solution of [PPN]Cl to give $[Et_3NH][HFe_4(CO)_{12}BH]$ or [PPN]-
$[HFe_4(CO)_{12}BH]$, respectively (*38*). The anion $[HFe_4(CO)_{12}BH]^-$
(16) may also be prepared by the cluster expansion reaction from
$[HFe_3(CO)_9BH_3]^-$ illustrated in Fig. 6 (*37,38*). In **15**, the *endo*-hydrogen
atoms are static on the NMR time scale, but in anion **16** the remaining
Fe—H—Fe and Fe—H—B protons undergo exchange. Multiple depro-
tonation of **15** may be brought about by treating **15** with butyllithium (*61*).
The sequence from $HFe_4(CO)_{12}BH_2$ to $[Fe_4(CO)_{12}B]^{3-}$ is shown in Fig. 10
and is compared with the deprotonation pattern observed for
$HRu_4(CO)_{12}BH_2$ (*62,63*) (see below). The observed difference in the site of
the second deprotonation in going from iron to ruthenium is in agreement
with prediction (*64*). The gradual exposure of the boron atom on deprotonation of **15** is accompanied by a shift to lower field for the ^{11}B-NMR

TABLE II
Comparison of Characteristic Geometrical Parameters of M_4B Butterfly Compounds

Compound	Height of the boron atom above the Fe_{wing}—Fe_{wing} vector (Å)	Internal dihedral angle of the M_4 butterfly (°)	Fe_{hinge}—B distances (Å)	Fe_{wing}—B distances (Å)	Ref.
$HFe_4(CO)_{12}BH_2$ (15)	0.31	114.0	2.044(6) 2.047(6)	1.966(6) 1.974(6)	59,60
$Fe_4(CO)_{12}Au_2(PPh_3)_2BH$	0.37	113.4(3)	2.13(1) 2.07(1)	2.01(1) 2.00(1)	68,69
$Fe_4(CO)_{12}Au_2(P(4-Me-C_6H_4)_3)_2BH$	0.38	113.3(3)	2.114(19) 2.126(19)	2.024(24) 2.008(24)	72
$HFe_4(CO)_{12}Au_2(PEt_3)_2B$ (17)	0.31	113.5(3)	2.065(11) 2.065(11)	1.989(3) 1.989(3)	72,73
$Fe_4(CO)_{12}Au_2(AsPh_3)_2BH$	0.32	113.6(3)	2.068(11) 2.114(13)	1.999(14) 2.031(14)	71
$Fe_4(CO)_{12}Au_3(PPh_3)_3B$	0.457(1)	116.9[a]	2.09(2) 2.16(2)	1.98(2) 2.13(2)	70
$[(HFe_4(CO)_{12}BH)_2Au]^-$ (18)	0.39[a]	116.6(3)	2.065(13) 2.082(12)	2.015(12) 1.990(12)	74
$HRu_4(CO)_{12}BH_2$ (20)	0.39[a]	118	2.195(5) 2.195(5)	2.111(6) 2.106(6)	63
$HRu_4(CO)_{12}Au_2(PPh_3)_2B$	0.37(1)	117.4(1)	2.259(13) 2.250(14)	2.130(11) 2.114(13)	62

[a] Calculated from atomic coordinates.

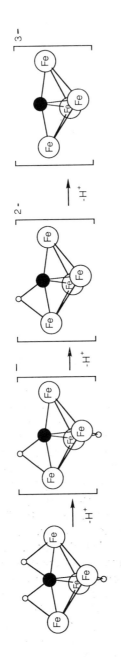
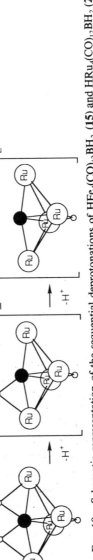

Fig. 10. Schematic representation of the sequential deprotonations of $HFe_4(CO)_{12}BH_2$ (15) and $HRu_4(CO)_{12}BH_2$ (20).

spectral resonance: δ +116 for $HFe_4(CO)_{12}BH_2$, δ +150 for $[HFe_4(CO)_{12}BH]^-$, and δ +153 for $[Fe_4(CO)_{12}BH]^{2-}$. The trianion is ^{11}B-NMR silent (61). Note that the change in chemical shift from $HFe_4(CO)_{12}BH_2$ to $[HFe_4(CO)_{12}BH]^-$ is considerably larger than that from $[HFe_4(CO)_{12}BH]^-$ to $[Fe_4(CO)_{12}BH]^{2-}$; this is consistent with the second proton being removed from the transition metal framework rather than from a B—H—Fe site.

The trianion $[Fe_4(CO)_{12}B]^{3-}$ is the first example of a discrete molecular homometallic boride and, moreover, one in which the boron atom is exposed. The reactivity of the partially exposed boron atom in both $[HFe_4(CO)_{12}BH]^-$ (16) and $[Fe_4(CO)_{12}BH]^{2-}$ toward electrophiles has been investigated. The $[PPN]^+$ salt of the dianion reacts with methyl iodide at room temperature to give $[HFe_4(CO)_{12}BMe]^-$. This anion is isoelectronic with $[Fe_4(CO)_{12}CMe]^-$ and is, by analogy with this and in keeping with PSEPT, proposed to possess a tetrahedral metal cluster core (61). The monoanion 16 reacts with $[Rh(CO)_2Cl]_2$ [formally a source of the $\{Rh(CO)_2\}^+$ electrophile] to give the closo-metalloboride cluster $[Fe_4Rh_2(CO)_{16}B]^-$ (see Sections V and VI) (65).

The reactions of 16 with gold(I) phosphine electrophiles have been exploited to generate a series of related metalloboride clusters of the types $HFe_4(CO)_{12}(AuPR_3)BH$, $HFe_4(CO)_{12}Au_2(PR_3)_2B$, and $Fe_4(CO)_{12}Au_3$ $(PPh_3)_3B$ (Fig. 11). The general methodology, based on the isolobal principle, of replacing an endo-hydrogen atom by a gold(I) phosphine fragment is well documented in transition metal cluster chemistry (66). The careful addition of an equimolar quantity of Ph_3PAuCl to 16 leads to the mono-gold derivative $HFe_4(CO)_{12}Au(PPh_3)BH$ (67). The addition of 2 or more equivalents of Ph_3PAuCl produces $Fe_4(CO)_{12}Au_2(PPh_3)_2BH$ as the major product (68,69). Over long reaction periods, another derivative, $Fe_4(CO)_{12}Au_3(PPh_3)_3B$, in which all three endo-hydrogen atoms have been replaced is isolated in low yield. A more efficient route to $Fe_4(CO)_{12}Au_3(PPh_3)_3B$ is to react anion 16 with $[(Ph_3PAu)_3O][BF_4]$, in which case the yield exceeds 70% (70). In $Fe_4(CO)_{12}Au_3(PPh_3)_3B$ the boron atom is completely encapsulated within a sheath of seven transition metal atoms (see Appendix II) and is therefore classed as an interstitial atom, that is, the cluster is a metalloboride (70). The space-filling diagram drawn in Fig. 12 along with a schematic representation of the Fe_4Au_3B core illustrates the extent to which the boron atom is protected. Indeed, the boron atom is not visible in any projection of a space filling model. Consequently, the air stability of $Fe_4(CO)_{12}Au_3(PPh_3)_3B$ is in marked contrast to the sensitivity of the precursors 15 and 16. In the solid-state structure, the three $AuPPh_3$ groups are nonequivalent, but, in solution, only one ^{31}P-NMR spectral resonance is observed at room temperature, implying that a fluxional process is operative.

Returning to $Fe_4(CO)_{12}Au_2(PPh_3)_2BH$, the crystallographically determined structure shows an asymmetrical positioning of the two gold atoms
with respect to the Fe_4B framework as depicted in the lower right-hand
scheme in Fig. 11. The cluster core is virtually unperturbed as two *endo*-
hydrogen atoms are replaced by two $AuPPh_3$ units (Table II) (*68,69*). A
similar structure is observed for $Fe_4(CO)_{12}Au_2(AsPh_3)_2BH$ (*71*). A series

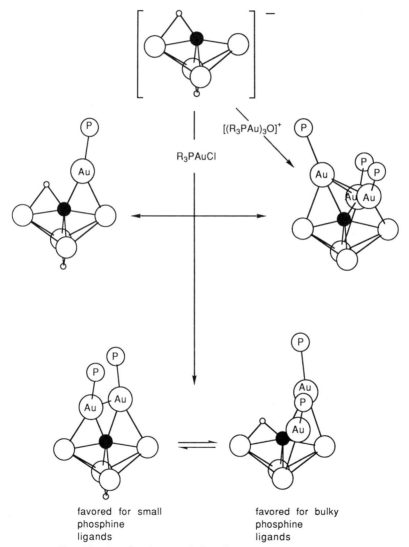

favored for small
phosphine
ligands

favored for bulky
phosphine
ligands

FIG. 11. Auraferraboranes derived from $[HFe_4(CO)_{12}BH]^-$ (**16**).

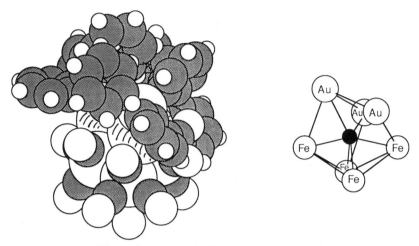

FIG. 12. Space-filling diagram and the core structure of the metalloboride
$Fe_4(CO)_{12}Au_3(PPh_3)_3B$.

of digold derivatives has been characterized, and an interesting isomeric
trend has been observed (72). The presence of a phosphine substituent
with a small cone angle allows the formation of a second isomer possessing
a symmetrical (C_2 axis) Fe_4Au_2B core (Fig. 11, lower left). The
second structure has been confirmed crystallographically for
$HFe_4(CO)_{12}Au_2(PEt_3)_2B$ (**17**) (72,73).

In the case of PMe_3 (Tollman cone angle, $\theta = 118°$), the symmetrical
system is preferred but is not exclusively present. For PPh_3 ($\theta = 145°$) only
the asymmetrical isomer is present. For $118° \leq \theta < 145°$ an isomeric
equilibrium (Fig. 11) is observed in solution and is evident in both the ^{11}B-
and ^1H-NMR spectra. The structural feature that makes the two isomers so
readily distinguishable is the position of the *endo*-hydrogen atom; it adopts
an Fe—H—Fe site in the symmetrical isomer and an Fe—H—B site in
the asymmetrical one. For example in dichloromethane solution the
bis(triethylphosphine) gold derivative exists with approximately 8% of
isomer $HFe_4(CO)_{12}Au_2(PEt_3)_2B$ (characterized by ^1H NMR $\delta -24.9$, ^{11}B
NMR $\delta +179$) and 92% of $Fe_4(CO)_{12}Au_2(PEt_3)_2BH$ (characterized by ^1H
NMR $\delta -9.1$, ^{11}B NMR $\delta +138$); the average ^1H-NMR resonance observed at room temperature is $\delta -10.4$, and the ^{11}B-NMR signal appears at
$\delta +142.2$. The change in *endo*-hydrogen location is forced by the reorienta-
tion of the carbonyl ligands on one hinge iron atom, and this in itself is
caused by the relocation of one $AuPR_3$ fragment from a $B—Fe_{wingtip}$
to a $B—Fe_{hinge}$ edge. The isomer system here is a novel example
of a metalloborane–boride equilibrium. Deprotonation of $Fe_4(CO)_{12}$-

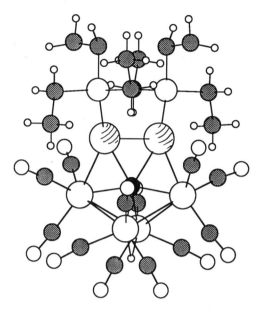

(17) HFe$_4$(CO)$_{12}$Au$_2$(PEt$_3$)$_2$B

Au$_2$(PR$_3$)$_2$BH (observed for R = Ph) presumably leads to a true boride cluster; at present only spectroscopic data are available (4).

For some phosphines, for example, PPhMe$_2$ and PPh$_2$Me, the formation of the derivatives HFe$_4$(CO)$_{12}$Au(PR$_3$)BH and HFe$_4$(CO)$_{12}$Au$_2$(PR$_3$)$_2$B is complicated by a pathway leading to the anion [{HFe$_4$(CO)$_{12}$BH}$_2$Au]$^-$ (18) (74). The full characterization of [{HFe$_4$(CO)$_{12}$BH}$_2$Au]-[Au(PPh$_2$Me)$_2$] illustrates a compound in which the empirical formula is the same as the simple monogold derivative HFe$_4$(CO)$_{12}$Au(PPh$_2$Me)BH and implies the formation of 18 via the latter with subsequent Au—P bond cleavage. Anion 18 comprises two {HFe$_4$(CO)$_{12}$BH} units fused "face-to-face" about a gold atom; the coordination geometry at Au deviates from square planarity by 30.9(5)°. If the {HFe$_4$(CO)$_{12}$BH} group is considered to act as a bidentate ligand, then the two ligands are mutually cis with respect to the central gold atom.

The electrochemistry of cluster compounds is still an underdeveloped area (75,76). A recent study has shown interesting differences in the redox behaviors of the isostructural compounds Fe$_4$(CO)$_{12}$Au$_2$(PPh$_3$)$_2$BH and Fe$_4$(CO)$_{12}$Au$_2$(AsPh$_3$)$_2$BH. Each undergoes a one-electron reduction and oxidation, but only the phosphine derivative can be oxidized further to the dication. All electrochemically generated species are short lived (71).

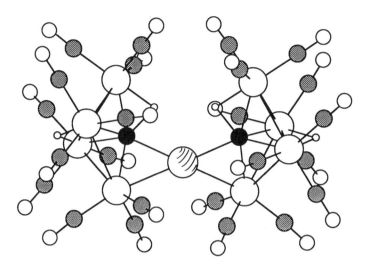

(18) $[\{HFe_4(CO)_{12}BH\}_2Au]^-$

Selected phosphine substitution chemistry of anion **16** has been studied; reaction of **16** with 1 or 2 equivalents of $PPhMe_2$ leads to $[HFe_4(CO)_{11}(PPhMe_2)BH]^-$ or $[Fe_4(CO)_{10}(PPhMe_2)_2BH_2]^-$, respectively (*38,77*). Spectroscopic data are consistent with each ligand substitution occuring at a (different) wingtip iron atom, each being in an equatorial site. Both *cis* and *trans* isomers of $[Fe_4(CO)_{10}(PPhMe_2)_2BH_2]^-$ have been detected at low temperature in the ^{31}P-NMR spectrum. The presence of phosphine in place of carbonyl ligands in the disubstituted derivative causes, not surprisingly, a significant change in the distribution of cluster electronic charge. In response to this, a reallocation of *endo*-hydrogen atoms occurs (Fig. 13) (*77*). The substitution reactions have been monitored by use of ^{11}B-NMR spectroscopy, and for each step an intermediate has been observed. An associative mechanism has been proposed (*38*). We noted above that in anion **16** the *endo*-hydrogen atoms are involved in a

(also *trans* isomer)

FIG. 13. *endo*-Hydrogen redistribution that accompanies the disubstitution reaction of $[HFe_4(CO)_{12}BH]^-$ to $[HFe_4(CO)_{10}(PPhMe_2)_2BH]^-$.

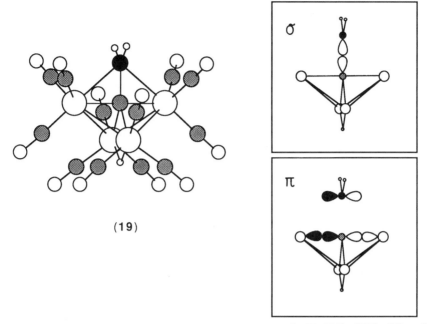

FIG. 14. The crystallographically determined structure of $HFe_4(CO)_{12}CBH_2$ (19) and schematic representations of the C—B σ and π bonding interactions.

fluxional process. However, in $[HFe_4(CO)_{11}(PPhMe_2)BH]^-$, there is no exchange of the corresponding cluster protons at room temperature (77).

One further cluster completes the section on Fe_4 metalloborane chemistry. $HFe_4(CO)_{12}CBH_2$ (19) (Fig. 14) forms in the reaction of $Fe_2(CO)_9$ and $BH_3 \cdot THF$ at 50°C, a system that also yields $HFe_3(CO)_9BH_4$. A structural determination of 19 illustrates a *carbide* cluster with the interstitial carbon atom supporting a terminal BH_2 unit. The $Fe_{wingtip}$—B distances of 2.427(3) Å are too long to be considered as σ bonds, but a reasonable bonding description is to consider the $\{HFe_4(CO)_{12}C\}$ cluster fragment as a source of both σ and π electron density as illustrated in Fig. 14 (78).

The first report of a tetraruthenaborane came in 1977 (47); a cluster of formula "$Ru_4(CO)_{12}BH_3$" (20) was reported to be a by-product of the borohydride reduction of $Ru_3(CO)_{12}$ and, on the basis of mass and ^1H-NMR spectroscopic data, was proposed to possess one of the two structures depicted in Fig. 15. The structural characterization in 1982 of the related ferraborane 15 (60) provided indirect support for cluster 20 being based on an Ru_4 butterfly framework. ^{11}B-NMR data reported in 1989 and 1990 were also consistent with this postulate (62,63). The butterfly-based struc-

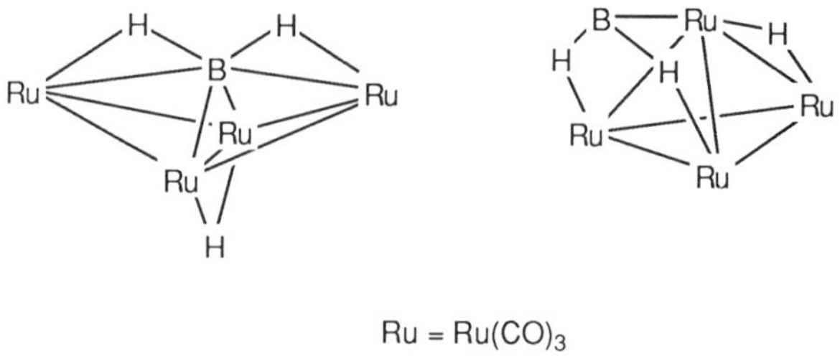

$Ru = Ru(CO)_3$

FIG. 15. Two proposed structures for $HRu_4(CO)_{12}BH_2$ (20); the left-hand structure has been confirmed by X-ray diffraction.

ture has now been confirmed crystallographically by Shore and co-workers (63). As in 15, the boron atom in 20 carries no terminal hydrogen atom and lies in direct contact with all four metal atoms (see Table II). Cluster 20 is a moderately air-stable yellow solid, and its ability to withstand limited exposure to air contrasts with the air sensitivity of its iron analog 15. Indeed, as might be expected, this distinction is also true for the pair of trimetal clusters $M_3(CO)_9BH_5$ with M equal to Fe and Ru, 6 and 7, respectively.

Two synthetic routes to 20 are available. The reaction of $Ru_3(CO)_{12}$ with an excess of $BH_3 \cdot THF$ and $Li[BHEt_3]$ in THF solution over a period of 1.5 hours at room temperature followed by protonation gives 20 in approximately 10% yield (62). The yield may be increased by reducing the molar equivalents of borane used (79). Alternatively, treatment of $H_4Ru_4(CO)_{12}$ in CH_2Cl_2 solution with $THF \cdot BH_3$ at 40°C for 4 days gives 20 in 60% yield (63).

The monodeprotonation of 20 mimics that of 15 (Fig. 10) with the removal of an M—H—B proton to give $[HRu_4(CO)_{12}BH]^-$ (21), and may be achieved either in a methanolic solution of [PPN]Cl at room temperature (62) or by using KH in Me_2O at −78°C (63). Treatment with KH at 30°C for a prolonged period (48 hours) leads to the salt $[HRu_4(CO)_{12}B]K_2$, and it is proposed on the basis of ^1H- and ^{11}B-NMR data that in the dianion the boron atom is completely exposed unlike that in the iron analog (Fig. 10). Interestingly, the exposed boron atom in the trianion $[Fe_4(CO)_{12}B]^{3-}$ is ^{11}B-NMR silent, whereas the dianion $[HRu_4(CO)_{12}B]^{2-}$ is characterized by a broad resonance at $\delta +159$ (63). The dianion is characterized in the ^1H-NMR spectrum at room temperature by a sharp signal at $\delta -20.09$ consistent with the presence of a Ru—H—Ru bridge. At 50°C this hydride resonance is replaced by a broad signal at $\delta -16.81$, suggesting that the single endo-hydrogen atom in $[HRu_4(CO)_{12}B]^{2-}$ is fluxional in solution at this temperature (63).

The reaction of anion **21** with Ph_3PAuCl in the presence of a halide abstractor (*e.g.*, $Tl[PF_6]$) gives both the mono- and digold derivatives $HRu_4(CO)_{12}Au(PPh_3)BH$ and $HRu_4(CO)_{12}Au_2(PPh_3)_2B$ (*62*). Remember that for the iron analog of the digold derivative an asymmetrical structure was imposed by the steric strain resulting from the presence of the two triphenylphosphine substituents. A crystallographic study of $HRu_4(CO)_{12}Au_2(PPh_3)_2B$ reveals a nearly symmetrical (but in fact not possessing a C_2 axis) geometry. Apparently, the boron-containing Ru_4 butterfly skeleton is sufficiently large compared to its Fe_4 counterpart to allow the two PPh_3 groups to adopt the more symmetrical structure (*62*). Thus, in $HRu_4(CO)_{12}Au_2(PPh_3)_2B$ the boron atom is more boridic in nature than in the analogous tetrairon compound.

The reactivity of neutral **20** toward diphenylacetylene has been investigated, and significant differences have been observed between the reaction pattern of $PhC{\equiv}CPh$ with $HRu_4(CO)_{12}BH_2$ compared to that with the isoelectronic $H_2Ru_4(CO)_{12}C$ or $HRu_4(CO)_{12}N$. When photolyzed in the presence of $PhC{\equiv}CPh$, **20** reacts smoothly to generate $HRu_4(CO)_{12}B(H)C(Ph)CHPh$ (**22**) (*80*). The insertion of the alkyne is

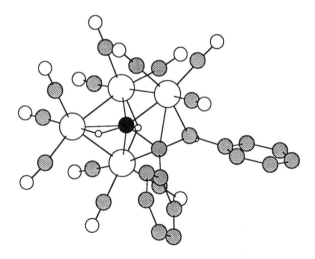

(**22**) $HRu_4(CO)_{12}B(H)C(Ph)CHPh$

accompanied by the transfer of one *endo*-hydrogen atom to give a {PhHCCPh} fragment, formation of a B—C bond, and cleavage of one $Ru_{wingtip}$—Ru_{hinge} bond of the initial butterfly framework. Figure 16 contrasts this pathway with those observed for the related carbide (*81*) and nitride (*82*) clusters. For the carbide, *endo*-hydrogen atom transfer occurs

(a) $HRu_4(CO)_{12}BH_2$

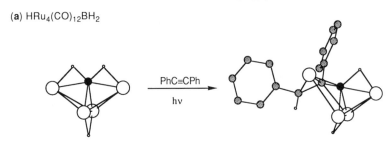

$\xrightarrow[h\nu]{PhC\equiv CPh}$

(b) $HRu_4(CO)_{12}CH$

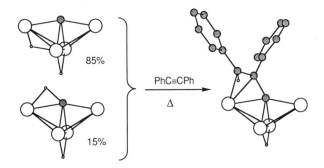

85%

15%

$\xrightarrow[\Delta]{PhC\equiv CPh}$

(c) $[Ru_4(CO)_{12}N]^-$

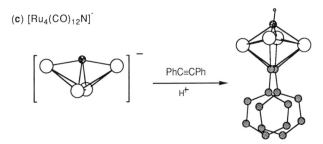

$\xrightarrow[H^+]{PhC\equiv CPh}$

Fig. 16. Comparative reaction pathways for isoelectronic Ru_4E clusters with diphenylacetylene: (a) E = B; (b) E = C; (c) E = N.

accompanied by C—C bond formation, but the Ru_4 butterfly is virtually unperturbed. In the case of the nitride, the nitrogen atom is drawn out of the cage to become a cluster vertex (and thus forms an *exo*-NH bond) as the alkyne inserts into the Ru_{hinge}—Ru_{hinge} edge.

Two tetracobaltaborane clusters have been reported to form under simi-

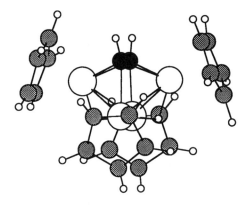

(23) $H_2Cp_4Co_4B_2H_2$

lar reaction conditions. In the first, $H_2Cp_4Co_4B_2H_2$ (**23**), the metal atoms are arranged in a butterfly geometry, and the cluster is structurally related to those of the general formula $Co_4(CO)_{10}C_2R_2$. Red, air-stable **23** is prepared by the reaction of $CpCo(PPh_3)(EtCCEt)$ or $CpCo(PPh_3)_3$ with $BH_3 \cdot THF$. The structure has been confirmed by X-ray diffraction, and one interesting feature is the rather long (unsupported by a bridging hydrogen atom) B—B bond [1.80(2) Å] (*83*). With careful tuning of the reaction conditions, the combination of reagents $CpCo(PPh_3)(EtCCEt)$ and $BH_3 \cdot THF$ leads to an alternative cobaltaborane, namely, green $Cp_4Co_4(PPh)B_2H_2$ (**24**) (*84*). The pentagonal bipyramidal geometry of **24**

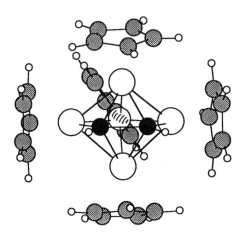

(24) $Cp_4Co_4(PPh)B_2H_2$

mimics that of $Ru_4(CO)_{11}(PPh)C_2Ph_2$ (85) and has been confirmed crystallographically; the two boron atoms lie in adjacent equatorial sites.

The cobalta- and nickelaboranes $Cp_4Co_4B_4H_4$ (**25**) (86) and $Cp_4Ni_4B_4H_4$ (**26**) (87) are synthesized from $Na[B_5H_8]$ and $CoCl_2$ with

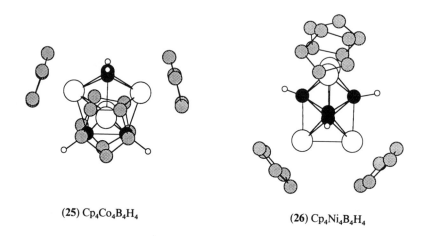

(**25**) $Cp_4Co_4B_4H_4$

(**26**) $Cp_4Ni_4B_4H_4$

$Na[C_5H_5]$ and from $Na[B_5H_8]$ with nickelocene in the presence of sodium amalgam, respectively. Structural characterization of **25** (88) and **26** (89) show that the clusters are related in that each cage is based on a dodecahedron. However, two points are significant. First, the positions of the metal and boron atoms are different in the two cases, and, second, despite possessing a common closed-polyhedral skeleton, the fragments in **26** contribute two more pairs of cluster-bonding electrons than do the fragments in **25**. Indeed, *neither* electron count (8 pairs for **25** and 10 pairs for **26** is consistent by PSEPT with the observed closo dodecahedral cage. The apparent problem may be rationalized in one of two ways. The first, put forward by Mingos and co-workers, illustrates that the M_4B_4 dodecahedral framework can be treated as being composed of two interpenetrating tetrahedra, one extended and one flattened. When only 8 electron pairs are available (i.e., in **25**), the metal atoms must define the flattened tetrahedral core; with 10 pairs, as in **26**, the metal atoms define the elongated tetrahedron (90). The second approach, discussed by Wade and O'Neill, shows that (1) the degeneracy of the highest occupied molecular orbitals in an ideal dodecahedral cluster (e.g., in $[B_8H_8]^{2-}$) leads to flexibility in the number of cluster-bonding electrons actually tolerated, and (2) since the {CpNi} fragment contributes more electrons than does the {CpCo} unit, the former will occupy sites of low skeletal connectivity (i.e., higher electron density) and the latter sites of high connectivity (91).

V

CLUSTERS WITH MIXED METAL FRAMEWORKS

There are still only a few examples of mixed metal–transition metal clusters containing boron atoms. The first to be reported was $Cp_2Co_2Fe(CO)_4B_3H_3$ (92). This cluster results from the photolysis of 2-$CpCoB_4H_8$ [a basally substituted isolobal analog of pentaborane(9)] with $Fe(CO)_5$. NMR spectroscopic data are consistent with the proposed structure shown in Fig. 17.

The expansion of the triruthenium cluster $Ru_3(CO)_9BH_5$ (7) to $HRu_3Fe(CO)_{12}BH_2$ proceeds smoothly when a dichloromethane solution of 7 is photolyzed with $Fe(CO)_5$. Both isomers of 7 shown in Fig. 3 are consumed equally during the reaction to generate one product in which the iron atom is proposed to reside in a wingtip site (32). ^{11}B- and ^1H-NMR spectroscopic data provide no evidence to suggest isomerism between the structural variants $HRu_3(Fe_{wingtip})(CO)_{12}BH_2$ and $HRu_3(Fe_{hinge})(CO)_{12}BH_2$. If this is indeed the case, then $HRu_3Fe(CO)_{12}BH_2$ parallels its neutral nitrido analog $HRu_3Fe(CO)_{12}N$ (93). Deprotonation of $HRu_3Fe(CO)_{12}BH_2$ by loss of a Fe—H—B proton appears from spectroscopic data to leave the Ru_3Fe framework unaffected; the Fe atom remains exclusively in a wingtip site (32). Significantly, however, deprotonation of $HRu_3Fe(CO)_{12}N$ by loss of the Ru—H—Ru proton causes a framework isomerization, and in the solid-state structure of $[Ru_3Fe(CO)_{12}N]^-$ the iron atom is found to be disordered over both wingtip and hinge atom sites (94).

The auraferraboranes and auraruthenaboranes of types $M_4(CO)_{12}$ $Au_x(PR_3)_xBH_{3-x}$ (M = Fe or Ru; x = 1, 2, or 3) may be included here for completeness, but a full discussion is more appropriate in Section IV. As described in detail above, each is formed from the interaction of the anion $[HM_4(CO)_{12}BH]^-$ with R_3PAuCl {or in the case of M = Fe and $x = 3$, the reaction of 16 with $[(Ph_3PAu)_3O][BF_4]$}.

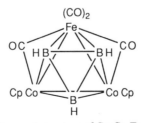

FIG. 17. Proposed structure of $Cp_2Co_2Fe(CO)_4B_3H_3$.

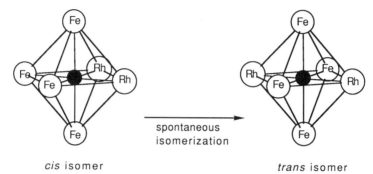

FIG. 18.   Isomerization of the heterometallic boride cluster $[Fe_4Rh_2(CO)_{16}B]^-$.

The reaction of the mono-, di-, or trianion of **15** with $[Rh(CO)_2Cl]_2$ is the means of preparing the first example of a *closo*-metalloborane cluster, namely, $[Fe_4Rh_2(CO)_{16}B]^-$. Monitoring the reaction by ^{11}B-NMR spectroscopy shows that the pathway proceeds via the formation of *cis*-$[Fe_4Rh_2(CO)_{16}B]^-$ (^{11}B NMR δ +205, $J_{RhB} = 23.3$ Hz) followed by isomerization to the trans isomer (^{11}B NMR δ +211, $J_{RhB} = 25.8$ Hz) as shown in Fig. 18. The cis and trans isomers are related isolobally to 1,2- and 1,6-$C_2B_4H_6$; a study of the kinetics of the cis to trans isomerization suggests that this process occurs (at 20°C) at a rate 3×10^4 times faster than does the 1,2- to 1,6-$C_2B_4H_6$ isomerization (at 250°C) *(65)*. A kinetic study of the skeletal isomerization has shown that the rate is first order in cluster and also depends on the concentration of CO in solution; in other words, the mechanism is an associative one *(95)*.

VI

CLUSTERS CONTAINING INTERSTITIAL BORON ATOMS

The area of solid-state metal borides is a large and important one *(96–98)*. There should be great scope, therefore, for investigations into molecular transition metal borides, and results in this area may well be of value in producing model systems for the bulk materials *(6,99)*. The number of discrete clusters in which a naked boron atom is encapsulated is still few, and most examples have been discussed in detail in Sections IV and V of this article. They are summarized as follows. $HFe_4(CO)_{12}BH_2$ *(59,60)* and $HRu_4(CO)_{12}BH_2$ *(47,62,63)* may be considered as metalloboride clusters since no terminal BH bond (the "conventional" hallmark of a borane) is present. However, the only true molecular boride cluster in which the boron atom is exposed is $[Fe_4(CO)_{12}B]^{3-}$ (61). Molecular construction on this framework produces the *closo*-boride $[Fe_4Rh_2(CO)_{16}B]^-$ (65,95). By using either $[HFe_4(CO)_{12}BH]^-$ or $[HRu_4(CO)_{12}BH]^-$ as precursors, a series of metalloboride clusters can be synthesized: $HFe_4(CO)_{12}Au_2$ $(PRR'_2)_2B$ (*e.g.*, R = R' = Et or Me; R = Me and R' = Ph; R = Ph and R' = Me) (72,73), $[Fe_4(CO)_{12}Au_2(PPh_3)_2B]^-$ (4), $Fe_4(CO)_{12}Au_3(PPh_3)_3B$ (70), and $HRu_4(CO)_{12}Au_2(PPh_3)_2B$ (62).

Three further borido clusters have been documented: $Co_6(CO)_{18}B$, $HRu_6(CO)_{17}B$, and $[Ru_6(CO)_{17}B]^-$. $Co_6(CO)_{18}B$ is reported to be a product in the reaction of $Co_2(CO)_8$ with BBr_3 at 60°C or after exposing $Co_2(CO)_8$ to 8–10 atm of B_2H_6. No structural data are available, and characterization is by elemental analysis and infrared spectroscopy only (22). The only closo homonuclear metalloboride cluster to be structurally characterized to date is $HRu_6(CO)_{17}B$ (**27**). It is produced thermally at 75°C when

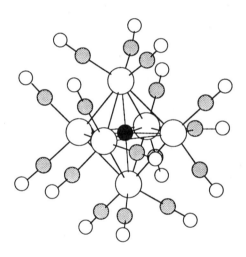

(**27**) $HRu_6(CO)_{17}B$

$Ru_3(CO)_{12}$ is combined with $BH_3 \cdot THF$ in toluene (100). The boron atom is equidistant within experimental error from all six metal atoms [Ru—B = 2.06(2), 2.05(2), 2.07(2), 2.06(2), 2.04(2), 2.06(2) Å], and this high degree of metal–boron interaction is reflected in the extreme downfield chemical shift of the ^{11}B-NMR signal ($\delta + 193.8$). The *endo*-hydrogen atom was not located, but the ^{1}H-NMR resonance at $\delta - 17.8$ is consistent with an Ru—H—Ru bridge. The boride cluster **27** also forms [along with $H_4Ru_4(CO)_{12}$ and some of the butterfly cluster **20**] when $Ru_3(CO)_9BH_5$ is photolyzed or is simply left to stand in solution (32). A second crystallographic study of **27** reveals a different structural isomer which exhibits a significantly distorted octahedral cage (*32*). Deprotonation of **27** occurs in a methanolic solution of [PPN]Cl, and retention of the Ru_6B core is proposed in the light of there being very little change in the ^{11}B-NMR spectral shift (*100*) (see Appendix I). The trimethylammonium salt $[Me_3NH][Ru_6(CO)_{17}B]$ is produced as the second boron-containing species in the reaction of $Ru_3(CO)_{12-x}(NCMe)_x$ ($x = 1$ or 2) with $BH_3 \cdot NMe_3$ (or $BH_3 \cdot THF$ in the presence of NMe_3), the other product being the adduct $HRu_3(CO)_{10}(\mu\text{-}COBH_2NMe_3)$ described in Section III (*25*).

VII

CONCLUDING REMARKS

Just 10 years ago, the chemistry of transition metal clusters containing boron atoms (defined as clusters in which the ratio of M and B atoms exceeds 1:1) was virtually undocumented. The recent unfolding of this area has brought with it clusters that are of structural and chemical interest in their own right, clusters that are models for, but may contrast with, transition metal organometallic systems (*7*), and clusters that are precursors for solid-state materials (*6*). By virtue of the electron-deficient nature of the boron atom, its environment in a cluster and its reactivity therein may differ quite dramatically from that of, for example, another first row atom in a structurally related cluster molecule. One pertinent example of this is the manner in which $Ru_4(CO)_{12}EH_x$ (E = B, $x = 3$; E = C, $x = 2$, E = N, $x = 1$) reacts with diphenylacetylene (*80*). New systems that are characterized often challenge the "cluster electron counter" and pose interesting problems as regards bonding rationales; borane chemists have long been used to the idea of boron atoms in sites of relatively high connectivity, for example, at the vertices of the icosahedral cluster $[B_{12}H_{12}]^{2-}$. Only in the last year or two, however, has the boron atom been structurally characterized *within* a discrete molecular cluster, for example, in $HRu_6(CO)_{17}B$ (*100*) or $[Fe_4Rh_2(CO)_{16}B]^-$ (*65*). These points are only a few of those which render this area of research a fascinating one.

APPENDIX I: NUCLEAR MAGNETIC RESONANCE SPECTROSCOPIC DATA AND AVAILABILITY OF STRUCTURAL DATA FOR METALLOBORANE CLUSTERS WITH A RATIO OF M AND B ATOMS OF AT LEAST 1:1

Compounds are arranged by metal and also in order of their ^{11}B-NMR downfield chemical shift. A shift to lower field is an indication of an increased degree of direct metal to boron bonding (δ). ^{11}B- and ^{1}H-NMR chemical shifts are with respect to $BF_3 \cdot OEt_2$ and tetramethylsilane (TMS), respectively, with downfield positive δ. For anionic metalloboranes and borides, ^{1}H data are provided for the anion only.

Cluster	^{11}B-NMR chemical shift[a]	^{1}H-NMR chemical shift[b]	NMR solvent	X-ray structure	Ref.
Tantalum					
(Cp*Ta)$_2$(μ-Br)$_2$B$_2$H$_6$ (major isomer)	+10.5	See text	C_6D_6	Yes	11
	+18.8				
(Cp*Ta)$_2$(μ-Br)$_2$B$_2$H$_6$ (minor isomer)	−19.9	See text	C_6D_6	No	11
(Cp*Ta)$_2$(B$_2$H$_6$)$_2$	−4.0	+4.4 (4H); +2.26 (30H); −10.5 (8H)	C_6D_6	No	11
Iron					
Fe$_2$(CO)$_6$B$_2$H$_6$	−24.2	+2.31 (2H); −2.44 (1H); −12.88 (1H); −15.57 (2H)	$^nC_6H_{14}$	No	16
[Fe$_2$(CO)$_6$B$_2$H$_5$][AsPh$_4$]	−17.4	+2.2 (2H); −2.6 (1H); −14.2 (2H)	CD_2Cl_2	No	16
[Fe$_2$(CO)$_6$B$_2$H$_5$][Et$_3$NH]	−17.4	+2.2 (2H); −2.6 (1H); −14.2 (2H)	CD_2Cl_2	No	16
HFe$_3$(CO)$_9$BH$_4$	+1.8	+3.2 (1H); −12.8 (1H); −15.8 (2H); −24.4 (1H)	C_6D_6 (^{11}B); $C_6D_5CD_3$, (^{1}H)	Yes	26,27
[HFe$_3$(CO)$_9$BH$_3$][PPN]	+6.2	+3.8 (1H); −13.1 (3H)	CD_2Cl_2	No[c]	27
HFe$_4$(CO)$_{12}$CBH$_2$	+9.6	+4.0 (2H); −26.1 (1H)	C_6D_6 (^{1}H); hexane (^{11}B)	Yes	78
HFe$_3$(CO)$_9$BH$_3$Me	+22.1	+1.03 (3H); −14.6 (3H); −24.0 (1H)	CD_2Cl_2	No	4,27
[HFe$_3$(CO)$_9$BH$_2$Me][AsPh$_4$]	+29.3	+1.09 (3H); −12.9 (3H)	$(CD_3)_2CO$	No	27
HFe$_3$(CO)$_9$(μ-CO)BH$_2$	+56	+5.9 (1H); −13.7 (1H); −25.6 (1H)	$C_6H_5CD_3$	No	33
[Fe$_3$(CO)$_8$(PPhMe$_2$)(μ-CO)BH$_2$][PPN]	+56.8	+7.49–7.37 (5H); +1.25 (6H); −11.6 (1H)	$(CD_3)_2CO$	No	36

(continued)

Cluster	^{11}B-NMR chemical shift^a	^{1}H-NMR chemical shift^b	NMR solvent	X-ray structure	Ref.
[Fe$_3$(CO)$_9$(μ-CO)BH$_2$][PPN]	+57.4	+6.0 (1H); −11.1 (1H)	(CD$_3$)$_2$CO	No	33
1,2,3-Fe$_3$(CO)$_9$B(H)C(H)C(Me)	+58.2	+7.82 (1H); +6.02 (1H); +2.53 (3H)	^{n}C$_6$H$_{14}$	Yes	53
[Fe$_3$(CO)$_8$(PPhMe$_2$)(μ-CO)BHMe][PPN]	+72.7	+7.4–7.2 (5H); +1.28 (6H); +1.10 (3H); −10.3 (1H)	(CD$_3$)$_2$CO	No	36
[Fe$_3$(CO)$_9$(μ-CO)BHMe][PPN]	+74.5	+1.10 (3H); −10.3 (1H)	(CD$_3$)$_2$CO	No	36
HFe$_3$(CO)$_9$(μ-CO)BHMe	+76.4	+1.09 (3H); −13.2 (1H); −25.6 (1H)	(CD$_3$)$_2$CO	No	36
HFe$_4$(CO)$_{12}$BH$_2$	+116.0	−11.9 (2H); −25.4 (1H)	C$_6$D$_6$	Yes	38,60
[Fe$_4$(CO)$_{10}$(PMe$_2$Ph)$_2$BH$_2$][PPN]	+117.9	+7.7–7.5 (10H); +1.90 (12H); −11.6 (2H)	(CD$_3$)$_2$CO	No	38,77
[HFe$_4$(CO)$_{12}$BMe][PPN]	+122	+2.4 (3H); −18.4 (1H)	THF	No	61
[(HFe$_4$(CO)$_{12}$BH]$_2$Au][Au(PMe$_2$Ph)$_2$]	+135.4	−6.5 (2H); −24.9 (2H)	CD$_2$Cl$_2$	Yes	74
HFe$_4$(CO)$_{12}$AuPPh$_3$BH	+137.3	+7.6–7.3 (15H); −7.4 (1H); −24.9 (1H)	CD$_2$Cl$_2$	No	67
Fe$_4$(CO)$_{12}$Au$_2$(AsPh$_3$)$_2$BH	+138.0	+7.7–7.5 (30H); −9.9 (1H)	CD$_2$Cl$_2$	Yes	71
Fe$_4$(CO)$_{12}$Au$_2$(P(4-MeC$_6$H$_4$)$_3$)$_2$BH	+138.0	+7.5–7.0 (30H); +2.34 (3H); −9.9 (1H)	CD$_2$Cl$_2$	Yes	72
Fe$_4$(CO)$_{12}$Au$_2$(PPh$_3$)$_2$BH	{ +141.3, +137.7	+7.57–7.25 (30H); −9.1 (1H)	(CD$_3$)$_2$CO, CD$_2$Cl$_2$	Yes	68,69
[HFe$_4$(CO)$_{11}$(PMe$_2$Ph)BH][PPN]	+141.7	+7.7–7.5 (5H); +1.89 (6H); −8.4 (1H); −23.8 (1H)	(CD$_3$)$_2$CO	No	38,77
[HFe$_4$(CO)$_{12}$BH][PPN]	+150.0	−8.5 (1H); −24.9 (1H)	CD$_2$Cl$_2$	No	37,38
[Fe$_4$(CO)$_{12}$BH][PPN]$_2$	+153	−8.7 (1H)	THF	No	61
HFe$_4$(CO)$_{12}$Au$_2$(PEt$_3$)$_2$B	+179	−25.0 (1H) (see text)	CD$_2$Cl$_2$	Yes	72,73
Fe$_4$(CO)$_{12}$Au$_3$(PPh$_3$)$_3$B	+183	+7.7–7.1 (45H)	CD$_2$Cl$_2$	Yes	70
[Fe$_4$(CO)$_{12}$Au$_2$(PPh$_3$)$_2$B][HNEt$_3$]	+192.2	+7.9–7.1 (30H)	(CD$_3$)$_2$CO	No	4
[cis-Fe$_4$Rh$_2$(CO)$_{16}$B][PPN]	+205		THF	No	65
[trans-Fe$_4$Rh$_2$(CO)$_{16}$B][PPN]	+211		THF	Yes	65
Ruthenium					
HRu$_3$(CO)$_9$BH$_4$ (isomer I)	+2.8	+3.5 (1H); −11.0 (1H); −12.2 (2H); −18.8 (1H)	CD$_2$Cl$_2$	No	28

Compound	δ(^{11}B)	^{1}H NMR	Solvent		Ref.
HRu₃(CO)₁₀(μ-COBH₂NMe₃)	+11.0	+2.56 (9H); +1.58 (2H); −14.31 (1H);	(CD₃)₂CO	Yes	25
HRu₃(CO)₉B₂H₅	+17.0	+4.5 (2H); −1.2 (1H); −12.3 (2H); −19.0 (1H)	CDCl₃	No	48
[Ru₃(CO)₉B₂H₅][PPN]	+18.4	+4.5 (2H); −0.4 (1H); −12.8 (2H)	CD₂Cl₂	No	48
H₂Ru₃(CO)₉BH₃ (isomer II)	+21.0	+4.0 (1H); −11.3 (2H); −18.4 (1H)	CD₂Cl₂	No	28
[HRu₃(CO)₉BH₃][PPN]	+22.5	+3.5 (1H); −12.1 (3H)	CD₂Cl₂	No	32
HRu₄(CO)₁₂B(H)C(Ph)CPhH	+93.7	+7.4–7.0 (10H); 5.04 (1H); −7.3 (1H); −19.06 (1H)	CDCl₃	Yes	80
HRu₄(CO)₁₂BH₂	{ +109.9	−8.4 (2H); −21.18 (1H)	CDCl₃	—	62
	+113.5	−8.5 (2H); −21.15 (1H)	(CD₃OCD₂)₂	Yes	63
HRu₃Fe(CO)₁₂BH₂	+114	−8.5 (1H); −11.3 (1H); −20.4 (1H)	CD₂Cl₂	No	32
HRu₄(CO)₁₂B(H)AuPPh₃	+137.2	+7.7–7.5 (15H); −4.7 (1H); −20.86 (1H)	(CD₃)₂CO	No	62
[HRu₄(CO)₁₂BH]K	+140.9	−6.6 (1H); −20.81 (1H)	(CD₃OCD₂)₂	No	63
[HRu₄(CO)₁₂BH][PPN]	+142.2	−6.7 (1H); −20.92 (1H)	(CD₃)₂CO	No	62
[HRu₃Fe(CO)₁₂BH][PPN]	+143.5	−6.5 (1H); −20.5 (1H)	CD₂Cl₂	No	32
[HRu₄(CO)₁₂B]K₂	+159	−20.09 (1H)	Not stated	No	63
HRu₄(CO)₁₂BAu₂(PPh₃)₂	+170.0	+7.8–7.6 (30H); −20.66 (1H)	(CD₃)₂CO	Yes	62
HRu₆(CO)₁₇B	+193.8	−17.8 (1H)	CDCl₃	Yes	100
[Ru₆(CO)₁₇B][PPN]	+196		Not stated	No	100
[Ru₆(CO)₁₇B][HNMe₃]	+202.0		(CD₃)₂CO	No	25
Osmium					
H₃Os₃(CO)₉CBCl₂(PPh₃)	+3.98	+1.62 (9H); −19.07 (3H)	CD₂Cl₂	No	42
H₃Os₃(CO)₉CBCl₂(PMe₃)	+13.1	+7.32 (3H); +7.11 (6H); +6.99 (6H); −19.00 (3H)	CD₂Cl₂	No	42
H₃Os₃(CO)₉BCO	+19.4	−19.8 (3H)	CD₂Cl₂	Yes	34
H₃Os₃(CO)₉CBCl₂(NMe₃)	+21.0	+3.18 (9H); −18.93 (3H)	CD₂Cl₂	No	42
H₃Os₃(CO)₉C(OBPhCl)BCl	+37.7 (BO) +18.5 (BCl)	+7.96 (2H); +7.54 (2H); +7.42 (1H); −11.69 (1H); −16.69 (1H); −21.71 (1H)	CD₂Cl₂	No	42,46
H₃Os₃(CO)₉CBBr₂	+52.5	−19.36 (3H)	CD₂Cl₂	No	42
H₃Os₃(CO)₈(PPh₃)CBCl₂	+53.4	+7.42 (15H); −18.41 (2H, J_{PH} = 10.5 Hz); −19.63 (1H)	CD₂Cl₂	No	42

(continued)

39

APPENDIX I (Continued)

Cluster	^{11}B-NMR chemical shift[a]	^{1}H-NMR chemical shift[b]	NMR solvent	X-ray structure	Ref.
$H_3Os_3(CO)_9BCH_2$	+53.5	+3.74 (1H); +3.49 (1H); −12.26 (1H); −13.45 (1H); −20.39 (1H)	CD_2Cl_2; $C_6H_5CD_3$	Yes	42,45
$H_3Os_3(CO)_9CBCl_2$	+57.4	−19.43 (3H)	CD_2Cl_2	Yes	42,43
$H_3Os_3(CO)_9C(OBC_8H_{14})BCl$	+58.8 (BO); +18.5 (BCl)	+1.90 (4H); +1.82 (8H); +1.39 (2H); −11.75 (1H); −16.18 (1H); −21.87 (1H)	CD_2Cl_2	Yes	42,46
$H_3Os_3(CO)_9BPMe_3$	+60.9	Not reported	CD_2Cl_2	Yes	34,40
$(H_3Os_3(CO)_9CO)_3B_3O_3$	Not given	−18.5 (9H)	CD_2Cl_2	Yes	35
Cobalt					
$Cp_2Co_2(\mu$-$PPh_2)B_2H_5$	+7.35; +24.2	+7.45−7.02 (10H); +6.47 (1H); +5.38 (1H); +4.92 (5H); 4.38 (5H); 3.01 (1H); −5.51 (1H); −21.15 (1H)	C_6D_6 (^{11}B); $C_6D_5CD_3$ (^{1}H)	Yes	20
4,6-Cp_2Co_2-3,5-$S_2B_2H_2$	+17.6	+5.94 (2H); +4.05 (10H)	C_6D_6	Yes	19
1,2,3-$Cp_3Co_3B_3H_5$	+62.7	+7.11 (3H); +4.90 (15H); −14.48 (2H)	$CDCl_3$	Yes	55,56
1,2,3-$Cp_2Co_2Fe(CO)_4B_3H_3$	+73.0 (2B); +87.5 (1B)	+9.46 (1H); +8.01 (2H); +4.98 (10H)	$CDCl_3$	No	92
1,2,3-$Cp_3Co_3(\mu_3$-$CO)B_3H_3$	+89.9	+9.07 (3H); +4.63 (15H)	CH_2Cl_2 (^{11}B); CS_2 (^{1}H)	Yes	57,58
$Cp_4Co_4(PPh)B_2H_2$	+103.8	+8.42−7.52 (5H); +4.62 (10H); +3.85 (10H) (BH not reported)	Not stated	Yes	84
$H_2Cp_4Co_4B_2H_2$	+114	+4.68 (10H); +4.46 (10H); −15.6 (2H) (BH not reported)	Not stated	Yes	83
$Cp_4Co_4B_4H_4$	+121.4	+11.83 (2H); +4.49 (20H)	$CDCl_3$	Yes	86,88
$Cp_3Co_3(\mu_3$-$BPh)(\mu_3$-$PPh)$	+143.7	+8.6−7.3 (10H); +4.45 (15H)	$CDCl_3$ (^{1}H); toluene (^{11}B)	Yes	54
Nickel					
$Cp_4Ni_4B_4H_4$	+56.2	+8.22 (4H); +5.34 (20H)	$CDCl_3$	Yes	87,89

[a] With respect to $BF_3 \cdot OEt_3$; downfield shifts are at positive δ.
[b] With respect to TMS; downfield shifts are at positive δ. Resonances for cations are omitted.
[c] Space group and cell dimensions are given; structure not solved.

40

APPENDIX II: SELECTED BOND DISTANCES FOR STRUCTURALLY CHARACTERIZED METALLOBORANE CLUSTERS WITH A RATIO OF M AND B ATOMS OF AT LEAST 1:1[a]

Cluster	M—M (Å)	M—B (Å)	B—B (Å)	B—H (Å)	M—H (Å)	Ref.
Tantalum						
$(Cp^*Ta)_2(\mu\text{-Br})_2B_2H_6$	2.839(1) 2.42(2) 2.40(2) 2.42(2)	2.37(2)	1.88(3)			11
Manganese						
$HMn_2(CO)_6(B_2H_6)Mn(CO)_4$	2.845(3)	Mean 2.30(2)	1.76(3)			13
Iron						
$HFe_3(CO)_9BH_4$	2.6026(21) 2.5923(19) 2.6732(22)	2.197(8) 2.176(8) 2.129(8)				26,27
$1,2,3\text{-}Fe_3(CO)_9B(H)C(H)C(Me)$	2.605(2) 2.552(2) 2.586(2)	2.010(11) 2.046(11)				53
$HFe_4(CO)_{12}CBH_2$	2.660(1) × 2 2.673(1) × 2 2.586(1)			1.04(5) 1.25(5)	1.56(5) 1.59(6)	78
$HFe_4(CO)_{12}BH_2$	2.637(1) 2.666(1) 2.672(1) 2.671(1) 2.662(1)	Fe—B 2.044(6) 2.047(6) 1.966(6) 1.974(6)		1.36(5) 1.38(5)	Fe—H—B 1.55(5) 1.58(5) Fe—H—Fe 1.62(4) 1.72(4)	59,60
$[(HFe_4(CO)_{12}BH)_2Au][Au(PMe_2Ph)_2]$	Fe—Fe 2.630(2) 2.649(2) 2.660(2) 2.650(2) 2.668(2) Fe—Au 2.614(1)	Fe—B 2.065(13) 2.082(12) 2.015(12) 1.990(12) Au—B 2.300(12)				74

(continued)

APPENDIX II (Continued)

Cluster	M—M (Å)	M—B (Å)	B—B (Å)	B—H (Å)	M—H (Å)	Ref.
$Fe_4(CO)_{12}Au_2(AsPh_3)_2BH$	Fe—Fe 2.586(2) 2.734(2) 2.668(3) 2.717(2) 2.685(2) Fe—Au 2.590(1) 2.613(1) Au—Au 2.931(1)	Fe—B 2.114(13) 2.068(11) 1.999(14) 2.031(14) Au—B 2.344(12) 2.341(13)				71
$Fe_4(CO)_{12}Au_2(PPh_3)_2BH$	Fe—Fe 2.720(2) 2.671(2) 2.578(2) 2.708(2) 2.655(3) Fe—Au 2.630(1) 2.606(1) Au—Au 2.943(1)	Fe—B 2.07(1) 2.00(1) 2.01(1) 2.13(1) Au—B 2.36(1) 2.35(1)				68,69
$Fe_4(CO)_{12}Au_2(P\{4\text{-}MeC_6H_4\}_3)_2BH$	Fe—Fe 2.741(4) 2.685(4) 2.580(4) 2.692(3) 2.674(5) Fe—Au 2.635(2) 2.627(2) Au—Au 2.975(1)	Fe—B 2.114(19) 2.008(24) 2.024(18) 2.126(19) Au—B 2.306(20) 2.368(18)				72

Compound			Ref.
HFe$_4$(CO)$_{12}$Au$_2$(PEt$_3$)$_2$B	Fe—Fe 2.621(2) 2.690(2) × 2 2.689(2) × 2 Fe—Au 2.615(1) × 2 Au—Au 2.880(1)	Fe—B 2.065(11) × 2 1.989(3) × 2 Au—B 2.262(11) × 2	72,73
Fe$_4$(CO)$_{12}$Au$_3$(PPh$_3$)$_3$B	Fe—Fe 2.599(4) 2.664(4) 2.713(5) 2.666(4) 2.709(6) Fe—Au 2.693(4) 2.711(3) 2.625(4) 2.616(3) Au—Au 2.877(1) 2.858(1)	Fe—B 2.09(2) 2.16(2) 1.98(2) 2.13(2) Au—B 2.34(2) 2.27(2) 2.32(2)	70
[*trans*-Fe$_4$Rh$_2$(CO)$_{16}$B][PPN]	Fe—Fe 2.748(2) 2.763(2) 2.748(2) 2.748(2) Fe—Rh 2.719(1) 2.719(2) 2.914(2) 2.834(2) 2.907(1) 2.881(2) 2.740(2) 2.743(2)	Fe—B 1.935(9) 1.952(9) 1.935(10) 1.951(10) Rh—B 2.013(11) 2.042(11)	65

(continued)

43

APPENDIX II (Continued)

Cluster	M—M (Å)	M—B (Å)	B—B (Å)	B—H (Å)	M—H (Å)	Ref.
Ruthenium						
$HRu_3(CO)_{10}(\mu\text{-}COBH_2NMe_3)$	2.823(1) 2.832(1) 2.835(1)			1.096(46) 1.028(49)	1.736(40) 1.784(43)	25
$HRu_4(CO)_{12}B(H)C(Ph)CPhH$	2.794(1) 2.922(1) 2.807(1) 2.950(1)	2.236(14) 2.178(13) 2.193(12) 2.152(13)				80
$HRu_4(CO)_{12}BH_2$	2.8220(4) × 2 2.8283(5) × 2 2.904(1)	2.111(6) 2.106(6) 2.195(5) × 2		1.3(2) 1.4	<u>Ru—H—B</u> 1.8(2) 1.7 <u>Ru—H—Ru</u> 1.73(5) × 2	63
$HRu_4(CO)_{12}BAu_2(PPh_3)_2$	<u>Ru—Ru</u> 2.871(1) 2.886(1) 2.913(1) 2.885(1) 2.864(1) <u>Ru—Au</u> 2.728(1) 2.730(1) <u>Au—Au</u> 2.849(1)	<u>Ru—B</u> 2.259(13) 2.250(14) 2.114(13) 2.130(11) <u>Au—B</u> 2.288(15) 2.272(13)				62
$HRu_6(CO)_{17}B$	2.832(2) 2.994(2) 2.925(2) 2.889(2) 2.912(3)	2.06(2) 2.05(2) 2.07(2) 2.06(2) 2.04(2)				100

44

Compound				Ref.
	2.914(2) 2.892(2) 2.910(2) 2.887(2) 2.867(2) 2.900(2) (one distance not reported)		2.06(2)	
Osmium				
H₃Os₃(CO)₉BCO	2.913(1) 2.917(1) 2.919(1)	2.155(9) 2.186(10) 2.130(10)		34
H₃Os₃(CO)₉BCH₂	2.929(1) 2.827(1) 2.815(1)	2.287(12) 2.270(11) 2.216(14)		45
H₃Os₃(CO)₉CBCl₂	2.875(1) 2.869(1) 2.879(1)			43
H₃Os₃(CO)₉C(OBC₈H₁₄)BCl	3.0454(6) 2.8532(6) 2.7974(5)	2.38(1) 2.36(1)		46
{H₃Os₃(CO)₉CO)₃B₃O₃	2.888(3) 2.886(3) 2.882(3) 2.889(3) 2.883(3) 2.891(3) 2.888(3) 2.892(3) 2.877(3)			35
Cobalt				
Cp₂Co₂(μ-PPh₂)B₂H₅	2.472(1)	2.110(9) 2.025(8) 2.138(9)	1.79(2)	20

APPENDIX II (Continued)

Cluster	M—M (Å)	M—B (Å)	B—B (Å)	B—H (Å)	M—H (Å)	Ref.
4,6-Cp$_2$Co$_2$-3,5-S$_2$B$_2$H$_2$	2.181(6) 2.148(6) 2.138(6) 2.181(7)	1.760(9)				19
Co$_3$(CO)$_9$(μ_3-COBH$_2$NEt$_3$)	2.498(2) 2.492(2) × 2					21
Co$_3$(CO)$_9$(μ_3-COBCl$_2$NEt$_3$)	2.476(3) 2.471(3) 2.470(3)					23
Co$_3$(CO)$_9$(μ_3-COBBr$_2$NEt$_3$)	2.489(4) 2.484(4) 2.478(4)					24
1,2,3-Cp$_3$Co$_3$B$_3$H$_5$	2.488(1) × 2 2.472(1)	2.051(3) × 2 2.047(3) × 2	1.716(6) 1.724(5) × 2 2.030(4) × 2	1.06(4) × 2 1.15(3)		56
1,2,3-Cp$_3$Co$_3$(μ_3-CO)B$_3$H$_3$	2.445(1) 2.442(1) 2.444(1)	2.040(11) 2.057(10) 2.049(10) 2.031(10) 2.048(10) 2.040(10)	1.712(11) 1.705(14) 1.722(13)	1.234(—) 1.209(—) 1.239(—)		57
Cp$_3$Co$_3$(μ_3-BPh)(μ_3-PPh)	2.553(1) 2.473(2) 2.561(1)	2.065(8) 2.018(8) 2.031(9)				54

Compound					Ref.
Cp$_4$Co$_4$(PPh)B$_2$H$_2$	2.517(1) 2.517(1) 2.520(1) 2.521(1)	1.937(9) 2.080(9) 2.057(9) 1.956(9) 2.069(9) 2.044(8)	1.79(1)		84
H$_2$Cp$_4$Co$_4$B$_2$H$_2$	2.456(1) × 2 2.461(2) × 2 2.501(2)	2.070(8) × 2 2.009(8) 2.019(8)	1.80(2)		83
Cp$_4$Co$_4$B$_4$H$_4$[b]	2.479(2) 2.477(2) 2.479(2) 2.481(2)	2.01(1) 2.03(1) 2.04(1) 2.03(1) 1.99(1) 2.01(1) 2.01(1) 2.03(1) 2.05(1) 1.99(1) 2.04(1) 2.04(1)	1.84(2) 1.89(2)		88
Nickel					
Cp$_4$Ni$_4$B$_4$H$_4$	2.354(1) × 2	2.037(4) × 2 2.031(3) × 2 2.018(4) × 2 2.063(4) × 2 2.028(4) × 2 2.045(4) × 2	1.868(7) 1.932(7) 1.955(5) × 2	0.89(3) × 2 1.10(3) × 2	89

[a] Parameters have been obtained from supplementary data via the Cambridge Crystallographic Data Center when not listed in published work.
[b] Parameters quoted are for molecules on general positions; of 12 molecules in the unit cell, 8 are on general and 4 on special positions.

ACKNOWLEDGMENTS

I should like to acknowledge the enthusiastic support of all my co-workers whose efforts have contributed toward our own investigations in the area described in this article, and I am grateful for the support of the Petroleum Research Fund, administered by the American Chemical Society, to the Royal Society for a University Research Fellowship (1987–1990) and to Johnson Matthey for generous loans of $RuCl_3$. The Cambridge Crystallographic Data Center is acknowledged for providing coordinates for many of the structural figures redrawn in this article.

REFERENCES

1. J. D. Kennedy, *Prog. Inorg. Chem.* **32**, 519 (1984).
2. J. D. Kennedy, *Prog. Inorg. Chem.* **34**, 211 (1986).
3. T. P. Fehlner, *New J. Chem.* **12**, 307 (1988).
4. C. E. Housecroft, *Polyhedron* **6**, 1935 (1987).
5. C. E. Housecroft, "Boranes and Metalloboranes: Structure, Bonding and Reactivity." Ellis Horwood Limited, Chichester, 1990.
6. T. P. Fehlner, *Adv. Inorg. Chem.* **35**, 199 (1990).
7. A. A. Aradi and T. P. Fehlner, *Adv. Organomet. Chem.* **30**, 189 (1990).
8. N. P. Rath and T. P. Fehlner, *J. Am. Chem. Soc.* **110**, 5345 (1988).
9. T. P. Fehlner, P. T. Czech, and R. F. Fenske, *Inorg. Chem.* **29**, 3103 (1990).
10. B. Martin, S. A. Cohen, R. E. Marsh, and J. E. Bercaw, *reported in Ref. 11.*
11. C. Ting and L. Messerle, *J. Am. Chem. Soc.* **111**, 3449 (1989).
12. P. H. Bird and M. G. H. Wallbridge, *J. Chem. Soc., Chem. Commun.,* 687 (1968).
13. H. D. Kaesz, W. Fellmann, G. R. Wilkes, and L. F. Dahl, *J. Am. Chem. Soc.* **87**, 2753 (1965).
14. E. L. Andersen, R. L. DeKock, and T. P. Fehlner, *Inorg. Chem.* **20**, 3291 (1981).
15. E. L. Andersen and T. P. Fehlner, *J. Am. Chem. Soc.* **100**, 4606 (1978).
16. G. B. Jacobsen, E. L. Andersen, C. E. Housecroft, F.-E. Hong, M. L. Buhl, G. J. Long, and T. P. Fehlner, *Inorg. Chem.* **26**, 4040 (1987).
17. S. W. Lee, W. D. Tucker, and M. G. Richmond, *J. Organomet. Chem.* **398**, C6 (1990).
18. X. Meng, A. K. Bandyopadhyay, and T. P. Fehlner, *J. Organomet. Chem.* **394**, 15 (1990).
19. R. P. Micciche, P. J. Carroll, and L. G. Sneddon, *Organometallics* **4**, 1619 (1985).
20. J. Feilong, T. P. Fehlner, and A. L. Rheingold, *J. Organomet. Chem.* **348**, C22 (1988).
21. F. Klanberg, W. B. Askew, and L. J. Guggenberger, *Inorg. Chem.* **7**, 2265 (1968).
22. G. Schmid, V. Bätzel, G. Etzrodt, and R. Pfeil, *J. Organomet. Chem.* **86**, 257 (1975).
23. V. Bätzel, U. Müller, and R. Allmann, *J. Organomet. Chem.* **102**, 109 (1975).
24. V. Bätzel, *Z. Naturforsch., B: Anorg. Chem., Org. Chem.* **31B**, 342 (1976).
25. A. K. Chipperfield, C. E. Housecroft, and P. R. Raithby, *Organometallics* **9**, 479 (1990).
26. J. C. Vites, C. Eigenbrot, and T. P. Fehlner, *J. Am. Chem. Soc.* **106**, 4633 (1984).
27. J. C. Vites, C. E. Housecroft, C. Eigenbrot, M. L. Buhl, G. J. Long, and T. P. Fehlner, *J. Am. Chem. Soc.* **108**, 3304 (1986).
28. A. K. Chipperfield and C. E. Housecroft, *J. Organomet. Chem.* **349**, C17 (1988).
29. T. J. Marks and J. R. Kolb, *Chem. Rev.* **77**, 263 (1977).
30. M. M. Lynam, D. M. Chipman, R. D. Barreto, and T. P. Fehlner, *Organometallics* **6**, 2405 (1987).
31. T. P. Fehlner, *Polyhedron* **9**, 1955 (1990).
32. A. K. Chipperfield, S. M. Draper, C. E. Housecroft, D. Matthews, and A. L. Rheingold, *200th ACS Natl. Meet. INORG,* **430** (1990).

33. J. C. Vites, C. E. Housecroft, G. B. Jacobsen, and T. P. Fehlner, *Organometallics* **3**, 1591 (1984).
34. S. G. Shore, D.-Y. Jan, L.-Y. Hsu, and W.-L. Hsu, *J. Am. Chem. Soc.* **105**, 5923 (1983).
35. S. G. Shore, D.-Y. Jan, W.-L. Hsu, L.-Y. Hsu, S. Kennedy, J. C. Huffman, T.-C. L. Wang, and A. G. Marshall, *J. Chem. Soc., Chem. Commun.,* 392 (1984).
36. C. E. Housecroft and T. P. Fehlner, *J. Am. Chem. Soc.* **108**, 4867 (1986).
37. C. E. Housecroft and T. P. Fehlner, *Organometallics* **5**, 379 (1986).
38. C. E. Housecroft, M. L. Buhl, G. J. Long, and T. P. Fehlner, *J. Am. Chem. Soc.* **109**, 3323 (1987).
39. T. P. Fehlner and C. E. Housecroft, *Organometallics* **3**, 764 (1984).
40. D.-Y. Jan, L.-Y. Hsu, and S. G. Shore, *188th ACS Natl. Meet. INORG,* **180** (1984).
41. R. D. Barreto, T. P. Fehlner, L.-Y. Hsu., D.-Y. Jan, and S. G. Shore, *Inorg. Chem.* **25**, 3572 (1986).
42. D. P. Workman, D.-Y. Jan, and S. G. Shore, *Inorg. Chem.* **29**, 3518 (1990).
43. D.-Y. Jan, L.-Y. Hsu, D. P. Workman, and S. G. Shore, *Organometallics* **6**, 1984 (1987).
44. M. M. Olmstead, P. P. Power, and K. J. Weese, *J. Am. Chem. Soc.* **109**, 2541 (1987).
45. D.-Y. Jan and S. G. Shore, *Organometallics* **6**, 428 (1987).
46. D. P. Workman, H.-B. Deng, and S. G. Shore, *Angew. Chem., Int. Ed. Engl.* **29**, 309 (1990).
47. C. R. Eady, B. F. G. Johnson, and J. Lewis, *J. Chem. Soc., Dalton Trans.,* 477 (1977).
48. A. K. Chipperfield, C. E. Housecroft, and D. M. Matthews, *J. Organomet. Chem.* **384**, C38 (1990).
49. T. P. Fehlner, *in* "Boron Chemistry" (R. W. Parry and G. Kodama, eds.), p. 95. Pergamon, Oxford, 1980.
50. C. E. Housecroft and D. M. Matthews, unpublished observations (1990).
51. N. N. Greenwood, *Pure Appl. Chem.* **49**, 791 (1977).
52. C. E. Housecroft, *Inorg. Chem.* **25**, 3108 (1986).
53. X. Meng, T. P. Fehlner, and A. L. Rheingold, *Organometallics* **9**, 534 (1990).
54. J. Feilong, T. P. Fehlner, and A. L. Rheingold, *Angew. Chem., Int. Ed. Engl.* **27**, 424 (1988).
55. V. R. Miller, R. Weiss, and R. N. Grimes, *J. Am. Chem. Soc.* **99**, 5646 (1977).
56. J. R. Pipal and R. N. Grimes, *Inorg. Chem.* **16**, 3255 (1977).
57. J. M. Gromek and J. Donohue, *Cryst. Struct. Commun.* **10**, 849 (1981).
58. G. J. Zimmerman, L. W. Hall, and L. G. Sneddon, *Inorg. Chem.* **19**, 3642 (1980).
59. T. P. Fehlner, C. E. Housecroft, W. R. Scheidt, and K. S. Wong, *Organometallics* **2**, 825 (1983).
60. K. S. Wong, W. R. Scheidt, and T. P. Fehlner, *J. Am. Chem. Soc.* **104**, 1111 (1982).
61. N. P. Rath and T. P. Fehlner, *J. Am. Chem. Soc.* **109**, 5273 (1987).
62. A. K. Chipperfield, C. E. Housecroft, and A. L. Rheingold, *Organometallics* **9**, 681 (1990).
63. F.-E. Hong, D. A. McCarthy, J. P. White, C. E. Cottrell, and S. G. Shore, *Inorg. Chem.* **29**, 2874 (1990).
64. T. P. Fehlner, *in* "Comments in Inorganic Chemistry," p. 326. Gordon & Breach Science Publishers, London, 1988.
65. R. Khattar, J. Puga, T. P. Fehlner, and A. L. Rheingold, *J. Am. Chem. Soc.* **111**, 1877 (1989).
66. I. M. Salter, *Adv. Organomet. Chem.* **29**, 249 (1989).
67. K. S. Harpp and C. E. Housecroft, *J. Organomet. Chem.* **340**, 389 (1988).
68. C. E. Housecroft and A. L. Rheingold, *J. Am. Chem. Soc.* **108**, 6420 (1986).

69. C. E. Housecroft and A. L. Rheingold, *Organometallics* **6**, 1332 (1987).
70. K. S. Harpp, C. E. Housecroft, A. L. Rheingold, and M. S. Shongwe, *J. Chem. Soc., Chem. Commun.,* 965 (1988).
71. C. E. Housecroft, M. S. Shongwe, A. L. Rheingold, and P. Zanello, *J. Organomet. Chem.* **408**, 7 (1991).
72. C. E. Housecroft, M. S. Shongwe, and A. L. Rheingold, *Organometallics* **8**, 2651 (1989).
73. C. E. Housecroft, A. L. Rheingold, and M. S. Shongwe, *Organometallics* **7**, 1885 (1988).
74. C. E. Housecroft, A. L. Rheingold, and M. S. Shongwe, *J. Chem. Soc., Chem. Commun.,* 1630 (1988).
75. P. Zanello, *in* "Stereochemistry of Organometallic and Inorganic Compounds" (I. Bernal, ed.), Vol. 5. Elsevier, Amsterdam, 1991.
76. S. R. Drake, *Polyhedron* **9**, 455 (1990).
77. C. E. Housecroft and T. P. Fehlner, *Organometallics* **5**, 1279 (1986).
78. X. Meng, N. P. Rath, and T. P. Fehlner, *J. Am. Chem. Soc.* **111**, 3422 (1989).
79. S. M. Draper and C. E. Housecroft, unpublished results (1990).
80. A. K. Chipperfield, B. S. Haggerty, C. E. Housecroft, and A. L. Rheingold, *J. Chem. Soc., Chem. Commun.,* 1174 (1990).
81. T. Dutton, B. F. G. Johnson, J. Lewis, S. M. Owen, and P. R. Raithby, *J. Chem. Soc., Chem. Commun.,* 1423 (1988).
82. M. Blohm and W. L. Gladfelter, *Organometallics* **5**, 1049 (1986).
83. J. Feilong, T. P. Fehlner, and A. L. Rheingold, *J. Am. Chem. Soc.* **109**, 1860 (1987).
84. J. Feilong, T. P. Fehlner, and A. L. Rheingold, *J. Chem. Soc., Chem. Commun.,* 1395 (1987).
85. J. Lunniss, S. A. MacLaughlin, N. J. Taylor, A. J. Carty, and E. Sappa, *Organometallics* **4**, 2066 (1985).
86. V. R. Miller and R. N. Grimes, *J. Am. Chem. Soc.* **98**, 1600 (1976).
87. J. R. Bowser and R. N. Grimes, *J. Am. Chem. Soc.* **100**, 4623 (1978).
88. J. R. Pipal and R. N. Grimes, *Inorg. Chem.* **18**, 257 (1979).
89. J. R. Bowser, A. Bonny, J. R. Pipal, and R. N. Grimes, *J. Am. Chem. Soc.* **101**, 6229 (1979).
90. D. N. Cox, D. M. P. Mingos, and R. Hoffmann, *J. Chem. Soc., Dalton Trans.,* 1788 (1981).
91. M. E. O'Neill and K. Wade, *Inorg. Chem.* **21**, 464 (1982).
92. R. Weiss, J. R. Bowser, and R. N. Grimes, *Inorg. Chem.* **17**, 1522 (1978).
93. M. L. Blohm, D. E. Fjare, and W. L. Gladfelter, *J. Am. Chem. Soc.* **108**, 2301 (1986).
94. D. E. Fjare and W. L. Gladfelter, *J. Am. Chem. Soc.* **106**, 4799 (1984).
95. A. K. Bandyopodhyay, R. Khattar, and T. P. Fehlner, *Inorg. Chem.* **28**, 4434 (1989).
96. R. Thompson, in *Prog. Boron Chem.* **2**, 173 (1970).
97. A. F. Wells, "Structural Inorganic Chemistry." Oxford Univ. Press, Oxford, 1984.
98. B. Ganem and J. O. Osby, *Chem. Rev.* **86**, 763 (1986).
99. W. N. Lipscomb, *J. Less-Common Met.* **82**, 1 (1981).
100. F.-E. Hong, T. J. Coffy, D. A. McCarthy, and S. G. Shore, *Inorg. Chem.* **28**, 3284 (1989).

ADVANCES IN ORGANOMETALLIC CHEMISTRY, VOL. 33

Organometallic Ions and Ion Pairs

JAY K. KOCHI and T. MICHAEL BOCKMAN

Department of Chemistry
University of Houston
Houston, Texas 77204

I

INTRODUCTION

The ability of organometallic species to undergo ready oxidation constitutes an important facet to their utility as ultimate reagents in organic synthesis and as reactive intermediates in catalytic organic reactions. Thus, the prototypical Grignard addition to carbonyl groups and Grignard coupling with alkyl halides represent, in their most fundamental constructs, the oxidation of the nucleophilic organomagnesium reagent (*1*). Indeed, the parallelism between reductant and nucleophile was originally pointed out by Edwards and Pearson a number of years ago (*2*). Furthermore, nucleophiles are often most effective as the negatively charged anions, and they are also referred to sometimes as Brønsted and Lewis bases or in terms of their softness on the hard and soft acid-base (HSAB) scale (*3,4*). Since each of these classifications relates in some way to a qualitative property that is considered vaguely in degrees of electron richness (*5,6*), the more inclusive description as electron donors, as originally defined by Mulliken (*7*), is preferred. The contrasting descriptions have also been applied to organometallic cations as electrophiles and oxidants in reference to their electron-acceptor behavior, as the direct comparison below emphasizes.

Electron donor (D)		Electron acceptor (A)
Anions		Cations
Reductant		Oxidant
Nucleophile		Electrophile
Base	(Brønsted, Lewis, HSAB)	Acid
(Electron rich)		(Electron poor)

The electron-donor property of an organometallic reagent, as indicated by the vertical ionization potential (IP), is strongly dependent on the ligands. For example, merely changing a methyl ligand to ethyl is sufficient to increase the donor property by almost 12 kcal/mol. In turn, a secondary isopropyl ligand is 8.5 kcal more effective than ethyl, and the *tert*-butyl ligand is 6.4 kcal/mol more effective than isopropyl (*8*). Metal hydrides are slightly less effective donors than the methyl analogs. Most importantly, the donor strengths of organometals are not related to the formal oxidation state of the metal centers, as indicated by the different IPs listed in Table I

TABLE I

Ionization Potentials of Organometals[a]

Organometal donor	IP (eV)	Organometal donor	IP (eV)
$Mo(CO)_6$	8.4	$Mn(CO)_5CH_3$	8.65
$Mo(CO)_2(dmpe)_2$	6.0	$Mn(CO)_5CF_3$	9.17
$Mo(CO)Cp_2$	5.9	$Mn(CO)_5H$	8.85
MoH_2Cp_2	6.4	$Mn(CO)_3Cp$	8.05
$Mo(CH_3)_2Cp_2$	6.1	$Mn(CO)_5Br$	8.83
		$Mn(CO)_5SnMe_3$	8.63
$Fe(CO)_5$	8.60		
$FeCp_2$	6.88	$Ni(CO)_4$	8.93
$Fe(CO)_2(CH_3)Cp$	7.7	$Ni(PF_3)_4$	8.82
$Fe(CO)_2(Br)Cp$	7.95	$Ni(bipy)Et_2$	6.4
$[Fe(CO)Cp]_4$	6.45	$Ni(allyl)_2$	7.76

[a] From Ref. *1*.

(*9*). The net charge on the complex is also an important factor, and organometallic anions are among the most effective electron donors currently available (*10*). Conversely, the positively charged organometallic cations are useful electron acceptors in measure with the magnitudes of their electron affinities (*11*).

Electron donors (D) and acceptors (A) constitute reactant pairs that are traditionally considered with more specific connotations in mind, such as nucleophiles and electrophiles in bond formation, reductant and oxidant in electron transfer, bases and acids in adduct production, and anion and cations in ion-pair annihilation (*12*). In the latter case, the preequilibrium formation of contact ion pairs (CIP), that is,

$$D^- + A^+ \underset{\text{}}{\overset{K}{\rightleftharpoons}} D^-,A^+ \tag{1}$$

has its counterparts that are variously described as an encounter complex, a nonbonded electron donor–acceptor (EDA) complex, a precursor complex, or a contact charge-transfer complex (*13,14*). In order to illustrate the utility of the electron donor–acceptor concept as an unifying theme in the delineation of the organometallic reactions of anions and cations, representative examples are presented in this article that revolve around the spectral identification of charge-transfer bands of the contact ion pair in Eq. (1).

II

CARBONYLMETALLATE ANIONS AS ELECTRON DONORS IN CHARGE-TRANSFER SALTS

Intermolecular interaction of ions has been roughly categorized in terms of contact ion pairs (CIP) and solvent-separated ion pairs (SSIP) (*14*),

$$D^-, A^+ \rightleftharpoons D^-//A^+ \qquad (2)$$
$$\text{(CIP)} \qquad \text{(SSIP)}$$

though doubtlessly a myriad of ion-pair types exist with continuously varying interionic separations. Of these, only the contact ion pair is the critical precursor to electrophile–nucleophile interactions (*15*). Unfortunately there are no techniques to establish CIP structures in solution, such as those routinely available for solids and gases. Electronic transitions of electron donor–acceptor complexes from both neutral and charged components have been delineated as charge-transfer (CT) excitations by Mulliken (*7*). The characteristic charge-transfer absorption bands have been identified in various pyridinium iodides that are constituted as acceptor–donor contact ion pairs in solution (*16*). Charge-transfer bands have also been reported for carbonylmetallate salts such as $Tl^+ Co(CO)_4^-$ (*17*), $PY^+ V(CO)_6^-$ (*18*), and $Co(CO)_3L_2^+ Co(CO)_4^-$ (*19*), where PY^+ stands for pyridinium cations and L for phosphines. Indeed, the strong and variable donor properties of carbonylmetallates (*20*) can be exploited in the quantitative study of contact ion pairs. Close examination of the charge-transfer characteristics of a series of salts, the structures of which have been established by X-ray crystallography, can be related to contact ion pairs in solution. Furthermore, the unique properties of anionic carbonylmetallate donors lead to CT excited states that result in a wide variety of productive photochemistry (*21*).

A. Isolation and Spectral Characterization of Charge-Transfer Salts of Carbonylmetallates

The carbonylmetallate anions tetracarbonylcobaltate $[Co(CO)_4^-]$, hexacarbonylvanadate $[V(CO)_6^-]$, and pentacarbonylmanganate $[Mn(CO)_5^-]$ can be prepared as the colorless sodium or bis(triphenylphosphoranyidene)ammonium $[(Ph_3P)_2N^+$ or PPN] salts (*22–24*) together with three distinct classes of colorless acceptor cations, namely, cobalticenium $(Cp_2Co^+$, where Cp is $\eta^5 - C_5H_5)$ (*25*), dibenzenechromium $[(C_6H_6)_2Cr^+$ or $\phi_2Cr^+]$ (*26*), and a series of N-methylpyridinium derivatives as either the halide or hexafluorophosphate salt. When a pair of colorless aqueous solutions containing 0.1 M Na^+ $Co(CO)_4^-$ and Cp_2Co^+ Cl^- are poured together, dark red crystals separate immediately (27). This remarkable visual display is also observed when an aqueous solution of Na^+ $Co(CO)_4^-$ is mixed with colorless saline solutions of N-methyl-4-cyanopyridinium (NCP^+), N-methyl-1-quinolinium (Q^+), N-methyl-4-phenylpyridinium (PP^+), and dibenzenechromium (ϕ_2Cr^+) chlorides to yield dark blue, burgundy, orange, and green crystals, respectively.

$$\text{[structure]}\ Cl^- + Na^+Co(CO)_4^- \rightarrow \text{[structure]}\ Co(CO)_4^- + Na^+Cl^- \qquad (3)$$

(burgundy)

Analogously, when a colorless aqueous solution of quinolinium chloride (Q^+ Cl^-) is mixed with an almost colorless aqueous solution of Na^+ $V(CO)_6^-$, the well-formed dark green crystals of the vanadate salt precipitate immediately. In each case, the spontaneous separation of the highly colored salts is made even more dramatic by the absence of color in the aqueous mother liquors throughout the course of precipitation.

The quantitative effects in these colored salts are observed as broad absorption bands in the spectral region from 350 to beyond 700 nm when they are dissolved in dichloromethane, or in the diffuse reflectance spectra of the crystalline salts. The correspondence of the band maximum (λ_{CT}) and breadth (fwhm) of the absorption and reflectance spectra indicates the close similarity of the various carbonylcobaltate salts in solution with those in the crystalline solid state (Table II). Significantly, the absorption bands of the various iodide salts in dichloromethane solution are also similar to those in the solid state. The absorption bands listed in Table II showed two important trends. For the series of carbonylcobaltate salts, the maxima λ_{CT} values are consistently red shifted with cation variation in the order $PP^+ > Cp_2Co^+ > Q^+ > \phi_2Cr^+ > NCP^+$. Furthermore, the same order applies to the corresponding carbonylvanadate, carbonylmanganate, and iodide salts. Such a progressive bathochromic shift of the absorption bands ($h\nu_{CT}$) with decreasing cathodic potentials E_c for cation reduction (as viewed down the rows in Table II) can be described by the linear correlation $h\nu_{CT} = -aE_c + \text{constant}$, with $a = 1.00$ for both series of carbonylcobaltate and iodide salts.

According to Mulliken charge-transfer theory (7), the separation of $\Delta h\nu_{CT} = 0.61$ eV for the two series represents the constant difference in the donor properties (ionization potential) of $Co(CO)_4^-$ and I^- in salt pairs with the same acceptor cation. The latter is a corollary of the Mulliken

TABLE II

VISIBLE ABSORPTION BANDS OF CHARGE-TRANSFER SALTS IN THE CRYSTALLINE SOLID STATE AND IN SOLUTION[a]

Cationic acceptor[c]	E_c (V versus SCE)[d]	Anionic donor[b]							
		$Co(CO)_4^-$		$V(CO)_6^-$		$Mn(CO)_5^-$		I^-	
PP+	−1.27	442	(494)	512	(578)		(562)[e]	(<380)	(390)
Cp$_2$Co+	−0.99	508	(520)	540	(630)	600	(570)[f]		
iQ+	−1.08	510	(516)				(580)[e]		
Q+	−0.90	520	(550)	520	(720)			465	(444)
Φ$_2$Cr+		516							
CMP+	−0.79		(590)[e]	600	(760)		(642)[e]	490	(454)
NCP+	−0.67	560	(620)[e]						

[a] Maximum (nm) of the electronic band obtained from the diffuse reflectance spectrum after 10% dispersion in silica (27).

[b] Absorption spectrum in 1 mM dichloromethane solution in parentheses.

[c] iQ+, N-Methyl-2-quinolinium; CMP+, N-methyl-4-carbomethoxypyridinium. See text for identification of other acronyms.

[d] Irreversible cathodic CV peak potential at 500 mV/second versus a standard calomel electrode.

[e] Prepared in situ from the PPN+ salt.

[f] Prepared in situ from the Na+ salt.

relationship for a series of carbonylmetallate salts in which the charge-transfer maxima viewed across the columns in Table II are consistently blue shifted in the order $Mn(CO)_5^- < V(CO)_6^- < Co(CO)_4^- < I^-$. Indeed, such a progressive increase in the energy of the absorption bands with increasing anodic potential E_a for anion oxidation represents the linear correlation $h\nu_{CT} = bE_a + $ constant, which applies to salts derived from a common cation and a series of donor anions. Both of the linear correlations derive from Mulliken theory, more commonly expressed (28) as $h\nu_{CT} = IP - EA + \omega + $ constant, where IP and EA are the ionization potential and electron affinity of the donor anions and acceptor cations, respectively, in the gas phase and ω represents the ion-pair interaction. (Note that the ionization potentials in the gas phase parallel the anodic potentials in solution for structurally related electron donors; the same interrelationship applies to electron affinities and cathodic potentials.) Accordingly, these colored crystals are also referred to as charge-transfer salts (29).

B. X-Ray Crystallographic Structures of Charge-Transfer Salts

The origin of the charge-transfer absorptions of the colored salts in Table II are established by X-ray crystallography of the tetracarbonylcobaltate salts of the representative cations Cp_2Co^+, Q^+, and NCP^+ [together with Cp_2Co^+ I^- and Cp_2Co^+ $V(CO)_6^-$ for comparison] as the expected $1:1$ ion pairs. In each salt, the tetracarbonylcobaltate moiety is present as a discrete tetrahedral anion with a slight distortion from ideal T_d to C_{2v} symmetry. The corresponding bond angles for $Co(CO)_4^-$ are listed in Table III, together with the critical bond distances (27). Indeed, the structural parameters, with only slight variations, are akin to those previously found in the colorless ionic salts of tetracarbonylcobaltate paired with alkali (30,31) and simple ammonium cations (32). Likewise, the molecular structures of the acceptor moieties Cp_2Co^+ and Q^+ exist as undistorted cations in the charge-transfer salts with respect to those found in other ionic salts (33). Most importantly, the X-ray crystallographic analyses of cobalticenium tetracarbonylcobaltate and iodide establish the interionic separations of the Cp_2Co^+–anion pairs relevant to the charge-transfer absorptions in Table II.

The molecular (space-filling) models in Fig. 1 illustrate the location of the anionic donors I^- and $Co(CO)_4^-$ relative to the cobalticenium acceptor for optimal orbital overlap with the LUMO in the equatorial plane (34). For the pyridinium salts of $Co(CO)_4^-$, the analogous charge-transfer interaction of the tetracarbonylcobaltate donor places it above the aromatic acceptor planes for optimal orbital overlap with the π-LUMOs of Q^+ and NCP^+. Such X-ray crystallographic structures indicate that these charge-transfer salts consist of contact ion pairs that are directionally constrained for optimum CT interaction in the crystal lattice.

TABLE III

BOND ANGLES AND DISTANCES FOR $Co(CO)_4^-$ IN
CHARGE-TRANSFER SALTS

	Cp_2Co^+	N⁺ (quinolinium)	NC—⟨⟩—N⁺—
C—Co—C(°)	106.7(4)[a]	107.7(6)[b]	107.5(5)[c]
	110.8(4)	111.2(8)	113.0(3)
Co—C(Å)	1.791(6)	1.747(5)	1.763(9)[c]
	1.779(6)	1.755(4)	1.772(9)
C—O(Å)	1.135(6)	1.151(4)	1.151(8)
	1.138(6)	1.157(5)	1.134(9)

[a] Tetracarbonylcobaltate in C_{2v} symmetry.
[b] Tetracarbonylcobaltate in C_{3v} symmetry.
[c] Tetracarbonylcobaltate in C_S symmetry.

A   B

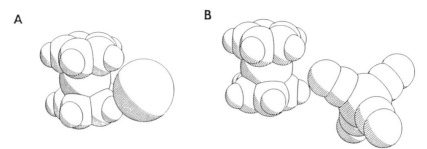

Fig. 1.   Space-filling models of (A) $Cp_2Co^+ I^-$ and (B) $Cp_2Co^+ Co(CO)_4^-$ based on X-ray crystallography showing the location of the anionic donors in the equatorial plane of the cobalticenium cation (27).

C. Infrared Spectra of Charge-Transfer Salts as Contact Ion Pairs in the Solid State and in Solution

The X-ray crystallographic structures emphasize the intimate contact that exists between the donor anion and the acceptor cation in charge-transfer salts. Such a close proximity of the anion–cation pair is sufficient to distort the normally tetrahedral $Co(CO)_4^-$ (Table III). The resulting decrease in symmetry is readily detected by changes in the carbonyl bands in the infrared spectrum. For example, Table IV includes the principal carbonyl bands (ν_{CO}) in the solid-state IR spectra of three charge-transfer salts, together with those of the Na^+ and PPN^+ salts for comparison. The single band at $\nu_{CO} = 1883$ cm^{-1} for the crystalline PPN^+ salt represents the T_2 mode of the undistorted tetrahedral $Co(CO)_4^-$, in accord with the X-ray crystallographic structure of $PPN^+ Co(CO)_4^-$ (35). In strong contrast, the IR spectra of the charge-transfer salts all show the symmetry-forbidden A_1 band at approximately 2005 cm^{-1}. Most importantly, the splitting of the major T_2 band in $Q^+ Co(CO)_4^-$ is akin to that previously observed in Na^+ $Co(CO)_4^-$ by Edgell and co-workers (22) and the IR spectra in Table IV are both in accord with the expected three (two A_1 plus E) bands for $Co(CO)_4^-$ in C_{3v} symmetry established by X-ray crystallography (36). Moreover, the four carbonyl bands in the crystalline cobalticenium and cyanopyridinium analogs are consistent with the C_{2v} and C_S symmetry of $Co(CO)_4^-$ in these salts, as established by X-ray crystallography (see above) (37). In other words, the splittings of the carbonyl IR bands provide reliable and sensitive measures of the tetracarbonylcobaltate distortions that are extant in crystalline charge-transfer salts.

Because the carbonyl IR bands are sensitive probes for $Co(CO)_4^-$ structure in the crystals, they are applicable to the direct interaction of oppo-

TABLE IV

SOLID-STATE AND SOLUTION IR SPECTRA OF $Co(CO)_4^-$ SALTS[a]

Cp_2Co^+		Q^+		NCP^+		Na^+		PPN^+	
2007	(2006)	2007	(2004)	2006	(2003)	2025	(2007)		
1907	(1906)	1928	(1910)	1911	(1916)	1935	(1910)[b]		
1872	(1886)	1895	(1887)	1878	(1886)	1868	(1853)[c]	1883	(1887)
1858	(1870)			1865	(1870)				

[a] ν_{CO} values in cm^{-1} from Ref. 27 for solid-state spectrum (10%) in KBr disk. Values for tetrahydrofuran solution (10 mM) are in parentheses.
[b] Resolved as two bands at 1899 and 1906 cm^{-1}.
[c] Resolved as 1846 and 1856 cm^{-1}, in addition to band at 1887 cm^{-1}.

sitely charged ions which persist as contact ion pairs on dissolution of the charge-transfer salts. Indeed comparison of the carbonyl IR bands in Table IV shows that the charge-transfer salts in tetrahydrofuran (THF) solution are strikingly akin to those found in the solid state. Furthermore, the difference between the ionic salts Na^+ $Co(CO)_4^-$ and PPN^+ $Co(CO)_4^-$ is maintained in THF solution. Since the latter derives from the undistorted $Co(CO)_4^-$, the structures of the crystalline charge-transfer salts as established by X-ray crystallography (see above) are closely related to the contact ion pairs extant when the salts are dissolved in THF. It is important to emphasize that all of these salts, *regardless of the cation*, when dissolved in a polar solvent such as acetonitrile show only a single band at $\nu_{CO} = 1892$ cm^{-1} in the carbonyl IR spectrum. Such a drastic simplification of the multiple splitting of the T_2 band of the charge-transfer salts with a simple change in solvent polarity is diagnostic of the return to a tetrahedral $Co(CO)_4^-$, most likely as a solvent-separated ion pair previously observed with Na^+ $Co(CO)_4^-$ (*38*) and Tl^+ $Co(CO)_4^-$ (*17*) and subsequently confirmed by conductivity measurements (*39*). Accordingly the solvent-dependent changes in the carbonyl IR bands can be related to the displacement of the contact ion pair, for example,

$$Cp_2Co^+ \ Co(CO)_4^- \underset{THF}{\overset{MeCN}{\rightleftharpoons}} Cp_2Co^+//Co(CO)_4^- \qquad (4)$$
$$\text{(CIP)} \qquad\qquad\qquad \text{(SSIP)}$$

where // denotes the solvent separation of the ion pair in which $Co(CO)_4^-$ is sufficiently unencumbered to adopt its most symmetric structure.

The close relationship in Fig. 2 between the diffuse reflectance spectra of crystalline salts and the absorption spectra of the salts in dichloromethane underscores the essential unity of the charge-transfer transitions in the solid state and in solution. The critical interionic separation (r_{A+D^-}) in Fig.

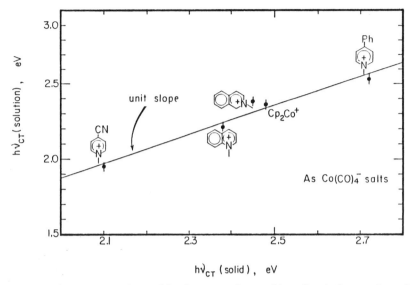

FIG. 2.   Direct correspondence of the charge-transfer transitions (hv_{CT}) of contact ion pairs (as indicated) in the solid state with those in dichloromethane solution. The line is arbitrarily drawn with a slope of unity (27).

1 is also pertinent to the charge-transfer transition (hv_{CT}) of Cp_2Co^+ $Co(CO)_4^-$ in solution. As such, the brightly colored solutions of charge-transfer salts derive directly from the contact ion pairs, which are closely related in kind to those defined by X-ray crystallography and IR spectroscopy in the crystalline solid state.

D. *Solvent and Salt Effects on Charge-Transfer Salts: Solvatochromism*

The marked changes in the carbonyl IR bands accompanying the solvent variation from THF to MeCN coincide with the pronounced difference in color of the solutions. For example, the charge-transfer salt Q^+ $Co(CO)_4^-$ is intensely colored violet in THF but imperceptibly orange in MeCN at the same concentration. The quantitative effects of such a solvatochromism are indicated by (1) the shifts of the absorption maxima and (2) the diminution in the absorbances at λ_{CT}. The concomitant bathochromic shift and hyperchromic increase in the charge-transfer bands follow the sizable decrease in solvent polarity from acetonitrile to THF as evaluated by the dielectric constants ($\mathscr{D} = 37.5$ and 7.6, respectively) (40). The same but even more pronounced trend is apparent in passing from butyronitrile and dichloromethane to diethyl ether ($\mathscr{D} = 26$, 9.1, and 4.3, respectively).

The marked variation in λ_{CT} with solvent polarity parallels the behavior of the carbonyl IR bands (see above), and the solvatochromism is thus readily ascribed to the same displacement of the CIP [Eq. (4)] and its associated charge-transfer band. As such, the reversible equilibrium between CIP and SSIP is described by

$$Q^+ Co(CO)_4^- \overset{K_{CIP}}{\rightleftharpoons} Q^+//Co(CO)_4^- \tag{5}$$

where the dissociation constant K_{CIP} applies to a particular solvent and temperature. The quantitative effects of solvent polarity on the dissociation constant are evaluated spectrophotometrically by measuring the change in the CT absorbance A_{CT} at various concentrations C of the charge-transfer salt from the relationship (41)

$$K_{CIP} = \frac{A_{CT}}{\epsilon_{CT}} + \frac{C^2\epsilon_{CT}}{A_{CT}} - 2C \tag{6}$$

where ϵ_{CT} is the extinction coefficient of the charge-transfer salt.

The CIP dissociation constants evaluated according to Eq. (6) are listed in Table V for some typical charge-transfer salts in both polar and nonpolar solvents. The values of K_{CIP} measured conductometrically (39) for the related ionic salts $PPN^+ Co(CO)_4^-$, $PPN^+ V(CO)_6^-$, and $Na^+ BPh_4^-$ are included in Table V for comparison. Two features in Table V are particularly noteworthy. First, the magnitudes of K_{CIP} in the nonpolar solvents (THF and CH_2Cl_2) are smaller by at least a factor of 100 compared with those in the polar MeCN. Thus, at the concentrations employed in the IR

TABLE V

CONTACT ION PAIR DISSOCIATION CONSTANTS OF CHARGE-TRANSFER SALTS IN POLAR AND NONPOLAR SOLVENTS[a]

CT salt	Solvent	λ_{mon}[b] (nm)	K_{CIP} (M)	ϵ_{CT} (M^{-1} cm^{-1})
$Cp_2Co^+ Co(CO)_4^-$	MeCN	520	1.1×10^{-2}	180
	CH_2Cl_2	520	1.5×10^{-4}	230
$Q^+Co(CO)_4^-$	MeCN	458	8.1×10^{-2}	227
	THF	550	1.2×10^{-4}	420
	CH_2Cl_2	560	1.5×10^{-5}	590
$PP^+ Co(CO)_4^-$	CH_2Cl_2	494	2.3×10^{-5}	380
$PPN^+ Co(CO)_4^-$	THF	—	9.4×10^{-5}	—
$PPN^+ V(CO)_6^-$	THF	—	1.2×10^{-4}	—
$PPN^+ BPh_4^-$	THF	—	9.1×10^{-5}	—

[a] At 25°C in the concentration range of 0.1 to 100 mM (27).
[b] Monitoring wavelength for the charge-transfer band.

studies (Table IV), the charge-transfer salts existed in THF primarily ($>90\%$) as the contact ion pair. Second, the extinction coefficient ϵ_{CT} evaluated at the maximum of the charge-transfer band λ_{CT} is relatively invariant with solvent polarity. This suggests that the same (or closely related) CIP is always formed, and the effect of solvent polarity largely resides with changes in K_{CIP}. Such a solvent effect on the thermodynamics of CIP dissociation is indeed consistent with the influence of added salts.

The addition of small amounts of inert salts such as tetrabutylammonium perchlorate (TBAP) or hexafluorophosphate (TBAH) to solutions of charge-transfer salts induces large changes in the intensity of the charge-transfer absorption bands. The magnitude of the salt effect is most pronounced in nonpolar solvents (THF, CH_2Cl_2). The monotonic decrease in the CT absorbance with increasing amounts of added TBAP is characteristic of the facile competition for the contact ion pair (42), namely,

$$Q^+ Co(CO)_4^- + Bu_4N^+ ClO_4^- \xrightleftharpoons{K_{ex}} Q^+ ClO_4^- + Bu_4N^+ Co(CO)_4^- \qquad (7)$$

Indeed, the single linear relationship between the charge-transfer absorbance and the mole fraction of the inert salt (TBAP) added to Cp_2Co^+ $Co(CO)_4^-$, $PP^+ Co(CO)_4^-$, $Q^+ Co(CO)_4^-$, and $\phi_2Cr^+ Co(CO)_4^-$ indicates that K_{ex} equals 1.0. Such a magnitude of K_{ex} points to the nonspecific interchange between contact ion pairs that effectively serves to disassemble the intimate charge-transfer ion pair $Q^+ Co(CO)_4^-$, without recourse to solvent variation as in SSIP formation (see above). As such, the primary salt effect in Eq. (7) is largely governed by ion-pair electrostatics. The charge-transfer spectrophotometry in Table V thus establishes the contact ion pairs to be the dominant species in nonpolar solvents (CH_2Cl_2, THF, etc.), and the solvatochromism in polar solvents (e.g., MeCN) mainly relates to CIP dissociation in Eq. (5), as given by the distinctive change in the carbonyl IR bands.

The quantitative treatment of primary salt effects demonstrates that the charge-transfer salts are purely ionic salts, indistinguishable from the more commonly used electrolytes such as TBAP or TBAH (43). Accordingly, the solvent effect on CIP dissociation to the solvent-separated ion pairs is the primary factor, since conductivity studies show that "free ions" are not particularly important in the aprotic solvents (43). Nonetheless "free ions" are not distinguishable from SSIP insofar as charge-transfer absorptions are observed in purely ionic salts. This spectral differentiation (or lack thereof) does not apply to other charge-transfer salts, of which $Tl^+ Co(CO)_4^-$ serves as the prime example. Thus, the elegant studies of Schramm and Zink demonstrate that a wide spectrum of charge-transfer entities exist, extending from the nonionic, highly covalent $TlCo(CO)_4$ at one extreme through a series of contact and solvent-separated ion pairs, the relative importance of which are strongly modulated by solvent polarity. In marked contrast, such a direct coordination of the metal to the carbonylmetallate anion is precluded in metallocenium and pyridinium salts owing to their coordinatively saturated character. Charge-transfer salts of the carbonylmetallates $Co(CO)_4{}^-$, $Mn(CO)_5{}^-$, and $V(CO)_6{}^-$ bear a striking resemblance to the well-studied pyridinium iodides. Thus, the solvatochromism of metallocenium and pyridinium salts of this study are fundamentally related to that of the pyridinium iodides, as seminally established by Kosower (16). The use of carbonylmetallate donors [$Co(CO)_4{}^-$, $Mn(CO)_5{}^-$, $V(CO)_6{}^-$, etc.] and metallocenium acceptors (Cp_2Co^+, ϕ_2Cr^+, etc.) successfully extends the range of donor–acceptor properties to expand the solvatochromic scale (40).

E. Photoexcitation of Charge-Transfer Salts in Solution

The red-brown solution of the contact ion pair Cp_2Co^+ $Co(CO)_4{}^-$ in dichloromethane shows no change, even on prolonged irradiation of the charge-transfer band at wavelengths beyond 520 nm (27). However, in the presence of triphenylphosphine the spontaneous evolution of carbon monoxide is observed together with the disappearance of $Co(CO)_4{}^-$, as judged by the diminution of its characteristic carbonyl IR band at $\nu_{CO} = 1887$ cm^{-1}. In its place a new band appears at $\nu_{CO} = 1958$ cm^{-1} for the dimeric $Co_2(CO)_6(PPh_3)_2$ (44) that can be isolated in 65% yield together with cobaltocene according to the stoichiometry

$$Cp_2Co^+ Co(CO)_4{}^- + PPh_3 \xrightarrow{h\nu_{CT}} Cp_2Co + \tfrac{1}{2} Co_2(CO)_6(PPh_3)_2 + CO \qquad (8)$$

Cobaltocene and the analogous cobalt dimer $Co_2(CO)_6(PMe_2Ph)_2$ are also obtained when triphenylphosphine is replaced with dimethylphenylphosphine. The stoichiometry in Eq. (8) corresponds formally to an oxidation–reduction process, in which the 1-electron reduction of Cp_2Co^+ to cobaltocene is accompanied by the 1-electron oxidation of $Co(CO)_4{}^-$ to the carbonylcobalt(0) dimer. Moreover, the analogous photoredox dimerization of the corresponding carbonylmanganate salt occurs only when tri-*n*-butylphosphine is present. Because these photoredox processes are specifically promoted by the CT excitation of the contact ion pair, they are referred to as *charge-transfer dimerizations.*

When the most basic phosphine *n*-Bu$_3$P is used as the additive in the charge-transfer photochemistry of the contact ion pair Cp_2Co^+ $Co(CO)_4{}^-$, it leads to CO evolution, and cobaltocene is isolated in 41% yield. However, a different carbonyl product is formed in 60% yield according to the stoichiometry

$$2\,Cp_2Co^+\,Co(CO)_4^- + 2\,PBu_3 \xrightarrow{h\nu_{CT}} 2\,Cp_2Co + Co(CO)_3(PBu_3)_2^+\,Co(CO)_4^- + CO \quad (9)$$

together with small amounts of the dimeric $Co_2(CO)_6(PBu_3)_2$, reminiscent of the stoichiometry in Eq. (8). The photoinduced process in Eq. (9) formally represents an oxidation–reduction of the charge-transfer salt. Accordingly this photochemical process is referred to simply as *charge-transfer disproportionation*.

The photoinduced reaction of the contact ion pair $Cp_2Co^+\,Co(CO)_4^-$ takes a third course when the triphenylphosphine or tributylphosphine additive is replaced with triphenylphosphite, namely,

$$Cp_2Co^+\,Co(CO)_4^- + P(OPh)_3 \xrightarrow{h\nu_{CT}} Cp_2Co^+\,Co(CO)_3[P(OPh)_3]^- + CO \quad (10)$$

which corresponds to an overall substitution of $Co(CO)_4^-$ by a single $P(OPh)_3$ ligand. Control experiments can establish that (1) neither PPh_3, PBu_3, nor $P(OPh)_3$ have any effect on the charge-transfer spectrum of $Cp_2Co^+\,Co(CO)_4^-$ and (2) no thermal reaction occurs in the absence of light. In order to differentiate the photoinduced process in Eq. (10) from that in either Eq. (8) or Eq. (9), it is referred to here as *charge-transfer substitution*.

The photochemical behavior of the other charge-transfer salts in Table II is comparable to that of $Cp_2Co^+\,Co(CO)_4^-$. Thus, the orange, purple, and green solutions of $PP^+\,Co(CO)_4^-$, $Q^+\,Co(CO)_4^-$, and $\phi_2Cr^+\,Co(CO)_4^-$, respectively, in dichloromethane are stable to visible radiation (with $\lambda > 550$ nm) for prolonged periods. Similarly, the carbonylvanadate and manganate salts $Cp_2Co^+\,V(CO)_6^-$ and $Cp_2Co^+\,Mn(CO)_5^-$ are unaffected by visible radiation with wavelengths greater than 520 nm. On the other hand, the actinic radiation of the charge-transfer salts in dichloromethane or THF containing the ligand L (phosphine or phosphite) results in the liberation of 1 mol of carbon monoxide, and the stoichiometry leading to either CT dimerization, disproportionation, or substitution as given in Eqs. (8), (9), and (10), respectively, varies with the additive L. It is important to emphasize that the photoinduced dimerization, disproportionation, and substitution arise directly from the charge-transfer excitation of only the contact ion pair. Thus, the use of visible light with wavelengths over 520 nm as the radiation source ensures the excitation of only the charge-transfer absorption bands. Since these CT absorptions relate specifically to the contact ion pair (see Table V), there is no ambiguity about the adventitious local excitation of either the separate anion or cation, or the photochemical generation of intermediates which do not arise from the CT excitation of the contact ion pair.

1. *Time-Resolved Spectra of Charge-Transfer Transients*

The reactive intermediates in the charge-transfer photochemistry of contact ion pairs can be identified by their time-resolved spectra immediately following the application of a 10-nsecond pulse consisting of the second harmonic at 532 nm of a mode-locked Nd^{3+}–YAG laser (*27*). The wavelength of this radiation source is ideally suited for the specific excitation of the contact ion pairs (see Table II). Accordingly, the time-resolved spectra from Cp_2Co^+ $Mn(CO)_5^-$ and Cp_2Co^+ $Co(CO)_4^-$ (Fig. 3) relate directly to the charge-transfer photochemistry. Most notably, the intense band centered at $\lambda_{max} = 800$ nm is identical to the 17-electron radical $Mn(CO)_5 \cdot$ that is independently generated from the direct homolytic cleavage of the dimeric $Mn_2(CO)_{10}$ (*45*). Other flash photolytic studies (*46*), as well as pulse radiolysis (*47*) and matrix isolation (*48*), verify the spectral assignment of $Mn(CO)_5 \cdot$ with a λ_{max} of approximately 800 nm. The analogous observation of the spectral transient absorbing at $\lambda_{max} = 780$ nm in Fig. 3A from the CT excitation of Cp_2Co^+ $Co(CO)_4^-$ can be similarly ascribed to the 17-electron radical $Co(CO)_4$. The laser-pulse excitation of the quinolinium and cyanopyridinium salts of $Co(CO)_4^-$ produces the same spectral transient. Most revealingly, the simultaneous appearance of two bands at λ_{max} values of 550 and about 380 nm (end absorption) from the CT excitation of Q^+ $Co(CO)_4^-$ in Fig. 3B can be readily assigned to the quinolinyl radical ($Q \cdot$) that can be independently generated (*49*).

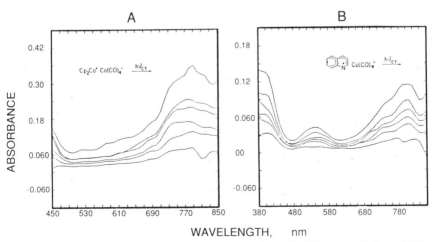

FIG. 3. Time-resolved difference spectra resulting from the 532-nm excitation of (A) 13 mM Cp_2Co^+ $Co(CO)_4^-$ at (top to bottom) 16, 20, 25, 28, 38, and 54 μseconds and (B) 10 mM Q^+ $Co(CO)_4^-$ at 60, 83, 106, 128, 150, and 217 μseconds in acetone (*27*).

Taken together, the time-resolved spectra in Fig. 3 provide strong support for the primary step in the photoactivation of tetracarbonylcobaltate salts, for example,

$$\text{(structure)} \; Co(CO)_4^- \xrightarrow{h\nu_{CT}} \text{(structure)} \; Co(CO)_4 \cdot \qquad (11)$$

The time-resolved spectroscopic studies thus show that the charge-transfer excitation of CIP results in the simultaneous production of the reactive 17-electron carbonylmetal radicals [$Mn(CO)_5 \cdot$, $Co(CO)_4 \cdot$, etc.] together with the reduced acceptor radical (Cp_2Co, $Q \cdot$, $NCP \cdot$, etc.). Furthermore, the time scale of the CT photoexcitation indicates that these radicals are initially formed as geminate pairs within the solvent cage. (50). The absence of productive photochemistry on steady-state irradiation of contact ion pairs alone in solution (i.e., without added phosphine) is consistent with the spontaneous return of the transient radicals to the ground state intact, for example,

$$Cp_2Co, Mn(CO)_5 \cdot \longrightarrow Cp_2Co^+ \, Mn(CO)_5^- \qquad (12)$$

Such a regeneration of the CIP from the radical pair accords with the invariance of the charge-transfer spectra even on prolonged irradiation. Back electron transfer from the radical pair is also supported in the time-resolved spectroscopic studies by the restoration of the transient absorbances to the original baselines in Fig. 3.

Kinetics for the return of the radical pair from $Cp_2Co^+ \, Mn(CO)_5^-$ are measured by following the absorbance change of $Mn(CO)_5 \cdot$ at the monitoring wavelength of $\lambda_{mon} = 800$ nm, and the second-order kinetics for the disappearance of $Mn(CO)_5 \cdot$ is demonstrated by the excellent fit of the smooth computed curve to the experimental decay. The second-order rate constant k_2 evaluated in this manner is insensitive to solvent variation. Disappearance of the radicals derived from such carbonylcobaltate salts as $Cp_2Co^+ \, Co(CO)_4^-$ also follows the same second-order kinetics. It is noteworthy that the CT excitation of the quinolinium salt $Q^+ \, Co(CO)_4^-$ allows a pair of second-order rate constants to be extracted from the absorbance decays at $\lambda_{mon} = 550$ and 780 nm. Assignment of the latter to the 17-electron radical $Co(CO)_4 \cdot$ then defines the second-order kinetics to derive from the mutual annihilation of the radicals, that is,

$$Q \cdot + Co(CO)_4 \cdot \xrightarrow{k_2} Q^+ \, Co(CO)_4^- \qquad (13)$$

Furthermore, the same kinetic relationship holds for the pair of radicals from the cyanopyridinium salt $NCP^+ Co(CO)_4^-$. The kinetic results also show that radical decay suffers only a minor retardation in the presence of added phosphine.

2. *Mechanism of Charge-Transfer Photochemistry of Contact Ion Pairs*

The charge-transfer salts of carbonylmetallates are thus distinguished from the iodide salts in the breadth of photochemistry that obtains on exposure to visible light. For example, continuous irradiation of the charge-transfer band ($\lambda \sim 350$–500 nm) of the pyridinium iodides inflicts no permanent change on either the crystalline salt or that dissolved in CH_2Cl_2 or THF (*51*). Moreover, the highly colored $Co(CO)_4^-$, $V(CO)_6^-$, and $Mn(CO)_5^-$ salts are similarly unaffected by visible light for prolonged periods in the solid state or in solution. However, merely the presence of small amounts of additives L is sufficient to activate the carbonylmetallate salts to undergo a variety of distinctive photochemical reactions in solution.

At first glance, all of the photochemical processes in Eqs. (8)–(10) are unique, with none showing any apparent stoichiometric relationship to the others. For example, the carbonylcobalt dimerization [Eq. (8)] and disproportionation [Eq. (9)] represent 1-electron oxidation processes of $Co(CO)_4^-$, whereas the formation of $Co(CO)_3L^-$ in Eq. (10) relates to a nonredox ligand substitution. The incorporation of the additive L into the carbonylmetal product, whether it be $Co_2(CO)_6L_2$, $Co(CO)_3L_2^+$, or $Co(CO)_3L^-$ in Eqs. (8), (9), and (10), respectively, is the sole feature that these photochemical processes have in common. Because the carbonylmetallates $Co(CO)_4^-$, $V(CO)_6^-$, and $Mn(CO)_5^-$ are all thermally substitution-stable anions (*52*), the introduction of L into the carbonylmetallate moiety must occur in some reactive intermediate.

The time-resolved spectroscopic studies of the transient intermediates provide the keys to understanding the charge-transfer photochemistry of contact ion pairs. Thus, the unambiguous identification of the carbonyl-metal radical $Mn(CO)_5\cdot$ immediately following the 10-nsecond laser-flash excitation of the contact ion pair $Cp_2Co^+ Mn(CO)_5^-$ at $\lambda = 532$ nm relates directly to the photoredox process,

$$Cp_2Co^+ Mn(CO)_5^- \xrightarrow{h\nu_{CT}} Cp_2Co, Mn(CO)_5\cdot \qquad (14)$$

in accord with Mulliken theory (*7*). The concomitant production of cobaltocene in Eq. (14) cannot be observed owing to its spectral overlap with the local absorptions of the separate ions Cp_2Co^+ and $Mn(CO)_5^-$. Nonetheless, the persistence of the transient 800-nm band at times as short as 100 nseconds following the laser excitation, coupled with the isolation of cobaltocene from the steady-state experiments, represents the reasonable experimental verification of the Mulliken charge-transfer formulation. Most importantly, the enhanced reactivity of the 17-electron radical $Mn(CO)_5 \cdot$ leads to rapid ligand substitution by tributylphosphine and homolytic dimerization, that is,

$$Mn(CO)_5 \cdot \; + PBu_3 \xrightarrow{k_S} Mn(CO)_4PBu_3 \cdot \; + CO$$

$$2\, Mn(CO)_4PBu_3 \cdot \xrightarrow{k_{dim}} Mn_2(CO)_8(PBu_3)_2$$

The large magnitudes of the rate constants k_S and k_{dim} (1.0×10^9 and $1.0 \times 10^8 \; M^{-1}$ second^{-1}, respectively) established by Brown and co-workers (53) are critical for the description of CT dimerization as given in the mechanistic pathway in Scheme 1.

$$Cp_2Co^+ \, Mn(CO)_5^- \underset{k_{ET}}{\overset{h\nu_{CT}}{\rightleftharpoons}} Cp_2Co + Mn(CO)_5 \cdot$$

$$Mn(CO)_5 \cdot \; + L \xrightarrow{k_S} Mn(CO)_4L \cdot \; + CO$$

$$2\, Mn(CO)_4L \cdot \xrightarrow{k_{dim}} Mn_2(CO)_8L_2$$

SCHEME 1

It is important to emphasize that the photoefficiency of CT dimerization according to Scheme 1 is generally dependent on the competition between back electron transfer (k_{ET}) and ligand substitution ($k_S L$). As such, the absence of photochemistry without added L derives from the facile back electron transfer of the initial radical pair, $Mn(CO)_5 \cdot$ and Cp_2Co, to regenerate the CIP. Thus, in order to account for the observed CT dimerization, a value of k_{ET} below 10^8 second^{-1} can be estimated in order to allow competition for phosphine scavenging of $Mn(CO)_5 \cdot$ with $k_S = 1.0 \times 10^9$ M^{-1} second^{-1} at roughly 0.1 M concentrations of added PBu_3. Furthermore, the absence of CT photochemistry when the added PBu_3 is replaced with the less nucleophilic PPh_3 is a direct consequence of the mechanism in Scheme 1, since the second-order rate constant for substitution with L equal to PPh_3 is roughly two orders of magnitude slower than that with PBu_3 ($k_S = 1.7 \times 10^7 \; M^{-1}$ second^{-1} (53).

The kinetic results from the time-resolved experiments show that back electron transfer leading to annihilation of $Mn(CO)_5 \cdot$ actually proceeds by *second*-order kinetics, with the rate constant k_{ET} being $3 \times 10^{10} \; M^{-1}$ second^{-1}. As such, any description of the photostationary state attained in the absence of additives must include the diffusive separation of the initial radical pair (54), for example,

$$Cp_2Co^+ Mn(CO)_5^- \underset{\kappa_3}{\overset{h\nu_{CT}}{\rightleftharpoons}} Cp_2Co, Mn(CO)_5 \cdot \underset{\kappa_2}{\overset{\kappa_1}{\rightleftharpoons}} Cp_2Co + Mn(CO)_5 \cdot$$

where the measured rate constant $k_{ET} \cong \kappa_2\kappa_3/\kappa_1$. Furthermore, the phosphine substitution of $Mn(CO)_5 \cdot$ to yield $Mn(CO)_4L \cdot$ serves to enhance the photochemical efficiency by minimizing the energy-wasting back electron transfer, that is,

$$Cp_2Co + Mn(CO)_4L \cdot \xrightarrow{slow} Cp_2Co^+ Mn(CO)_4L^-$$

owing to the significantly attenuated reduction potentials of the phosphine-substituted radicals (55).

The time-resolved spectra are also in accord with the Mulliken formulation for the photochemical activation of carbonylcobaltate salts. As such, the charge-transfer mechanism outlined for $Mn(CO)_5 \cdot$ is generally applicable to the kindred 17-electron radical (Scheme 2).

$$Cp_2Co^+ Co(CO)_4^- \underset{\kappa_3}{\overset{h\nu_{CT}}{\rightleftharpoons}} Cp_2Co, Co(CO)_4 \cdot \underset{\kappa_2}{\overset{\kappa_1}{\rightleftharpoons}} Cp_2Co + Co(CO)_4 \cdot \qquad (15)$$

$$Co(CO)_4 \cdot + L \xrightarrow{k_s} Co(CO)_3L \cdot + CO \qquad (16)$$

SCHEME 2

Indeed, the cage mechanism in Eq. (15) is confirmed by the limiting value of the quantum yield, $\Phi_p(limit)$, that is observed at high concentrations of added phosphine, as illustrated in Fig. 4. According to Scheme 2, the efficiency of CT photochemistry arises from the carbonylcobalt radicals $Co(CO)_4 \cdot$ that successfully undergo cage escape (κ_1) in competition with back electron transfer (κ_3). Importantly, the trend in the values of $\Phi_p(limit)$ in the order $\phi_2Cr^+ > Q^+ > Cp_2Co^+$ derives from the relative ease of back electron transfer with the acceptor radicals $\phi_2Cr < Q \cdot < Cp_2Co$, as based on the values of the cathodic potentials at $E_c = -0.80, -0.90$, and -0.95 V, respectively. Most importantly, however, there are specific elaborations to Scheme 2 that must be included to account for the unique product selectivities in charge-transfer dimerization, substitution, and disproportionation as they obtain with different phosphines described separately below.

a. Charge-Transfer Dimerization. The second-order rate constant for back electron transfer (k_{ET}) with $Co(CO)_4 \cdot$ is roughly an order of magnitude slower than that for $Mn(CO)_5 \cdot$ in Scheme 1. Second, the formation of the dimeric $Co_2(CO)_6L_2$ with L being PPh_3 and PMe_2Ph undoubtedly arises from the efficient scavenging of the carbonylcobalt radical [Eq. (16)],

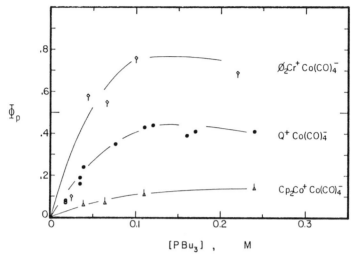

FIG. 4.   Limiting quantum yields for charge-transfer photochemistry of $Co(CO)_4^-$ salts of various acceptor cations at increasing concentrations of the additive $L = PBu_3$ (27).

followed by dimerization, that is,

$$2\ Co(CO)_3L \cdot \xrightarrow{k_{dim}} Co_2(CO)_6L_2 \qquad (17)$$

with the second-order rate constant k_S exceeding $10^5\ M^{-1}$ second^{-1} and k_{ET} approximating $10^7\ M^{-1}$ second^{-1}. Moreover, the energy-wasting back electron transfer of $Co(CO)_4 \cdot$ (with Cp_2Co) is obviated by the conversion to $Co(CO)_3PPh_3^-$ whose annihilation, namely,

$$Cp_2Co + Co(CO)_3PPh_3 \cdot \xrightarrow{slow} Cp_2Co^+\ Co(CO)_3PPh_3^-$$

is minimized as the result of its significantly reduced potential for reduction (56). Indeed, the quantum yield of $\Phi p = 0.1$ for the CT dimerization of $Cp_2Co^+\ Co(CO)_4^-$ with PPh_3 indicates that $Co(CO)_4 \cdot$ undergoes ligand substitution roughly 10 times slower than back electron transfer in Eq. (15).

b. Charge-Transfer Substitution.   Carbonylcobalt radicals are also critical intermediates when the phosphines PPh_3 and PMe_2Ph are replaced by the phosphite $P(OPh)_3$. The formation of $Co(CO)_3[P(OPh)_3]^-$ does not derive in Eq. (10) from direct ligand substitution, since $P(OPh)_3$ is too weak a nucleophile to effect CO replacement in the substitution-inert $Co(CO)_4^-$. However, the additives $L = PPh_3$, PMe_2Ph, and $P(OPh)_3$ are

strongly differentiated in the redox properties of the 17-electron intermediates $Co(CO)_3L\cdot$ as formed in Eq. (16) (Scheme 2). For example, the
phosphite-substituted radical $Co(CO)_3[P(OPh)_3]\cdot$ is significantly more
readily reduced than its phosphine analog $Co(CO)_3[PMe_2Ph]\cdot$, the cathodic peak potentials E_c at -0.05 and -0.45 V, respectively (56), represent-
ing a difference of 9 kcal/mol. Coupled with the oxidation potential of
-0.95 V for cobaltocene (57), the driving force for back electron transfer

$$Cp_2Co + Co(CO)_3[P(OPh)_3]\cdot \xrightarrow{k_{ET}} Cp_2Co^+ \, Co(CO)_3[P(OPh)_3]^- \qquad (18)$$

is estimated to be exergonic ($\Delta G_{ET} \sim -20$ kcal/mol). Although the direct
measure of the rate of electron transfer is not available by time-resolved
spectroscopy, it can be inferred that k_{ET} is greater than $10^8 \, M^{-1}$ second^{-1}
from the large driving force for Eq. (18). Such large values of k_{ET} leading
to $Cp_2Co^+ \, Co(CO)_3[P(OPh)_3]^-$ could thus be sufficient to obviate the
formation of $Co_2(CO)_6[P(OPh)_3]_2$ via the bimolecular coupling of
$Co(CO)_3[P(OPh)_3]\cdot$ in Eq. (17). The extent to which any of the dimeric
$Co_2(CO)_6[P(OPh)_3]_2$ is formed will be subsequently reduced to the salt,

$$Co_2(CO)_6[P(OPh)_3]_2 + 2\, Cp_2Co \longrightarrow 2\, Cp_2Co^+ \, Co(CO)_3[P(OPh)_3]^-$$

c. Charge-Transfer Disproportionation. The role of carbonylcobalt radicals in the conversion of $Co(CO)_4^-$ to $Co(CO)_3L_2^+$ is delineated by independent electrochemical studies (58), and the charge-transfer disproportionation of $Cp_2Co^+ \, Co(CO)_4^-$ can be considered in two distinct parts,
namely, (i) the 2-electron oxidation of the anionic donor,

$$Co(CO)_4^- + 2\, PBu_3 \xrightarrow{-2e} Co(CO)_3(PBu_3)_2^+ + CO \qquad (19)$$

coupled with (ii) the reduction of 2 equivalents of Cp_2Co^+, so that the
combination of (i) and (ii) represents the experimental stoichiometry in
Eq. (9). The facile ligand substitution of carbonylcobalt radicals constitutes
the pathway for the 2-electron transformation in Eq. (19) (Scheme 3),
which represents an electrochemical ECE mechanism. As such, transient
electrochemical techniques differentiate phosphine ligands L = PPh_3 and
PBu_3 by their effect on $Co(CO)_3L_2^+$, the redox potential of
$Co(CO)_3(PBu_3)_2^+$ being 0.61 V or 14 kcal/mol more favorable than that for
$Co(CO)_3(PPh_3)_2^+$ (56).

$$Co(CO)_4^- \xrightarrow{-e} Co(CO)_4\cdot$$

$$Co(CO)_4\cdot + L \longrightarrow Co(CO)_3L\cdot + CO$$

$$Co(CO)_3L\cdot + L \xrightarrow{-e} Co(CO)_3L_2^+$$

SCHEME 3

The importance of such a redox property is also illustrated in the photodissociation (59) of $Co_2(CO)_6(PBu_3)_2$ in the presence of both $Q^+ PF_6^-$ and PBu_3. Separately, neither Q^+ nor PBu_3 has any perceptible effect on the photostationary state of $Co_2(CO)_6(PBu_3)_2$ on irradiation of the σ,σ^* band (60) at wavelengths of approximately 380 nm. However, when both are present, essentially quantitative yields of $Co(CO)_3(PBu_3)_2^+ PF_6^-$ are obtained by a photoredox process that can be described as

$$Co_2(CO)_6(PBu_3)_2 \xrightarrow{h\nu} 2\ Co(CO)_3(PBu_3)\cdot$$

$$Co(CO)_3(PBu_3)_2\cdot + PBu_3 \xrightarrow{[e]} Co(CO)_3(PBu_3)_2^+$$

where $[e]$ represents the redox couple $[A^+ \rightarrow A\cdot]$ for the pyridinium acceptor A^+ (61). The importance of the latter is underscored by the acceptor cations which favor CT disproportionation in the order $A^+ = \phi_2Cr^+ > Q^+ > Cp_2Co^+ > PP^+$. It is noteworthy that this also represents the decreasing trend in the CV reduction potentials of the cations in the same order, namely, $E_c = -0.80, -0.90, -0.95,$ and -1.22 V, respectively. A particularly dramatic influence of the cation is shown by the entirely different products obtained from the CT activation of $Cp_2Co^+ Co(CO)_4^-$ and $Q^+ Co(CO)_4^-$ in the presence of the same phosphine additive PMe_2Ph.

Finally, care must be exercised in the photochemical activation of contact ion pairs to irradiate only the charge-transfer absorption bands, and not those of the (colored) products. For example, the irradiation of either $Cp_2Co^+ Co(CO)_4^-$ or $Q^+ Co(CO)_4^-$ in the presence of PBu_3 at wavelengths beyond 510 nm leads only to the dimeric $Co_2(CO)_6(PBu_3)_2$, despite the fact that the CT photochemistry of the same solution at wavelengths below 550 nm leads smoothly to only the disproportionation products. In fact, control experiments demonstrate that the carbonylcobalt dimer arises from a secondary process by the adventitious excitation of the disproportionation salt, namely,

$$Co(CO)_3(PBu_3)_2^+ Co(CO)_4^- \xrightarrow{h\nu_{CT}} Co_2(CO)_6(PBu_3)_2 + CO$$

III

THERMAL ANNIHILATION OF CARBONYLMETALLATE ANIONS BY CARBONYLMETAL CATIONS

The photoinduced activation of organometallic ion pairs as presented in Section II has its counterpart in the purely thermal (adiabatic) processes for metal–metal bond formation. Thus, the dimeric metal carbonyls can be synthesized via various reductive procedures, including the use of carbon monoxide, metals, and alkylmetals as reagents (62). The oxidation of carbonylmetallate anions is also known to lead to carbonylmetal dimers (63). Heterobimetallic carbonyls result from the interaction of carbonyl-metallates with different types of metal halides and homodimers (64). The latter can be generally classified as nucleophilic substitution processes, although mechanistic studies of such metal–metal bond formations are generally lacking.

Most importantly, the heterolytic coupling of carbonylmetal cations and anions is indicated in the treatment of tetracarbonylcobaltate(−I) with hexacarbonylrhenium(I) to afford the mixed carbonyl (65). Similarly, the treatment of $Co(CO)_4^-$ with the phenanthroline-substituted tetracarbonyl-manganese(I) cation leads to the substituted heterobimetallic carbonyl (66), and the highly efficient coupling of the carbonylmanganese anion and cation yields the homodimer (67),

$$Mn(CO)_5^- + Mn(CO)_6^+ \longrightarrow Mn_2(CO)_{10} + CO \qquad (20)$$

The availability of various substituted carbonylmanganese cations and anions presents an opportunity to examine the mechanism of the coupling processes in detail, especially with regard to the products and stoichiometry for the general representation:

$$Mn(CO)_5L^+ + Mn(CO)_4P^- \longrightarrow Mn_2(CO)_8(L)(P) + CO \qquad (21)$$

where L and P represent various phosphorus- and nitrogen-centered ligands associated with the carbonylmanganese cation and anion, respectively.

A. Coupling of Anionic Carbonylmanganates(−I) and Carbonylmanganese(I) Cations

The composition of the reaction mixture from the ion-pair interaction of $Mn(CO)_5L^+$ and $Mn(CO)_4P^-$ in Eq. (21) differs according to the nature of the ligands L and P. In order to facilitate the descriptions of the product mixtures, the interactions of the carbonylmanganates with each of the cations are described separately.

1. Pentacarbonylmanganate(−I) and Hexacarbonylmanganese(I)

The ion pair pentacarbonylmanganate(−I) and hexacarbonylmangan-ese(I) reacts on mixing (<5 minutes in THF solutions to afford dimangan-ese decacarbonyl as the sole carbonyl product (67). Essentially the same results are obtained in THF solution with the acetonitrile and pyridine derivatives,

$$Mn(CO)_5^- + Mn(CO)_5L^+ \longrightarrow Mn_2(CO)_{10} + L \qquad (22)$$

where L = CO, MeCN, and pyridine. When pentacarbonylmanganate(I) is treated with the triarylphosphine-substituted cations, high yields of $Mn_2(CO)_{10}$ are formed together with minor amounts of the bis-substituted dimer $Mn_2(CO)_8L_2$ and the substituted hydride $HMn(CO)_4L$,

$$Mn(CO)_5^- + Mn(CO)_5L^+ \longrightarrow Mn_2(CO)_{10} + Mn_2(CO)_8L_2 + HMn(CO)_4L$$

where L = phosphines. Treatment of pentacarbonylmanganate with the alkyl- and aralkylphosphine derivatives also affords variable amounts of $Mn_2(CO)_{10}$ and the analogous carbonylmanganese products $Mn_2(CO)_8L_2$ and $HMn(CO)_4L$. In addition, small but distinctive amounts of the mono-substituted dimanganese carbonyl $Mn_2(CO)_9L$ are detected. Furthermore, the coupling processes of $Mn(CO)_5^-$ with $Mn(CO)_5PPh_3^+$, etc. often proceed at significantly attenuated rates. The material balance indicates that all of the phosphine (L) included in $Mn(CO)_5L^+$ is not retained in the carbonylmanganese products, with the deficit being present as free phosphine.

2. Tetracarbonyl(triphenylphosphite)manganate(−I) and Cationic Hexacarbonylmanganese(I)

The ion pair tetracarbonyl(triphenylphosphite)manganate(−I) and hexacarbonylmanganese(I) rapidly yields two manganese dimers $Mn_2(CO)_{10}$ and $Mn_2(CO)_8[P(OPh)_3]_2$ in a roughly 4:3 ratio, namely,

$$Mn(CO)_4P^- + Mn(CO)_6^+ \longrightarrow [Mn_2(CO)_{10} + Mn_2(CO)_8P_2] + CO$$

where $P = P(OPh)_3$. When the acetonitrile- and pyridine-substituted cations are employed, the same pair of dimanganese carbonyls is obtained on mixing, together with significant amounts of the monosubstituted analog,

$$Mn(CO)_4P^- + Mn(CO)_5L^+ \longrightarrow [Mn_2(CO)_{10} + Mn_2(CO)_9P + Mn_2(CO)_8P_2] + L \quad (23)$$

where $P = P(OPh)_3$ and L = MeCN and py. The absence of any other carbonyl-containing product in the reaction mixture indicates that the nitrogen ligands (MeCN and py) are rapidly replaced. In marked contrast, the reaction of the phosphine-substituted cations $Mn(CO)_5L^+$ with $Mn(CO)_4[P(OPh)_3]^-$ affords no $Mn_2(CO)_{10}$ or $Mn_2(CO)_9P$; the most important products are the other dimanganese carbonyls, namely,

$$Mn(CO)_4P^- + Mn(CO)_5L^+ \longrightarrow Mn(CO)_8P_2 + Mn(CO)_8L_2 + Mn_2(CO)_8(L)(P)$$

where $P = P(OPh)_3$. In addition, minor amounts of the manganese hydrides $HMn(CO)_4L$ and $HMn(CO)_4P$ are observed. With $Mn(CO)_5PEt_3^+$

and $Mn(CO)_5PEt_2Ph^+$, the IR spectrum of the solution taken immediately after mixing with $Mn(CO)_4P(OPh)_3^-$ shows the presence ($\sim 20\%$) of the reactant ion pair, the complete disappearance of which requires about 1 hour.

3. Effect of Added Phosphine

The ion-pair couplings can be examined with the anionic carbonylman-ganates($-I$) and carbonylmanganese(I) cations in the presence of added ligands since they are all substitution stable. The effect of added phosphine on the course of dimanganese formation shows three important changes that accompany the presence of triphenylphosphine on the ion-pair coupling of $Mn(CO)_5^-$ and $Mn(CO)_5(NCMe)^+$ in Eq. (22). First, with increasing amounts of PPh_3, the yield of $Mn_2(CO)_{10}$ drops precipitously, and the deficit is made up with a pair of new products, $Mn_2(CO)_8(PPh_3)_2$ and $HMn(CO)_4PPh_3$,

$$Mn(CO)_5^- + Mn(CO)_5(NCMe)^+ \xrightarrow{PPh_3} Mn_2(CO)_8(PPh_3)_2 + HMn(CO)_4PPh_3, \text{ etc.}$$

Small amounts of the monosubstituted dimer $Mn_2(CO)_9PPh_3$ are observed only at the highest concentrations of added PPh_3.

4. Mechanism of Mn–Mn Bond Formation by Ion-Pair Annihilation

The formulation in Scheme 4 derives from the ion-pair annihilations via the known behavior of 19- and 17-electron carbonylmanganese radicals (67). Accordingly, the initiation by electron transfer is included in the generalized mechanism for Mn–Mn bond formation (Scheme 4). The

$$Mn(CO)_4P^- + Mn(CO)_5L^+ \longrightarrow [Mn(CO)_4P\cdot, Mn(CO)_5L\cdot] \qquad (24)$$

$$[Mn(CO)_4P\cdot, Mn(CO)_5L\cdot] \left\{ \begin{array}{l} \xrightarrow{\text{collapse}} Mn_2(CO)_8(P)(L) + CO \qquad (25) \\ \\ \xrightarrow{\text{diffuse}} Mn(CO)_4P\cdot + Mn(CO)_5L\cdot \qquad (26) \end{array} \right.$$

$$Mn(CO)_5L\cdot \xrightarrow{k_L, k_{CO}} Mn(CO)_4X\cdot + L(CO) \qquad (27)$$

$$2 Mn(CO)_4X\cdot \xrightarrow{k_{dim}} Mn_2(CO)_8X_2 \qquad (28)$$

SCHEME 4. $X = CO, P, L.$

first-order rate constants k_L and k_{CO} represent the ligand dissociation from the 19-electron radical, and k_{dim} is the second-order rate constant for the couplings of pairs of 17-electron radicals. According to Scheme 4, electron transfer is the first step [Eq. (24)] in the annihilation of the carbonylmanganese cations by anions. The nonstatistical distribution among the homocoupled and cross-coupled dimers derives primarily from the competition between cage collapse [Eq. (25)] and diffusive separation [Eq. (26)] of the initially formed radical pair. The extent to which cage collapse occurs faster than diffusive separation will determine the amounts of cross-coupled products obtained. Alternatively, the efficiency with which added phosphines divert the products to $Mn_2(CO)_8P_2$ reflects the extent to which 17-electron carbonylmanganese radicals have escaped and consequently are subject to ligand substitution. In the same way, the trapping of 19-electron $Mn(CO)_5L\cdot$ by hydrogen transfer reflects cage escape. Indeed the multiple substitution of phosphine in the hydride products suggests that the 19-electron carbonylmanganese precursors are longer lived.

A further indication of this difference is shown in the carbonylmanganate $Mn(CO)_4(\eta^1\text{-DPPE})^-$ in which the corresponding 17-electron radical will be susceptible to intramolecular trapping (68), namely

$$(OC)_4\overset{\cdot}{Mn}PPh_2\diagup\diagdown PPh_2 \xrightarrow{-\sigma} (OC)_4\overset{\cdot}{Mn}\begin{array}{c}\overset{Ph_2}{P}\\ \diagup\diagdown \\ P\\ Ph_2\end{array} \rightarrow (OC)_3\overset{\cdot}{Mn}\begin{array}{c}\overset{Ph_2}{P}\\ \diagup\diagdown \\ P\\ Ph_2\end{array} + CO \quad (29)$$

The facile annihilation of $Mn(CO)_4(\eta^1\text{-DPPE})^-$ by an equivalent amount of either $Mn(CO)_5(NCMe)^+$ or $Mn(CO)_5py^+$ affords three principal products, namely, $Mn_2(CO)_{10}$ (30%), $Mn_2(CO)_9(\eta^1\text{-DPPE})$ (15%), and $HMn(CO)_3(\eta^2\text{-DPPE})$ (35%). The formation of the chelated $HMn(CO)_3(\eta^2\text{-DPPE})$ is consistent with the trapping of the 19-electron intermediate, as indicated by hydride formation. The observation of the monohapto dimanganese carbonyl $Mn_2(CO)_9(\eta^1\text{-DPPE})$ suggests that the cross-coupling involves either a concerted process or a cage collapse in Eq. (25) which is faster than intramolecular trapping [Eq. (29)]. From a purely operational point of view, there are two pathways for the formation of Mn–Mn carbonyls by ion-pair annihilation, namely, (i) a process in which pairs of carbonylmanganese radicals behave more or less independently as discrete species as a result of diffusive separation and (ii) a minor pathway in which the cation and anion more or less maintain their identity by either cage collapse of the radical pair [Eq. (25)] or by concerted action.

B. Coupling of Carbonylcobaltate(−I) Anions with Carbonylcobalt(I) Cations

The ready disproportionation of dicobalt octacarbonyl induced by Lewis bases has been known for more than three decades (69), but the process still presents an intriguing mechanistic knot. With nitrogen- and oxygen-centered bases (B), the disproportionation of $Co_2(CO)_8$ usually affords a $Co(II)/Co(-I)$ salt with the general stoichiometry (70)

$$3\ Co_2(CO)_8 + 12\ B \longrightarrow 2\ CoB_6^{2+} \{Co(CO)_4^-\}_2 + 8\ CO$$

On the other hand, with phosphorus-centered bases P (especially in polar solvents), dicobalt octacarbonyl disproportionates even at room temperature to produce the $Co(I)/Co(-I)$ salt with a simpler stoichiometry (71,72) that is,

$$Co_2(CO)_8 + 2\ P \rightarrow Co(CO)_3P_2^+Co(CO)_4^- + CO \qquad (30)$$

Interestingly, the same phosphines react with dicobalt octacarbonyl in nonpolar solvents at higher temperatures by effecting an overall ligand substitution (72,73)

$$Co_2(CO)_8 + 2\ P \longrightarrow Co_2(CO)_6P_2 + 2\ CO$$

It is known that this disubstitution product can also be obtained by heating the $Co(I)/Co(-I)$ salt (73):

$$Co(CO)_3L_2^+ + Co(CO)_4^- \longrightarrow Co_2(CO)_6L_2 + CO$$

Such an ion-pair annihilation can be formally considered as the microscopic reverse of the disproportionation in Eq. (30).

Among the various carbonylcobalt(I) cations extant, the bisphosphine-substituted derivative $Co(CO)_3L_2^+$ is the most common, with only a few examples of the monophosphine-substituted cation $Co(CO)_4L^+$ having been reported (74). Accordingly the previous mechanistic investigations of cobalt-centered ion pairs (75) have focused on the annihilation of the disproportionation ion pair $Co(CO)_3L_2^+ Co(CO)_4^-$ that is obtained directly from dicobalt octacarbonyl by treatment with phosphines (76). However, this problem can be approached more generally from a consideration of carbonylcobalt cations as electrophilic electron acceptors and carbonylcobalt anions as nucleophilic electron donors in a manner similar to that established in the annihilation of carbonylmanganese ion pairs in Section III,A. Because phosphine ligands are known to strongly modulate the redox properties of carbonylmetals (77), the substituted carbonylcobalt(I) cations $Co(CO)_3L_2^+$ and carbonylcobaltate($-I$) anions $Co(CO)_3P^-$ that contain PPh_3 and PBu_3 can be selected as the prototypical aryl- and alkylphosphines, respectively (78).

1. Dimers from Carbonylcobalt Ion Pairs

When equimolar amounts of $Co(CO)_3(PPh_3)_2^+$ and $Co(CO)_3PPh_3^-$ in THF solutions are mixed at $25°C$, a quantitative reaction results immediately (<5 minutes) according to the stoichiometry in Eq. (31) (79). The homodimer $Co_2(CO)_6(PPh_3)_2$ largely precipitates as a red solid in 98% yield, and it is characterized by its distinctive carbonyl band at 1943 cm^{-1} in the IR spectrum (73). An equivalent amount of free triphenylphosphine is also extracted from the reaction mixture in accord with the stoichiometry in Eq. (31). When the corresponding ion pair that contains only tri-n-butylphosphine as the ligand is treated in a similar manner, high yields (93%) of the homodimer $Co_2(CO)_6(PBu_3)_2$ are obtained. However, this coupling occurs at a significantly diminished rate (~ 30 minutes), and the more soluble homodimer $Co_2(CO)_6(PBu_3)_2$ is extracted in 93% yield [Eq. (32)]. When $Co(CO)_3(PPh_3)_2^+$ is treated with an equimolar amount of $Co(CO)_3PBu_3^-$ in THF solution at 25°C, the initial IR spectrum of the reaction mixture that is recorded within 5 minutes shows no carbonyl band of either starting material (79). Immediate separation of the red solid yields the homodimer $Co_2(CO)_6(PPh_3)_2$ in 35% yield, and thin-layer chromatography (TLC) of the mother liquor affords a pure crystalline sample of the other homodimer, $Co_2(CO)_6(PBu_3)_2$, in 28% yield. The remaining carbonyl-containing product is identified as the cross-dimer Co_2-$(CO)_6(PPh_3)(PBu_3)$ (29% yield).

$$Co(CO)_3(PPh_3)_2^+ + Co(CO)_3PPh_3^- \xrightarrow[\text{THF}]{\text{fast}} Co_2(CO)_6(PPh_3)_2 + PPh_3 \qquad (31)$$

$$Co(CO)_3(PBu_3)_2^+ + Co(CO)_3PBu_3^- \xrightarrow[\text{THF}]{\text{slow}} Co_2(CO)_6(PBu_2)_2 + PBu_3 \qquad (32)$$

2. Effect of Added Phosphine on Ion-Pair Coupling

The rate of the homocoupling of $Co(CO)_3(PPh_3)_2^+$ and $Co(CO)_3PPh_3^-$ in Eq. (31) is materially unaffected when the THF solution contains up to 15 equivalents of added tri-n-butylphosphine. However, no evidence of $Co_2(CO)_6(PPh_3)_2$ can be found ($< 5\%$), the only carbonyl-containing product being the replaced homodimer $Co_2(CO)_6(PBu_3)_2$ isolated in 84% yield, that is,

$$Co(CO)_3(PPh_3)_2^+ + Co(CO)_3PPh_3^- \xrightarrow[\text{(excess)}]{\text{PBu}_3} Co_2(CO)_6(PBu_3)_2 + 3\ PPh_3 \qquad (33)$$

Control experiments show that neither the carbonylcobalt cation nor the anion undergoes prior ligand substitution by the added PBu_3 under the reaction conditions. Moreover, the homodimer $Co_2(CO)_6(PPh_3)_2$ and the cross-dimer $Co_2(CO)_6(PPh_3)(PBu_3)$ are not subject to subsequent ligand substitution by tributylphosphine to yield the homodimer in Eq. (33).

3. Effect of Solvent Polarity on Ion-Pair Annihilation

When the homocoupling of $Co(CO)_3(PPh_3)_2^+$ and $Co(CO)_3PPh_3^-$ in Eq. (31) is carried out in a more polar solvent such as acetonitrile, the formation of $Co_2(CO)_6(PPh_3)_2$ in 84% yield is noticeably retarded. Moreover, the solvent effect is unmistakable in the homocoupling of the tributylphosphine-substituted ion pair [Eq. (32)], which is essentially inert in acetonitrile for periods exceeding 10 hours. The crossed ion pair is similarly unreactive in acetonitrile, with only a small amount ($\sim 10\%$) of the anion $Co(CO)_3PPh_3^-$ diminished after 10 hours, and the concentration of the more stable cation $Co(CO)_3(PBu_3)_2^+$ is unchanged.

4. Inhibitory Effects of Added Salts on Ion-Pair Coupling

The homocoupling of $Co(CO)_3(PPh_3)_2^+$ and $Co(CO)_3PPh_3^-$ in THF solution is visibly retarded by the presence of 0.3 M tetra-n-butylammonium perchlorate (TBAP), but the yield of the homodimer $Co_2(CO)_6(PPh_3)_2$ is unaffected (79). The negative salt effect in THF is dramatic in the case of the less reactive $Co(CO)_3(PBu_3)_2^+/Co(CO)_3PBu_3^-$ pair of ions. Thus, the presence of 0.3 M TBAP is sufficient to retard the conversion in Eq. (32) to less than 5% over a 10-hour period. Similarly, the cross-coupling pair that consists of $Co(CO)_3(PBu_3)_2$ and $Co(CO)_3PPh_3^-$ yields less than 5% $Co_2(CO)_6(PBu_3)_2$ in THF solution over a 10-hour period when 0.3 M TBAP is added. IR analysis indicates that the concentration of the reactant ion pair is essentially undiminished.

C. Spectral Study of Contact Ion Pairs from Carbonylcobalt Cations and Carbonylcobaltate Anions

The ion-pair interactions of carbonylcobalt(I) cations with carbonylcobaltate($-$I) anions is apparent in both their electronic and IR spectra taken in solution and in the solid state.

1. Electronic Spectra of Precursor Salts

Both $Co(CO)_3(PBu_3)_2^+$ ClO_4^- and $Co(CO)_4^-$ PPN^+ show essentially no absorptions beyond 350 nm. However, the crystalline carbonylcobalt salt $Co(CO)_3(PBu_3)_2^+$ $Co(CO)_4^-$ obtained by disproportionation [Eq. (30), $P = n$-Bu_3P] is a bright yellow color which appears in the diffuse reflectance spectrum as a broad absorption with a shoulder ($\lambda_{max} \sim 400$ nm) tailing out to beyond 500 nm (79). By comparison, the color of a solution of 3.5 mM $Co(CO)_3(PBu_3)_2^+$ $Co(CO)_4^-$ dissolved in THF is also yellow, and it shows the same absorption band. Moreover, this band appears when the separate ions $Co(CO)_3(PBu_3)_2^+$ and $Co(CO)_4^-$ as the ClO_4^- and PPN^+ salts, respectively, are mixed in THF. As such, the new spectral band with λ_{max} around 400 nm is assigned to the charge-transfer absorption ($h\nu_{CT}$) of the contact ion pair (27):

$$Co(CO)_3(PBu_3)_2^+ Co(CO)_4^- \xrightarrow{h\nu_{CT}} Co(CO)_3(PBu_3)_2 \cdot Co(CO)_4 \cdot$$

In order to confirm this assignment, the effects of solvent polarity and added salts are examined on the behavior of the disproportionation salt $Co(CO)_3(PBu_3)_2^+ Co(CO)_4^-$ in solution.

Solvent polarity strongly affects the nature of $Co(CO)_3(PBu_3)_2^+$ $Co(CO)_4^-$ in solution as evidenced by the color change from orange in diethyl ether, yellow-orange in chlorobenzene, yellow in THF, pale yellow in dichloromethane, and finally to colorless in acetonitrile. The corresponding shift in the electronic absorption bands in Fig. 5A parallels the trend in the dielectric constant. As such, the complete bleaching of the color is consistent with the disappearance of the contact ion pair (and its associated charge-transfer band at ~ 400 nm) by stabilization of the solvent-separated ions in polar solvents such as acetonitrile, that is,

$$Co(CO)_3(PBu_3)_2^+Co(CO)_4^- \underset{\longleftarrow}{\overset{MeCN}{\longrightarrow}} Co(CO)_3(PBu_3)_2^+ // Co(CO)_4^-$$

By the same token, the spectral red shift observed in diethyl ether can be attributed to the interionic contraction induced by this least polar solvent (80).

Added salts like TBAP strongly perturb the intensity of the color of $Co(CO)_3(PBu_3)_2^+ Co(CO)_4^-$ in THF solution. Figure 5B shows the pronounced hypochromic effect on only the charge-transfer band with λ_{max} around 400 nm by incremental amounts of added TBAP. Indeed the presence of merely 1 equivalent of TBAP is sufficient to reduce the concentration of the contact ion pair by one half. Ion-pair exchange in Eq. (34) effectively separates the carbonylcobalt cation from the carbonylcobaltate anion in the contact ion pair. As such, the disappearance of the CT band in Fig. 5B is reminiscent of the situation achieved with highly polar solvents. It is important to emphasize that the equimolar combination of salts in Eq. (34) is equivalent to the mixture of the separate carbonylcobalt salts such as $Co(CO)_3(PBu_3)_2^+ ClO_4^-$ and $n\text{-}Bu_4N^+ Co(CO)_4^-$ that are employed above in ion-pair annihilation, and these expectations are borne out by the presence of the well-resolved charge-transfer band (with $\lambda_{CT} = 520$ nm) of $Co(CO)_3(PBu_3)_2^+ Co(CO)_3PBu_3^-$. This contact ion pair is instantaneously generated on mixing the separate carbonylcobalt salts that are individually transparent beyond 400 nm. Moreover the suppression of the CT absorption band by added TBAP is the same as that presented for the disproportionation salt in Fig. 5B. Indeed, the generality of the negative salt effect on such charge-transfer spectra emphasizes its validity as the experimental

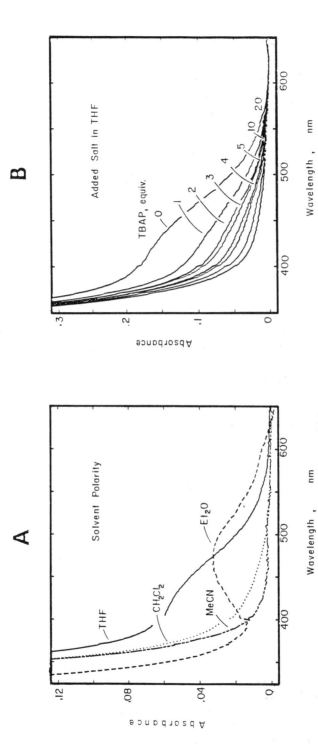

Fig. 5. Charge-transfer absorption spectrum of 3.5 mM Co(CO)$_3$(PBu$_3$)$_2$$^+$ Co(CO)$_4$$^-$. (A) Solvent effect in THF, CH$_2$Cl$_2$, MeCN, and Et$_2$O. (B) Salt effect of added TBAP (top to bottom): none, 1, 2, 3, 4, 5, 10, and 20 equivalents, relative to 10 mM Co(CO)$_3$(PBu$_3$)$_2$$^+$ Co(CO)$_4$$^-$ in THF (27).

criterion for the existence of contact ion pairs (14).

$$Co(CO)_3(PBu_3)_2^+ Co(CO)_4^- + Bu_4N^+ ClO_4^-$$
$$\rightleftharpoons Co(CO)_3(PBu_3)_2^+ ClO_4^- + Bu_4N^+ Co(CO)_4^- \quad (34)$$

2. Infrared Spectra of Disproportionation Salts

Both $Co(CO)_3(PBu_3)_2^+ Co(CO)_4^-$ and $Co(CO)_3(PPh_3)_2^+ Co(CO)_4^-$ consist by and large of the spectral composite of the individual ionic components. Thus, the IR absorptions of the carbonylcobalt cation in the carbonyl region occur at $v_{CO} = 2000-2800$ cm^{-1} and those for the carbonylcobaltate anionic moiety at $v_{CO} = 1870-1900$ cm^{-1}. However, a closer examination of the IR absorption bands reveals a systematic variation of v_{CO} with solvent polarity. The carbonyl band of $Co(CO)_3(PPh_3)_2^+$ in acetonitrile and dichloromethane consists of a single band with $v_{CO} = 2010$ cm^{-1} for the E' mode in D_{3h} symmetry. This band in a less polar THF medium is split into two components appearing as a shoulder and a weak A_1' band which are forbidden in D_{3h} symmetry. Moreover, the salt in either a KBr disk or a Nujol mull also shows well-resolved splittings into three distinct bands symptomatic of cation distortion to C_{2v} symmetry (81).

A change in the carbonyl bands of the carbonylcobaltate anion also accompanies the transition from solution to the crystal. The carbonyl band of $Co(CO)_4^-$ in either acetonitrile, dichloromethane, or THF consists of a single band at approximately 1890 cm^{-1}, as expected for the tetrahedral anion with T_d symmetry. In the crystal of $Co(CO)_3(PPh_3)_2^+ Co(CO)_4^-$, however, this triply degenerate band is split into three distinct components diagnostic of a configurational change to C_{3v} symmetry (17). Owing to an adventitious overlap with the cation absorptions, the weak A_1 band, which is forbidden in T_d symmetry, is not identified. Such simultaneous distortions of both the carbonylcobalt cation and carbonylcobaltate anion (as indicated by IR studies) thus mirror the solvent-induced changes in the amount of ion pairing as deduced from the charge-transfer absorptions). Both spectroscopic probes point to the relevance of the intimate molecular interactions extant in contact ion pairs.

D. Formation of Carbonylcobalt Dimers from Contact Ion Pairs

The existence of the charge-transfer absorption bands is characteristic of important electron donor–acceptor contributions to the contact ion pair that is the direct precursor in the formation of metal–metal dimers by the mutual annihilation of carbonylcobalt(I) cations and carbonylcobaltate(−I) anions (79). The diverse results cannot be explained by any single process in which the metal–metal bond for the dimer is formed by the *direct* replacement of a phosphine ligand (L) on the cationic $Co(CO)_3L_2^+$ by the nucleophilic anion $Co(CO)_3P^-$. Indeed, the experimental observations are (i) the homo coupling with $L = P = PPh_3$ in the presence of added PBu_3 produces only the substituted dimer $Co_2(CO)_6(PBu_3)_2$ under conditions in which neither the anion nor cation alone is subject to ligand substitution [Eq. (33)], and (ii) the rate of the substitutive coupling in Eq. (33) is essentially unaffected by the presence of 2–10 equivalents of foreign phosphine (PBu_3). Factors (*i*) and (*ii*) together demand that one or more reactive intermediates intervene between the rate-limiting activation process and the product-determining step.

In order to facilitate the mechanistic analysis, the strong correlation of the solvent and salt effect on the *rate* of ion-pair coupling is directly related to their effect on the *formation* of contact ion pairs in Fig. 5. Accordingly, the rapid equilibria leading up to the rate-determining step can be readily summarized as shown in Scheme 5 for the homocoupling where L and P are identical. Values of the formation constant K greater than $10^4 \, M^{-1}$ together with the configurational distortion of contact ion pairs in THF and CH_2Cl_2 point to a first-order activation process (k_1) in Eq. (36). The extensive scrambling of the ligands during ion-pair annihilation further indicates that Eq. (36) in Scheme 5 must proceed via labile intermediates, such as those formed in the rate-limiting activation by an electron-transfer process. On this basis the sequence of reactions leading to the metal–metal dimers evolve spontaneously from a pair of 19- and 17-electron carbonylcobalt radicals, as presented in Scheme 6 for the homocoupling process where L and P are identical.

$$Co(CO)_3P_2^+ + Co(CO)_3P^- \xrightleftharpoons{K} Co(CO)_3P_2^+ \, Co(CO)_3P^- \qquad (35)$$

$$Co(CO)_3P_2^+ + Co(CO)_3P^- \xrightarrow{k_1} Co_2(CO)_6P_2 + P \qquad (36)$$

SCHEME 5

$$Co(CO)_3P_2^+ \, Co(CO)_3P^- \xrightarrow{k_{ET}} Co(CO)_3P_2\cdot \, + Co(CO)_3P\cdot$$

$$Co(CO)_3P_2\cdot \longrightarrow Co(CO)_3P\cdot + P$$

$$2 \, Co(CO)_3P\cdot \longrightarrow Co_3(CO)_6P_2$$

SCHEME 6

E. Driving Force for Thermal Annihilation of Contact Ion Pairs by Electron Transfer

The coupling of the carbonylmanganese cation/anion pairs and the carbonylcobalt cation/anion pairs both follow a common pattern involving (a) the prior association to form contact ion pairs, followed by (b) interionic electron transfer to afford (c) the carbonylmetal radical pairs that ultimately lead to (d) their coupling for the metal–metal bond. In this general formulation, the rate-limiting activation process is largely represented by the adiabatic electron transfer within the contact ion pair.

1. *Electron Transfer in Formation of Mn–Mn Bonds*

Ion-pair annihilation via electron transfer is exemplified by the reaction of the carbonylmanganese(I) cation $Mn(CO)_6^+$ and the carbonylmanganate(−I) anion $Mn(CO)_5^-$ on mixing in THF solution [Eq. (22)] to afford high yields of dimanganese decarbonyl. Similarly, the substituted cations $Mn(CO)_5L^+$ with L = py, MeCN, aryl- and alkylphosphines and the substituted anions $Mn(CO)_4P^-$ with P = PPh$_3$ and P(OPh)$_3$ lead to mixtures of dimanganese carbonyls labeled with the P and L tracers [Eq. (23)]. The extensive (if not complete) scrambling of the carbonylmanganese moieties during ion-pair annihilation is ascribed to the 17- and 19-electron radicals $Mn(CO)_4P\cdot$ and $Mn(CO)_5L\cdot$, respectively, as the reactive intermediates. This conclusion is strongly supported by the known behavior of both types of radicals when they are independently generated by the anodic oxidation of $Mn(CO)_4P^-$ and the cathodic reduction of $Mn(CO)_5L^+$. The reversible addition of ligands to 17-electron radicals provides a ready means for interconversion with their 19-electron counterparts. Thus, the effect of added phosphine in altering the course of Mn–Mn coupling *via* ligand substitution and the formation of the hydrido by-products $HMn(CO)_4L$ *via* hydrogen transfer provide compelling evidence for carbonylmanganese radicals since neither the cation nor the anion is susceptible to additives on the time scale of the coupling experiments.

These experiments also rule out any type of concerted displacement mechanism as a dominant route in the formation of $Mn_2(CO)_{10}$ from ion pairs such as $Mn(CO)_5(NCMe)^+$ $Mn(CO)_5^-$ or $Mn(CO)_5py^+$ $Mn(CO)_5^-$ in which the carbonylmanganese moieties are not labeled. The participation of specific cross-coupling is included in the unified mechanism as a minor pathway stemming from cage collapse [Eq. (25)] of radicals immediately following an initial electron transfer. The latter is consistent with the reactivity trends for carbonylmanganese ion pairs which qualitatively parallel the differences ΔE in the oxidation and reduction potentials of $Mn(CO)_4P^-$ and $Mn(CO)_5L^+$ as follows. Since the radical-pair mechanism in Scheme 4 commences with an electron-transfer initiation, a critical part of the driving force for Mn–Mn bond formation is the oxidation–reduction of the anion–cation pair. As such, the energetics for the production of the 19- and 17-electron carbonylmanganese radical pair derives from the reversible electrode potentials for $Mn(CO)_5L^+$ and $Mn(CO)_4P^-$.

(a) For the *reduction* of the parent cation $Mn(CO)_6^+$, a value of E_{red}^0 of -1.0 V can be estimated from the cyclic voltammetric (CV) peak potentials (82). The wide span in potentials indicates a large ligand dependence on the ease of reduction of $Mn(CO)_5L^+$ in the order $L = py > MeCN > CO > PPh_3 > PPh_2Et > PPhEt_2 > PEt_3 \gg \eta^2\text{-DPPE}$. (b) For the anodic *oxidation* of the anionic $Mn(CO)_4PPh_3^-$, E_{ox}^0 equals -0.40 V versus a standard calomel electrode (SCE). The reversible potentials for the other carbonylmanganates lead to the conclusion that the ligand effect on the ease of oxidation is opposite to that in the cation in the order $L = PPh_3 \gg P(OPh)_3 > CO$. On this basis, the driving force for electron transfer will be the most favorable for the ion pair $Mn(CO)_5py^+$ $Mn(CO)_4PPh_3^-$ ($\Delta E \sim 0.5$ V) and the least favorable for $Mn(CO)_2\text{-}(DPPE)_2^+ Mn(CO)_5^-$ ($\Delta E \sim 1.6$ V). Indeed the latter is an unreactived ion pair, and it can be readily isolated as a simple salt from THF solution, namely,

$$Na^+ Mn(CO)_5^- + Mn(CO)_2(DPPE)_2^+BF_4^-$$
$$\longrightarrow Mn(CO)_2(DPPE)_2^+Mn(CO)_5^- + NaBF_4 \quad (37)$$

owing to the insolubility of $NaBF_4$.

Although the reaction rates are too fast to be measured quantitatively, the ion-pair reactivities qualitatively follow the trend in the driving forces. Thus, the ion pairs $Mn(CO)_5py^+ Mn(CO)_4PPh_3^-$ and $Mn(CO)_5(NCMe)^+$ $Mn(CO)_4PPh_3^-$ are the most reactive. At the other extreme, the couplings of the parent anion $Mn(CO)_5^-$ with the phosphine-substituted cations require the longest times. Since ΔE is approximately 1.2 V for $Mn(CO)_5PEt_3^+ Mn(CO)_5^-$, it appears that the threshold in the driving force for ion-pair annihilation lies somewhere between 1.2 and 1.6 V in THF solution. If the rate of electron transfer in Eq. (24) is taken in the outer-sphere context of Marcus theory (83), two other factors must also be considered. Thus, the interaction of oppositely charged ions in contact ion pairs will be aided considerably by electrostatics. Indeed such a positive work term is in accord with the strong solvent dependence and the negative salt effect. The contribution from the reorganization energies of $Mn(CO)_4P^-$ and $Mn(CO)_5L^+$ of these large highly polarizable five- and six-coordinate ions are likely to be small owing to the minor differences in the basic structural changes. If so, the driving force ΔE will be the dominant factor in determining the ease with which ion pairs are annihilated to form metal–metal bonds.

2. Annihilation Rates of Isomeric Contact Ion Pairs

The cation $Mn(CO)_2(DPPE)_2^+$ and the anion $Mn(CO)_2(DPPE)_2^-$ are critical to the successful mechanistic elucidation of the reduction that involves the facile annihilation of the ion pair. For a reversible process, the second-order rate constants are related to the driving force by the equilibrium relationship (84) $RT \ln(k_{ET}/k_d) = n\mathscr{F} \Delta E$ in which the magnitude of ΔE is given by $(E_{ox}^0 + E_{red}^0) = 0.11$ V and k_{ET} is the optimized second-order rate constant for electron transfer of 2×10^2 M^{-1} second^{-1}. The calculated value for the second-order rate constant for radical disproportionation $(k_d = 0.4\ M^{-1}$ second$^{-1})$ is indeed within the range predicted by the CV simulation (82). The rate of ion-pair annihilation of cis-$Mn(CO)_2(DPPE)_2^+$ by $Mn(CO)_2(DPPE)_2^-$ with a second-order rate constant $(k_{ET} = 2 \times 10^2\ M^{-1}$ second$^{-1})$ is a factor of 6 faster than the annihilation of the isomeric trans-$Mn(CO)_2(DPPE)_2^+$ with the same anion $(k_{ET} = 35\ M^{-1}$ second$^{-1})$ examined earlier (68). Such a trend in the electron-transfer rates parallels the difference in the driving force of 3.4 kcal/mol evaluated from the reduction potentials of -1.61 and -1.76 V for cis- and trans-$Mn(CO)_2(DPPE)_2^+$, respectively. The latter, together with the value of E_{ox}^0 of -1.45 V for the oxidation of the anionic $Mn(CO)_2(DPPE)_2^-$, indicates that the reductions of both cis- and trans-$Mn(CO)_2(DPPE)_2^+$ by $Mn(CO)_2(DPPE)_2^-$ have driving forces of 3.7 and 7.1 kcal/mol respectively.

These driving forces are exergonic and considerably more favorable than those involved in the electron-transfer reactions of the simple, monosubstituted carbonylmanganese cations $Mn(CO)_5L^+$ and anions $Mn(CO)_4P^-$ (where L and P are both monodentate phosphines and phosphites). Nonetheless, the rate constants k_{ET} for cis- and trans-$Mn(CO)_2(DPPE)_2^+$ with $Mn(CO)_2(DPPE)_2^-$ are considerably slower than those qualitatively observed between $Mn(CO)_5L^+$ and $Mn(CO)_4P^-$ (67). Such large rate differences that belie thermodynamics can be attributed to steric hindrance in the tetrasubstituted carbonylmanganese cations and the anion which are absent in the simpler ions. Such structural effects, even in these apparently outer-sphere electron transfers, merit a further quantitative evaluation as in the application of Marcus theory (83).

3. Rates of Ion-Pair Annihilation

The fate of the carbonylcobalt ions $Co(CO)_3L_2^+$ and $Co(CO)_3P^-$ are highly dependent on the phosphine ligand, the rate with $L = P = PPh_3$ in Eq. (31) taking place substantially faster than that with $L = P = PBu_3$ in Eq. (32). As such, the rate of ion-pair annihilation will parallel the driving force for electron transfer in Scheme 6, as given by the relationship $\Delta G_{ET} \cong -\mathscr{F}(E_c + E_a)$ where E_c and E_a are the CV peak potentials for $Co(CO)_3L_2^+$ and $Co(CO)_3P^-$, respectively (79). Evaluated in this way, the driving force for electron transfer of the PPh$_3$-substituted ion pair is substantially (~ 13 kcal/mol) greater than that for the PBu$_3$-substituted ion pair. This conclusion thus accounts for the reactivity pattern established for homocoupling {i.e., the homocoupling with L = P = PPh$_3$ is quantitative and rapid [Eq. (31)], whereas the same process with L = P = PBu$_3$ is slow [Eq. (32)]} and for cross-coupling [i.e., the cross coupling with L = PPh$_3$ and P = PBu$_3$ is rapid, and it yields a more or less equimolar mixture of the three dicobalt carbonyls; however, the same process for the reverse combination with L = PBu$_3$ and P = PPh$_3$ is slow and preferentially leads to Co$_2$(CO)$_6$L$_2$ with lesser amounts of Co$_2$(CO)$_6$LP with no Co$_2$(CO)$_6$P$_2$].

IV
ELECTRON-TRANSFER EQUILIBRIA FOR CONTACT ION PAIRS

The establishment of electron-transfer equilibria is allowed by the unique stabilizations inherent in the organometallic anion TpMo(CO)$_3$$^-$ [where Tp = hydridotris(3,5-dimethylpyrazolyl)borate] (85), in which the 17-electron radical (m$_2$·) is persistent (86). Since other members of the group VIB metals, namely, TpW(CO)$_3$$^-$ and TpCr(CO)$_3$$^-$, are also available (87), this triad of anions enables the structural effects to be systematically examined,

$$\text{TpM(CO)}_3{}^- \underset{}{\overset{E^0_{ox}}{\rightleftharpoons}} \text{TpM(CO)}_3\cdot + e^- \tag{38}$$

Electron-transfer equilibria of the organometallic anions TpM(CO)$_3$$^-$ can be examined by coupling them with a graded series of triarylpyrylium cations (TaP$^+$) for which 1-electron reduction potentials are known to be strongly dependent on the substituents (88). The evaluation of the constant K' for the electron-transfer equilibrium in Eq. (39) requires the quantitative analysis of the anionic organometallic redox couple in Eq. (38), as well as that of triarylpyrylium cation.

$$\text{TpM(CO)}_3{}^- + \text{TaP}^+ \overset{K'}{\rightleftharpoons} \text{TPM(CO)}_3\cdot + \text{TaP}\cdot \tag{39}$$

A. Spectral Characterization of TpM(CO)$_3$$^-$ Anions and Their 17-Electron Radicals, Where M Is Mo, W, and Cr

Each TpM(CO)$_3$$^-$ anion is readily identified by its characteristic IR spectrum consisting of two carbonyl bands corresponding to the symmetric A_1 and the antisymmetric E stretching frequencies for idealized C_{3v} symmetry (37). The isostructural molybdenum, tungsten, and chromium derivatives all show a single sharp (fwhm ~ 10 cm^{-1}) A_1 band at approximately 1880 cm^{-1} and a broadened (fwhm ~ 30 cm^{-1}) E band at about 1740 cm^{-1} in acetonitrile solutions of the tetra-n-butylammonium (TBA$^+$) and N-methyl-4-cyanopyridinium (NCP$^+$) salts.

The molybdenum radical TpMo(CO)$_3\cdot$ can be isolated from the oxidation of TpMo(CO)$_3^-$ with ferrocenium salt (85) and the parent pyrazoyl derivative characterized by X-ray crystallography (86). The cyclic voltammetry of TpMo(CO)$_3^-$ is characterized by a well-defined reversible voltammogram obtained on the initial-positive scan at sweep rates (v) of 0.5 and 20 V second^{-1}, and the reversible oxidation potential E^0_{ox} for Eq. (38) is obtained as the average of the anodic and cathodic CV peak potentials, ($E_a + E_c$)/2 (89). The anodic peak current, calibrated against the ferrocene standard, corresponds to 1-electron oxidation of TpMo(CO)$_3^-$, and chemical reversibility is indicated by the current ratio of the cathodic and anodic peaks (i_c/i_a) of unity. In order to confirm the production of the 17-electron radical, a solution of TpMo(CO)$_3^-$ is subjected to bulk electrolysis at a constant potential of 0 V versus SCE in dichloromethane containing 0.2 M tetrabutylammonium perchlorate, and the course of the anodic oxidation is followed by IR spectrophotometry for the growth of a pair of new carbonyl bands for the 17-electron radical TpMo(CO)$_3\cdot$ The clean oxidative conversion of TpMo(CO)$_3^-$ is indicated in the visible absorption spectrum by the appearance of a single isobestic point at 380 nm. Coulometry shows that the anodic current remains invariant until the requisite passage of charge (calculated for 0.99 e), whereon it drops abruptly to nil. The clear golden yellow anolyte shows the characteristic carbonyl bands at 2001 and 1860 cm^{-1} of TpMo(CO)$_3\cdot$ The solution of TpMo(CO)$_3\cdot$ in dichloromethane shows no change in its IR spectrum on prolonged standing, but it is extremely oxygen-sensitive and turns dark red immediately on exposure to air. The remarkable stability of TpMo(CO)$_3\cdot$ is underscored by its failure to react with σ donors such as triphenylphosphine even when present in a large 20-fold excess (90).

The electrochemical reduction of the triarylpyrylium cations can be examined in dichloromethane solution by cyclic voltammetry. The reversible one-electron potentials for the cation reduction,

$$(40)$$

are consistent with those of a variety of pyrylium cations examined earlier (88) and include a small constant contribution from the reversible dimerization of the triarylpyranyl radicals (91),

$$2 \quad \underset{\substack{Ar \\ Ar}}{\overset{Ar}{\bigodot}} \quad \xrightarrow{K_{dim}} \quad (41)$$

as judged by the K_{dim} that varies only slightly from 600 M^{-1} for the trianisylpyranyl radical (TAP·) (92) to 800 M^{-1} for the triphenylpyranyl radical.

B. Electron-Transfer between TpM(CO)$_3^-$ Anions and Pyrylium Cations

When a dilute (millimolar) solution of TpMo(CO)$_3^-$ as the tetrabutylammonium salt in dichloromethane is treated with 0.2 equivalent of 2,4,6-triphenylpyrylium triflate (TPP$^+$OTf$^-$), the pale yellow solution immediately acquires a dark orange color (93). Inspection of the visible absorption spectrum reveals the presence of three characteristic bands with λ_{max} values of 512,552, and 780 nm of the 2,4,6-triphenylpyranyl radical (91). The infrared spectral changes accompanying the successive addition of the TPP$^+$ salt to Bu$_4$N$^+$ TpMo(CO)$_3^-$ in dichloromethane reveals the diminution in the carbonyl bands of the anion ($\nu_{CO} = 2001$ and 1865 cm^{-1}), in a manner indistinguishable from that observed during anodic oxidation. Such a concomitant appearance of the 17-electron organometallic radical and the pyranyl radical, attendant on mixing of the molybdenum anion and the pyrylium cation, is due to the reversible electron-transfer process, namely,

$$TPP^+ + TpMo(CO)_3^- \rightleftharpoons TPP\cdot + TpMo(CO)_3\cdot \qquad (42)$$

Indeed the same color change occurs when the molybdenum anion and pyrylium cation are mixed in acetonitrile. However, the significantly diminished intensity of the orange color (and the reduced absorbance of the visible bands at $\lambda_{max} = 508$, 548 and 780 nm) indicates that lower concentrations of TPP· (by roughly a factor of 10) are formed in acetonitrile relative to dichloromethane containing the same salt concentrations. This observation, coupled with the lower concentrations of the 17-electron radical is ascribed to the equilibrium in Eq. (42), which is displaced toward the left in the more polar solvent (40).

Similarly, exposure of the tungsten anion TpW(CO)$_3^-$ to TPP$^+$ in dichloromethane immediately results in the same qualitative color change, but the deeper red-orange solution together with intense visible bands at 512,

552, and 780 nm indicates that higher concentrations of the pyranyl radical TPP· are present. Moreover, the presence of strong carbonyl bands at 1983 and 1856 cm^{-1} for the 17-electron radical TpW(CO)$_3$· (90) suggests that the electron-transfer equilibrium is displaced further to the right than that observed in Eq. (42) for the molybdenum anion. Indeed, the presence of merely 1.2 equivalents of TPP$^+$ OTf$^-$ in a dilute solution of Bu$_4$N$^+$ TpW(CO)$_3$$^-$ in dichloromethane is sufficient for the complete conversion of TpW(CO)$_3$$^-$.

The presence of 1 equivalent of TPP$^+$ OTf$^-$ with the chromium anion TpCr(CO)$_3$$^-$ as the tetrabutylammonium salt in dichloromethane results in the loss of the carbonyl bands of the anion at 1890 and 1740 cm^{-1}. Their complete replacement by the sharp A_1 band at 2018 cm^{-1} and the broad E band (1898 and 1838 cm^{-1}) of the 17-electron radical TpCr(CO)$_3$· indicates that the ion-pair annihilation proceeds to completion. Variation of the pyrylium cation, by the replacement of TPP$^+$ with a weaker acceptor such as tri-p-anisylpyrylium triflate (TAP$^+$ OTf$^-$), consistently results in lower conversions of the carbonylmetal anions. For example, the treatment of TpMo(CO)$_3$ with the TAP$^+$ salt leads to a light red solution of TAP· (λ_{max} 560 nm) (92) and a greatly diminished concentration of TpMo(CO)$_3$· as judged by the reduced carbonyl absorbances in comparison with that obtained from TPP$^+$ at the same concentration. Even with this weaker acceptor cation, however, the strong chromium anionic donor TpCr(CO)$_3$$^-$ is completely oxidized by 1 equivalent of TAP$^+$ to form TpCr(CO)$_3$· in essentially quantitative yields.

The tropylium cation (C$_7$H$_7$$^+$) with a reduction potential of -0.18 V versus SCE in acetonitrile is only a slightly better oxidant than TPP$^+$. Nonetheless, both TpMo(CO)$_3$$^-$ and TpW(CO)$_3$$^-$ are quantitatively converted in dichloromethane solution to the 17-electron radicals (as assayed by their carbonyl bands). The rapid and irreversible dimerization of tropyl radicals is undoubtedly responsible for driving the electron transfer to completion,

$$TpM(CO)_3^- + C_7H_7^+ \longrightarrow TpM(CO)_3 \cdot + \tfrac{1}{2}(C_7H_7)_2 \qquad (43)$$

Indeed, the 17-electron radicals in Eq. (43) with M = Mo and W can both be readily collected as red and orange crystals merely when the ion-pair annihilation is carried out in acetonitrile solutions.

The simultaneous spectral observation of the organometallic redox couple TpM(CO)$_3$$^-$ and TpM(CO)$_3$· in dichloromethane allows their concentrations to be accurately determined in the presence of various stoichiometric amounts of the pyrylium cations. The establishment of the electron-transfer equilibrium as presented in Eq. (39) is indicated by the constant value of K', irrespective of the initial molar ratio of the ion pair.

Although the constant K' includes a contribution from the reversible dimerization of the pyranyl radical in Eq. (41), the trend generally shows an increase with acceptor strength (E_{red}^0) of the pyrylium cation and donor strength (E_{ox}^0) of the organometallic anion.

The influence of added salt on the electron-transfer equilibria is examined by the addition of an inert salt, tetrabutylammonium perchlorate (TBAP), to dichloromethane solutions of $TpM(CO)_3^-$ and TaP^+. The dramatic decrease in the concentration of the 17-electron radical (quantitatively measured by its carbonyl absorbances) occurs in the presence of as little as 1 equivalent of added TBAP. A slightly smaller salt effect is shown on the concentration of $TpMo(CO)_3 \cdot$ obtained from the weaker pyrylium acceptor (diphenylanisylpyrylium, DAP$^+$) containing one p-methoxy substituent. The large salt effects induced by relatively small amounts of added TBAP, coupled with the sharp leveling off at slightly elevated concentrations, are typical of the special salt effect resulting from the ion-pair exchange examined earlier (94). Such an ionic exchange effectively serves to disassemble the active ion pair that is directly pertinent to the electron-transfer equilibrium, that is

$$TaP^+TpM(CO)_3^- + Bu_4N^+ ClO_4^- \xrightleftharpoons{K_{ex}} TaP^+ ClO_4^- + Bu_4N^+ TpM(CO)_3^-$$

(active ion pair)   (inactive ion pair)   (inactive ion pair)

The equilibrium constant K_{ex} of essentially unity points to the nonspecific interchange between contact ion pairs (27).

C. Contact Ion Pairs of TpM(CO)₃⁻ Anions as Charge-Transfer Salts

Direct evidence for the involvement of the organometallic anions $TpM(CO)_3^-$ in contact ion pairs obtains from the electronic transitions associated from proximate cation–anion pairs and delineated as charge-transfer excitations by Mulliken (7). This consideration is brought out by the dramatic colors obtained on the exposure of the $TpM(CO)_3^-$ anions to selected pyridinium cations with reduction potentials E_{red}^0 that are sufficiently negative to obviate spontaneous electron transfer. Thus, the mixing of a dilute, pale yellow solution of $Bu_4N^+ TpMo(CO)_3^-$ in dichloromethane with a colorless solution of N-methyl-4-phenylpyridinium (MPP$^+$) triflate with an E_{red}^0 value of -1.3 V versus SCE immediately produces a deep purple coloration. The quantitative effect of this color change is indicated by the appearance of a new absorption band with a λ_{max} of 540 nm covering a broad span (fwhm 7500 cm^{-1}) characteristic of an intermolecular charge-transfer excitation (30). Examination of the IR spectrum of the purple solution reveals the low-energy E band with $\nu_{CO} =$

1742 cm^{-1} in the Bu$_4$N$^+$ salt to be split into a pair of bands with $\nu_{CO} =$ 1744 and 1728 cm^{-1}, owing to the presence of the cationic MPP$^+$ in the contact ion pair (27). (The high energy A band is unaffected by added MPP$^+$.)

Most importantly, the new absorption band undergoes a progressive red (bathochromic) shift when the molybdenum anion is replaced by successively better anionic donors, such as those with tungsten (purple, λ_{max} 562 nm) and chromium (green, λ_{max} 636 nm). In accord with the expectations based on Mulliken theory (7), the new absorption band also undergoes a progressive red shift as the cationic acceptor MPP$^+$ is replaced with increasingly more effective acceptors. With the most powerful pyridinium acceptor, namely, N-methyl-4-cyanopyridinium (NCP$^+$) triflate with an E^0_{red} value of -0.64 V versus SCE, the ion-pair interaction with TpMo(CO)$_3^-$ leads to the most red-shifted CT band (λ_{max} 726 nm). Inspection of the IR spectrum of the green solution shows prominent carbonyl bands of TpMo(CO)$_3^-$ with $\nu_{CO} =$ 1880, 1744, and 1728 cm^{-1} in addition to a minor absorbance at 2001 cm^{-1}, indicative of the admixture of the 17-electron radical. Indeed, the best anionic donor TpCr(CO)$_3^-$ reacts rapidly with NCP$^+$ on mixing to form a golden yellow solution, the IR spectrum of which is bereft of the diagnostic carbonyl bands of the chromium anion. It is also interesting to note that all of the colored solutions of the charge-transfer salts in dichloromethane are highly air-sensitive; typically, that from TpMo(CO)$_3^-$ is bleached within 30 seconds of exposure to air under conditions in which the solution of Bu$_4$N$^+$ TpMo(CO)$_3^-$ itself is unchanged.

The addition of the cyanopyridinium salt NCP$^+$ OTf$^-$ to a THF solution of Bu$_4$N$^+$ TpMo(CO)$_3^-$ immediately results in a green solution from which a deep green solid precipitates on standing to leave a colorless solution. The IR spectrum of the salt in a KBr disk shows the characteristic carbonyl bands at 1882, 1757, and 1736 cm^{-1} similar to that obtained in dichloromethane solution (see above). A single crystal of the dark green powder suitable for X-ray crystallography can be grown from a solution of diethyl ether and acetonitrile. The structure of NCP$^+$ TpMo(CO)$_3^-$ is solved in the monoclinic space group $P2_1/n$, and the ORTEP diagram in Fig. 6 shows a perspective of the relevant interionic separation that is responsible for the charge-transfer absorption band with a λ_{max} of 726 nm. Most importantly, the intimate nature of the contact ion pair shows up in the charge-transfer salt with a cationic acceptor which is strongly oriented toward one of the carbonyls of TpMo(CO)$_3^-$ at an oxygen distance of 3.13 Å to the pyridinium centroid. Such an asymmetric orientation of the cationic acceptor relative to the Mo(CO)$_3$ moiety is sufficient to break the degeneracy of the E band in the fac-tricarbonyl IR spectrum (37). The accompanying distor-

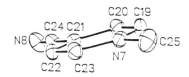

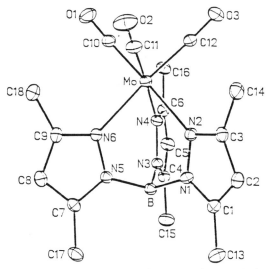

FIG. 6. ORTEP diagram of the charge-transfer salt showing the relative disposition of the cationic pyridinium acceptor P(NCP$^+$) to the anionic TpMo(CO)$_3^-$ donor. The thermal ellipsoids are 20% equiprobability envelopes with hydrogens omitted for clarity (93).

tion of the tricarbonylmolybdenum moiety from C_{3v} symmetry is exemplified by three distinct OC—Mo—CO bond angles of 84.7(2), 85.9(2), and 86.6(2)°, and it is translated to the trispyrazolylborate ligand binding with N—Mo—N bond angles of 80.3, 81.2, and 82.4°. This slightly distorted octahedral structure of the anion contrasts with the strict C_{3v} symmetry of the corresponding uncharged 17-electron radical TpMo(CO)$_3\cdot$ (86) and is a direct consequence of the contact ion pairing (see above).

D. Evaluation of Electron-Transfer Equilibria in TpM(CO)$_3^-$ Contact Ion Pairs

The isostructural pyrazolylborate complexes TpM(CO)$_3^-$ thus form a graded series of organometallic anions with donor strengths increasing in the order $M = Mo < W < Cr$, as judged by the E^0_{ox} values. The highly persistent nature of the 17-electron radicals $TpM(CO)_3 \cdot$ derived by 1-electron oxidation (and their inertness to ligand substitution) enable these anions to be particularly useful in electron-transfer studies. By comparison, most 17-electron carbonylmetal radicals rapidly undergo irreversible dimerizations (95); and even the stable $V(CO)_6$ is subject to ready ligand substitution (96). For these reasons, the redox reactions of $TpM(CO)_3^-$ with a similarly graded series of triarylpyrylium cations TaP^+ can be used to establish the structural effects on electron-transfer equilibria. Before proceeding, however, it is important to quantitatively assess the contribution from the reversible dimerization of pyranyl radicals (91). Furthermore, the central importance of contact ion pairs to the establishment of electron-transfer equilibria is underscored by the pronounced salt effects arising from ion-pair exchange. Accordingly, the reversible pyranyl dimerization and ion-pair exchange are specifically designated by the constants K_{dim} and K_{ex}, respectively, and they are included with the electron-transfer equilibrium (K_{ET}) as shown in Scheme 7, where X^- and Y^+ are generic representations of all the inert (spectator) ions (Bu_4N^+, ClO_4^-, $O_3SCF_3^-$, etc.). According to Scheme 7, the added salt $X^- Y^+$ depresses the concentration of the redox-active ion pair $TaP^+ TpM(CO)_3^-$ and qualitatively shifts the electron-transfer equilibrium away from the radical species. The inclusion of the (special) salt effect thus provides a steady-state analogy to the kinetic salt effect that affects the dynamics of transient redox-active ion pairs (97).

$$TaP^+ X^- + Y^+ TpM(CO)_3^- \xrightleftharpoons{K_{ex}} TaP^+ TpM(CO)_3^- + Y^+ X^-$$

$$TaP^+ TpM(CO)_3^- \xrightleftharpoons{K_{ET}} TaP \cdot + TpM(CO)_3 \cdot$$

$$2 \, TaP \cdot \xrightleftharpoons{K_{dim}} (TaP)_2$$

SCHEME 7

In order to determine the quantitative effects of the pyranyl dimerization and the salt effect on the electron-transfer equilibrium, an operational equilibrium constant K' can be defined so that it is evaluated solely from the reliable quantification of $TpM(CO)_3^-$ and $TpM(CO)_3 \cdot$ by infrared spectrophotometry,

$$K' = [TpM(CO)_3 \cdot]^2/[TpM(CO)_3^-](a - [TpM(CO)_3 \cdot]) \qquad (44)$$

where a is the formal (initial) concentration of the pyrylium salt. The applicability of Eq. (44) is shown by the values of K' which are more or less invariant at various formal concentrations of $TpM(CO)_3^-$ and TaP^+. Since the study in Eq. (7) indicates the absence of preferential ion pairing, that is, K_{ex} is approximately 1 (27), it can be shown that the set of equilibria in Scheme 7 reduces to a direct relationship between the operational equilibrium constant K' and the equilibrium constant K_{ET} for electron transfer (93), given as

$$K_{ET} \cong K'\{a + b - [TpM(CO)_3 \cdot]/(2K_{dim}[TpM(CO)_3 \cdot])^{1/2}\} \quad (45)$$

where a and b are the formal concentrations of P^+ and $TpM(CO)_3^-$, respectively.

The equilibrium constants for electron transfer obtained with the aid of Eq. (45) with K_{dim} equal to 6×10^2 M (91) are included in Table VI for various redox-active ion-pair combinations. Most notably, the large factor between these equilibrium constants (i.e., $K'/K_{ET} \cong 10^2$) relates to the pyranyl dimerization that drives the electron-transfer equilibrium by roughly a hundredfold. Otherwise the magnitudes of the equilibrium constant for electron transfer follow the expected trend, with K_{ET} for the better anionic donor $TpW(CO)_3^-$ being a factor of 6.5 times larger than that for $TpMo(CO)_3^-$. Likewise, the values of K_{ET} parallel the variation in the reduction potentials E_{red}^0 of the cationic pyrylium acceptors. However, the results in the last column of Table VI indicate that K_{ET} is not simply related to the thermodynamic free-energy change based on the potentials of the redox couples, that is, $\Delta G = -\mathscr{F}(E_{ox}^0 + E_{red}^0)$, where $\dot{E}_{ox}^0$ for $TpM(CO)_3^-$ and E_{red}^0 for TaP^+ are given in Eqs. (38) and (40), respectively, and $\mathscr{F}$ is the Faraday constant. An important factor in the discrepancy undoubtedly lies

TABLE VI

EQUILIBRIUM CONSTANTS FOR ELECTRON TRANSFER FROM CONTACT ION PAIRS P^+ $TpM(CO)_3^-$ IN DICHLOROMETHANE SOLUTIONS

$TpM(CO)_3^-$ M	TaP^{+a}	K_{ET} (M)	ΔG_{ET} (kcal/mol)[b]	ΔG_{calc} (kcal/mol)[c]
Mo	TPP$^+$	0.15	1.1	0
	DAP$^+$	4.9×10^{-3}	3.0	2.0
	DVP$^+$	7.5×10^{-3}	2.8	1.7
	TAP$^+$	6.2×10^{-5}	5.6	5.8
W	TAP$^+$	2.3×10^{-2}	2.2	3.7
Cr	TAP$^+$	$> 10^2$		-1.4

[a] TPP$^+$, Triphenylpyrylium; DAP$^+$, diphenylanisylpyrylium; DVP$^+$, phenyldiveratrylpyrylium; TAP$^+$, trianisylpyrylium.
[b] From the value of $-RT \ln K_{ET}$ at 294 K from Ref. 93.
[c] From $(E_{ox}^0 + E_{red}^0)$ as described in the text.

in the extensive ion pairing in the form of contact ion pairs that must be specifically taken into account, especially in nonpolar solvents such as dichloromethane.

An indication of the magnitude of this problem is obtained from the intimate structure of $NCP^+ TpMo(CO)_3^-$, as illustrated by the ORTEP diagram in Fig. 6. That this structure of the crystalline charge-transfer salt also represents the contact ion pair in dichloromethane is strongly indicated by the diagnostic carbonyl stretching bands that are the same in the crystal and in solution. In particular the characteristic splitting of the (otherwise degenerate) antisymmetric E band ($v_{CO} = 1758$ and $1737 cm^{-1}$) relates directly to the distortion of the $M(CO)_3$ moiety from ideal C_{3v} symmetry. Indeed the X-ray crystal structure of $NCP^+ TpMo(CO)_3^-$ (Fig. 6) shows the origin of the distortion to lie with the unsymmetrical disposition of the pyridinium cation, which is closest to one of the carbonyl ligands with a C—O bond length that is 0.024 Å (5σ) longer and OC—Mo—CO bond angles that are larger (86.6 and 85.9 *versus* 84.7°, 10σ) than the others. Structural studies of other types of crystalline charge-transfer salts and their contact ion pairs in solution will provide further quantitative insight into such electron-transfer equilibria.

V

NUCLEOPHILIC AND ELECTRON-TRANSFER PROCESSES IN ION-PAIR ANNIHILATION

Nucleophilic addition to unsaturated ligands, such as olefins, acetylenes, and arenes, coordinated to various metal centers is a useful strategy in organic synthesis (*98*). The ion-pair interaction of organometallic cations and anions is facile and can successfully lead to the nucleophilic activation of ethylene, acetylene, and benzene (*99*), for example,

$$m_1(\eta^2\text{-}C_2H_4)^+ + m_2^- \rightarrow m_1 \diagdown\diagup m_2 \qquad (46)$$

where $m_1 = CpMo(CO)_3$ and $CpW(CO)_2(PPh_3)$ and $m_2 = CpMo(CO)_3$, $Re(CO)_5$, and $CpW(CO)_2(PPh_3)$ (*100*). Many organometallic cations are also readily convertible to 19-electron radicals and, likewise, anions to 17-electron radicals by 1-electron reduction and oxidation as described above. The latter raises the question as to whether the nucleophilic additions to coordinated ligands (L) are actually preceded by an electron-transfer step, namely,

$$m_1L^+ + m_2^- \longrightarrow m_1L\cdot + m_2\cdot, \text{ etc.} \qquad (47)$$
$$(19\,e^-) \quad (17\,e^-)$$

This mechanistic possibility is supported by sporadic reports of the isolation of dimeric by-products and other types of evidence of the radical intermediates $m_1L\cdot$ and $m_2\cdot$ during the annihilation of organometallic ion pairs (101). Indeed, the detailed comparison of a series of the structurally related alkenyliron cations (102),

reveal the divergent products that stem from apparently concurrent radical and nucleophilic processes (103). Each $Fe(CO)_3L^+$ as the hexafluorophosphate salt (102) and the nucleophilic reagent (104) PPN$^+$ CpMo(CO)$_3^-$ are sufficient to establish two distinctive courses of ion-pair reaction, as described individually below.

A. Ion-Pair Annihilation to Nucleophilic Adducts

When the pentadienyliron cation **1** is combined with CpMo(CO)$_3^-$ at 25°C, only a single yellow adduct **5a** is formed in 95% yield, as immediately judged by the growth of the characteristic carbonyl stretching band at 2010 cm^{-1} in the IR spectrum. The bright yellow adduct **5a** is also the sole product when a slurry of the pentadienyliron salt, rather insoluble in nonpolar solvents such as dichloromethane or THF, is treated under otherwise the same reaction conditions. The homologous hexadienyliron cation **2** is similarly converted to the analogous yellow adduct **5b** with $\nu_{CO} = 2012$ cm^{-1} as the predominant product (75% yield),

Since the structure of **5b** is established by X-ray crystallography (103), the series of related products **5** are hereafter designated as nucleophilic adducts.

B. Ion-Pair Annihilation to Radical Dimers

Radicals are formed spontaneously on the addition of $CpMo(CO)_3^-$ to the cyclohexadienyliron cation **3**, as observed spectroscopically at 25°C by the rapid disappearance of both carbonylmetal ions and the simultaneous formation of the corresponding pair of homodimers, namely,

$$\text{--Fe(CO)}_3^+ + CpMo(CO)_3^- \xrightarrow{25°C}$$

3

$$(OC)_3Fe\diagdown\diagup Fe(CO)_3 + [CpMo(CO)_3]_2 \quad (49)$$

6a   **7**

The formation of **6a** is particularly diagnostic, since this unique carbon–carbon bonded reductive dimer was demonstrated by Wrighton and co-workers to arise via the transient 19-electron radical $(\eta^5$-cyclohexadienyl)Fe(CO)$_3\cdot$ by regiospecific coupling at a ligand center (*105*). Furthermore the 17-electron radical $CpMo(CO)_3\cdot$ is the precursor to the accompanying oxidative dimer $[CpMo(CO)_3]_2$ (**7**), as described in the earlier anodic studies of $CpMo(CO)_3^-$ (*106*). Accordingly these products (**6** and **7**) of ion-pair annihilation are referred to hereafter as radical (homo) dimers.

The singular absence of a nucleophilic adduct from either **3** or **4** suggests that much milder conditions be employed in the ion-pair annihilation. When a THF slurry of the cyclohexadienyliron cation **3** is treated at −78°C with $CpMo(CO)_3^-$, it gradually dissolves to afford a bright yellow mixture. No evidence of the red intermediate that is readily apparent at higher temperatures (see above) can be discerned. Instead, after the separation of the colorless precipitate of $PPN^+ PF_6^-$ at −50°C, the IR spectrum of the chilled solution is found to be essentially identical to that of the nucleophilic adducts (**5**) (see above), that is,

$$\text{--Fe(CO)}_3^+ + CpMo(CO)_3^- \xrightarrow[\text{THF}]{-78°C} (OC)_3Fe\text{---}\diagdown_{Mo(CO)_3Cp} \quad (50)$$

3   **5c**

Upon warming the clear yellow solution to $-20°C$, it turns red and deposits dark purple crystals of the molybdenum dimer **7** in essentially quantitative yields, and the accompanying mother liquor also affords an equimolar amount of the (cyclohexadiene)iron dimer **6a**,

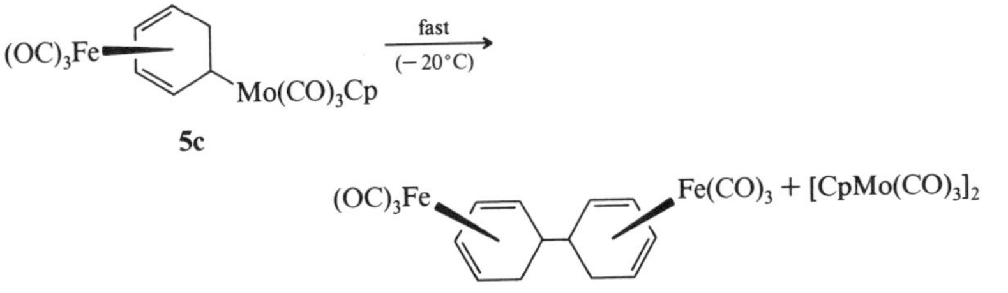

$$(OC)_3Fe \overset{\text{fast}}{\underset{(-20°C)}{\longrightarrow}} Mo(CO)_3Cp$$

5c

$$(OC)_3Fe \qquad Fe(CO)_3 + [CpMo(CO)_3]_2 \quad (51)$$

6a

Careful workup of the labile yellow solution by solvent removal *in vacuo* at $-30°C$, yields the thermally unstable, bright yellow complex **5c** in high yields. A similar yellow solid **5d** is also obtained when the homologous cycloheptadienyliron cation **4** is treated with $CpMo(CO)_3^-$ at $-78°C$, but rapid decomposition to an equimolar mixture of the molybdenum and iron dimers **7** and **6b** at $-40°C$, indicates that **5d** is even less stable than **5c**. The ready homolysis of the labile intermediate in Eq. (51) raises the parallel question as to the corresponding fate of the nucleophilic adducts **5a** and **5b**.

C. Thermal Decomposition of Nucleophilic Adducts

Solutions of the nucleophilic adducts **5a** and **5b** are indeed thermally unstable and slowly degrade, even at room temperature in the absence of air. Visually, the bright yellow solution of **5b** in hexane gradually darkens merely on standing in the dark. After several hours, it deposits the insoluble purple crystals of the molybdenum dimer **7** in essentially quantitative yields, and an equimolar amount of the iron dimer **6d** can be isolated from the mother liquor,

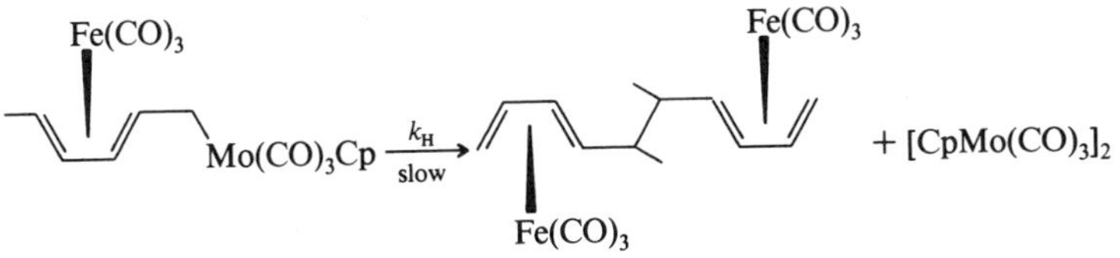

$$Fe(CO)_3 \qquad\qquad Fe(CO)_3$$
$$\overset{}{\underset{Mo(CO)_3Cp}{\quad}} \xrightarrow[\text{slow}]{k_H} \qquad + [CpMo(CO)_3]_2 \quad (52)$$
$$Fe(CO)_3$$

5b   **6d**   **7**

The kinetics of the decomposition are measured in various solvents by

simultaneously following the disappearance of the nucleophilic adduct and
the appearance of the iron and molybdenum dimers by quantitative IR
spectrophotometry of their characteristic carbonyl stretching bands,
namely, at 2011 (2031), 2044, and 1912 cm^{-1} for **5b, 6d,** and **7,** respectively, in THF. Clean first-order kinetics are followed by the decay of the
reactant and the growth of both products. The effect of solvent polarity on
the first-order decomposition of adduct **5b** is illustrated in Fig. 7. Indeed
the limited variation ($\pm 10\%$) in the first-order rate constant k_H for ther-
molyses carried out in media differing as widely as acetonitrile, THF, and
hexane underscore the minor role of solvent polarity in adduct decomposition (*107*). Significantly more pronounced in Fig. 7 is the effect of solvent
viscosity, as indicated by the approximate 500% decrease in k_H on proceeding from hexane to dodecane to Nujol (*108*). From the temperature dependence of the rate constant k_H for decomposition of the nucleophilic
adducts **5a** and **5b** in THF solvent, the activation parameters can be
determined as $\Delta H^{\ddagger} = 31$ and 28 kcal/mol and $\Delta S^{\ddagger} = 20$ and 14 cal/mol/
K, respectively. Transient ESR studies coupled with the extensive use of

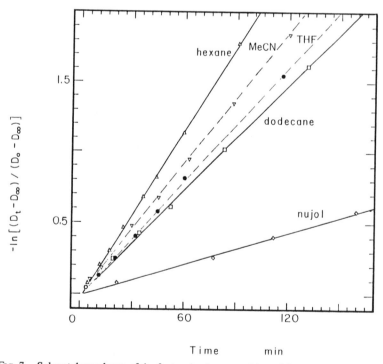

FIG. 7.   Solvent dependence of the first-order decomposition of the nucleophilic adduct **5b**
at 30°C in hexane, acetonitrile, THF, dodecane, and Nujol (*103*).

free radical (spin) traps are consistent with transient paramagnetic species as reactive intermediates (103), that is,

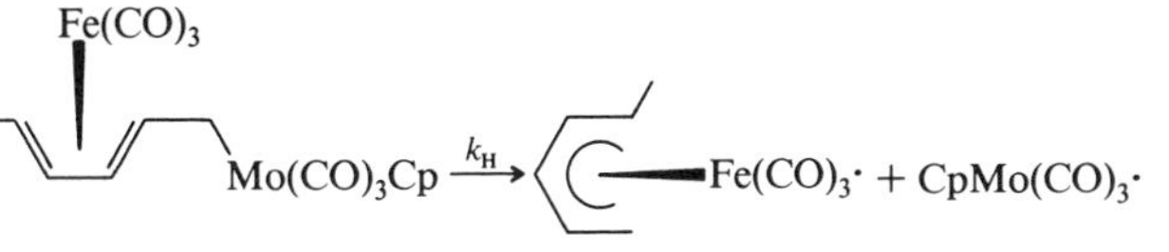

$$\text{(53)}$$

5b

during the decomposition of the nucleophilic adduct according to the stoichiometry in Eq. (52).

D. Nucleophilic Addition versus Electron Transfer as Competing Pathways

The alkenyliron cations **1–4** represent an optimum series of electrophiles for the study of nucleophilic interactions as ion pairs since two distinctive processes are clearly delineated, despite minimal changes in reactant structures. The pentadienyl and hexadienyl derivatives **1** and **2** cleanly yield the nucleophilic adducts (**5**), whereas the cyclic analogs **3** and **4** yield only the homodimers (**6**) according to the stoichiometries in Eqs. (48) and (49), respectively. Indeed, these processes occur *concurrently* in the case of (hexadienyl)Fe(CO)$_3^+$ which in combination with CpMo(CO)$_3^-$ produces both the nucleophilic adduct **5b** in 75% yield together with the homodimers in 25% yields. It is important to emphasize that **5b** is not an intermediate in the formation of **6d** and **7** since its rate of decomposition [Eq. (52)] is much too slow to account for the amounts of homodimers observed.

The foregoing concurrence of pathways from ion-pair interactions indicates that the rate-limiting step leading to nucleophilic addition is highly competitive with that leading to homodimers. Thus, the activation barriers for both processes are closely coupled. Such divergent products as nucleophilic adducts and homodimers in Eqs. (48) and (49) are usually considered to arise via two distinctive pathways. On one hand, the formation of adducts is mostly associated with the direct nucleophilic attack with rate constant k_N at the site of unsaturation on the electrophile in a one-step mechanism (Scheme 8), where fe and mo refer to Fe(CO)$_3$ and CpMo(CO)$_3$, respectively, and L to an unsaturated ligand. On the other hand, the homodimers often derive from a two-step process in which the radical couplings are preceded by electron transfer with the rate constant k_{ET} (Scheme 9).

$$\text{feL}^+ + \text{mo}^- \xrightarrow{k_N} \text{feLmo}$$

SCHEME 8

$$feL^+ + mo^- \xrightarrow{k_{ET}} feL\cdot + mo\cdot$$

$$2\,feL\cdot \longrightarrow (feL)_2$$

$$2\,mo\cdot \longrightarrow mo_2$$

SCHEME 9

In a more general context, Schemes 8 and 9 represent the duality of electrophile–nucleophile interactions that are more commonly described in terms of 2-electron (concerted) and 1-electron (electron-transfer) processes, respectively. However, the rigorous distinction between these apparently disparate processes is more subtle, as the homolytic scission of the nucleophilic adduct,

$$feLmo \xrightarrow{k_H} feL\cdot + mo\cdot \qquad (54)$$

affords the pair of radicals leading to the homodimers in Eq. (52). The combination of Eq. (54) with Scheme 8 is tantamount to the electron transfer in Scheme 9; stated alternatively, the activation barrier for electron transfer is equivalent to that for nucleophilic addition as modified by the feL—mo bond energy. However, the sizable magnitude of the experimentally determined feL—mo bond energy ΔH of approximately 30 kcal/mol for the nucleophilic adducts **5** is not consistent with the concurrence of the two pathways (Schemes 8 and 9), which leads to the different conclusion that log $k_N \cong$ log k_{ET} (see above). This mechanistic dilemma can be addressed in several ways: (a) rate-limiting electron transfer in Scheme 9 is a common step leading to nucleophilic adducts, (b) rate-limiting nucleophilic addition in Scheme 8 is a common step leading to homodimers, and (c) simultaneous rate-limiting electron transfer and nucleophilic addition are fortuitous. Let us consider each of these possibilities in more detail.

1. Rate-Limiting Electron Transfer

The rate-limiting electron transfer leading to both nucleophilic adducts and homodimers formally involves the distinction between the coupling of unlike and like radicals, respectively. Such a mechanism is traditionally formulated (*109*) as shown in Scheme 10, where the brackets denote the solvent cage. According to Scheme 10, the competition leading to the nucleophilic adduct and the homodimers arises largely from the relative rates of cage collapse (k_{cage}) and diffusive separation (k_{diff}) of the radical pair formed as the successor complex in electron transfer. Such a formulation, however, is hard pressed to account for the very high selectivity to feLmo shown by cations **1** and **2**, certainly in relation to the structurally related cations **3** and **4**.

$$feL^+ + mo^- \xrightarrow{k_{ET}} [feL\cdot, mo\cdot]$$

$$[feL\cdot, mo\cdot] \overbrace{}^{\begin{array}{l}\xrightarrow{k_{cage}} feLmo\\[1em]\xrightarrow{k_{diff}} feL\cdot + mo\cdot\end{array}}$$

$$2\ feL\cdot \longrightarrow (feL)_2$$

$$2\ mo\cdot \longrightarrow mo_2$$

$$feL\cdot + mo\cdot \longrightarrow feLmo$$

SCHEME 10

2. Rate-Limiting Nucleophilic Addition

The rate-limiting nucleophilic addition leading to homodimers formally includes the reversible homolysis of the nucleophilic adduct (*110*) (e.g., Scheme 11). Indeed the successful isolation of the labile intermediates **5c** and **5d** from both cyclic cations **3** and **4** indicates that the nucleophilic adduct (feLmo) is the critical intermediate leading to the homodimers **6** [compare Eq. (51) with Eq. (49)]. Bond homolysis (*111*) as the mechanism for the decomposition of the nucleophilic adduct is experimentally supported by kinetics, ESR studies, as well as spin and chemical trapping. Furthermore, the low sensitivity of the first-order rate constant k_H (Fig. 7) on solvent polarity and the marked effect of solvent viscosity are both earmarks of the homolytic bond scission (*107,108*).

$$feL^+ + mo^- \xrightarrow{k_N} feLmo \tag{55}$$

$$feLmo \xrightarrow{k_H} feL\cdot + mo\cdot \tag{56}$$

$$feL\cdot + mo\cdot \longrightarrow (feL)_2, mo_2, feLmo \tag{57}$$

SCHEME 11

The mechanism for ion-pair annihilation in Scheme 11 taken in a more general context predicts the relative amounts of nucleophilic adduct and homodimers to be modulated by the homolytic rate constant, namely, whether k_H is small or large. Such a conclusion requires the high selectivity to derive from a strong dependence of the feL—mo bond strength on the structure of the nucleophilic adduct. Thus, a closer consideration of feLmo structures reveals that the persistent ones (**5a** and **5b**) involve the binding of the ligand L to mo at a primary carbon ($-CH_2$) site, whereas the ligands in the transient feLmo (**5c** and **5d**) can only be bonded to mo at a secondary carbon $\left(\diagdown CH \diagup\right)$ site. The estimated difference in bond energy of approximately 5 kcal/mol (*112*) indeed predicts a half-life τ of only about 10 seconds for **8** relative to the measured value of 10^5 seconds for **5** under the same conditions. Further evidence of this relationship between structure and reactivity is included in the detailed behavior of the cation **2** in the earliest stages of ion-pair annihilation, since the unsymmetrical hexadienyl ligand offers both a primary and a secondary site for nucleophilic attack. Most revealing is the spectral (IR, ^{1}H NMR) observation of the first-formed, highly transient intermediate **A** at low temperatures, followed by the sequential appearance of metastable species **B** and **C** as well as the nucleophilic adduct (**5b**) (*103*).

A

B

C

In order to facilitate the presentation, Scheme 12 is included to illustrate how **A, B,** and **C** are interrelated as critical intermediates in the conversion of (η^5-hexadienyl)Fe(CO)$_3{}^+$, with the ligand originally bound to iron in a cis configuration, to the all-trans hexadiene ligand present in the final nucleophilic adduct by a series of standard organometallic transformations (*113*). In Scheme 12, step (i) represents ion-pair annihilation by nucleophilic addition to the methylene terminus of the ligand to yield the initially observed intermediate **A**, which is transformed by a 1,2-shift of mo to the iron center in step (ii) to generate the fluxional species **B** and **B**' (step iii) (*114*). In both cases, rotation about the single bond, followed by the retroshift of mo, leads to the nucleophilic adduct **5b** (step iv) and its isomer **C** (step v). Finally the disappearance of **C** on thermal equilibration at

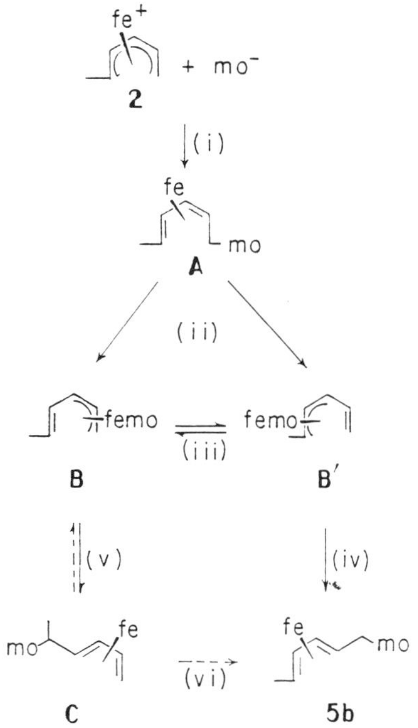

fe$^+$

$+$ mo$^-$

2

(i)

fe

mo

A

(ii)

femo   femo

(iii)

B   **B'**

(v)   (iv)

mo   fe

fe   mo

(vi)

C   **5b**

Scheme 12. Ion-pair annihilation to the nucleophilic adducts (*103*).

higher temperatures to the more stable adduct **5b** can occur via the reversal of step (v). Alternatively, the bond homolysis–recombination in step (vi) relates the intermediate **C**, with its weak secondary (CH—mo) bond (*112*), directly to the labile intermediate **8** (see above).

The generalized mechanism for ion-pair annihilation as presented in Scheme 11 involves the rather circuitous route for radical-pair production [involving Eqs. (55) and (56), certainly in comparison with the direct electron-transfer pathway (Scheme 8)]. In other words, why do ion pairs first make a bond and then break it, when the simple electron transfer directly from anion to cation would achieve the same end? The question thus arises as to whether electron transfer between $Fe(CO)_3L^+$ and $CpMo(CO)_3^-$ is energetically disfavored. The evaluation of the driving force for the electron transfer process obtains from the separate redox couples, namely,

$$feL^+ + e^- \xrightarrow{E^0_{red}} feL \cdot$$

and

$$mo^- \xrightarrow{E^0_{ox}} mo \cdot + e^-$$

The overall driving forces for electron transfer $[-\Delta G_{ET} = \mathscr{F}(E^0_{red} + E^0_{ox})]$ for ion-pair annihilation in THF between mo$^-$ and feL$^+$ (i.e., cations **1**, **2**, **3**, and **4**) are ΔG_{ET} = 3.2, 3.9, 5.3, and 6.0 kcal/mol, respectively, based on the electrochemical measurements (*115*). Such driving forces all easily lie within the isoergonic bounds for the facile electron transfer between feL$^+$ and mo$^-$. Moreover, the differences in driving forces are not sufficient to strongly distinguish the cations **1** and **2** from their cyclic analogs **3** and **4** for the consideration of *simultaneous nucleophilic addition and electron transfer,* as presented in possibility (c) above.

We are thus left with the inescapable conclusion that nucleophilic addition is the favored process for the ion-pair annihilation of feL$^+$ and mo$^-$, despite the favorable driving force for electron transfer. Some of the factors that discourage the latter may be related to the large intrinsic reorganization energy required to convert the organometallic cation (feL$^+$) to the corresponding 19-electron radical (feL$\cdot$) (*116*). Thus, the facile self-reaction of feL$\cdot$ to couple only at the ligand site (*105*) suggests that electron transfer to feL$^+$ is accompanied by a large ligand reorganization to place the spin density at a terminal carbon. The "slippage" of the unsaturated ligand would lead to a 17-electron radical (*117*), for example,

Such a change in ligand hapticity is probably a general problem associated with the supersaturated nature of 19-electron radicals (*118*). In the context of Marcus theory (*84*), the sizable barrier for outer-sphere electron transfer posed by a large reorganization energy may be circumvented by a configurational change to favor an inner-sphere activated complex (*119*). In the latter regard, the distinction between nucleophilic addition and inner-sphere electron transfer is sufficiently blurred to obscure any further meaningful mechanistic insight into ion-pair annihilation (*120*).

VI

BIS(ARENE)IRON(II) DICATIONS AS ELECTRON ACCEPTORS

The relevance of the charge on an organometallic complex is dramatically underscored by the comparison of ferrocene (Cp$_2$Fe), which is a powerful electron donor (*121*), with its isoelectronic analog bis(benzene)iron(II) dication (BZ$_2$Fe^{2+}), which is an effective electron acceptor (*122*).

A. Charge-Transfer Complexes of Bis(arene)iron(II) and Ferrocene: Spectral Characterization and X-Ray Crystallography

When dilute solutions of ferrocene and bis(hexamethylbenzene)iron(II) hexafluorophosphate [(HMB)$_2$Fe^{2+}(PF$_6^-$)$_2$] in acetonitrile are mixed, the pale yellow-orange mixture instantly takes on a dark brown coloration. The spectral change accompanying this dramatic visual transformation is accompanied by the obscuration of the visible absorption spectra of ferrocene and (HMB)$_2$Fe^{2+} by an enveloping new absorption at lower energy. Spectral (digital) subtraction reveals the latter as a very broad absorption band with a λ_{max} of 626 nm. A similar treatment of ferrocene with the analogous durene dication, (DUR)$_2$Fe^{2+}, yields a corresponding difference spectrum with a λ_{max} of 647 nm. The bathochromic shift of 510 cm^{-1} parallels the relative ease of reduction of the bis(arene)iron(II) dications as given by their reversible redox potentials. The charge-transfer complex spontaneously separates from acetonitrile solution as dark brown, almost black crystals when ferrocene and (DUR)$_2$Fe^{2+} are employed at higher (0.1 M) concentrations, and single crystals suitable for X-ray crystallography are obtained by the slow diffusive mixing of these solutions (*123*).

The structure of the charge-transfer complex can be solved in the tetragonal space group $P4_2/nmc$, with $a = 11.200(3)$ Å and $c = 13.283(5)$ Å, as a discrete 1:1 mixture of individual ferrocene and bis(durene)iron(II) hexafluorophosphate units, that is,

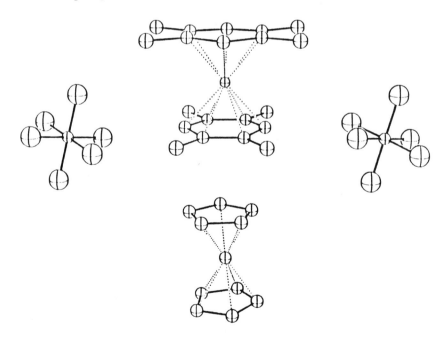

The ferrocene component maintains its structural integrity within the crystalline complex, and it is observed with an unexceptional disorder of the Cp rings about the rotational axis (*124*). By way of contrast, in the $(DUR)_2Fe^{2+}$ acceptor the coplanar durene rings are rigidly locked in the mutually orthogonal orientation shown above. Otherwise the iron–ring (centroid) distance of 1.58 Å in the dication is only slightly less than that (1.66 Å) extant in ferrocene (*125*). Importantly, the diagram of the unit cell consists of the stacked alignment of alternating donor–acceptor interactions of Cp_2Fe and $(DUR)_2Fe^{2+}$ units that are separated by the interannular Cp–DUR distance of 3.43 Å. The overall Fe–Fe separation of 6.64 Å within the stack compares with the closest nonbonded Fe–Fe distance of 7.92 Å in the horizontal direction between stacks.

In order to generalize the spectral complexation of the bis(arene)iron(II) acceptor, it is exposed to various arenes whose donor characteristics in charge-transfer complexes are already established (*126*). When a solution of $(DUR)_2Fe^{2+}$ $(PF_6^-)_2$ in acetonitrile is added to various arene donors, colors ranging from bright yellow for the benzene donors, to red for the naphthalene donors, and finally to purple for the anthracene donors are observed. The monotonic red shift of the new absorption band with λ_{max} values of 368, 412, 462, and 534 nm coincides with the progressively decreasing ionization potentials of 8.05, 7.83, 7.72, and 7.25 eV, respectively, for the following aromatic donors: durene, 1,3,5-trimethoxybenzene, 1-methoxynaphthalene, and 9-methylanthracene. Such a parallel trend in the absorption band with the donor property of the arenes, as measured by the ionization potential, relates directly to the formation of the charge-transfer complexes,

$$(DUR)_2Fe^{2+} + Ar \xrightleftharpoons{K} [(DUR)_2Fe^{2+}, Ar] \qquad (58)$$

in accord with the predictions of the Mulliken theory (*7*).

The formation constant for the charge-transfer complex between 9-methylanthracene and bis(mesitylene)iron(II) is evaluated as $K = 2.5$ M^{-1} (ϵ_{max} 400 M^{-1} cm^{-1}) in acetonitrile, which places it in the weak category (*30*). Nonetheless, single crystals of various charge-transfer complexes suitable for X-ray crystallography can be grown, and they are typified by the bright orange-red crystals from $(HMB)_2Fe^{2+}$ and durene in the triclinic space group $P\bar{1}$ with $a = 9.270(4)$ Å, $b = 11.040(5)$ Å, $c = 12.130(5)$ Å, $\alpha = 71.72(3)°$, $\beta = 67.30(3)°$, and $\gamma = 77.41(3)°$ (*123*). The ORTEP diagram of the 1:1 donor–acceptor pair identifies the durene donor to lie below the hexamethylbenzene ligand (and coplanar with it) at an intermolecular separation of 3.65 Å. The slight displacement of the quasi C_6 axes is ascribed to the methyl groups, whose closest nonbonded H–H separation of 2.5 Å is reduced to only 1.5 Å in the symmetric structure, which is clearly too close for the allowable van der Waals separation of 2.4 Å. Importantly, the diagram of the unit cell shows the same stacked alignment of alternating donor–acceptor interactions of durene and $(HMB)_2Fe^{2+}$ units that are found in the direct structural counterpart of the ferrocene complex with $(DUR)_2Fe^{2+}$ depicted above.

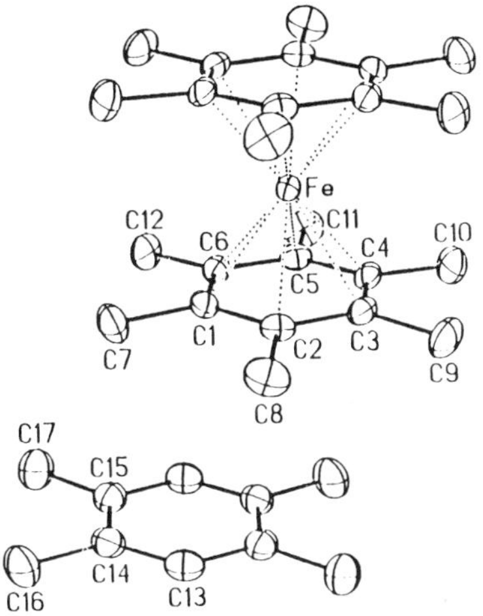

B. *Photoinduced Electron Transfer in Bis(arene)iron(II) Complexes with Ferrocene and Aromatic Donors*

The charge-transfer character of the new absorption bands is directly examined by the selective electronic excitation of only the complex. For example, an acetonitrile solution of ferrocene and $(HMB)_2Fe^{2+}$ can be irradiated with the output from a 450-W xenon lamp passed through a 620-nm interference filter. The use of such a monochromatic irradiation ensures the photoexcitation of only the charge-transfer band of the complex $[Cp_2Fe, (HMB)_2Fe^{2+}]$, and not the local bands of either the ferrocene donor or the bis(arene)iron(II) acceptor, as established by the comparison of their absorption spectra. The absorption of the 620-nm light is accompanied by the gradual bleaching of the charge-transfer band. Analysis of the photolysate indicates the presence of 2 molar equivalents of free hexamethylbenzene and an equivalent of ferrous iron $Fe(MeCN)_6^{2+}$ identified by complexation with 2,2′-bipyridine followed by the spectrophotometric determination of $Fe(bipy)_3^{2+}$ at its characteristic absorption at a λ_{max} of 520 nm with $\epsilon = 8560 \ M^{-1} \ cm^{-1}$ (*127*). Since the ferrocene is recovered largely intact, the photochemical conversion of the charge-transfer complex corresponds to the stoichiometry

$$[Cp_2Fe, (HMB)_2Fe^{2+}] \xrightarrow[\text{MeCN}]{hv_{CT}} Cp_2Fe + 2\, HMB + Fe(MeCN)_6^{2+} \qquad (59)$$

The reactive intermediates leading to the charge-transfer deligation of the bis(arene)iron(II) acceptors in Eqs. (59) and (60) are examined by time-resolved picosecond spectroscopy immediately following the application of an 18-psecond laser pulse. The time-resolved spectrum from the $(HMB)_2Fe^{2+}$ complex with ferrocene consists of the superposition of the absorption spectrum of the monocationic $(HMB)_2Fe^+$ PF_6^- with a λ_{max} of 580 nm (ϵ_{max} 604 M^{-1} cm^{-1}) (128) and the spectrum of the ferrocenium ion with a λ_{max} of 619 nm (ϵ_{max} 360 M^{-1} cm^{-1}) (129). Analogously, the charge-transfer excitation of a solution of $(HMB)_2Fe^{2+}$ and 9-methylanthracene yields the transient spectrum of the 9-methylanthracene cation radical (MeAn$^+\cdot$) in Fig. 8 that corresponds to the one from the γ-radiolysis, anodic, or photoinduced oxidation of 9-methylanthracene (130). The

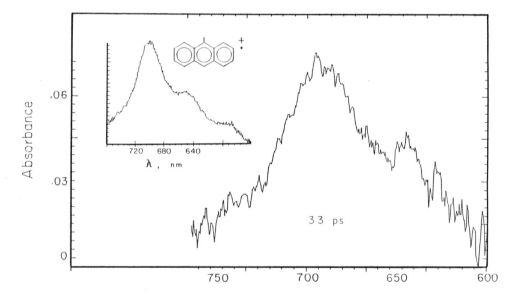

FIG. 8.   Transient spectrum (33 pseconds) obtained from the charge-transfer excitation at 532 nm of a solution of 0.2 M 9-methylanthracene and 3.4 mM $(HMB)_2Fe^{2+}$. The inset shows the time-resolved spectrum of the 9-methylanthracene cation radical for comparison (123).

specific photoexcitation of the charge-transfer band of the ferrocene complex with $(DUR)_2Fe^{2+}$ similarly leads to the degradation of the bis(arene)iron(II) acceptor. Since each cationic acceptor is characterized by the loss of the ligand, the photoprocess is referred to hereafter as *charge-transfer deligation.*

In order to establish the deligation in Eq. (59) to be a direct consequence of charge-transfer excitation, the ferrocene donor is replaced by a series of arenes with varying donor properties (i.e., ionization potentials). Initially, 9-methylanthracene is chosen as the donor to be used in conjunction with the $(DUR)_2Fe^{2+}$ acceptor, owing to the clean spectral separation of the charge-transfer absorption band from the local absorption of the cationic acceptor. The use of monochromatic light of 620 ± 5 nm ensures the selective excitation of only the charge-transfer band. Under these conditions, the clean liberation of 2 molar equivalents of durene is observed (¹H-and ¹³C-NMR analysis), and the 9-methylanthracene donor is recovered unchanged, together with quantitative yields of $Fe(MeCN)_6^{2+}$ according to the stoichiometry

$$[(DUR)_2Fe^{2+}, \quad MeAn] \xrightarrow[CH_3CN]{h\nu_{CT}} MeAn + 2\,DUR + Fe(CH_3CN)_6^{2+} \qquad (60)$$

The quantum yield for the charge-transfer deligation of $(DUR)_2Fe^{2+}$ by 9-methylanthracene is measured by Reineckate actinometry and found to be independent of the donor concentration, and independent of the CT excitation energy. However the photoefficiency of the charge-transfer deligation of bis(arene)iron(II) acceptors is strongly dependent on the aromatic donor strength, and the trend in the values of ϕ_{CT} for different donors is quite remarkable, in that the weaker donor induces markedly higher quantum efficiencies.

C. *Charge-Transfer Structures of Bis(arene)iron(II) Acceptors with Ferrocene and Arene Donors*

The 1:1 charge-transfer complex of $(DUR)_2Fe^{2+}$ and ferrocene, as described above, compares with the green 1:1 complex of ferrocene with tetracyanoethylene that exhibits its charge-transfer absorption at 900 and 1075 nm in cyclohexane solution (*131*). The location of the ferrocene donor poised over the bis(arene)iron(II) acceptor is consistent with the charge-transfer interaction in the molecular complex to occur principally *via* the contiguous placement of cyclopentadienyl and arene ligands on the donor and acceptor, respectively. Indeed the intimate van der Waals contact of these coplanar ligands, with coincident (quasi) C_5 and C_6 axes, optimizes the overlap of the filled cyclopentadienyl π orbital with the vacant e_{1g} arene orbital.

However, such an orbital correlation of charge transfer in this pair of sandwich structures does not represent the usual HOMO–LUMO transition in other molecular complexes (132). The interaction diagram in Fig. 9 (133) of ferrocene [and also applicable to bis(arene)iron(II) with equivalent orbital topology] identifies a set of three close-lying HOMOs centered on iron (134). The charge transfer from the subjacent ligand π orbital of the ferrocene donor to the bis(arene)iron(II) acceptor is thus analogous to that previously presented in the ferrocene complex with the π acceptor tetracyanoethylene (135). Basically the same charge-transfer formulation is therefore applicable to the structural situation with the reverse polarity, namely, that found in the charge-transfer complex of the bis(arene)iron(II) acceptor with the arene π donors (see above). The alternative HOMO–LUMO interaction stemming from the iron-centered orbitals is structur-

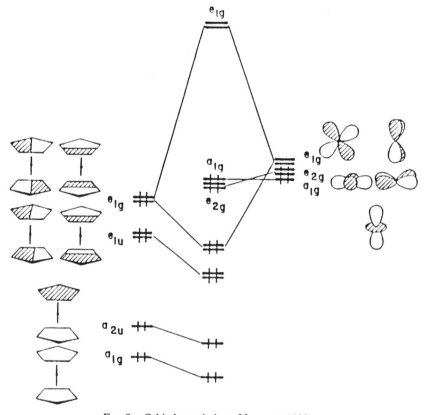

FIG. 9.   Orbital correlation of ferrocene (133).

ally provided by the side-by-side placement of the ferrocene donor and bis(arene)iron(II) acceptor, in which the nonbonded Fe–Fe separation of 7.92 Å occurs between the heterosoric stacks. Although this intermolecular separation is only slightly greater than the iron-to-iron distance of 6.64 Å within the stack, the HOMO–LUMO transition does not provide a consistent formulation for the structures of the [ferrocene–tetracyanoethylene] and [arene–bis(arene)iron(II)] complexes, in which the possibility of such an orbital correlation does not exist. On the other hand, charge-transfer transitions involving the subjacent orbitals as described above provide the unifying formulation for all the ferrocene and bis(arene)iron(II) complexes in which the proximal juxtapositions of ligand π orbitals are unmistakable.

Photoactivation of the bis(arene)iron(II) complexes with ferrocene and arene donors (D) by the selective irradiation of the charge-transfer absorption bands uniformly results in the deligation of the acceptor moiety according to the stoichiometry in Eqs. (59) and (60), respectively. Time-resolved picosecond spectroscopy establishes the transient cation radical of the donor (i.e., $D^{+\cdot}$) to be the spectral feature that both complexes share in common on charge-transfer excitation ($h\nu_{CT}$). Thus, the fleeting appearance of the oxidized donors, ferrocenium ion (Cp_2Fe^+) and 9-methylanthracene cation radical ($MeAn^{+\cdot}$), immediately following the application of the 18-psecond laser pulse is in accord with the Mulliken formulation of the charge-transfer photoactivation of the molecular complexes, namely,

$$[Cp_2Fe, Ar_2Fe^{2+}] \xrightarrow{h\nu_{CT}} [Cp_2Fe^+, Ar_2Fe^+]$$

and

$$[MeAn, Ar_2Fe^{2+}] \xrightarrow{h\nu_{CT}} [MeAn^{+\cdot}, Ar_2Fe^+]$$

in which the reduced acceptor as the 19-electron radical Ar_2Fe^+ is a common partner. Since the transient electrochemical studies of bis(arene)iron(II) reduction identify the rapid fragmentation of the 19-electron cation Ar_2Fe^+, the charge-transfer deligation is readily formulated as shown in Scheme 13.

$$Ar_2Fe^{2+} + D \rightleftharpoons [D, Ar_2Fe^{2+}] \tag{61}$$

$$[D, Ar_2Fe^{2+}] \underset{k_{-1}}{\overset{h\nu_{CT}}{\rightleftharpoons}} [D^{+\cdot}, Ar_2Fe^+] \tag{62}$$

$$[D^{+\cdot}, Ar_2Fe^+] \xrightarrow{k_1} [D^{+\cdot}, 2\ Ar, Fe^+] \tag{63}$$

$$[D^{+\cdot}, 2\ Ar, Fe^+] \xrightarrow{fast} [D + 2\ Ar + Fe^{2+}] \tag{64}$$

SCHEME 13

According to Scheme 13, the photochemical process proceeds from the selective excitation of the molecular complex to the charge-transfer ion pair in Eq. (62). As such, the fragmentation (k_1) of the labile 19-electron radical Ar_2Fe^+ in Eq. (63) leads to the deligation of the acceptor. The overall photoefficiency of the process would be largely determined by the competition between the fragmentation (k_1) and back electron transfer (k_{-1}). The mechanism in Scheme 13 is indeed consistent with the various experimental facets of charge-transfer deligation. For example, the photoefficiency of charge-transfer deligation, as measured by the quantum yield, relates directly to the ratio of rate constants in Scheme 13, that is,

$$\Phi_{CT} = k_1/(k_1 + k_{-1}) \qquad (65)$$

Thus, in the methylanthracene-induced deligation, the trend in the quantum yields for $(DUR)_2Fe^{2+} > (HMB)_2Fe^{2+}$ follows predictably from the lifetimes $(k_1{}^{-1})$ of the labile 19-electron radicals $(DUR)_2Fe^+ <$ $(HMB)_2Fe^+$, as evaluated by transient electrochemical methods (136). Furthermore, the remarkable trends in the quantum yields to decrease with the increasing strength of the arene donor must take specific cognizance of the rate of back electron transfer (k_{-1}). Since the latter results in the annihilation of the radical ion pair $Ar_2Fe^+/D^{+\cdot}$, it is readily evaluated from the separate redox couples,

$$Ar_2Fe^{2+} + e^- \xrightarrow{E^0_{red}} Ar_2Fe^+$$

and

$$D \xrightarrow{E^0_{ox}} D^{+\cdot} + e^-$$

by the driving force $-\Delta G_{BET} = \mathscr{F}\Delta E_{BET}$, where $\Delta E_{BET} = -(E^0_{ox} + E^0_{red})$ and $\mathscr{F}$ is the Faraday constant. Based on the linear free energy relationship, $-\Delta G_{BET} = (RT/\alpha) \ln k_{-1} + $ constant (α is the Brønsted coefficient), the photoefficiency in Eq. (65) can be reexpressed as

$$\Delta E_{BET} = (RT/\alpha\mathscr{F})[\ln(1 - \Phi_{CT})/\Phi_{CT} + \ln k_1] + \text{constant} \qquad (66)$$

Indeed, the monotonic trend observed in Fig. 10 for the quantum yield data represents strong support for the mechanism in Scheme 13, since $\ln k_1$ is invariant for a given bis(arene)iron(II) acceptor. The listings for the various arene donors in Table VII show the driving force for back electron transfer to be highly exergonic (40–60 kcal/mol). It is noteworthy that such large magnitudes of $-\Delta G_{BET}$ are likely to lie in the Marcus inverted region of the free energy relationship for electron transfer (137). The net result is an increase in the rate of back electron transfer with increasing donor strength (138).

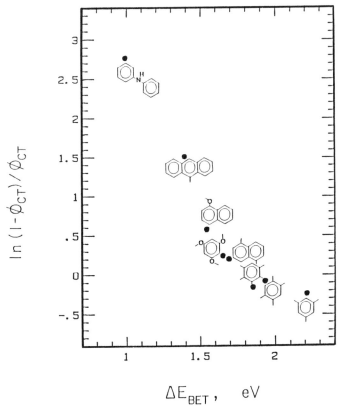

FIG. 10. Variation of the photoefficiency for charge-transfer deligation with the driving force for back electron transfer according to Eq. (66) (123).

The majority of the charge-transfer complexes of the bis(arene)iron(II) acceptors persist for prolonged periods—provided the solutions are protected from light, even adventitious room light. In some instances, however, the bis(arene)iron(II) complexes react spontaneously in the dark. For example, the characteristically broad charge-transfer band from the mesitylene dication $(MES)_2Fe^{2+}$ and ferrocene is fleeting, and it disappears in less than 15 min after mixing the solutions. The resultant spectrum shows a residual absorbance centered at 620 nm which is diagnostic of the ferrocenium ion. By way of contrast, the weaker acceptors $(DUR)_2Fe^{2+}$ and $(HMB)_2Fe^{2+}$ produce charge-transfer complexes that persist for prolonged periods under the same dark conditions. In an analogous manner, the solutions of the charge-transfer complexes of $(DUR)_2Fe^{2+}$ with such strong amine donors as N-methylacridane and N,N-dimethylaniline on standing

TABLE VII

Driving Force for Back Electron Transfer $(-\Delta G_{\text{BET}})$ of Charge-Transfer Ion Pairs from Bis(arene)iron(II) Complexes with Ferrocene and Arene Donors[a]

Donor	E^0_{ox} $(V)^b$	$-\Delta G_{\text{BET}}$ (kcal/mole) for $(\text{Ar})_2\text{Fe}^{2+c}$			
		BZ	MES	DUR	HMB
Benzene	2.40	55	57	59	61
Mesitylene	2.05	47	49	51	53
Durene	1.77	41	42	45	47
Pentamethylbenzene	1.69	39	40	43	45
Hexamethylbenzene	1.56	36	37	40	42
1,5-Dimethylnaphthalene	1.53	35	36	39	41
1,3,5-Trimethoxybenzene	1.49	34	36	38	40
1-Methoxynaphthalene	1.38	32	33	36	38
9-Bromoanthracene	1.38	32	33	36	38
2,6-Dimethoxynaphthalene	1.33	30	32	34	36
9-Methylanthracene	1.23	28	30	31	34
N-Methylacridane	0.86	20	21	24	26
Diphenylamine	0.83	19	21	23	25
N,N-Dimethylaniline	0.68	15	17	19	22
Ferrocene	0.41	9	11	13	15

[a] In acetonitrile solution at 25°C, from Ref. *123*.

[b] Relative to SCE.

[c] $E^0_{\text{red}} = 0.01$, -0.06, -0.16, and -0.26 V versus SCE for BZ, MES, DUR, and HMB derivatives of $\text{Ar}_2\text{Fe}^{2+}$.

in the dark liberate durene with essentially the same stoichiometry as that given in Eq. (60). The dark reactions derive from the charge-transfer complexes that clearly involve the most electron-rich donors and electron-poor acceptors, as judged by their oxidation and reduction potentials in Table VII. Inspection of Table VII thus reveals that the driving force for electron transfer between the bis(arene)iron(II) acceptor and the various donors, when evaluated as $-\Delta G_{\text{ET}} = \mathcal{F}(E^0_{\text{ox}} + E^0_{\text{red}})$, is the largest for ferrocene, N-methylacridane, and N,N-dimethylaniline, especially when those strong donors are paired with the electron-poor mesitylene derivative of the bis(arene)iron(II) acceptor. As such, the dark reactions may arise via essentially the same multistep mechanism as that proposed in Scheme 13 for charge-transfer deligation, with the difference arising from an adiabatic electron transfer in the contact ion pair, for example,

$$\text{Cp}_2\text{Fe}, (\text{MES})_2\text{Fe}^{2+} \xrightarrow{\Delta} \text{Cp}_2\text{Fe}^+, (\text{MES})_2\text{Fe}^+, \text{ etc.} \qquad (67)$$

that is thermally allowed when the driving force $-\Delta G_{\text{ET}}$ is sufficient to surmount the barrier that is otherwise overcome by charge-transfer activation in Eq. (62). Such a conclusion provides a relevant link to the unifying theme for the photochemical and thermal activations of electron transfer in various other organometallic processes (*139*).

VII

CONCLUSIONS

The recurring theme in the facile reactions of organometallic ion pairs is the prior charge-transfer (CT) interaction stemming from the preequilibrium formation of contact ion pairs (CIP). The contact ion pairs are often visually apparent by their colors and are unequivocally detected by the appearance of new absorption bands in the electronic spectra. They are strongly influenced by polar solvents, and they are subject to special salt effects. The chemical activation of contact ion pairs to geminate radical pairs by electron transfer is photochemically effected by the deliberate irradiation of the charge-transfer absorption band. The variety of ensuing photoinduced reactions arises from the labile character of 17- and 19-electron paramagnetic intermediates from the diamagnetic anion and cation, respectively. Thermal activation is thus dependent on the electron donor strength (E_{ox}^0) of the organometallic anion and the electron acceptor strength of the organometallic cation (E_{red}^0), as determined by the driving force for electron transfer, namely, $\mathscr{F}(E_{ox}^0 + E_{red}^0)$, when an inner-sphere contribution is included. The strong influence of ligands on the charge-transfer behavior of organometallic anions and cations is included with other electron donors and acceptors in the direct comparison in Table VIII,

TABLE VIII

Isoelectronic Charge-Transfer Pairs

Donor	Acceptor
$Pb^{IV}(CH_2CH_3)_4$	$Pb^{IV}(O_2CCH_3)_4$
$Co^I(DMG)_2^-$	$Co^I(CO)_3P_2^+$
$Hg^{II}(CCl_3)_2$	$Hg^{II}(O_2CCCl_3)_2$
$Cr^{IV}(bornyl)_4$	$Cr^{IV}{=}O(Salen)$
$Cr^{II}(en)_2^{2+}$	$Cr(C_6H_6)_2^{2+}$
$Fe^{II}Cp_2$	$Fe^{II}(CO)_3(C_4H_6)^+$
$W^{VI}(CH_3)_6$	$W^{VI}F_6$

in which a metal in a given oxidation state is transformed from an electron-rich donor to an electron-poor acceptor.

ACKNOWLEDGMENTS

We thank our co-workers, especially R. E. Lehman, K. Y. Lee, and J. D. Korp, for creative and diligent efforts in delineating the various aspects of organometallic ion and ion-pair interactions described in this article. We also thank the National Science Foundation, the Robert A. Welch Foundation, and the Texas Advanced Research Program for financial support.

REFERENCES

1. J. K. Kochi, "Organometallic Mechanisms and Catalysis," p. 504ff. Academic Press, New York, 1978.
2. J. O. Edwards, *J. Am. Chem. Soc.* **76**, 1540 (1954); J. O. Edwards and R. G. Pearson, *J. Am. Chem. Soc.* **84**, 16 (1962).
3. R. G. Pearson, *J. Am. Chem. Soc.* **85**, 3533 (1963); see also A. Ahrland, J. Chatt, and N. R. Davies, *Q. Rev. Chem. Soc.* **12**, 265 (1958); V. Gutmann, *Coord. Chem. Rev.* **18**, 225 (1976).
4. For a review, see T.-L. Ho, *Chem. Rev.* **75**, 1 (1975).
5. See, for example, G. Klopman, *J. Am. Chem. Soc.* **90**, 223 (1968); G. Klopman, "Chemical Reactivity and Reaction Paths." Wiley, New York, 1974; I. Fleming, "Frontier Orbitals and Organic Chemical Reactions." Wiley, New York, 1976.
6. W. B. Jensen, *Chem. Rev.* **78**, 1 (1978); see also W. B. Jensen, "The Lewis Acid–Base Concepts." Wiley, New York, 1980; W. B. Jensen, "Nucleophilicity" (J. M. Harris and S. P. McManus, eds.), p. 215. Adv. Chem. Ser., American Chemical Society, Washington, D.C., 1987.
7. R. S. Mulliken, *J. Am. Chem. Soc.* **74**, 811 (1952); R. S. Mulliken and W. B. Person, "Molecular Complexes." Wiley, New York, 1969.
8. See T. P. Fehlner, J. Ulman, W. A. Nugent, and J. K. Kochi, *Inorg. Chem.* **15**, 2544 (1976).
9. From Ref. *1*, p. 501.
10. Compare R. B. King, *Adv. Organomet. Chem.* **2**, 157 (1964); G. N. Schrauzer and E. Deutsch, *J. Am. Chem. Soc.* **91**, 3341 (1969); for a summary, see Ref. *1*, p. 156ff.
11. For example, T. J. Alavosus, and D. A. Sweigart, *J. Am. Chem. Soc.* **107**, 985 (1985).
12. J. D. Atwood, *Inorg. Chem.* **26**, 2918 (1987); P. Szabo, L. Fekete, and G. Bor, *J. Organomet. Chem.* **12**, 245 (1968); A. J. Pearson, S. L. Kole, and T. Ray, *J. Am. Chem. Soc.* **106**, 6060 (1984).
13. N. Sutin, *in* "Inorganic Biochemistry" (G. L. Eichorn, ed.), Vol. 2, p. 611. Elsevier, Amsterdam, 1973; A. Haim, *Acc. Chem. Res.* **8**, 264 (1975).
14. R. M. Fuoss and F. Accascina, "Electrolytic Conductance." Wiley, New York, 1959; C. W. Davis, "Ion Association." Butterworth, London, 1962; M. Szwarc, ed., "Ions and Ion Pairs in Organic Reactions," Vols. 1 and 2. Wiley, New York, 1972 and 1974; J. E. Gordon, "Organic Chemistry of Electrolyte Solutions." Wiley, New York, 1975.
15. For example, see E. H. Cordes and R. B. Dunlap, *Acc. Chem. Res.* **2**, 329 (1969); H. Kessler and M. Feigel, *Acc. Chem. Res.* **15**, 2 (1982); E. B. Troughton, K. E. Molter, and E. M. Arnett, *J. Am. Chem. Soc.* **106**, 6726 (1984); E. D. Hughes and C. K. Ingold, *J. Chem. Soc.*, 244 (1935); E. D. Hughes, C. K. Ingold, and C. S. Patel, *J. Chem. Soc.*, 526

(1933); C. K. Ingold, "Structure and Mechanism in Organic Chemistry," 2nd Ed. Cornell University, Ithaca, New York, 1969; for the microscopic reverse, see S. Winstein and G. C. Robinson, *J. Am. Chem. Soc.* **80**, 169 (1958); S. Winstein, E. Clippinger, A. H. Fainberg, and G. C. Robinson, *J. Am. Chem. Soc.* **76**, 2597 (1954); J. M. Harris, *Prog. Phys. Org. Chem.* **11**, 89 (1974); V. J. Shiner, Jr., *in* "Isotope Effects in Chemical Reactions" (C. J. Collin and N. S. Bowman, eds.), p. 250. Van Nostrand, New York, 1970.

16. E. M. Kosower, *J. Am. Chem. Soc.* **80**, 3253 (1958); E. M. Kosower and J. A. Skorcz, *J. Am. Chem. Soc.* **82**, 2195 (1960); E. M. Kosower and M. Mohammad, *J. Phys. Chem.* **74**, 1153 (1970).

17. C. Schramm and J. I. Zink, *J. Am. Chem. Soc.* **101**, 4554 (1979); see also S. E. Pedersen and W. R. Robinson, *Inorg. Chem.* **14**, 2360 (1975).

18. F. Calderazzo, G. Pampaloni, M. Lanfranchi, and G. Pelizzi, *J. Organomet. Chem.* **296**, 1 (1985); F. Calderazzo, G. Pampaloni, G. Pelizzi, and F. Vitali, *Organometallics* **7**, 1083 (1988).

19. A. Vogler and H. Kunkely, *Organometallics* **7**, 1449 (1988).

20. W. E. Geiger and N. G. Connelly, *Adv. Organomet. Chem.* **23**, 1 (1984).

21. See M. A. Fox, *Adv. Photochem.* **13**, 237 (1986); G. J. Kavarnos and N. J. Turro, *Chem. Rev.* **86**, 401 (1986); J. Mattay, *Angew. Chem., Int. Ed. Engl.* **26**, 825 (1987); H. Hennig, D. Rehorek, and R. D. Archer, *Coord. Chem. Rev.* **61**, 1 (1985).

22. D. P. Schussler, W. R. Robinson, and W. F. Edgell, *Inorg. Chem.* **13**, 153 (1974).

23. D. Rehder, *J. Organomet. Chem.* **37**, 303 (1972).

24. R. A. Faltynek and M. S. Wrighton, *J. Am. Chem. Soc.* **100**, 2701 (1978).

25. Y. S. Sohn, D. N. Hendrickson, and H. B. Gray, *J. Am. Chem. Soc.* **93**, 3603 (1971).

26. E. O. Fischer, F. Scherer, and H. O. Stahl, *Chem. Ber.* **93**, 2065 (1960).

27. T. M. Bockman and J. K. Kochi, *J. Am. Chem. Soc.* **111**, 4669 (1989).

28. M. W. Hanna and J. L. Lippert, "Molecular Complexes" (R. Foster, ed.), Vol. 1, p. 1ff. Elek Science, London, 1973.

29. R. Foster, "Organic Charge-Transfer Complexes." Academic Press, New York, 1969.

30. P. Klüfers, *Z. Kristallogr.* **167**, 275 (1984).

31. P. Klüfers, *Z. Kristallogr.* **167**, 253 (1984); P. Klüfers, *Z. Kristallogr.* **165**, 217 (1983).

32. F. Calderazzo, G. Fachinetti, F. Marchetti, and P. F. Zanazzi, *J. Chem. Soc., Chem. Commun.*, 181 (1981).

33. P. E. Riley and R. E. Davis, *J. Organomet. Chem.* **152**, 209 (1978); H. Kobayashi, F. Marumo, and Y. Saito, *Acta Crystallogr.* **B27**, 373 (1971).

34. T. A. Albright, J. K. Burdett, and M. H. Whangbo, "Orbital Interactions in Chemistry." Wiley, New York, 1985.

35. H. B. Chin and R. Bau, *J. Am. Chem. Soc.* **98**, 2434 (1976).

36. W. F. Edgell, J. Lyford, A. Barbetta, and C. I. Jose, *J. Am. Chem. Soc.* **93**, 6403 (1971).

37. P. S. Braterman, "Metal Carbonyl Spectra." Academic Press, New York, 1975; for other examples of distorted carbonylmetallates, see K. H. Pannell and D. Jackson, *J. Am. Chem. Soc.* **98**, 4443 (1976); G. B. McVicker, *Inorg. Chem.* **14**, 2087 (1975); M. Y. Darensbourg, D. J. Darensbourg, and H. L. C. Barros, *Inorg. Chem.* **17**, 297 (1978); M. Y. Darensbourg, *Prog. Inorg. Chem.* **13**, 221 (1985).

38. W. F. Edgell, S. Hegde, and A. Barbetta, *J. Am. Chem. Soc.* **100**, 1406 (1978).

39. M. Darensbourg, H. Barros, and C. Borman, *J. Am. Chem. Soc.* **99**, 1647 (1977); see also Ref. *37*.

40. C. Reichardt, "Solvent Effects in Organic Chemistry," 2nd Ed. VCH, Weinheim, 1988.

41. R. S. Drago and N. J. Rose, *J. Am. Chem. Soc.* **81**, 6138 (1959).

42. See J. M. Masnovi and J. K. Kochi, *J. Am. Chem. Soc.* **107**, 7880 (1985).

43. See J. E. Gordon in Ref. *14*, p. 55ff.

44. A. R. Manning, *J. Chem. Soc. (A)* 1135 (1968).
45. L. J. Rothberg, N. J. Cooper, K. S. Peters, and V. Vaida, *J. Am. Chem. Soc.* **104,** 3536 (1982).
46. H. W. Walker, R. S. Herrick, R. J. Olsen, and T. L. Brown, *Inorg. Chem.* **23,** 3748 (1984).
47. W. L. Waltz, O. Hackelberg, L. M. Dorfman, and A. Wojcicki, *J. Am. Chem. Soc.* **100,** 7259 (1978).
48. S. P. Church, M. Poliakoff, J. A. Timney, and J. J. Turner, *J. Am. Chem. Soc.* **103,** 7515 (1981).
49. R. F. Cozzens and T. A. Gover, *J. Phys. Chem.* **74,** 3003 (1970).
50. Compare E. F. Hilinski, J. M. Masnovi, J. K. Kochi, and P. M. Rentzepis, *J. Am. Chem. Soc.* **106,** 8071 (1984).
51. See E. M. Kosower and L. Lindqvist, *Tetrahedron Lett.,* 4481 (1965); T. W. Ebbesen and G. Ferraudi, *J. Phys. Chem.* **87,** 3717 (1983); Ref. *49.*
52. J. A. S. Howell and P. M. Burkinshaw, *Chem. Rev.* **83,** 557 (1983); A. Davison and J. E. Ellis, *J. Organomet. Chem.* **31,** 239 (1971); F. Ungváry and A. Wojcicki, *J. Am. Chem. Soc.* **109,** 6848 (1987).
53. T. R. Herrinton and T. L. Brown, *J. Am. Chem. Soc.* **107,** 5700 (1985); see also Ref. *46.*
54. Compare, e.g., T. Koenig and H. Fischer, *in* "Free Radicals" (J. K. Kochi, ed.), Vol. 1, p. 157ff. Wiley, New York, 1973.
55. D. J. Kuchynka, C. Amatore, and J. K. Kochi, *J. Organomet. Chem.* **328,** 133 (1987).
56. K. Y. Lee and J. K. Kochi, *Inorg. Chem.* **28,** 567 (1989).
57. U. Koelle, *J. Organomet. Chem.* **152,** 225 (1978).
58. P. Reeb, Y. Mugnier, C. Moise, and E. Laviron, *J. Organomet. Chem.* **273,** 247 (1984); see also K. Y. Lee and J. K. Kochi in Ref. *56.*
59. See R. W. Wegman and T. L. Brown, *Inorg. Chem.* **22,** 183 (1983).
60. G. L. Geoffroy and M. S. Wrighton, "Organometallic Photochemistry." Academic Press, New York, 1979; T. J. Meyer and J. V. Caspar, *Chem. Rev.* **85,** 187 (1985).
61. I. Carelli, M. E. Cardinali, A. Casini, and A. Arnone, *J. Org. Chem.* 41, 3967 (1976); S. Kato, J. Nakaya, and E. Imoto, *J. Electrochem. Soc. Jpn.* **46,** 708 (1972).
62. H. E. Podall, J. H. Dunn, and H. Shapiro, *J. Am. Chem. Soc.* **82,** 1325 (1960); W. Hieber and H. Fuchs, *Z. Anorg. Allg. Chem.* **248,** 256 (1941); A. Davison, J. A. McCleverty, and G. Wilkinson, *J. Chem. Soc.* 1133 (1963); W. Hieber, H. Schulten, and R. Marin, *Z. Anorg. Allg. Chem.* **240,** 261 (1939); J. A. McCleverty and G. Wilkinson, *Inorg. Synth.* **8,** 211 (1966); C. Ungurenasu and M. Palie, *J. Chem. Soc., Chem. Commun.,* 388 (1975).
63. R. B. King and F. G. A. Stone, *Inorg. Synth.* **7,** 193 (1963); R. Birdwhistell, P. Hackett, and A. R. Manning, *J. Organomet. Chem.* **157,** 239 (1978); J. A. Armstead, D. J. Cox, and R. Davis, *J. Organomet. Chem.* **236,** 213 (1982); R. E. Dessy, P. M. Weissman, and R. L. Pohl, *J. Am. Chem. Soc.* **88,** 5117 (1966).
64. R. E. Dessy and P. M. Weissman, *J. Am. Chem. Soc.* **88,** 5129 (1966); V. N. Pandey, *Inorg. Chim. Acta* **23,** L26 (1977); S. Fontana, O. Orama, E. O. Fischer, U. Schubert, and F. R. Kreissl, *J. Organomet. Chem.* **149,** C57 (1978); A. Davison and J. E. Ellis, *J. Organomet. Chem.* **36,** 113 (1972); D. A. Roberts, W. C. Mercer, S. M. Zahurak, G. L. Geoffroy, C. W. DeBrosse, M. E. Cass, and C. G. Pierpont. *J. Am. Chem. Soc.* **104,** 910 (1982); M. A. Bennett and D. J. Patmore, *Inorg. Chem.* **10,** 2387 (1971); U. Anders and W. A. G. Graham, *J. Am. Chem. Soc.* **89,** 539 (1967); J. K. Ruff, *Inorg. Chem.* **7,** 1818 (1968); E. O. Fischer, T. L. Linder, F. R. Kreissl, and P. Braunstein, *Chem. Ber.* **110,** 3139 (1977); G. D. Michels and H. J. Svec, *Inorg. Chem.* **20,** 3445 (1981); L. Carlton, W. E. Lindsell, K. J. McCullough, and P. N. Preston, *J. Chem. Soc., Chem. Commun.,*

1001 (1982); compare also J. A. Hriljac, P. N. Swepston, and D. Shriver, *Organometallics* **4**, 158 (1985); R. E. Dessy and P. M. Weissman, *J. Am. Chem. Soc.* **88**, 5124 (1966); see also D. L. Morse and M. S. Wrighton, *J. Organomet. Chem.* **125**, 71 (1977).

65. T. Kruck and M. Höfler, *Angew. Chem., Int. Ed. Engl.* **3**, 701 (1964); see also K. K. Joshi and P. L. Pauson, *Z. Naturforsch.* **17**, 565 (1962); A. N. Nesmeyanov, K. N. Anisimov, N. E. Kolobova, and I. S. Kolomnokov, *Izv. Akad. Nauk SSSR, Otdel, Khim. Nauk,* 194 (1963); P. Szabo, L. Fekete, G. Bor, Z. Nagy-Magos, and L. Marko, *J. Organomet. Chem.* **12**, 245 (1968).

66. T. Kruck and M. Höfler, *Chem. Ber.,* **97**, 2289 (1964).

67. K. Y. Lee, D. J. Kuchynka, and J. K. Kochi, *Organometallics* **6**, 1886 (1987).

68. D. J. Kuchynka and J. K. Kochi, *Inorg. Chem.* **27**, 2574 (1988); D. J. Kuchynka and J. K. Kochi, *Organometallics* **8**, 677 (1989).

69. J. E. Ellis, *J. Organomet. Chem.* **86**, 1 (1975); T. A. Manuel, *Adv. Organomet. Chem.* **3**, 181 (1965); R. D. W. Kemmitt and D. R. Russell, *in* "Comprehensive Organometallic Chemistry" (G. Wilkinson, ed.), Vol. 5, p. 1. Pergamon, Oxford, 1982; A. L. Pidcock *in* "Transition Metal Complexes of Phosphorus, Arsenic and Antimony Ligands" (C. A. McAuliffee, ed.), p. 88. Wiley, New York, 1973.

70. I. Wender, H. W. Sternberg, and M. Orchin, *J. Am. Chem. Soc.* **74**, 1216 (1952); E. R. Tucchi and G. H. Gwynn, *J. Am. Chem. Soc.* **86**, 4838 (1964); W. Hieber, J. Sedlmeier, and W. Abeck, *Chem. Ber.* **86**, 700 (1953); W. Hieber and J. Sedlmeier, *Chem. Ber.* **87**, 25 (1954); W. Hieber and R. Wiesboeck, *Chem. Ber.* **91**, 1146 (1958).

71. A. Sacco and M. Freni, *J. Inorg. Nucl. Chem.* **8**, 566 (1958); A. Sacco and M. Freni, *Ann. Chim.* **48**, 218 (1958); O. Vohler, *Chem. Ber.* **91**, 1235 (1958); S. Attali and R. Poilblanc, *Inorg. Chim. Acta* **6**, 475 (1972).

72. W. Hieber and W. Freyer, *Chem. Ber.* **93**, 462 (1960); W. Hieber and W. Freyer, *Chem. Ber.* **91**, 1230 (1958).

73. J. A. McCleverty, A. Davison, and G. Wilkinson, *J. Chem. Soc. (A)* 2610 (1969); A. R. Manning, *J. Chem. Soc. (A),* 1135 (1968).

74. W. Hieber and H. Duchatsch, *Chem. Ber.* **98**, 1744 (1965); see I. Wender *et al.* in Ref. *70.*

75. M. F. Mirbach, M. J. Mirbach, and R. W. Wegman, *Organometallics* **3**, 900 (1984); compare Ref. *73* and R. B. King, *Inorg. Chem.* **2**, 936 (1963); J. D. Atwood, *Inorg. Chem.* **26**, 2918 (1987).

76. R. F. Heck, *J. Am. Chem. Soc.* **85**, 657 (1963); M. Absi-Halabi, J. D. Atwood, N. P. Forbus, and T. L. Brown, *J. Am. Chem. Soc.* **102**, 6248 (1980); see also R. L. Sweany and T. L. Brown, *Inorg. Chem.* **16**, 415 (1977).

77. J. W. Hershberger and J. K. Kochi, *Polyhedron* **2**, 919 (1983).

78. R. B. King, *Inorg. Chem.* **2**, 936 (1963); A. R. Manning, Ref. *73*; J. A. McCleverty *et al.,* Ref. *73*; see M. S. Arabi, A. Maisonnat, S. Attali, and R. Poilblanc, *J. Organomet. Chem.* **67**, 109 (1974).

79. K. Y. Lee and J. K. Kochi, *Inorg. Chem.* **28**, 567 (1989).

80. K. M. C. Davis, *Mol. Assoc.* **1**, 151 (1975).

81. W. Hieber and H. Duchatsch, *Chem. Ber.* **98**, 2530 (1965).

82. D. J. Kuchynka and J. K. Kochi, *Inorg. Chem.* **28**, 855 (1989).

83. R. A. Marcus, *J. Chem. Phys.* **24**, 966 (1956); for a review, see R. D. Cannon, "Electron Transfer Mechanisms." Butterworth, London, 1980.

84. A. J. Bard and L. R. Faulkner, "Electrochemical Methods." Wiley, New York, 1980.

85. K. B. Shiu and L. Y. Lee, *J. Organomet. Chem.* **348**, 357 (1988).

86. M. D. Curtis, K. B. Shiu, W. M. Butler, and J. C. Huffman, *J. Am. Chem. Soc.* **108**, 3335 (1986).

87. S. Trofimenko, *J. Am. Chem. Soc.* **89**, 3170 (1967); S. Trofimenko, *J. Am. Chem. Soc.* **91**, 588 (1969); S. Trofimenko, *Prog. Inorg. Chem.* **34**, 115 (1986).
88. F. Pragst, *Electrochim. Acta* **21**, 497 (1976); F. Pragst, R. Ziebeg, U. Seydewitz, and G. Driesel, *Electrochim. Acta* **25**, 341 (1980).
89. J. O. Howell, J. M. Goncalves, C. Amatore, L. Klasinc, R. M. Wightman, and J. K. Kochi, *J. Am. Chem. Soc.* **106**, 3968 (1984).
90. See M. C. Baird, *Chem. Rev.* **88**, 1217 (1988).
91. V. Wintgens, J. Pouliquen, J. Kossanyi, and M. Heintz, *New J. Chem.* **10**, 345 (1986); A. T. Balaban, C. Bratu, and C. N. Rentea, *Tetrahedron* **20**, 265 (1964).
92. H. Kawata and S. Niizuma, *Bull. Chem. Soc. Jpn.* **62**, 2279 (1989).
93. T. M. Bockman and J. K. Kochi, *New J. Chem.* 1991, **15**, 0000 (1991).
94. See J. M. Masnovi and J. K. Kochi, in Ref. *42*; compare with Eq. (7).
95. T. L. Brown, *Ann. N.Y. Acad. Sci.* **80**, 333 (1980); see also M. Chanon, M. Julliard, and J. C. Poite, eds., "Paramagnetic Organometallic Species in Activation/Selectivity, Catalysis." Kluwer Academic Publishers, Dordrecht, The Netherlands, 1989; W. C. Trogler, ed., "Organometallic Radical Processes." Elsevier, New York, 1990.
96. Q.-Z. Shi, T. G. Richmond, W. C. Trogler, and F. Basolo, *J. Am. Chem. Soc.* **106**, 71 and 76 (1984).
97. J. M. Masnovi, S. Sankararaman, and J. K. Kochi, *J. Am. Chem. Soc.* **111**, 2263 (1989); S. Sankarararaman, W. A. Haney, and J. K. Kochi, *J. Am. Chem. Soc.* **109**, 7824 (1987).
98. A. J. Birch, B. M. R. Bandara, K. Chamberlain, B. Chauncey, D. Dahler, A. I. Day, I. D. Jenkins, L. F. Kelly, T. C. Khor, G. Kretschmer, A. J. Liepa, A. S. Narula, W. D. Raverty, E. Rizzardo, C. Sell, G. R. Stephenson, D. J. Thompson, and D. H. Williamson, *Tetrahedron* **37**, 289 (1981); A. J. Birch and A. J. Pearson, *Tetrahedron Lett.*, 2379 (1975); L. A. P. Kane-Maguire, E. D. Honig, and D. A. Sweigart, *Chem. Rev.* **84**, 525 (1984); Y. K. Chung, H. S. Khoi, D. A. Sweigart, and N. G. Connelly, *J. Am. Chem. Soc.* **104**, 4245 (1982); J. W. Faller and K.-H. Chao, *Organometallics* **3**, 927 (1984); A. J. Pearson *et al.*, in Ref. *12*; W. E. Van Arsdale, R. E. K. Winter, and J. K. Kochi, *J. Organomet. Chem.* **296**, 31 (1985); Y.-L. Lai, W. Tam, and K. P. Vollhardt, *J. Organomet. Chem.* **216**, 97 (1981); A. M. Madonik, D. Mandon, P. Michaud, C. Lapinte, and D. Astruc, *J. Am. Chem. Soc.* **106**, 3381 (1984); for reviews, see P. S. Braterman, ed., "Reactions of Coordinated Ligands." Plenum, New York, 1986; B. M. Trost and T. R. Verhoeven, *in* "Comprehensive Organometallic Chemistry" (G. Wilkinson, F. G. A. Stone, and E. W. Abel, eds.), Vol. 8, Chap. 58, p. 799ff. Pergamon, New York, 1982; A. J. Pearson, *in* "Comprehensive Organometallic Chemistry" (G. Wilkinson, F. G. A. Stone, E. W. Abel, eds.), Vol. 8, Chap. 58, p. 939ff. Pergamon, New York, 1982; W. Watts, *in* "Comprehensive Organometallic Chemistry" (G. Wilkinson, F. G. A. Stone, and E. W. Abel, eds.), Vol. 8, Chap. 59, p. 1013ff. Pergamon, New York, 1982.
99. W. Beck, K. Raab, U. Nagold, and W. Sacher, *Angew. Chem.* **97**, 498 (1985); W. Beck and B. Olgemoeller, *J. Organomet. Chem.* **127**, C45 (1977); A. D. Cameron, D. E. Laycock, V. H. Smith, and M. C. Baird, *J. Chem. Soc., Dalton Trans.*, 2857 (1987); H.-J. Mueller, U. Nagel, and W. Beck, *Organometallics* **6**, 193 (1987); W. Beck, H.-J. Mueller, and U. Nagel, *Angew Chem., Int. Ed. Engl.* **25**, 734 (1986); H.-T. Mueller and W. Beck, *J. Organomet. Chem.* **330**, C13 (1987); M. Green, N. C. Norman, and A. G. Orpen, *J. Am. Chem. Soc.* **103**, 1271 (1981); B. Niemer, M. Steinmann, and W. Beck, *Chem. Ber.* **121**, 1767 (1988); W. Beck, B. Niemer, and B. Wagner, *Angew. Chem.* **101**, 1699 (1989).
100. B. Olegmoeller and W. Beck, *Chem. Ber.* **114**, 867 (1981); K. Raab, U. Nagel, and W. Beck, *Z. Naturforsch. B,* **38B**, 1466 (1983).
101. A. J. Pearson and J. Yoon, *Tetrahedron Lett.* **26**, 2399 (1985); B. R. Reddy, V.

Vaughan, and J. S. McKennis, *Tetrahedron Lett.* **21,** 3639 (1980); M. Moll, P. Wurstl, H. Behrens, and P. Merbach, *Z. Naturforsch. B,* **33B,** 1304 (1978); E. W. Abel, M. A. Bennett, and G. Wilkinson, *J. Chem. Soc.* 4559 (1958); D. Ciapanelli and M. Rosenblum, *J. Am. Chem. Soc.* **91,** 3673 (1969); A. N. Nesmeyanov, L. G. Makarova, and N. A. Ustynyuk, *J. Organomet. Chem.* **23,** 517 (1970); P. M. Treichel and R. L. Shubkin, *Inorg. Chem.* **6,** 1328 (1967); P. V. Bonnesen, A. T. Baker, and W. J. Hersh, *J. Am. Chem. Soc.* **108,** 8304 (1986); J. D. Atwood, *Inorg. Chem.* **26,** 2918 (1987); P. Szabo, L. Fekete, and G. Bor, *J. Organomet. Chem.* **12,** 245 (1968); A. J. Pearson, *et al.,* in Ref. *12*; M. Airoldi, G. Deganello, G. Dia, P. Saccone, and J. Takats, *Inorg. Chim. Acta* **41,** 171 (1980); H. Adams, N. A. Bailey, J. T. Gauntlett, and M. J. Winter, *J. Chem. Soc., Chem. Commun.* 1360 (1984); M. Green and R. P. Hughes, *J. Chem. Soc., Chem. Commun.,* 862 (1975); W. E. Van Arsdale, R. E. K. Winter, and J. K. Kochi, *Organometallics* **5,** 645 (1986); D. Mandon and D. Astruc, *Organometallics* **8,** 2372 (1989).

102. J. E. Mahler and R. Pettit, *J. Am. Chem. Soc.* **85,** 3955 (1963); J. E. Mahler, D. H. Gibson, and R. Pettit, *J. Am. Chem. Soc.* **85,** 3959 (1963); M. Anderson, A. D. H. Clague, L. P. Blaauw, and P. A. Couperus, *J. Organomet. Chem.* **56,** 307 (1973); E. O. Fischer and R. D. Fischer, *Angew. Chem.* **72,** 919 (1960); H. J. Dauben, Jr., and D. J. Bertelli, *J. Am. Chem. Soc.* **83,** 497 (1961); A. J. Birch, *Org. Synth.* **57,** 107 (1979).

103. R. E. Lehman and J. K. Kochi, *Organometallics* **10,** 190 (1991).

104. E. W. Abel, A. Singh, and G. Wilkinson, *J. Chem. Soc.,* 1321 (1960); E. W. Abel, A. Singh, and G. Wilkinson, in "Organometallic Synthesis" (J. J. Eisch and R. B. King, eds.), Vol. 1, p. 114. Academic Press, New York, 1965; R. B. King, *Acc. Chem. Res.* **3,** 417 (1970); P. Inkrott, R. Goetze, and S. G. Shore, *J. Organomet. Chem.* **154,** 337 (1978); R. G. Hayter, *Inorg. Chem.* **2,** 1031 (1963).

105. C. Zou, K. J. Ahmed, and M. S. Wrighton, *J. Am. Chem. Soc.* **111,** 1133 (1989).

106. P. Lemoine, A. Giraudeau, and M. Gross, *J. Chem. Soc., Chem. Commun.,* 77 (1980); K. M. Kadish, D. A. Lacombe, and J. E. Anderson, *Inorg. Chem.* **25,** 2246 (1986).

107. J. C. Martin, in "Free Radicals" (J. K. Kochi, ed.), Vol. 2, p. 493ff. Wiley, New York, 1975; J. Halpern, *Acc. Chem. Res.* **15,** 238 (1982); M. K. Geno and J. Halpern, *J. Chem. Soc., Chem. Commun.,* 1052 (1987), and references therein.

108. T. Koenig and H. Fischer, in "Free Radicals" (J. K. Kochi, ed.), Vol. 2, p. 157ff. Wiley, New York, 1973; H. Keifer and T. G. Traylor, *J. Am. Chem. Soc.* **89,** 6667 (1967).

109. T. Koenig and R. G. Finke, *J. Am. Chem. Soc.* **110,** 2657 (1988).

110. R. M. Noyes, *J. Am. Chem. Soc.* **77,** 2042 (1955); R. M. Noyes, *J. Am. Chem. Soc.* **78,** 5486 (1956); R. M. Noyes, *Prog. React. Kinet.* **1,** 129 (1961); Ref. *109.*

111. M. K. Geno and J. Halpern, *J. Am. Chem. Soc.* **109,** 1238 (1987); B. P. Hay and R. G. Finke, *J. Am. Chem. Soc.* **109,** 8012 (1987); and related papers.

112. See J. Halpern, *Bull. Chem. Soc. Jpn.* **61,** 13 (1988).

113. J. D. Atwood, "Inorganic and Organometallic Reaction Mechanisms." Brooks/Cole, Monterey, California, 1985.

114. Compare H. Maltz and B. A. Kelly, *J. Chem. Soc., Chem. Commun.,* 1390 (1971); L. K. K. Li Shing Man, and J. Takats, *J. Organomet. Chem.* **117,** C104 (1976).

115. Compare J. L. Hughey, C. R. Bock, and T. J. Meyer, *J. Am. Chem. Soc.* **97,** 4440 (1975); and M. L. Olmstead, R. G. Hamilton, and R. S. Nicholson, *Anal. Chem.* **41,** 260 (1969).

116. R. J. Klingler and J. K. Kochi, *J. Am. Chem. Soc.* **103,** 5846 (1981).

117. See P. J. Krusic and J. San Filippo, Jr., *J. Am. Chem. Soc.* **104,** 2645 (1982); W. J. Bowyer, J. W. Merkert, W. E. Geiger, and A. L. Rheingold, *Organometallics* **8,** 191 (1989).

118. For the highly labile nature of 19-electron radicals, see D. J. Kuchynka and J. K. Kochi in Ref. *82* and W. C. Trogler in Ref. *95.*

119. See Figure 9 in S. Fukuzumi, C. L. Wong, and J. K. Kochi, *J. Am. Chem. Soc.* **102,** 2928 (1980).

120. See J. K. Kochi, *Angew. Chem., Int. Ed. Engl.* **27,** 1227 (1988); S. S. Shaik, *Acta Chem. Scand.* **44,** 205 (1990).

121. J. W. Rabalais, L. O. Werme, T. Bergmark, L. Karlsson, M. Hussain, and K. J. Siegbahn, *J. Chem. Phys.* **57,** 1185 (1972); in acetonitrile, see A. M. Stolzenberg and M. T. Stershic, *J. Am. Chem. Soc.* **110,** 6391 (1988); A. Haaland, *Acc. Chem. Res.* **12,** 415 (1979); E. S. Yang, M. S. Chan, and A. C. Wahl, *J. Phys. Chem.* **84,** 3094 (1980).

122. D. M. Braitsch, *J. Chem. Soc., Chem. Commun.,* 460 (1974); J. F. Helling, S. L. Rice, D. M. Braitsch, and T. Mayer, *J. Chem. Soc., Chem. Commun.,* 930 (1971); J. F. Helling and D. M. Braitsch, *J. Am. Chem. Soc.* **92,** 7207 (1970); D. Mandon and D. Astruc, *J. Organomet. Chem.* **369,** 383 (1989); see also E. O. Fischer and R. Böttcher, *Chem. Ber.* **89,** 2397 (1956).

123. R. E. Lehmann and J. K. Kochi, *J. Am. Chem. Soc.* **113,** 501 (1991).

124. See P. Seiler and J. D. Dunitz, *Acta Crystallogr.* **B35,** 1068 (1979).

125. P. Seiler and J. D. Dunitz, *Acta Crystallogr.* **B35,** 2020 (1979).

126. See for example, J. M. Wallis and J. K. Kochi, *J. Am. Chem. Soc.* **110,** 8207 (1988); Y. Takahashi, S. Sankararaman, and J. K. Kochi, *J. Am. Chem. Soc.* **111,** 2954 (1989).

127. D. H. Busch and J. C. Bailar, Jr., *J. Am. Chem. Soc.* **78,** 1137 (1956).

128. E. O. Fischer and F. Röhrscheid, *Z. Naturforsch.* **17B,** 483 (1962).

129. G. Wilkinson, M. Rosenblum, M. C. Whiting, and R. B. Woodward, *J. Am. Chem. Soc.* **74,** 2125 (1952).

130. See T. Shida, "Electronic Absorption Spectra of Radical Ions." Elsevier, New York, 1988; H. Miyasaka, S. Ojima, and N. Mataga, *J. Phys. Chem.* **93,** 3380 (1989); see also N. Mataga, *Pure Appl. Chem.* **56,** 1255 (1984).

131. M. Rosenblum, R. W. Fish, and C. Bennett, *J. Am. Chem. Soc.* **86,** 5166 (1964); R. L. Collins and R. Pettit, *J. Inorg. Nucl. Chem.* **29,** 503 (1967).

132. C. K. Prout and B. Kamenar, *in* "Molecular Complexes" (R. Foster, ed.), Vol. 1, p. 151ff. Crane, Russak & Co., New York, 1973; F. H. Herbstein, *in* "Perspectives in Structural Chemistry" (J. A. Ibers, ed.), p. 166ff. Wiley, New York, 1971.

133. T. A. Albright *et al.* in Ref. *34*; We thank T. A. Albright for kindly permitting this reproduction.

134. J. Green, *Struct. Bonding (Berlin)* **43,** 37 (1981).

135. E. Adman, M. Rosenblum, S. Sullivan, and T. N. Margulis, *J. Am. Chem. Soc.* **89,** 4540 (1967).

136. Compare D. Astruc, *Tetrahedron* **39,** 4027 (1983); P. Michaud, D. Astruc, and J. H. Ammeter, *J. Am. Chem. Soc.* **104,** 3755 (1982); A. Darchen, *J. Chem. Soc., Chem. Commun.,* 768 (1983); A. N. Nesmeyanov, N. A. Vol'kenau, L. S. Shilovtseva, and V. A. Petrakova, *J. Organomet. Chem.* **61,** 329 (1973); C. Moinet, E. Roman, and D. J. Astruc, *Electroanal. Chem.* **121,** 241 (1981); J.-R. Hamon, D. Astruc, and P. Michaud, *J. Am. Chem. Soc.* **103,** 758 (1981).

137. R. A. Marcus, *J. Chem. Phys.* **24,** 966 (1956); R. A. Marcus, *Discuss. Faraday Soc.* **29,** 21 (1960); G. L. Closs and J. R. Miller, *Science* **240,** 440 (1988); I. R. Gould, R. Moody, and S. Farid, *J. Am. Chem. Soc.* **110,** 7242 (1988).

138. T. Ohno, A. Yoshimura, and N. Mataga, *J. Phys. Chem.* **94,** 4871 (1990).

139. See J. K. Kochi, *Acta Chem. Scand.* **44,** 409 (1990).

ADVANCES IN ORGANOMETALLIC CHEMISTRY, VOL. 33

Structure and Bonding in the Parent Hydrides and Multiply Bonded Silicon and Germanium Compounds: from MH_n to $R_2M = M'R_2$ and $RM \equiv M'R$

ROGER S. GREV

Center for Computational Quantum Chemistry
University of Georgia
Athens, Georgia 30602

I

INTRODUCTION

The 1980s were an exciting decade for those working with the heavier main group analogs of carbon, particularly silicon and germanium. The classic double bond rule was decisively overturned with the isolation and characterization of stable $Si = C$ (*1–9*), $Si = Si$ (*10–20*), $Ge = C$ (*21–24*), and $Ge = Ge$ (*25–30*) doubly bonded compounds. Evidence for the formation of a $Ge = Si$ intermediate has only recently been obtained (*31*). The burst of activity on the experimental side has been matched by theoretical interest in these compounds, and a consensus is slowly emerging concerning the structures and relative energies of the simplest of the parent compounds, particularly silene, $H_2Si = CH_2$ (*32–39*), disilene, $H_2Si = SiH_2$ (*39–50*), silyne, $HSi \equiv CH$ (*51–57*), and disilyne, $HSi \equiv SiH$ (*58–66*). The germanium compounds germene, $H_2Ge = CH_2$ (*67–70*), digermene, $H_2Ge = GeH_2$ (*28,71–74*), germasilene, $H_2Ge = SiH_2$ (*74*), and digermyne $HGe \equiv GeH$ (*75*) have also been studied, but at somewhat less definitive levels of theory.

125

Experimentalists have been particularly well served by numerous reviews from some of the leading workers in the field. For example, Brook and Baines (76) have reviewed silenes, Wiberg (77) discussed M=C and M=N double bonds (M = Si and Ge), Cowley and Norman (78) have reviewed M=M' (M, M' = group 14 and 15 elements) double bonds, Raabe and Michl (79,80) have provided complementary reviews of multiple bonds to silicon, West (81) has reviewed disilene chemistry, Masamune (82) has reviewed the work of his group on Si=Si and Ge=Ge compounds, and most recently Barrau et al. (83) have summarized multiple bonds to germanium. Studies from the reactive intermediates era of this field are beautifully summarized by Gusel'nikov and Nametkin (84).

On the theoretical side, Schaefer reviewed the early studies on silene that contributed greatly to the theoretical interest in these compounds (35). The review by Raabe and Michl (79) on multiply bonded silicon included a comprehensive examination of the early theoretical work. Gordon (85,86) has summarized multiply bonded silicon, and the studies of Luke et al. deserve mention as a comprehensive study of multiply bonded silicon at a uniform level of theory (87,88). Nagase et al. (89) have reviewed their studies on doubly bonded silicon and germanium, with particular emphasis on reactivity. Most recently, Apeloig (90) has provided encyclopedic coverage of theoretical studies of silicon compounds through the middle of 1987.

With that background, the reader is probably wondering, as did the author, what could possibly be said that has not already been rehashed countless times in the 1000 plus pages of review articles on multiply bonded silicon and germanium compounds from the past decade. The number of theoretical studies that have appeared since the reviews of Raabe and Michl (79), Apeloig (90), and Gordon (85,86) is in fact relatively small, and many of them are fiddling at the kilocalorie per mole level with some important isomeric energy differences. One notable exception is the beautiful series of papers by Trinquier and co-workers on trans bending and bridging isomers in doubly bonded compounds (91–96). The one aspect that does appear to be lacking is an overview of the field like that employed in typical organic chemistry textbooks. That is, to start out with simple compounds, like the parent hydrides MH_n, discuss their bond energies and geometries and the like, and show how they relate to the properties of the bigger, unsaturated compounds. With the exception of the work of Trinquier et al. (91–96), this approach has been largely ignored to date. In all the excitement to discover new features and quantify trends in the unsaturated compounds, many interesting parallels with the properties of the MH_n compounds have been neglected. As we shall see, most of the major differences between alkenes and metallenes—such as the emergence of stable divalent isomers, the ease of transbending, and the small dissociation energies of Si=Si, Ge=Ge, and Si=Ge doubly bonded compounds —are qualitatively, and sometimes quantitatively, predictable from the properties of the MH_n compounds and standard M—H and M—M' bond energies.

Predictions of isomeric energy differences and double bond dissociation energies from properties of the parent hydrides should not be expected to yield quantitatively accurate results. It is true that heats of formation of quite complex organic substances can be accurately predicted from tables of group increments or component enthalpies using additivity rules (97,98). Unfortunately, the situation in silicon and germanium chemistry is not so advanced, and the simple trends we shall highlight are among those compounds that are considered references in the modern machinery of organic thermochemistry. Our approach is dictated by what is available, not by what would be best, and falls somewhere between the hierarchical schemes described by Benson (97) as "additivity of bond properties" and "additivity of group properties."

Before delving into the subject we should provide a bit of justification for the approach taken in this article. It might be abundantly obvious why it is important to relate properties of parent hydrides and saturated compounds to the properties of the unsaturated compounds, but to some it may seem like a menial exercise in qualitative bonding theory. It is, in fact, vitally important that we attempt this exercise. The principal reason is that we know very little about the effects of substituents on such quantities as isomeric energy differences and double bond dissociation energies in unsaturated silicon and germanium compounds. Put another way, we are not likely to see accurate theoretical studies of tetrakis[bis(trimethylsilyl)methyl]digermene, $[(Me_3Si)_2CH]_2Ge=Ge[HC(SiMe_3)_2]_2$, anytime in the very near future. However, break that molecule in two and replace terminal methyl groups by hydrogens, and there is a respectable chance that someone with a lot of CPU time and adequate disk space could do a semirespectable theoretical study of $[(H_3Si)_2CH]_2Ge$ in the next few years. If we knew how to relate the properties of this fragment to the properties of the digermene, we might be able to say something intelligent about one of the most fascinating and puzzling molecules yet synthesized (28). In relative terms, we know much about things like Si—H, Si—X (87,88,99–109), and Ge—H (110–113) bond dissociation energies, substituent effects on Si—H bond strengths (99–109), SiR_2 (114–126) and GeR_2 (127,128) singlet–triplet splittings, and geometries and barriers to planarity of substituted silyl radicals

(115,129–131). We want to use this knowledge to tell us something about $R_2M=MR_2$ and $RM\equiv MR$, about which we know considerably less. Maybe this information will help chemists make digermenes and distannenes that will retain their structures in solution or the gas phase. Maybe it will aid in the synthesis of stable triply bonded compounds of silicon and germanium, or hexasilabenzene!

The intent of this article, then, is to provide a synopsis of the current understanding of the structure and bonding of compounds containing multiple bonds between carbon, silicon, and germanium, from the point of view of what has been learned of the simpler compounds like the hydrides MH_n. The emphasis is on qualitative bonding aspects from a molecular orbital theory viewpoint, a reflection of the author's biases. When we received the generous offer to write this review, the instructions were that "it should be written for ordinary preparative chemists, rather than for theoreticians. What our readers will be interested in is the results and predictions which can be made from them." Thus, details of the theoretical studies cited are omitted for the most part, except where it is necessary to make a point. The interested reader can find these details in the original literature cited. Furthermore, no attempt has been made to be comprehensive in covering the early literature as this information can be obtained from previous review articles, and much of the quantitative data resulting from the early theoretical studies has been superseded by subsequent higher-level attacks. The relative paucity of theoretical studies of the germanium compounds necessitates heavy dependence on the results of silicon chemistry as a guide. Fortunately, it appears that germanium chemistry *is* only a minor perturbation on silicon chemistry, at least for the hydrides. Ethylene and acetylene are discussed only for comparative purposes, since their characteristics are covered extensively in textbooks and are quite well known. To those whose work has been unintentionally slighted, we offer our apologies for remaining at the computer terminal when the library has so much to offer.

This article starts with a discussion of the parent hydrides, MH_n ($M = C$, Si, and Ge, $n = 2-4$), highlighting the major differences between the methane series and the silane and germane series. Some attention is paid to substituent effects on bond energies and geometries, and we introduce the important notion of the divalent state stabilization energy (DSSE). After laying this foundation, $Si=C$ and $Ge=C$ compounds are reviewed. This is followed by $Si=Si$, $Ge=Ge$, and $Ge=Si$ doubly bonded structures and their isomers. Finally, triply bonded $Si\equiv C$, $Si\equiv Si$, and $Ge\equiv Ge$ compounds are briefly considered.

II

PRELIMINARY CONSIDERATIONS

A. *Bond Dissociation Energies and Divalent State Stabilization Energies*

In Tables I, II, and III we have assembled some important bond dissociation energies (BDEs) that will aid us in our discussions. Table I contains a collection of C—H, Si—H, and Ge—H BDEs in the parent hydrides, as well as all possible M—M' (M, M' = C, Si, and Ge) single bond dissociation energies. The most complete source of BDE data for these compounds is from the kinetic studies of Walsh and co-workers (99–101,110,111). Berkowitz, Ruscic, and collaborators (104,112) have determined sequential BDEs of SiH$_4$ and GeH$_4$ from photoionization mass spectrometric studies that are in good agreement with the values of Walsh. Theoretical studies (105–109) have recently become an important source of thermochemical data, with accuracies that rival, and sometimes exceed, those from experiment. Comparison of the theoretically determined BDEs in Table I with the experimental values exhibits well the power of quantum mechanically derived BDEs, and provides confidence in theoretically derived values in cases where experiments do not exist. The values we shall use in subsequent analysis are given in bold in Table I.

A dominant theme in the chemistry of the lower group 14 elements is the enhanced stability of the divalent state relative to carbon. This is often noted to be a somewhat trivial consequence of overall weaker bonding. Thus the trend toward the "inert *s*-pair effect," observed through the decreasing endothermicity (increasing exothermicity) of the reaction

$$MX_4 \rightarrow MX_2 + X_2$$

as M goes from C to Si, to Ge, to Sn, to Pb, can be considered to be a general reflection of weaker M—X bonds (132,133). Of course, such trends could also be used to argue for increased stability of *all* the lower oxidation states.

A more direct indication of the increased stability of the divalent state relative to the others derives from the patterns in the individual bond dissociation energies. Specifically, it is generally observed that the second BDE in an SiX$_4$ compound is considerably smaller than the first. Walsh (99) was the first to point this out, and he has defined the difference between the first and second BDE to be the divalent state stabilization energy (DSSE). That is, for

$$SiX_4 \rightarrow SiX_3 + X \qquad \Delta H_1$$

$$SiX_3 \rightarrow SiX_2 + X \qquad \Delta H_2$$

TABLE I

M—H AND M—M' BOND DISSOCIATION ENERGIES[a]

Bond	Walsh	Berkowitz	NBS[b]	Theory
H_3C—H			103.2	103.6[h]
H_2C—H			108.8[c]	110.2[h]
H_3Si—H	90.3[d]	88.8[f]		91.6[i]
H_2Si—H	71.0[d]	69.6[f]		70.1[i]
H_3Ge—H	82.7[e]	82[g]		84.8[j]
H_2Ge—H	56.9[e]	59[g]		58.0[j]
H_3C—CH_3			87.8	
H_3C—SiH_3	88.2[d]			
H_3Si—SiH_3	73.6[d]			
H_3C—GeH_3	78.2[e]			
H_3Si—GeH_3				70[k]
H_3Ge—GeH_3	65.7[e]			
$MeSiH_2$—H	89.6[d]			
$MeSiH$—H	70.7[d,l]			
Me_2SiH—H	89.5[d]			
Me_2Si—H	67.8[d,l]			
H_3SiSiH_2—H	86.3[d]			
H_3SiSiH—H	73.4[d]			

[a] BDEs are in kcal/mol. Values given in bold type are used in the text.

[b] D. D. Wagman, W. H. Evans, V. B. Parker, R. H. Schumm, T. Halow, S. M. Bailey, K. L. Churney, and R. L. Nuttall, *J. Phys. Chem. Ref. Data* **11** (Suppl. 2) (1982).

[c] Using heat of formation of CH_2 from Ref. *148*.

[d] Ref. *101*.

[e] Ref. *111*.

[f] Ref. *104*.

[g] Ref. *112*.

[h] Ref. *105*.

[i] Ref. *108*.

[j] Ref. *113*.

[k] Estimated from Ref. *74*.

[l] The heats of formation of MeHSi and Me_2Si listed in reference *101* are in dispute. We have employed the theoretically obtained values of M. S. Gordon and J. A. Boatz, *Organometallics* **8**, 1978 (1989), which are in good agreement with experimental results of H. E. O'Neal, M. A. Ring, W. H. Richardson, and G. F. Licciardi, *Organometallics* **8**, 1968 (1989). These values lead to significantly smaller DSSEs for MeHSi and Me_2Si than the older, apparently erroneous values used in reference *74*, but cannot be considered definitive.

TABLE II

SEQUENTIAL BOND DISSOCIATION ENERGIES, D, AND DIVALENT STATE STABILIZATION ENERGIES, DSSE, OF MONOSUBSTITUTED SILANES AND METHANES[a]

X	$D(XH_2Si-H)$	$D(XHSi-H)$	DSSE(XHSi)	$D(XH_2C-H)$	$D(XHC-H)$	DSSE(XCH)
H[b]	90.3	71.0	19.3	103.2	108.8	−5.6
Li	78.3	78.8	−0.5	93.8	91.2	2.6
BeH	82.3	76.3	6.0	94.2	99.3	−5.1
BH$_2$	78.1	75.4	2.7	90.8	93.4	−2.6
CH$_3$	91.1	69.5	21.6	99.0	106.7	−6.8
NH$_2$	89.4	49.6	39.8	91.0	68.5	22.5
OH	91.1	55.2	35.9	94.6	75.2	19.4
F	93.8	58.2	35.6	98.8	87.2	11.6

[a] See text for details. D and DSSE values are in kcal/mol.
[b] See Table I.

the DSSE of SiX$_2$ is given by

$$DSSE(SiX_2) = D(X_3Si-X) - D(X_2Si-X) = \Delta H_1 - \Delta H_2 \quad (1)$$

Equivalently, the DSSE is equal to the exothermicity of the disproportionation reaction

$$2\ SiX_3 \rightarrow SiX_2 + SiX_4$$

TABLE III

PREFERRED VALUES FOR DIVALENT STATE STABILIZATION ENERGIES[a] OF RR′M

RR′M	DSSE(RR′M)	RR′M	DSSE(RR′M)
H$_2$Si	19.3	H$_2$Ge	25.8
(H$_3$C)HSi	18.9	H$_2$C	−5.6
(H$_3$C)$_2$Si	21.7	LiHC	2.6
(H$_3$Si)HSi	12.9	(HBe)HC	−5.1
LiHSi	−0.5	(H$_2$B)HC	−2.6
(HBe)HSi	6.0	(H$_3$C)HC	−6.8
(H$_2$B)HSi	2.7	(H$_2$N)HC	22.5
(H$_2$N)HSi	39.8	(HO)HC	19.4
(HO)HSi	35.9	FHC	11.6
FHSi	33.8	F$_2$C	41.7
F$_2$Si	54.1	H$_2$C=C	28.7
ClHSi	30.8	HN≡C	78.0
Cl$_2$Si	43.5	O≡C	72.8

[a] DSSEs are in kcal/mol.

From Table I, the DSSE of SiH_2 is seen to be $90.3 - 71.0 = 19.3$ kcal/mol. For GeH_2, the DSSE is 25.8 kcal/mol.

The reader appreciates, of course, that the second BDE being smaller than the first is exactly the opposite of the pattern found in the methane series (see Table I), which has a negative DSSE, -5.6 kcal/mol. The familiar explanation for the trend in the CH_n BDEs is that the methyl radical is planar and has stronger sp^2 hybridized bonds than the sp^3 hybrids in CH_4. Given that doubly bonded carbon is regarded as sp^2 hybridized, and that no special stability can be attributed to the analogous trivalent silicon and germanium hydrides, the inverted order of sequential BDEs in the SiH_n and GeH_n series relative to CH_n is a big hint that the traditional stability of multiply bonded compounds found in organic chemistry does not hold further down the periodic table.

Divalent state stabilization energies are not easy to come by, as they require knowledge of both the first *and* second BDE, and the reactive intermediates MR_2 are not trivially characterized. Quantum mechanical studies are certainly ahead of experiment in this area, and we can combine the results of two separate studies, one by Coolidge and Borden (*109*) and the other by Luke *et al.* (*88*), to assemble a small list of DSSEs for monosubstituted carbenes and silylenes. Specifically, Coolidge and Borden determined the effects of substituents, X, on the stability of methyl and silyl radicals through determination of the heat of reaction

$$X-AH_2\cdot + AH_4 \to X-AH_3 + AH_3\cdot$$

We can combine this with $D(H_3C-H)$ and $D(H_3Si-H)$ in Table I to obtain an absolute value for $D(XH_2A-H)$, with A = C and Si. These are listed in Table II.

In the study of Luke *et al.*, stabilization energies for singlet and triplet silylenes and carbenes were assessed through isodesmic bond separation equations

$$XSiH(singlet) + SiH_4 \to H_3SiX + H_2Si(singlet)$$

$$XSiH(triplet) + SiH_4 \to H_3SiX + H_2Si(triplet)$$

and analogous reactions for singlet and triplet carbenes. This is equivalent to determination of the difference of the *sum* of the first two A—H BDEs in AH_4 and XAH_3, and we can combine this with $D(XH_2A-H)$ determined from the study of Coolidge and Borden to determine the $D(XHA-H)$, and thus DSSE(XHA), with one catch. For those XHA that have different ground state multiplicities than AH_2, we require knowledge of the exact singlet–triplet splitting of AH_2, which are in fact known with good precision (*117*) to be 9.0 kcal/mol for CH_2, the triplet being the ground state, and 20.9 kcal/mol for SiH_2, with the singlet being the ground state. The result is the set of $D(XHSi{-}H)$ values listed in Table II and, through subtraction, the accompanying set of DSSEs. The trend in DSSEs for silicon is obvious; the more electronegative and, especially, the more π donating the substituent, the more positive the DSSE. Clearly, divalent silicon with amino, hydroxy, and fluorine substituents is quite stable. For the carbon series, the trend in DSSEs from Table II is a bit more erratic, but similar to that in the silane series. We should stress the fact that substituent effects on DSSEs are quite a bit larger than the effects of substituents on first bond dissociation energies.

The series of molecules ethylene, methyleneimine ($H_2C{=}NH$), and formaldehyde ($H_2C{=}O$) provides an interesting parallel with silanes that we shall use later. All of these have positive DSSEs, and in the latter two cases *huge* DSSEs. For ethylene, $D(H_2CCH{-}H)$ is 110 kcal/mol and $D(H_2CC{-}H)$ is 81 kcal/mol, yielding a DSSE for vinylidene of 29 kcal/mol (*134*). The second BDE being so much smaller than the first was criticized as being nonsensical by Benson (*135*), but an avalanche of evidence (*134,136–138*) suggests it is correct. For methyleneimine and formaldehyde, one might suggest that we are almost cheating, as the products of two sequential $C{-}H$ bond dissociations yields structures that we ordinarily draw with triple bonds, namely, isocyanide ($HN{\equiv}C$) and carbon monoxide ($C{\equiv}O$). The difference, however, is more one of degree than of kind; Lewis dot structures without formal charges and including lone pairs on N and O are equally plausible zeroth-order descriptions to those with triple bonds. For methyleneimine, we obtain the DSSE by combining heats of formation (*139*) of $H_2C{=}NH$, HNC, and HCN with a theoretically obtained value (*140*) for the hydrogen addition reaction to HCN to yield *trans*-HCNH. This analysis yields $D(HNCH{-}H) = 99$, $D(HNC{-}H) = 21$, and $DSSE(HNC) = 78$ kcal/mol. Experimental values for formaldehyde (*137*) are $D(OCH{-}H) = 87$, $D(OC{-}H) = 14$, and $DSSE(OC) = 73$ kcal/mol.

For the DSSE to be a useful tool, a requirement is that it truly depend only on the substituents X and Y in the silylene SiXY. Strictly, the assumption that $DSSE(MX_2)$ be the same whether we determine it from MX_4 or MX_2Y_2 is equivalent to the statement that the heat of reaction for

$$MX_4 + 2\,MX_2Y \rightarrow 2\,MX_3 + MX_2Y_2$$

be identically zero. That is, the law of bond additivity must be obeyed. This dependence has recently been tested (*74*) using the quantum mechanically determined heats of formation of all possible SiH_nCl_m compounds in the study of Ho *et al.* (*106*). The $DSSE(SiH_nCl_{2-n})$, with n equal to 0–2, was found to be similar whether the two bonds being broken were both hydrogen or chlorine. For example, from $D(ClH_2Si{-}H) = 90.5$ kcal/mol and $D(ClHSi{-}H) = 59.7$ kcal/mol, one obtains DSSE(ClHSi) = 30.8 kcal/mol, whereas from $D(Cl_2HSi{-}Cl) = 110.0$ kcal/mol and $D(ClHSi{-}Cl) = 79.2$ kcal/mol the result is DSSE(ClHSi) = 31.8 kcal/mol.

Analysis of heats of formation of SiH_nF_m determined by Ignacio and Schlegel (*108*) shows greater dependences of the DSSEs on the exact method used to determine them than those from the chlorine compounds. In the worst case, that of SiH_2, the DSSE differs by 9 kcal/mol depending on whether one uses sequential Si—F BDEs in SiH_2F_2 or sequential Si—H BDEs in SiH_4. With only two complete sets of data on which to judge the soundness of the above hypothesis, this might be considered damning evidence. We suspect, however, that this is just another example of the perversity of fluorine chemistry. In particular, it appears to be a manifestation of the "*gem*-difluoro effect," a term used to describe enhanced stabilization in compounds containing more than one fluorine attached to the same atom. Thus, in what follows we assume that the DSSE(SiXY) is a property of SiXY. Table III contains a list of the DSSEs that we shall use later. In order of preference, these are from Walsh's experimental values derived from Table I, Ho *et al.*'s study of chlorinated silanes, Ignacio and Schlegel's study of fluorinated silanes, and values from Table II.

Another important piece of data for the multiply bonded structures is the π bond energy. A common definition for the π bond energy, and one that is often assessable from experiment, is provided by the barrier to internal rotation about the double bond. Schmidt *et al.* (*141*) have determined π bond energies for a number of systems using this method. Owing to its formal connection to isomeric energy differences (see below), a better definition, proposed by Benson (*97*), is the difference in the energy of hydrogen atom addition to the doubly bonded compound and the free radical that results. For doubly bonded compounds of the form $H_2M{=}M'H_2$, there is the additional proviso that the addition be to the same atom (M or M') in each step. This amounts to the determination of the energy of the reaction

$$H_2M{=}M'H_2 + H_3M{-}M'H_3 \rightarrow H_3M{-}M'H_2\cdot + H_3M'{-}MH_2\cdot$$

This method, as well as the internal rotation barrier technique, has been used by Dobbs and Hehre to yield C=X π bond energies for X = C, Si, Ge, and Sn (*142*). Schleyer and Kost (*143*) compared an X=Y double bond to two single bonds via isodesmic reactions and used standard single bond energies to obtain an estimate for the π bond energies of first and second row systems. The Ge=Ge and Ge=Si π bond energies were

TABLE IV

π Bond Energies[a]

Bond	Bond energy
$H_2C{=}CH_2$	**65**,[b] 70,[c] 64–68[d]
$H_2Si{=}CH_2$	**38**,[b] 36,[c] 35–36[d]
$H_2Si{=}SiH_2$	**25**,[b,e] 24[c]
$H_2Ge{=}CH_2$	**31**,[d] 29[f]
$H_2Ge{=}SiH_2$	**25**[e]
$H_2Ge{=}GeH_2$	**25**[e]
$H_2Sn{=}CH_2$	**19**[d]

[a] Bond energies are in kcal/mol. Values given in bold type are used in the text.
[b] Ref. *141*.
[c] Ref. *143*.
[d] Ref. *142*.
[e] Ref. *74*.
[f] Ref. *68*.

obtained from rotational barrier data. Overall, the various methods yield similar π bond energies, which are close to experimentally determined values in substituted compounds (*13,14,16,18,49*). The resulting data are collected together in Table IV, with the values we shall use later given in bold. The π bond strengths involving silicon and germanium range from about 35 to 60% of the 65 kcal/mol barrier in ethylene.

B. *Qualitative Bonding Considerations in MR₃ and MR₂ Compounds*

The DSSE for SiH_2 and GeH_2 being positive while that of CH_2 is negative is only part of the story. It is not just the ordering of the sequential BDEs that differ from CH_4 to SiH_4 and GeH_4, but the individual compounds MH_3 and MH_2 differ in qualitative ways as well. Thus, whereas CH_3 is planar, both SiH_3 and GeH_3 are pyramidal, with inversion barriers of approximately 5.8 kcal/mol for SiH_3 (*115*) and apparently slightly less (maybe 5.5 kcal/mol) for GeH_3 (*144*). This can be rationalized by a simple orbital-mixing argument known as the second-order Jahn–Teller effect (SOJT) (*145*). Consider the highest lying occupied molecular orbital (HOMO) of a planar MH_3 radical. It is, of course, the singly occupied out-of-plane p orbital, represented schematically in Fig. 1. The lowest unoccupied molecular orbital (LUMO), an antibonding σ orbital (shown at the top of Fig. 1) is orthogonal to the HOMO at planar geometries and cannot mix. However, on pyramidalization, the two orbitals are of the

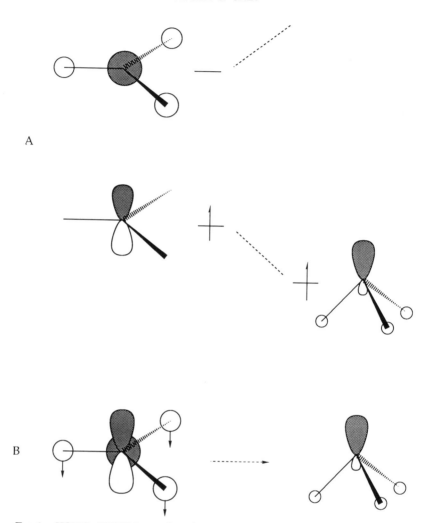

FIG. 1.   HOMO–LUMO interaction diagram for MR_3 radicals. (A) Interaction of a singly occupied p orbital (HOMO) with the σ^* (LUMO) of planar MR_3 that results in stabilization of the HOMO on pyramidalization. (B) Superposition of the HOMO and LUMO from (A). Arrows show the distortion that results in a stabilizing interaction.

same symmetry and can therefore interact. The stabilization energy that results (145) is the decrease of the HOMO energy multiplied by the number of electrons, n, in the HOMO, that is,

$$\Delta E = n \langle \Psi_{HOMO} | H | \Psi_{LUMO} \rangle^2 / (E_{HOMO} - E_{LUMO}) \qquad (2)$$

Aside from the number of electrons, there are thus two important factors that determine the magnitude of ΔE. First, it depends inversely on the orbital energy difference $E_{HOMO} - E_{LUMO}$. Second, it depends on the square of the interaction integral, which will be large only if Ψ_{HOMO} and Ψ_{LUMO} have substantial electron density in the same regions. Electronegativity differences play a crucial role in both of these factors. Substitution of hydrogen by a more electronegative element such as fluorine both reduces the HOMO–LUMO splitting *and* leads to σ-bonding orbitals that are substituent based and σ-antibonding orbitals that are predominantly on the central atom. Hence, the interaction integral between HOMO and LUMO will be greater, and the result is a larger ΔE, a more pyramidal radical. Hydrogen is more electronegative than silicon and germanium but less electronegative than carbon, so on that basis alone we would expect SiH_3 and GeH_3 to be more pyramidal than CH_3. The result of mixing some LUMO character into the HOMO after pyramidalization is shown at right in Fig. 1. Figure 1B is a cartoon shorthand where we have superimposed the HOMO and LUMO and put arrows on the terminal atoms to show the direction of the distortion that provides a stabilizing interaction. An alternative, but less insightful, analysis of this problem is to resort to Bent's rules (*146*), which is really the same thing in hybridization language. We shall see later that the occurrence of trans-bent double bonds in disilene, digermene, and germasilene can be reduced to an identical analysis.

Concerning MH_2 compounds, the differences between SiH_2 and GeH_2 on the one hand and CH_2 on the other hand are even greater than those between the corresponding MH_3 compounds. To begin with, CH_2 has a triplet ground state, approximately 9.0 kcal/mol below the lowest lying singlet state (*117,147,148*), whereas both SiH_2 (*104,117*) and GeH_2 (*127,128*) are ground state singlets, 20.9 and approximately 23.0 kcal/mol below their lowest lying triplet states. Second, the H—M—H angles for SiH_2 and GeH_2 are significantly smaller than those for the corresponding state of CH_2. For CH_2 the bond angles are approximately 134° for the triplet and 102° for the singlet, compared to 118° and 93° for triplet and singlet SiH_2 and 120° and 92° for triplet and singlet GeH_2, respectively. These effects are also evident in the related barriers to linearity, which are approximately 6 and 28 kcal/mol for the $^3B_1$ and $^1A_1$ states of CH_2 (*147*), but are nearer 25 and 68 kcal/mol for SiH_2 (*122,123*). Note that the barrier to linearity in CH_2 is only a fraction of a $C{=}C$ π bond energy, whereas triplet SiH_2 has a barrier to linearity that is equal to an $Si{=}Si$ π bond energy. Imagine replacing one of the hydrogens in SiH_2 by an SiH group to form a hypothetical HSi—SiH linkage isomer and recoupling the spins of the two adjacent triplet silylenes to form disilyne, $HSi{\equiv}SiH$. Clearly, one of the π bonds *at least* is going to be severely stressed by the inherent desire of silylenes to bend.

The trends in the angles and barriers to linearity in MH_2 can be rationalized by the SOJT effect in a similar way to that used to rationalize the pyramidalization of SiH_3 and GeH_3 relative to CH_3. The pertinent orbital interaction diagram for the singlet state of linear H—M—H is shown in Fig. 2. The diagram is slightly simplified in that we show only one component of the degenerate set of π orbitals. The actual electronic state of linear singlet H—M—H, which lies above the triplet state at linear geometries, has one electron in each of the degenerate π orbitals, spin coupled to yield a $^1\Delta$ state; however, the degeneracy is broken on bending, and then the electrons both principally occupy the in-plane component, so we focus on that orbital. With this simplification, the HOMO is now a pure p orbital, and the LUMO is once again a σ-antibonding orbital. For H—Si—H and H—Ge—H, the σ-antibonding orbital is again more localized on the central atom and lower lying than it is for H—C—H, because of electronegativity differences. Thus, as for MH_3, we expect much stronger interaction (stabilization) on bending H—Si—H and H—Ge—H than for H—C—H. The qualitative result of mixing the two orbitals is shown at right in Fig. 2, and the superimposed HOMO–LUMO cartoon and arrows indicating the distortion required for a stabilizing interaction are shown in Fig. 2B. The higher barrier to planarity in the singlet states compared to the triplets is trivially explained by the number of electrons, n, in Eq. (2) being equal to two for the HOMO of the singlets and equal to one for the triplets, for which one of the electrons remains in the out-of-plane p orbital. For SiH_2 and GeH_2 singlet ground states, the mixing of HOMO and LUMO on bending is so great that most authors note that the lone pair has mainly s character.

The discussion above raises two general questions. First, how can we make carbon chemistry more like silicon and germanium chemistry? The second, and more important, question is how can we make silicon and germanium mimic the fantastic diversity of carbon chemistry? With the ready admission that we shall never completely accomplish this task, we can at least provide pointers in the right direction. To make carbon chemistry more like silicon and germanium, we can perfluorinate the carbon compounds. The qualitative differences between CH_n on the one hand and SiH_n and GeH_n on the other evaporate entirely if we replace hydrogen by fluorine. In this case, the sequential BDEs all have the second smaller than the first, all the MF_3 compounds are pyramidal, and all MF_2 compounds are ground state singlets. The correspondences go further still, as we demonstrate later, to include isomeric energy difference and double-bond BDE trends.

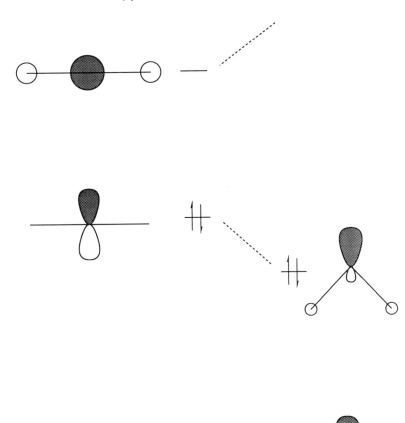

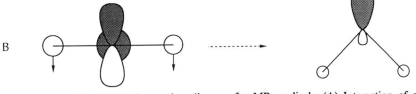

Fig. 2. HOMO–LUMO interaction diagram for MR_2 radicals. (A) Interaction of a doubly occupied p orbital (HOMO) with the σ^* (LUMO) of linear MR_2 that results in stabilization of the HOMO on bending. (B) Superposition of the HOMO and LUMO from (A). Arrows show the distortion that results in a stabilizing interaction.

To make silicon and germanium chemistry at least somewhat more like carbon we can, for example, use silyl substituents. To wit, H_3SiSiH has a reported DSSE of 12.9 kcal/mol, compared to 19.3 for SiH_2. The DSSE of H_3SiSiH is not negative like CH_2, but at least it is smaller than in SiH_2. Silyl groups are also known to make amines planar, and this presumably extends to silyl radicals as well. Furthermore, disilylsilylene, $(H_3Si)_2Si$, has been determined (125) to have a singlet-triplet splitting around 5 kcal/mol compared to 20.9 for SiH_2. Again, silyl substitution is not enough to produce a ground state triplet like that of CH_2, but it is at least a trend in the right direction. Recent evidence from theoretical studies ($47,48$) suggests that these similarities *do* carry over to disilenes, for which silyl substituted disilenes have increased stability toward dissociation *and* are less prone to trans-bending than $H_2Si{=}SiH_2$.

A fascinating rationalization of the apparent lack of hybridization in elements beyond the first row, Li–Ne, was provided by Kutzelnigg (149). He noted that the principal reason was not due to inherently greater $s-p$ energy differences, but rather to an orbital size mismatch. For Li–Ne, the valence orbitals are $2s$ and $2p$, with a $1s^2$ core. The requirement that the $2s$ orbitals be orthogonal to the $1s$ core increases their radial extent. The result is that the $2s$ orbitals are localized in the same spatial regions as the $2p$ orbitals, for which there are no corresponding core orbitals, and thus $s-p$ hybridization is efficient. In the second row, Na–Ar, the valence $3p$ orbitals are now required to be orthogonal to the core $2p$, with the result being that the valence $3s$ and $3p$ orbitals now have significantly different radial extents and hybridization is not efficient. These arguments also rationalize the tendency of first row elements to have weak σ bonds and strong π bonds, whereas second row elements show the opposite trend of strong σ bonds and weak π bonds.

In summary, the differences between the CH_n series and the SiH_n and GeH_n series are considerable. Bond strengths are not just weaker in the silane and germane compounds; they exhibit entirely different trends. This is neatly summarized by the DSSEs. Furthermore, the individual compounds differ in important respects. Thus, silyl and germyl radicals are pyramidal, whereas methyl is planar; silylene and germylene are ground state singlets, whereas methylene is a ground state triplet; and, finally, barriers to linearity of both the singlet and triplet electronic states of SiH_2 (and presumably GeH_2 as well) are large, rivaling π bond energies, whereas CH_2 has smaller barriers to linearity that pale in comparison to the $C{=}C$ π bond energy. The effect of substituting a hydrogen atom by an X group to form XAH_3 appears to have similar qualitative effects for both series of compounds, with π donors making the methane series behave more like the silane and germane series, and π acceptors (most importantly, silyl) leading the silane series in the direction of methanes. Finally, we have a simple bonding model, the second-order Jahn–Teller effect, that accounts qualitatively for many of these trends. We next show how to relate these phenomena to that which is known in the fascinating world of the unsaturated multiply bonded compounds, and speculate a bit on that which is not.

III

DOUBLY BONDED COMPOUNDS AND THEIR ISOMERS

A. *Isomeric Energy Differences, Bond Energies, and Divalent State Stabilization Energies*

Recall that Benson's definition of the π bond energy in ethylene (97) is the difference between the first BDE of ethane and the second BDE for cleavage of a hydrogen β to the radical center, that is,

$$H_3C—CH_3 \rightarrow H_3C—CH_2\cdot + H \qquad \Delta H = D(H_3CCH_2—H)$$
$$H_3C—CH_2\cdot \rightarrow H_2C{=}CH_2 + H \qquad \Delta H = D(H_3CCH_2—H) - D_\pi$$

Consider now, the reaction

$$H_3C—CH_2\cdot \rightarrow H_3CCH + H \qquad \Delta H = D(H_3CCH—H)$$

We have already defined $D(H_3CCH—H)$ to be $D(H_3CCH_2—H) -$ DSSE(H_3CCH). Thus, assuming Benson's definition of the π bond energy and Walsh's definition of the DSSE, ΔH for the following reaction is given by:

$$H_3CCH \rightarrow H_2C{=}CH_2 \qquad \Delta H = DSSE(H_3CCH) - D_\pi$$

It is an assumption that the π bond energy, thus defined, is equal to the internal rotation barrier. For ethylene, the thermochemical π bond energy and the barrier to internal rotation are within experimental error of one another. Note that the isomerization enthalpy is strictly valid only if the DSSE(H_3CCH) is determined from H_3CCH_3.

In Fig. 3, we show how the isomeric energy differences of unsaturated derivatives of C_2H_6 are related to the BDEs, DSSEs, and the π bond energy, D_π. These relationships follow strictly from the definitions of these quantities and Hess' Law, provided that the DSSEs are determined from the compounds indicated, without approximation. As such, this diagram is equally valid for any M_2X_6 system. An analogous diagram for H_3SiCH_3 is shown in Fig. 4, and is necessarily more complicated. Figure 4 is equally valid for any $X_3MM'X_3$ system, without approximation. As a mere statement of thermodynamic identities, Figs. 3 and 4 contain nothing new; their potential utility is as an aid in making reasonable approximations to obtain isomeric energy differences and double bond dissociation energies.

There are very few systems M_2X_6 for which the quantities in Fig. 3 are all known, C_2H_6 being the most obvious example. The reasons are many. Dissociation energies are difficult to determine accurately from first principles quantum mechanical studies, although methods are improving. Most

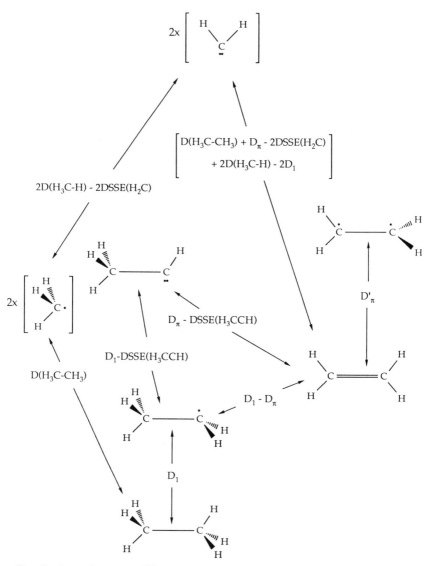

FIG. 3. Isomeric energy differences of ethane and derived species in terms of bond dissociation energies (*D*), π bond energies (*D*$_π$), and divalent state stabilization energies (DSSE).

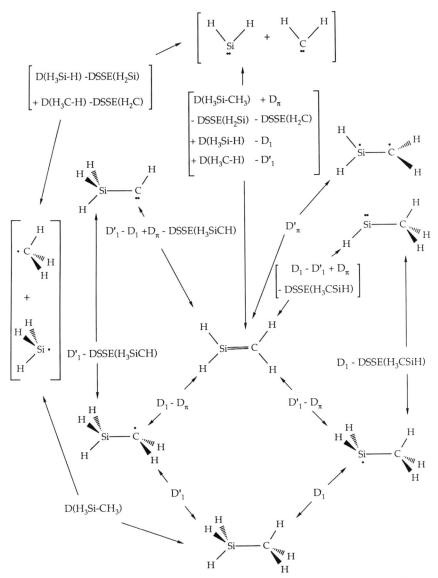

FIG. 4. Isomeric energy differences of H_3SiCH_3 and derived species in terms of bond dissociation energies (D), π bond energies (D_π), and divalent state stabilization energies (DSSE).

often studied are isomeric energy differences, such as ethylene–methylmethylene, disilene–silylsilylene, or digermene–germylgermylene, which we see is equivalent to determining the quantity D_π DSSE(X_3MMX). Experimentally, the most accessible π bond energy is certainly that determined from barriers to internal rotation, but unsymmetrically substituted XYM=MXY compounds are required for this to work. Nonetheless, in the absence of other alternatives, we can approximate D_π by D'_π (see Fig. 3) and, as is usually necessary, assume that D'_π is independent of X. BDEs from thermochemical kinetics become increasingly difficult to obtain as the size of the molecule, and thus the number of reaction pathways, increases. The parent hydrides, MH_n, are better characterized, although even there it is an arduous task. The efforts of quantum chemists, spectroscopists, and kineticists to determine the properties of methylene alone probably represents the activities of a major university for a 5- or 10-year period!

For the unsymmetrical system in Fig. 4, the change in enthalpy for the reaction

$$X_2M{=}M'X_2 \rightarrow X_3MM'X \tag{3}$$

is given by

$$\Delta H(3) = D(X_3MM'X_2{-}X) - D(X_3M'MX_2{-}X)$$
$$+ D_\pi - \text{DSSE}(X_3MM'X)$$

or alternatively

$$\Delta H(3) = D(X_3MM'X{-}X) - D(X_3M'MX_2{-}X) + D_\pi$$

where we have simply replaced $D(X_3MM'X_2{-}X) - \text{DSSE}(X_3MM'X)$ by $D(X_3MM'X{-}X)$. In other words, the isomeric energy difference is determined by D_π plus the difference between a first substituted M—X BDE and the second substituted M'—X BDE. In cases where we do not have BDEs for the substituted compounds, we can reasonably approximate this by

$$\Delta H(3) \cong D(HX_2M'{-}X) - D(HX_2M{-}X) + D_\pi - \text{DSSE}(M'XH)$$

For the carbon, silicon, and germanium compounds considered here, this should not lead to exceptionally large errors, because H for M substitution generally changes BDEs by 4 kcal/mol or less.

Another approximation that deserves special mention concerns the dimerization energy of the doubly bonded structures. From Fig. 3, we see that the enthalpy required to break the double bond for the reaction

$$X_2M{=}MX_2 \rightarrow 2\,MX_2 \tag{4}$$

is given by

$$\Delta H(4) = D(X_3M—MX_3) + D_\pi - 2DSSE(X_2M) + 2D(X_3M—X)$$
$$- 2D(X_3MMX_2—X)$$

For M = C, Si, and Ge, the approximation

$$D(X_3M—X) = D(X_3MMX_2—X)$$

will almost certainly introduce errors of less than 10 kcal/mol. Thus, to a reasonable approximation, the heat of the above reaction is given by

$$\Delta H(4) \cong D(X_3M—MX_3) + D_\pi - 2DSSE(X_2M)$$

A similar set of approximations can be applied to the case of an unsymmetrical doubly bonded isomer, $X_2M{=}M'X_2$, as in Fig. 4. Thus, if we assume that

$$D(X_3M—X) = D(X_3M'MX_2—X)$$

and

$$D(X_3M'—X) = D(X_3MM'X_2—X)$$

then the enthalpy for the reaction

$$X_2M{=}M'X_2 \rightarrow X_2M + M'X_2 \tag{5}$$

will be given by

$$\Delta H(5) \cong D(X_3M—M'X_3) + D_\pi - DSSE(X_2M) - DSSE(M'X_2)$$

Within this approximation, the DSSE is seen to play a prominent role in the thermochemical stability of the doubly bonded isomers toward dissociation. Fortunately, we know a bit about the effects of substitution on DSSEs for silicon, and we can assume that they will be similar for germanium.

B. Silene, $H_2Si{=}CH_2$, and Germene, $H_2Ge{=}CH_2$

The lowest lying isomers of C_2H_4, $SiCH_4$, and $GeCH_4$ are represented schematically in Fig. 5. A number of qualitative differences between ethylene and the silicon- and germanium-substituted isomers are apparent, the most striking difference being the tremendous stabilization of the singlet state divalent isomers methylsilylene, H_3CSiH, and methylgermylene, H_3CGeH, relative to the doubly bonded isomers silene, $H_2Si{=}CH_2$, and germene, $H_2Ge{=}CH_2$. Thus, whereas triplet methylmethylene (ethylidene), H_3CCH, is nearly 70 kcal/mol above ethylene (88), methylsilylene

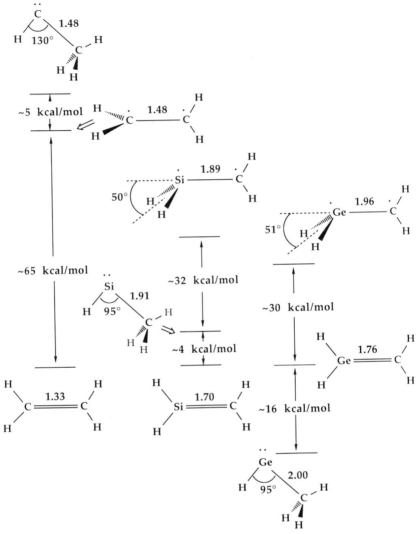

FIG. 5.   Structures and relative energies (kcal/mol) of the lowest lying isomers of ethylene, silene, and germene. Bond distances are in angstroms.

is about 4 kcal/mol above silene (*38,39*), and methylgermylene lies approximately 15–18 kcal/mol *below* germene (*68,69*). From the discussion of the parent hydrides and the analysis in Fig. 4, the stability of the divalent isomers should come as no surprise.

Another qualitative difference between the ethylene isomers and those of silene and germene is that the twisted diradical states, both singlets and triplets, are not planar about the heavy atoms as in ethylene, but instead are pyramidal, with out-of-plane angles near $50°$. The angles about silicon and germanium are, in fact, quite close to those in the corresponding SiH_3 and GeH_3 radicals, as we might expect. Thus, the nonplanarity is a trivial consequence of the electronic structure corresponding to two weakly coupled radical centers. We show only a single geometry and relative energy for both the singlet and triplet diradical states as they are nearly identical for both electronic states. The singlet and triplet diradical states do differ in one important aspect, however, as the triplet state represents a potential energy minimum whereas the singlet state is the transition state for internal rotation about the $M=C$ bond (141). The energy difference between the transition state and the double bonded minimum is the source of numerous π bond energy estimates, which we have already commented on. The best *ab initio* transition state barriers are 65 kcal/mol for ethylene, 36 kcal/mol for silene, and 29–31 kcal/mol for germene.

Substituent effects on the $Si=C$ bond length have been extensively studied, mainly to clear up initial discrepancies between theoretical and experimental results. An early electron diffraction (ED) study (8) of 1,1-dimethyl-1-silaethylene yielded a distance $r(Si=C)$ of 1.83 Å, which was strongly criticized (35). In a recent study of the microwave spectra of the same compound by H. S. Gutowsky *et al.* (9) that appeared with the title "The Silicon–Carbon Double Bond: Theory Takes a Round," it was demonstrated that the rotational spectra were consistent with the theoretical values near 1.70 Å, and that the ED work must be in error. Apeloig and Karni (33) demonstrated that the substituents in Brooks' silene $(1-3)$ were the cause of the elongated distance, 1.764 Å, compared to that in $H_2Si=CH_2$, which can be reliably given as 1.70 ± 0.02 Å $(35,38)$. Wiberg *et al.* $(4-7)$ have synthesized and obtained crystal structures of $Me_2Si=C(SiMe_3)(SiMe\text{-}t\text{-}Bu_2)$ and its tetrahydrofuran (THF) adduct. The THF-free compound has an $Si=C$ bond distance of 1.702 Å—in near perfect agreement with the quantum mechanical results for $H_2Si=CH_2$—whereas the fascinating THF adduct has a slightly elongated bond distance of 1.747 Å. The structurally characterized germanium–carbon doubly bonded compounds $(21-24)$ have bond distances of 1.80 and 1.83 Å, slightly longer than self-consistent-field (SCF) values of 1.75–1.78 Å $(67-70)$ for $H_2Ge=CH_2$. It seems that we can generalize and say that planar double bonds are usually 0.2 Å shorter than single bonds, a constancy that is often obscured by using percentage shortenings. One difference between the hydrides and the experimentally observed doubly bonded compounds is that, as a rule, the known $Si=C$ and $Ge=C$ structures are twisted by varying degrees about the double bond.

The relative energies of silene and methylsilylene have been studied at sufficiently high levels of theory (38,39) that the relative energies can be confidently stated to be near 4 kcal/mol. This is in good agreement with the analysis of Walsh (150). From Fig. 4 [see also reaction (3)] ΔH for the reaction

$$H_2Si{=}CH_2 \rightarrow HSiCH_3$$

is given by

$$\Delta H = D(H_3CSiH_2{-}H) - D(H_3SiCH_2{-}H) + D_\pi - DSSE(H_3CSiH)$$

$$= 89.6 - 99.2 + 38 - 18.9$$

$$= 9.5 \ kcal/mol$$

with values from Tables I, II, and IV, except for $D(H_3SiCH_2{-}H)$, which is taken to be equal to the value of $D[(H_3C)_3SiCH_2{-}H]$ from Walsh's studies. Curiously, a better estimate of this reaction enthalpy can be obtained by using BDEs from the parent hydrides [i.e., using the approximate form of Eq. (3)] instead of from the actual silyl-substituted methanes and methylsilanes, that is,

$$\Delta H = D(H_3Si{-}H) - D(H_3C{-}H) + D_\pi - DSSE(SiH_2)$$

$$= 90.3 - 103.2 + 38 - 19.3$$

$$= 5.8 \ kcal/mol$$

This is probably a result of uncertainties in heats of formation of methylsilylenes (see footnote l of Table I). An easy way to remember the various component BDEs that enter into this isomerization enthalpy is to consider breaking the π bond, requiring D_π energy, then breaking a second Si—H bond, requiring $D(H_2Si{-}H) = D(H_3Si{-}H) - DSSE(SiH_2)$, and finally making a first C—H bond, giving back energy equal to $D(H_3C{-}H)$.

It is fortunate that the simplified scheme using BDEs from the parent hydrides works so well, because the requisite data for the analogous substituted germanium compounds are nonexistent. Thus, using the approximate form of $\Delta H(3)$ as above, we can predict the enthalpy change for the germene–methylgermylene rearrangement as

$$\Delta H = D(H_3Ge{-}H) - D(H_3C{-}H) + D_\pi - DSSE(GeH_2)$$

$$= 82.7 - 103.2 + 31 - 25.8$$

$$= -15.3 \ kcal/mol$$

in good agreement with the available quantum mechanical values of -17.6 and -15.0 kcal/mol. Clearly, the main reason these simplified ΔH values work so well is that silyl and germyl only slightly perturb C—H BDEs and methyl only slightly perturbs Si—H and Ge—H BDEs. This procedure fails (74) for systems such as $H_2Si=NH \rightleftharpoons HSiNH_2$ since the second Si—H BDE is dramatically influenced by the amino substituent, as seen in Table II.

With such a small energy difference between silene and methylsilylene, it is not surprising that the relative stabilities can be shifted around by proper choice of substituent. One way to stabilize the silene isomer is to disubstitute with electronegative groups like fluorines (56). The principal reason for this effect is the dramatically increased stability of Si—F bonds relative to C—F, with average values of 142.3 and 116.4 kcal/mol in SiF_4 and CF_4, respectively. We can estimate the relative energies of $F_2Si=CH_2$ and $FSi—CH_2F$ using the simplified scheme above as

$$\Delta H = D(H_2FSi—F) - D(H_3C—F) + D_\pi - DSSE(SiHF)$$

$$= 160.3 - 112.8 + 38 - 33.8$$

$$= 51.7 \text{ kcal/mol}$$

compared to SCF level predictions by Gordon of 33.0 kcal/mol, which is probably around 10 kcal/mol smaller than the actual value. Monosubstitution by fluorine has the opposite effect of stabilizing the silylene isomer, $FSiCH_3$, which Gordon predicts to be about 10 kcal/mol more stable than $FHSi=CH_2$. The stabilization of the silylene isomer upon monofluorination by 10 to 15 kcal/mol relative to that in $SiCH_4$ on monofluorination is quantitatively related to the fact that the DSSE of SiHF is about 14 kcal/mol greater than that of SiH_2. Presumably, OR substituents will have similar effects on the silene–silylene isomer energies, namely, monosubstitution favoring silylene isomers and disubstitution favoring silene isomers, as they share with fluorine the features of stronger bonds to silicon than carbon and increasing silylene DSSEs.

Substituent effects on germene–germylene isomerization energies have not been studied. It may be that no combination of substituents can be employed to yield a ground state germene isomer. Substitution by two fluorines is unlikely to have the same effect of stabilizing the doubly bonded isomer as it did in the silene–silylene case because no differential advantage exists for Ge—F bonds compared to C—F bonds as occurs for Si—F bonds relative to C—F bonds; experimental heats of formation (139) for GeF_4 yield an average Ge—F bond of only 112.5 kcal/mol. In fact, this suggests that the exact opposite effect—stabilization of the germylene isomer—occurs, since both stronger C—F than Ge—F bonds and larger DSSE(GeHF) than DSSE(GeH_2) values should tend to stabilize the germylene isomer relative to the germene isomer. One possibility is to use bulky substituents, as the longer bonds to germanium should result in smaller destabilizing steric interactions. Of course, this is exactly the usual experimental approach to creating stable doubly bonded isomers in the first place. That germenes exist at all as stable structures is attributable to the large barrier (37 kcal/mol) for the germene to germylene rearrangement (*69*).

C. Disilene, $H_2Si = SiH_2$, Digermene, $H_2Ge = GeH_2$, and Germasilene, $H_2Ge = SiH_2$

Figures 6, 7, and 8 show the structures and relative energies of disilene, digermene, and germasilene isomers, respectively, obtained from theoretical studies. The various isomers of Si_2H_4 have been studied by numerous authors (*39–50,74*), whereas Ge_2H_4 has been studied to a lesser degree (*28,71–74*); only one published study of $GeSiH_4$ exists (*74*). In common with silene and germene, all of these systems exhibit low lying (Si_2H_4 and Ge_2H_4) or ground state ($GeSiH_4$) divalent isomers. One new feature that arises in these molecules that was not evident in $SiCH_4$ and $GeCH_4$ is that the doubly bonded isomers are no longer planar, but instead adopt trans-bent geometries, with the planar structures being transition states. This trans-bending phenomenon has attracted a great deal of attention (*28,48,71,91–96*). For disilene, the potential energy surface is very flat along this coordinate, and subtle changes in basis set at the SCF level of theory can have a dramatic effect on the out-of-plane distortion angle. Still, the barrier to planarity is usually very small (~0.1 kcal/mol) or the planar structure is found to be a minimum. On the other hand, studies including the effects of electron correlation invariably find the structure to be transbent, with a barrier to planarity that is near 1 kcal/mol. For digermene and germasilene, the doubly bonded structures are significantly pyramidal even at the SCF level of theory.

Another curious feature of disilene and digermene is the existence of stable doubly bridged isomers as in Fig. 9, of both the cis and trans form. For disilene, the trans- and cis-dibridged isomers have been found to be about 23 and 25 kcal/mol above $H_2Si = SiH_2$, respectively, in the most recent studies (*93*). For digermene, the trans- and cis-dibridged isomers are only about 9 and 12 kcal/mol above the doubly bonded isomer, respectively, and in distannene, Sn_2H_4, and diplumbene, Pb_2H_4, the transbridged isomer is predicted to be the ground state (*93*)! For all of the M_2H_4 compounds (M = Si, Ge, Sn, and Pb) the energy of these dibridged structures relative to two MH_2 appears to be a relatively constant -30 ± 3 kcal/mol. The interested reader should consult the fascinating studies of Trinquier *et al.* (*91,92,94*) for further discussion of this topic.

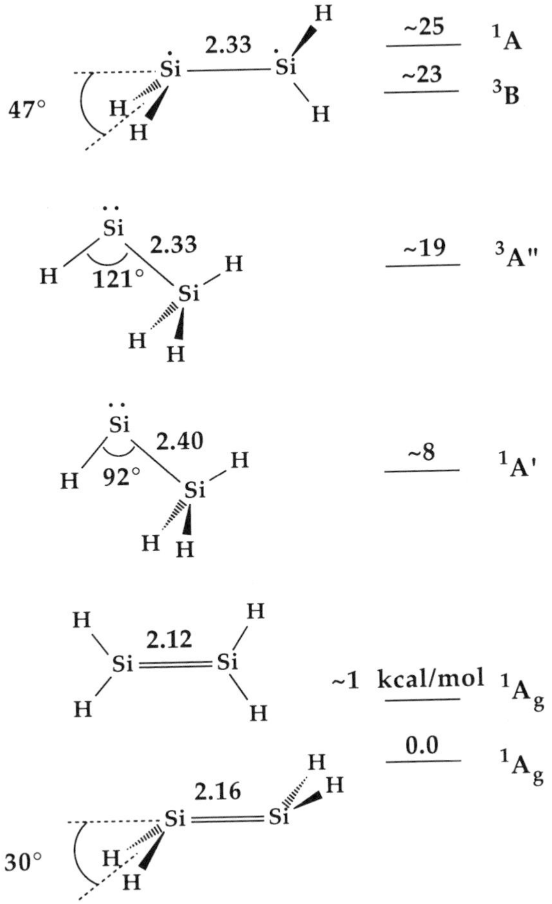

FIG. 6. Structures and relative energies (kcal/mol) of the lowest lying isomers of disilene. Bond distances are in angstroms.

Of all the Si_2H_4 isomers (Figs. 6 and 9), the trans-bent disilene structure is the most stable, being around 8 kcal/mol more stable than the singlet silylsilylene isomer according to the best available theoretical studies (39). The experimental studies of Becerra and Walsh (151) suggested a lower bound for this energy difference of 12 kcal/mol, and White et al. (152) concluded it was in excess of 10.9 kcal/mol. The discrepancy between these results is not large and may be due to a number of factors. From Fig. 3 this energy difference is identically $D_\pi - DSSE(H_3SiSiH)$, and small revisions in any of the quantities involved could bring the theoretical and experimental results into better agreement. Given the large multireference char-

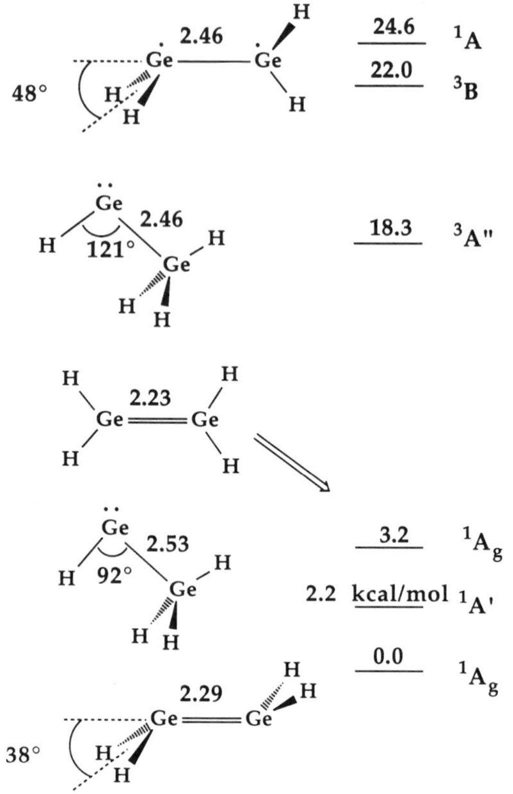

FIG. 7. Structures and relative energies (kcal/mol) of the lowest lying isomers of digermene. Bond distances are in angstroms.

acter of disilene (*40,45,46*), it would not be surprising if better theoretical treatments would result in a value larger than 8 kcal/mol. This author's suspicion is that the experimental value for DSSE(H_3SiSiH) may also be too small, and, given the history of such problems, this might be more precisely aimed at inaccuracies in the heat of formation of silylsilylene. A value for the disilene–silylsilylene energy difference of about 10 kcal/mol is unlikely to be more than 2–3 kcal/mol from the truth.

The double bond distance in disilene, around 2.16 Å, is in good agreement with experimental results for the stable, and heavily substituted, structures that are known (*11,12,14,17,19*), which range from 2.14 to 2.16 Å. It is interesting that the bulky functional groups on the stable disilenes do not result in longer Si=Si bonds than that in $H_2Si=SiH_2$. This may well be tied to the fact that the experimentally known structures

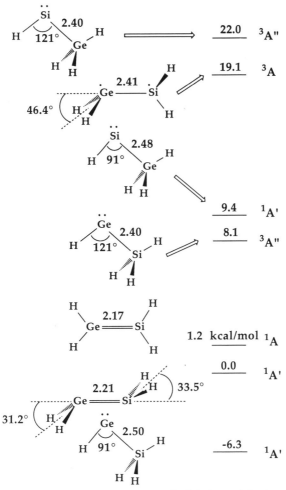

FIG. 8.   Structures and relative energies (kcal/mol) of the lowest lying isomers of germasilene. Bond distances are in angstroms.

are all significantly less pyramidal than theoretical studies of Si_2H_4 indicate. There is clearly a strong correlation between the out-of-plane angle and the $Si=Si$ distance in the published theoretical studies, with more pyramidal structures also having longer $Si=Si$ bonds. It is also curious that the barrier to internal rotation about the $Si=Si$ bond in Si_2H_4, around 25 kcal/mol (*50,74,141*), agrees so well with the various experimentally determined values, which range from 23 to 31 kcal/mol (*13,14,16,18*) in the highly substituted disilenes. By comparison, substi-

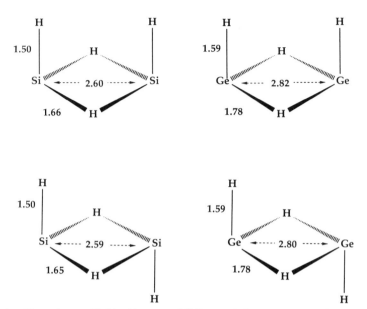

FIG. 9.   Cis- and trans-dibridged isomers of disilene and digermene. Bond distances are in angstroms.

tuted ethylenes have internal rotation barriers that vary over an incredibly wide range, namely, from near zero all the way up to 65 kcal/mol (*153*).

The published studies of digermene (*28,71–74*) are not as comprehensive as those on disilene. We (*154*) have refined some of our earlier work (*74*) on this molecule, and these results are shown in Fig. 7. The barrier to planarity of digermene and the energy relative to singlet germylgermylene, H_3GeGeH, are in excellent agreement with the most recent study of Trinquier (*93*). The digermene–germylgermylene isomers are nearly isoenergetic, trans-bent digermene being only about 2 kcal/mol lower than H_3GeGeH. The barrier to planarity is now over 3 kcal/mol, compared to nearer 1 kcal/mol for disilene. The energies of the rest of the structures (triplet H_3GeGeH, triplet $H_2Ge—GeH_2$, and the digermene internal rotation transition state), relative to trans-bent digermene, all bear an eerie resemblance to those of disilene.

Using configuration interaction methods, the Ge=Ge distance in planar digermene is 2.23 Å, and it increases by 0.06 Å to 2.29 Å in the minimum energy trans-bent structure, for which the folding angle is found to be 38° (*154*). The structurally characterized digermenes exhibit tremendous variability in Ge=Ge bond distances. Lappert's digermene, Ge_2R_4, with $R = CH(SiMe_3)_2$, has a Ge=Ge distance of 2.347 Å and is trans-bent by 32° (*25,26,28*). It is sufficiently unstable that it does not retain its digermene structure in solution (*25*) or in the gas phase (*155,156*). Masamune has reported crystal structures of two digermenes. The first, with R = 2,6-diethylphenyl, has $r(Ge{=}Ge) = 2.213$ Å and a folding angle of only 12° (*18,30*). Most recently, he has reported the structure of the first unsymmetrical digermene, (*Z*)-1,2-bis(2,6-diisopropylphenyl)-1,2-dimesityldigermene, with a bond distance of 2.301 Å and a folding angle of 36° (*18*). The $E \rightleftharpoons Z$ isomerism of the latter compound has also been analyzed (*18*), and barriers of 20.0 and 22.2 kcal/mol have been deduced, in reasonable agreement with the theoretical results for Ge_2H_4, which predict the transition state energy to be 24.6 kcal/mol above trans-bent digermene (*74,154*). Unlike Lappert's digermene, both of Masamune's digermenes retain their structures in solution.

Only one theoretical study of the isomers of $GeSiH_4$ has been published (*74*), and the results of that study are summarized in Fig. 8. For this system, the singlet silylgermylene isomer is the lowest energy isomer, similar to $GeCH_4$ and in contrast to Si_2H_4 and Ge_2H_4. The asymmetry of the system, in particular the differences in Si—H and Ge—H BDEs, is responsible for this result. As in the analysis of $GeCH_4$, the enthalpy change for the germasilene–silylgermylene rearrangement can be approximated by

$$\Delta H = D_\pi + D(H_3Ge{-}H) - DSSE(GeH_2) - D(H_3Si{-}H)$$

$$= 25 + 82.7 - 25.8 - 90.3$$

$$= -8.4 \text{ kcal/mol}$$

which is close to the theoretically determined value of −6.3 kcal/mol. For the germasilene–germylsilylene(singlet) isomeric system, a similar analysis predicts

$$\Delta H = D_\pi + D(H_3Si{-}H) - DSSE(SiH_2) - D(H_3Ge{-}H)$$

$$= 25 + 90.3 - 19.3 - 82.7$$

$$= 13.0 \text{ kcal/mol}$$

compared to 9.4 kcal/mol from theoretical studies. It is likely that more elaborate studies will increase the stability of the doubly bonded isomers relative to the others. Bridged isomers analogous to those in Fig. 9 for Si_2H_4 and Ge_2H_4 have not been investigated, and it is at least possible that the asymmetry of this system will render them unstable, but this remains to be investigated.

A difference between disilenes, digermenes, and germasilenes relative to alkenes that is of vital importance to the experimental community is the incredibly small energies required to dissociate the doubly bonded isomers to fragments. It is well known that the dissociation energy of ethylene to two triplet methylenes is highly endothermic, requiring about 170 kcal/mol in energy, compared to 88 kcal/mol to break the $C-C$ single bond in ethane. By contrast, dissociation of disilene to two ground state singlet silylenes requires only about 65 kcal/mol according to the best available theoretical studies (39), compared to an $Si-Si$ single bond energy of 74 kcal/mol in disilane. This curious fact appears to have first been recognized by Gordon *et al.* (43). Digermene and germasilene also require *less* energy to dissociate to fragments than do the analogous single-bonded compounds digermane and germylsilane. The best available theoretical data yield $D(H_2Ge{=}GeH_2) = 37$ kcal/mol compared to an experimental value of $D(H_3Ge-GeH_3) = 66$ kcal/mol, and $D(H_2Ge{=}SiH_2) = 45$ kcal/mol compared to $D(H_3Ge-SiH_3) = 67$ kcal/mol, both from theoretical studies. Considering the basis sets used and levels of theory employed, the values of $D(H_2Ge{=}GeH_2)$, $D(H_2Ge{=}SiH_2)$, and $D(H_3Ge-SiH_3)$ are all likely to be a few kilocalories per mole smaller than the exact values.

We demonstrated above [see Fig. 3 and 4 and Eqs. (4) and (5)] that the dissociation energies for multiply bonded compounds were intimately connected to the DSSEs of the resulting fragments. Using the approximate form of ΔH for reactions (4) and (5), we predict

$$D(H_2Ge{=}GeH_2) \cong D(H_3Ge-GeH_3) + D_\pi - 2DSSE(GeH_2)$$

$$= 65.7 + 25 - 2(25.8)$$

$$= 39.1 \text{ kcal/mol}$$

in good agreement with the lower bound of 37 kcal/mol quoted above. For disilene, the analogous prediction yields

$$D(H_2Si{=}SiH_2) \cong D(H_3Si-SiH_3) + D_\pi - 2DSSE(SiH_2)$$

$$= 73.6 + 25 - 2(19.3)$$

$$= 60.0 \text{ kcal/mol}$$

which, by comparison with the exact $\Delta H(4)$, is too low by exactly twice the difference $D(H_3Si-H) - D(H_3SiSiH_2-H) \cong 3$ kcal/mol, or about 6 kcal/mol. For germasilene, one obtains

$$D(H_2Ge{=}SiH_2) \cong D(H_3Ge-SiH_3) + D_\pi$$

$$- DSSE(GeH_2) - DSSE(SiH_2)$$

$$= 70 + 25 - 25.8 - 19.3$$

$$= 49.9 \text{ kcal/mol}$$

from the approximate form for $\Delta H(5)$, compared to a lower bound of 45 kcal/mol from theoretical studies.

From the DSSEs in Table III and their relationships to double-bond dissociation energies it is clear that π donor substituents on silicon and germanium will lead to double-bonded structures that may be only weakly stable, or unstable, with respect to dissociation. If one includes entropic effects and the possibility of stable bridged structures, it is probably unwise to choose substituents that yield dissociation energies below 20 kcal/mol or so. These predictions are in general agreement with the observation that germylenes with N-, O-, S-, and halogen-centered ligands yield monomeric crystals (157–162). Given that $H_2Ge{=}GeH_2$ is only stable by around 40 kcal/mol (with a likely maximum around 45 kcal/mol), it is obvious that great care should be exercised in the proper choice of substituents. Ideally, π acceptors should be used, as these are known to decrease DSSEs.

A curious connection between the small dissociation energies of Si=Si, Si=Ge, and Ge=Ge compounds and cumulenic carbon compounds deserves comment. According to Eq. (4), the C=C dissociation energies of the cumulenic structures $H_2C{=}C{=}C{=}CH_2$, $HN{=}C{=}C{=}NH$, and $O{=}C{=}C{=}O$ can be approximated by the sum of a standard C—C σ bond energy and a C=C π bond energy, minus twice the DSSE of the resulting $H_2C{=}C$, $HN{=}C$, or $O{=}C$ fragment. From the DSSEs in Table III, it is clear that the latter two compounds should be very unstable with respect to dissociation, as twice the DSSE is comparable to a $\sigma + \pi$ bond energy, and that the first should have a dissociation energy near 100 kcal/mol.

Small double bond dissociation energies have been discussed by Carter and Goddard (163–165), who have advocated that they are associated with ground state singlet character in the dissociation products. Specifically, they predict that the dissociation energy of a substituted ethylene, $XYC{=}CX'Y'$, will be equal to that in $H_2C{=}CH_2$ (which they call the intrinsic C=C bond energy, D_{int}) if the two carbene fragments CXY and CX'Y' have triplet ground states, and that if either CXY or CX'Y' have singlet ground states, then the dissociation energy will be lower than that in ethylene by the singlet–triplet splitting in the ground state singlet fragment. For example

$$D(XYC{=}CX'Y') = D_{int}(C{=}C) - [\Delta E_{ST}(CXY) + \Delta E_{ST}(CX'Y')]$$

for the dissociation energy in a substituted ethylene for which the CXY and CX'Y' fragments have singlet ground states. This relationship between carbene singlet–triplet splittings and double bond dissociation energies does seem to work reasonably well for $F_2C{=}CF_2$ but is in error by 25 to 30 kcal/mol in other cases such as FHC=CHF (74) and $Cl_2C{=}CCl_2$ (166).

It should be noted that, thus far, this relationship has only been tested for substituents that preferentially stabilize singlet carbenes. It seems counter-intuitive that substituents which stabilize triplet carbenes, such as CN, SiH_3, and other π acceptors, should not also lead to smaller dissociation energies than that in ethylene. A back-of-the-envelope calculation of $D[(H_3Si)_2C=C(SiH_3)_2]$ suggests the dissociation energy may be as low as 130 kcal/mol, which is in serious disagreement with the above model (which would predict 170 kcal/mol) but obviously requires rigorous verification.

Karni and Apeloig (47) have determined $Si=Si$ dissociation energies for a number of substituted disilenes that show strong correlation with the sum of singlet–triplet splittings, as postulated by Carter and Goddard, although the slope of the plot of $D(Si=Si)$ with the sum of the singlet–triplet splittings does not exhibit a slope of -1 as predicted from their model. Instead, a slope of -0.62 is found (47). The data show the expected trend, from the DSSE connection, that π donor substituents such as NH_2, OH, and F yield smaller dissociation energies and π acceptors like Li, BeH, BH_2, and SiH_3 help stabilize the double-bonded structure with respect to dissociation. The theoretically determined values appear to be in good agreement with values that we would predict using an appropriate form of $\Delta H(4)$ and DSSEs from Table III. Dramatic effects on Si–Si distances and trans-bending angles that paralleled the dissociation energies were also noted: π donors led to longer Si–Si distances and more pyramidal structures, with the opposite for π acceptors. The observation that silyl substituents stabilize disilenes with respect to dissociation may be of greatest benefit for digermenes, for which the dissociation energies are much smaller.

The trans-bending phenomenon observed in disilene, digermene, and germasilene has also been related to the singlet–triplet splittings of the fragments involved. Specifically, Trinquier and Malrieu (91,92,95) have proposed, and tested, a simple model which predicts that double bonds will be unstable with respect to this distortion whenever the sum of ΔE_{ST} values of the fragments is greater than one-half of the $X=Y$ double-bond energy. This model has been extensively reviewed and will not be commented on further. From a pedagogical perspective, the primary conclusion of their studies is that the trans-bending observed is a direct manifestation of the desire for the fragments to preserve, as much as possible, their own identities. In this case, the identity is that of a singlet state monomer, and one possible interpretation of the resulting trans-bent structure is that of two datively bonded singlet carbenelike species.

Another way to rationalize the trans-bent geometries is through a $\pi-\sigma^*$ mixing model (28,95,145), using the second-order Jahn–Teller effect as was employed above to rationalize the pyramidal structure of silyl and germyl radicals, and the more angular geometries of silylenes and germylenes compared to carbenes. The relevant orbital interaction diagram is shown in Fig. 10A, which contains a schematic of the doubly occupied HOMO, a π bonding orbital, and the unoccupied σ^* orbital. The LUMO of these systems is, of course, the π^* orbital, but it is not of the correct symmetry to interact with the HOMO under a trans-bending distortion. The π^* orbital can mix in with the σ bonding orbital on trans-bending, but the $\pi-\sigma$ energy difference is smaller than $\sigma-\pi^*$ and is therefore the dominant effect. In Fig. 10B, we have superimposed the π and σ^* orbitals and drawn in arrows on the terminal atoms to show the direction of distortion that allows for a stabilizing interaction, namely, trans-bending.

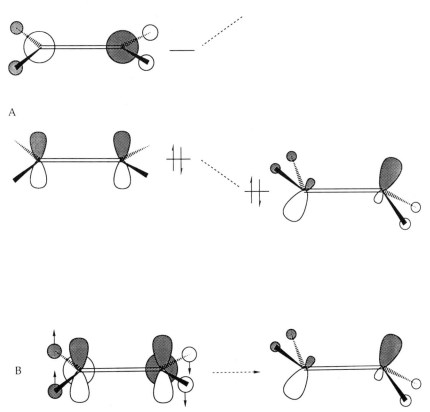

FIG. 10.   Interaction diagram for analysis of trans-bending in disilene, digermene, and germasilene. (A) Interaction of the π HOMO and the σ^* orbital of planar doubly bonded structures that is responsible for the trans-bending. (B) Superimposed π and σ^* orbitals from (A) with arrows indicating the stabilizing distortion.

The $\pi-\sigma^*$ mixing argument can, in fact, be derived from the analysis of SOJT effects for the silyl and germyl radicals. In other words, the trans-bending in these systems can also be regarded as a direct consequence of their characterization as two weakly coupled X_2M^- radical centers. This analysis is shown in Fig. 11. Starting from the right-hand side in Fig. 11A and considering it to be a radical center, we have drawn in a $p-\sigma^*$ interaction picture as in Fig. 1B. From the perspective of the left-hand radical center, the phase of the p orbital is determined by the fact that the two adjacent p orbitals must form a bonding π set, and the phases on the left-hand terminal atoms are required to be out of phase with that of the left-hand central atom as in a σ^* orbital. This is shown in Fig. 11B. Comparison with Fig. 10 shows that the two results are identical. In other words, the $\pi-\sigma^*$ mixing argument as applied to the trans-bending of doubly bonded compounds is nothing more than the $p-\sigma^*$ mixing argument for the pyramidal structures of the trivalent radicals. This connection

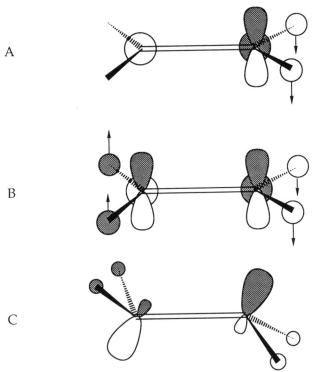

A

B

C

FIG. 11.   Trans-bending of doubly bonded compounds viewed as a consequence of MR_3 pyramidalization (compare with Figs. 1 and 10). See text for details.

has been overlooked in previous analyses and provides another example of the persistence of fragment properties in molecules.

IV

TRIPLY BONDED COMPOUNDS

The triply bonded systems that have been investigated by quantum mechanical methods are $SiCH_2$ (51–57,88), Si_2H_2 (46,58–66,88), and Ge_2H_2 (75). Dimethyldisilyne, $Si_2(CH_3)_2$, has also been investigated (62,66) at reliable levels of theory, as evidence for its intermediacy in the thermolysis of bis(7-silanorbornadiene) has been presented by West's group (167–169). A comprehensive high-level theoretical study of disilyne by Colegrove and Schaefer, using large basis sets and including the effects of electron correlation, has recently been published (65). This study included an extensive review of previous work on Si_2H_2. No new studies of the $SiCH_2$ potential energy surface have appeared recently, and the previous studies have been reviewed (85,86,88,90); thus, we shall just point out the salient features. If there is one conclusion to be drawn from the studies on these fascinating molecules, it is that the more unsaturated the silicon and germanium compounds are, the stranger the potential energy surface becomes. From a computational point of view, there is a second conclusion: electron correlation is essential in describing these systems.

SCF studies of the singlet states of $SiCH_2$ reveal only one minimum, corresponding to silylidene $H_2C=Si$:. At correlated levels of theory, at least with small (by today's standards) basis sets, a trans-bent minimum exists corresponding to a triply bonded isomer. It is quite high lying, around 50 kcal/mol above silylidene, and the barrier to rearrangement to silylidene is only about 6 kcal/mol. It is not clear that this structure will exist as a minimum if larger basis sets and more extensive electron correlation procedures are employed.

Disilyne, Si_2H_2, and digermyne, Ge_2H_2, reveal a structural complexity that is unparalleled in unsaturated carbon chemistry. Colegrove and Schaefer (65) located *four* singlet state minima within 20 kcal/mol of each other, and three transition states within 15 kcal/mol of the ground state. Since the work of Lischka and Kohler (58), quantum chemists have universally agreed that the ground state of disilyne is a dibridged $Si(H_2)Si$ structure. This stunning theoretical prediction of a dibridged ground state structure for disilyne has, in fact, been confirmed by microwave spectroscopic detection of Si_2H_2 in a silane plasma (170). The r_0 structure obtained by Bogey *et al.* from analysis of the rotational spectrum is in good agreement with the theoretically determined r_e structure, and observation of the ^{29}Si and ^{30}Si isotopomer spectra leaves no doubt concerning its identification (*170*).

The four minima on the Si$_2$H$_2$ surface located by Colegrove and Schaefer are shown in Fig. 12, as is the linear structure. For comparison, the Ge$_2$H$_2$ minima and the linear structure obtained by Grev *et al.* are shown (*75*). Obviously, the potential energy surfaces of Si$_2$H$_2$ and Ge$_2$H$_2$ are very similar to one another. It is possible that a singly bridged isomer of Ge$_2$H$_2$ similar to that found for Si$_2$H$_2$ may also exist, but this was not

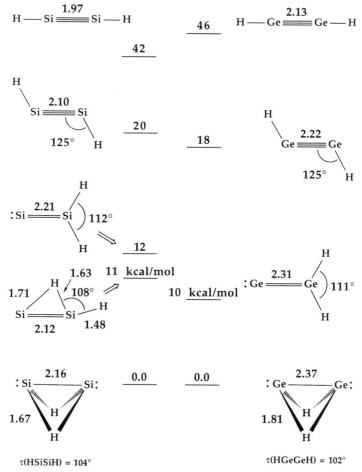

FIG. 12. Structures and relative energies (kcal/mol) of the singlet state minima of Si$_2$H$_2$ and Ge$_2$H$_2$. The linear structures are not minima. Bond distances are in angstroms.

investigated. For both molecules, the dibridged isomers are the ground states.

The transition states connecting the three lowest minima of Si_2H_2 were studied by Colegrove and Schaefer (65), and the results show that the surface is exceptionally flat. Specifically, the transition state between the disilavinylidene isomer, $H_2Si=Si:$, and the singly bridged isomer lies only 3 kcal/mol above $H_2Si=Si:$, and the transition state barrier connecting the singly bridged isomer and the $Si(H_2)Si$ ground state is only 2 kcal/mol. Thus, it is questionable whether disilavinylidene will have any stability to rearrangement to the ground state dibridged isomer.

The triply bonded linear structures of disilyne and digermyne are not minima, and are over 40 kcal/mol above the dibridged ground state structure. Following the normal mode corresponding to the imaginary frequency of the linear structure leads to the trans-bent structures which, for both Si_2H_2 and Ge_2H_2, are near 20 kcal/mol above the ground state. At the SCF level of theory, the trans structures are not minima. Depending on the basis set used, they may have one or two imaginary frequencies. However, on inclusion of electron correlation effects the trans-bent disilynes and digermynes *are* minima. Analysis of the correlated wavefunctions shows that there is a substantial amount of diradical character in these states. Unfortunately, transition states connecting *trans*-disilyne and the other, lower lying, minima have not been located, so the stability of this isomer to rearrangement is unknown. Given the flatness of the rest of the surface, it would be surprising if a substantial barrier exists. We leave it as an exercise for the reader to show that the trans-bending of the linear structures can be derived from the second-order Jahn–Teller effects in linear SiH_2 and GeH_2 (Fig. 2) in an analogous fashion to that used to demonstrate that trans-bending in disilenes and digermenes (Fig. 11) was derivable from analysis of SiH_3 and GeH_3 (Fig. 1).

Triplet states of Si_2H_2 have also been investigated (46,58–60,63,88). For a given structural type, the triplets appear always to be higher in energy than the singlets. The lowest lying of the triplet states is the disilavinylidene structure. This isomer has been found to be either planar or pyramidal, with the most recent studies finding it to be planar, although the results are very sensitive to the choice of method (46,63). Trans-bent, cis-bent, and possibly even twisted disilyne isomers exist as well, but all are higher lying. There appears to be agreement that all of the triplet structures are higher in energy than any of the singlet state minima in Fig. 12.

The relationship between the electronic structure and stability of the various disilyne and digermyne isomers is, with hindsight, at least rationalizable: it is a direct consequence of the number of lone pairs (64,65,75). The dibridged isomers are most easily viewed as in Fig. 12, namely, as singly bonded M—M structures, each heavy atom also containing a lone pair, and with two three-center–two-electron $(3c-2e)$ bonds to the hydrogens (64). Thus, it can be considered to be the result of the union of two protons with M_2^{2-}, represented as $:M{\equiv}M:$, and each proton forms a $3c-2e$ bond with one of the π bonds, π_x and π_y, at right angles to one another. On the other hand, the $H_2M{=}M:$ isomers have only one lone pair, and the trans-bent structure has none, although it does have diradical character. In resonance language, we might consider the trans-bent isomer to be of the form $HM^-{=}M^+H \rightleftharpoons HM^+{=}M^-H$. Regardless, it is clear that the structures with the most lone pairs are more stable.

For dimethyldisilyne, the dibridged structure is high lying and is not a minimum on the potential energy surface (62). Instead, the dibridged structure has two imaginary vibrational frequencies leading to a disilavinylidene structure and to a trans-bent disilyne structure. The disilavinylidene structure is the ground state for this molecule, about 12 kcal/mol lower in energy than the trans-bent $(62,66)$. The high energy of the dibridged structure (about 32 kcal/mol above disilavinylidene) suggests that a single bridging methyl group is also unstable, and thus it appears likely that the trans-bent structure will be stable toward rearrangement. It also seems reasonable that functional groups larger than methyl should preferentially stabilize the trans-structure relative to disilavinylidene for steric reasons.

Substituent effects on the structures and relative energies of triply bonded compounds obviously deserve more attention. Studies of silyl-substituted disilynes and digermynes are certainly called for. Since silyl substituents stabilize disilenes toward dissociation, and stabilize radicals through their π-accepting abilities, a preferential stabilization of the disilyne isomer may result. On the other hand, silyl is also a better bridging ligand than methyl, which suggests that the bridged isomers may reemerge. Nonetheless, bulky silyl groups are an attractive ligand in the quest for a stable disilyne.

One class of substituents that could yield a new breed of structures is π donors. We have seen numerous times already that π donors stabilize

Fig. 13. A disilylene. Structures such as this may emerge as low-lying minima if R is a strong π-donor.

divalent isomers. If two π donor substituents are used, a structure such as that in Fig. 13, which we shall call a disilylene, may emerge as the ground state. Isomers like this have been seen in theoretical studies of Si_2H_2 and $Si_2(CH_3)_2$ when small basis sets are used, or if electron correlation is neglected. However, these disilylene isomers have always been shown to be artifacts of the theoretical method, and disappear entirely as stable minima when the methods are improved. We suspect that they will return as minima, if not ground states, if substituents such as NR_2 or OR are used.

V

CONCLUSIONS

Experimental and theoretical studies of unsaturated silicon and germanium compounds have produced a wealth of new and unexpected phenomena, and more is sure to come. Foremost among those already recognized is the enhanced stability of divalent isomers in these systems, but we should include, as well, the trans-bent nature of double- and triple-bonded systems, the curious inversion of single- and double-bond dissociation energies, and the emergence of low lying stable or ground state bridged isomers.

Many of these new phenomena are foreshadowed by the properties of the parent hydrides. The reversal of the patterns in the sequential bond dissociation energies of CH_n compared to SiH_n and GeH_n already gives a clue to the lack of any special stability associated with sp^2 hybridized bonding environments. Furthermore, the pyramidal nature of silyl and germyl radicals, and the large barriers to linearity in SiH_2 and certainly GeH_2 as well, presages the instability of the planar and linear bonding environment customarily seen in alkenes and alkynes, respectively.

The divalent state stabilization energy (DSSE) is seen to be a particularly useful concept. It arises naturally from the comparison of $X_2M{=}MX_2 \rightleftharpoons X_3MMX$ energy differences and leads to simple and reasonably accurate approximation schemes for the determination of double-bond dissociation energies and isomeric energy differences.

Many of the results reviewed here suggest that a replacement of the usual alkyl or aryl substituents by silyl substituents in unsaturated silicon and germanium compounds may be rewarding. As we noted, silyl substituents do tilt the properties of silylenes, silyl radicals, and sequential BDE trends toward those in carbon chemistry. They have already been shown to stabilize disilenes with respect to dissociation to two silylenes, and this may be crucial to the further development of digermene and distannene chemistry.

ACKNOWLEDGMENTS

This research was supported by National Science Foundation Grant CHE-8718469. The author wishes to thank the silicon and germanium community as a whole for their unending encouragement of theoretical investigations.

REFERENCES

1. A. G. Brook, F. Abdesaken, B. Gutekunst, G. Gutekunst, and R. K. Kallury, *J. Chem. Soc., Chem. Commun.* 191 (1981).
2. A. G. Brook, S. C. Nyburg, F. Abdesaken, B. Gutekunst, G. Gutekunst, R. K. M. R. Kallury, Y. C. Poon, Y.-M. Chang, and W. Wong-Ng, *J. Am. Chem. Soc.* **104**, 5667 (1982).
3. S. C. Nyburg, A. G. Brook, F. Abdesaken, G. Gutekunst, and W. Wong-Ng, *Acta Crystallogr.* **C41**, 1632 (1985).
4. N. Wiberg, G. Wagner, and G. Muller, *Angew. Chem., Int. Ed. Engl.* **24**, 229 (1985).
5. N. Wiberg, G. Wagner, G. Muller, and J. Riede, *J. Organomet. Chem.* **271**, 381 (1984).
6. N. Wiberg, G. Wagner, J. Riede, and G. Muller, *Organometallics* **6**, 32 (1987).
7. N. Wiberg, G. Wagner, G. Reber, J. Riede, and G. Muller, *Organometallics* **6**, 35 (1987).
8. P. G. Mahaffy, R. Gutowsky, and L. K. Montgomery, *J. Am. Chem Soc.* **102**, 2854 (1980).
9. H. S. Gutowsky, J. Chen, P. J. Hajduk, J. D. Keen, and T. Emilsson, *J. Am. Chem. Soc.* **111**, 1901 (1989).
10. R. West, M. J. Fink, and J. Michl, *Science* **214**, 1343 (1981).
11. M. J. Fink, M. J. Michalczyk, K. J. Haller, R. West, and J. Michl, *J. Chem. Soc., Chem. Commun.* 1010 (1983).
12. M. J. Fink, M. J. Michalczyk, K. J. Haller, R. West and J. Michl, *Organometallics* **3**, 793 (1984).
13. M. J. Michalczyk, R. West, and J. Michl, *Organometallics* **4**, 826 (1985).
14. B. D. Shepherd, D. R. Powell, and R. West, *Organometallics* **8**, 2664 (1989).
15. S. Masamune, Y. Hanzawa, S. Murakami, T. Bally, and J. F. Blount, *J. Am. Chem. Soc.* **104**, 1150 (1982).
16. S. Murakami, S. Collins, and S. Masamune, *Tetrahedron Lett.* **25**, 2131 (1984).
17. S. Masamune, S. Murakami, J. T. Snow, H. Tobita, and D. J. Williams, *Organometallics* **3**, 333 (1984).
18. S. A. Batcheller, T. Tsumuraya, O. Tempkin, W. M. Davis, and S. Masamune, *J. Am. Chem. Soc.* **112**, 9394 (1990).
19. H. Watanabe, K. Takeuchi, N. Fukawa, M. Kato, M. Goto, Y. Nagai, *Chem. Lett.* 1341 (1987).
20. B. D. Shepherd, C. F. Campana, and R. West, *Heteroatom Chem.* **1**, 1 (1990).
21. C. Couret, J. Escudie, J. Satge, and M. Lazraq, *J. Am. Chem. Soc.* **109**, 4411 (1987).
22. M. Lazraq, J. Escudie, C. Couret, J. Satge, M. Drager, and R. Dammel, *Angew. Chem., Int. Ed. Engl.* **27**, 828 (1988).
23. H. Meyer, G. Baum, W. Massa, and A. Berndt, *Angew. Chem., Int. Ed. Engl.* **26**, 798 (1987).
24. A. Berndt, H. Meyer, G. Baum, W. Massa, and S. Berger, *Pure Appl. Chem.* **59**, 1011 (1987).
25. P. J. Davidson, D. H. Harris, and M. F. Lappert, *J. Chem. Soc., Dalton Trans.*, 2268 (1976).

26. P. B. Hitchcock, M. F. Lappert, S. J. Miles, and A. J. Thorne, *J. Chem. Soc., Chem. Commun.*, 480 (1984).
27. P. Bleckman, R. Minkwitz, W. P. Neumann, M. Schriewer, M. Thibud, and B. Watta, *Tetrahedron Lett.* **25**, 2467 (1984).
28. D. E. Goldberg, P. B. Hitchcock, M. F. Lappert, K. M. Thomas, A. J. Thorne, T. Fjeldberg, A. Haaland, and B. E. R. Schilling, *J. Chem. Soc., Dalton Trans.*, 2387 (1986).
29. S. Masamune, Y. Hanzawa, and D. J. Williams, *J. Am. Chem. Soc.* **104**, 6136 (1982).
30. J. T. Snow, S. Murakami, S. Masamune, and D. J. Williams, *Tetrahedron Lett.* **25**, 4191 (1984).
31. K. M. Baines, J. A. Cooke, R. J. Groh, and B. Joseph, presented at the 24th Organosilicon Symposium, University of Texas at El Paso, April 1991.
32. P. Rosmus, H. Bock, B. Solouki, G. Maier, and G. Mihm, *Angew. Chem., Int. Ed. Engl.* **20**, 598 (1981).
33. Y. Apeloig, and M. Karni, *J. Am. Chem. Soc.* **106**, 6676 (1984).
34. H.-J. Kohler and H. Lischka, *J. Am. Chem. Soc.* **104**, 5884 (1982).
35. H. F. Schaefer, *Acc. Chem. Res.* **15**, 283 (1982).
36. M. W. Schmidt, M. S. Gordon, and M. Dupuis, *J. Am. Chem. Soc.* **107**, 2585 (1985).
37. S. K. Shin, K. K. Irikura, J. L. Beauchamp, and W. A. Goddard, *J. Am. Chem. Soc.* **110**, 24 (1988).
38. R. S. Grev, G. E. Scuseria, A. C. Scheiner, H. F. Schaefer, and M. S. Gordon, *J. Am. Chem. Soc.* **110**, 7337 (1988).
39. J. A. Boatz and M. S. Gordon, *J. Phys. Chem.* **94**, 7331 (1990).
40. K. Somasundram, R. D. Amos, and N. C. Handy, *Theor. Chim. Acta* **70**, 393 (1986); K. Somasundram, R. D. Amos, and N. C. Handy, *Theor. Chim. Acta* **72**, 69 (1987).
41. K. Krogh-Jespersen, *Chem. Phys. Lett.* **93**, 327 (1982).
42. K. Krogh-Jespersen, *J. Am. Chem. Soc.* **107**, 537 (1985).
43. M. S. Gordon, T. N. Truong, and E. K. Bonderson, *J. Am. Chem. Soc.* **108**, 1421 (1986).
44. G. Olbrich, *Chem. Phys. Lett.* **130**, 115 (1986).
45. H. Teramae, *J. Am. Chem. Soc.* **109**, 4140 (1987).
46. A. F. Sax and J. Kalcher, *J. Mol. Struct. (THEOCHEM)* **208**, 123 (1990).
47. M. Karni and Y. Apeloig, *J. Am. Chem. Soc.* **112**, 8589 (1990).
48. C. Liang and L. C. Allen, *J. Am. Chem. Soc.* **112**, 1039 (1990).
49. G. Olbrich, P. Potzinger, B. Reimann, and R. Walsh, *Organometallics* **3**, 1267 (1984).
50. D. A. Hrovat, H. Sun, and W. T. Borden, *J. Mol. Struct. (THEOCHEM)* **163**, 51 (1988).
51. J. N. Murrell, H. W. Kroto, and M. F. Guest, *J. Chem. Soc., Chem. Commun.*, 619 (1977).
52. A. C. Hopkinson and M. H. Lien, *J. Chem. Soc., Chem. Commun.*, 107 (1980).
53. A. C. Hopkinson, M. H. Lien, and I. G. Csizmadia, *Chem. Phys. Lett.* **95**, 232 (1983).
54. M. S. Gordon and R. D. Koob, *J. Am. Chem. Soc.* **103**, 2939 (1981).
55. M. S. Gordon and J. A. Pople, *J. Am. Chem. Soc.* **103**, 2945 (1981).
56. M. S. Gordon, *J. Am. Chem. Soc.* **104**, 4352 (1982).
57. M. R. Hoffmann, Y. Yoshioka, and H. F. Schaefer, *J. Am. Chem. Soc.* **105**, 1084 (1983).
58. H. Lischka and H.-J. Kohler, *J. Am. Chem. Soc.* **105**, 6646 (1983).
59. J. S. Binkley, *J. Am. Chem. Soc.* **106**, 603 (1984).
60. J. Kalcher, A. Sax, and G. Olbrich, *Int. J. Quantum Chem.* **25**, 543 (1984).
61. H.-J. Kohler and H. Lischka, *Chem. Phys. Lett.* **112**, 33 (1984).
62. B. S. Thies, R. S. Grev, and H. F. Schaefer, *Chem. Phys. Lett.* **140**, 355 (1987).

63. S. Koseki and M. S. Gordon, *J. Phys. Chem.* **92**, 364 (1988).
64. S. Koseki and M. S. Gordon, *J. Phys. Chem.* **93**, 118 (1989).
65. B. T. Colegrove and H. F. Schaefer, *J. Phys. Chem.* **94**, 5593 (1990).
66. B. T. Colegrove and H. F. Schaefer, *J. Am. Chem. Soc.* **113**, 1557 (1991).
67. T. Kudo and S. Nagase, *Chem. Phys. Lett.* **84**, 375 (1981).
68. G. Trinquier, J.-C. Barthelat, and J. Satge, *J. Am. Chem. Soc.* **104**, 5931 (1982).
69. S. Nagase and T. Kudo, *Organometallics* **3**, 324 (1984).
70. K. D. Dobbs and W. J. Hehre, *Organometallics* **5**, 2057 (1986).
71. G. Trinquier, J.-P. Malrieu, and P. Riviere, *J. Am. Chem. Soc.* **104**, 4529 (1982).
72. S. Nagase and T. Kudo, *J. Mol. Struct.* **103**, 35 (1983).
73. T. Fjeldberg, A. Haaland, M. F. Lappert, B. E. R. Schilling, R. Seip, and A. J. Thorne, *J. Chem. Soc., Chem. Commun.*, 1407 (1982).
74. R. S. Grev, H. F. Schaefer, and K. M. Baines, *J. Am. Chem. Soc.* **112**, 9458 (1990).
75. R. S. Grev, B. J. DeLeeuw, and H. F. Schaefer, *Chem. Phys. Lett.* **165**, 257 (1990).
76. A. G. Brook and K. M. Baines, *Adv. Organomet. Chem.* **25**, 1 (1986).
77. N. Wiberg, *J. Organomet. Chem.* **273**, 141 (1984).
78. A. H. Cowley and N. C. Norman, *Prog. Inorg. Chem.* **34**, 1 (1986).
79. G. Raabe and J. Michl, *Chem. Rev.* **85**, 419 (1985).
80. G. Raabe and J. Michl, *in* "The Chemistry of Organic Silicon Compounds" (S. Patai and Z. Rappoport, eds.), p. 1015. Wiley, New York, 1989.
81. R. West, *Angew. Chem., Int. Ed. Engl.* **26**, 1201 (1987).
82. S. Masamune, *in* "Silicon Chemistry" (J. Y. Corey, E. R. Corey, and P. P. Gaspar, eds.), p. 257. Wiley, New York, 1988.
83. J. Barrau, J. Escudie, and J. Satge, *Chem. Rev.* **90**, 283 (1990).
84. L. E. Gusel'nikov and N. S. Nametkin, *Chem. Rev.* **79**, 529 (1979).
85. M. S. Gordon, *in* "Molecular Structure and Energetics" (J. F. Liebman and A. Greenberg, eds.), Vol. 1, p. 101. VCH, Deerfield Beach, Florida, 1986.
86. K. K. Baldridge, J. A. Boatz, S. Koseki, and M. S. Gordon, *Annu. Rev. Phys. Chem.* **38**, 211 (1987).
87. B. T. Luke, J. A. Pople, M.-B. Krogh-Jespersen, Y. Apeloig, J. Chandrasekhar, and P. v. R. Schleyer, *J. Am. Chem. Soc.* **108**, 260 (1986).
88. B. T. Luke, J. A. Pople, M.-B. Krogh-Jespersen, Y. Apeloig, M. Karni, J. Chandrasekhar, and P. v. R. Schleyer, *J. Am. Chem. Soc.* **108**, 270 (1986).
89. S. Nagase, T. Kudo, K. Ito, *in* "Applied Quantum Chemistry" (V. H. Smith, H. F. Schaefer, and K. Morokuma, eds.), p. 249. Reidel, Dordrecht, The Netherlands, 1986.
90. Y. Apeloig, *in* "The Chemistry of Organic Silicon Compounds" (S. Patai and Z. Rappoport, eds.), p. 57. Wiley, New York, 1989.
91. G. Trinquier and J.-P. Malrieu, *J. Am. Chem. Soc.* **109**, 5303 (1987).
92. J.-P. Malrieu and G. Trinquier, *J. Am. Chem. Soc.* **111**, 5916 (1989).
93. G. Trinquier, *J. Am. Chem. Soc.* **112**, 2130 (1990).
94. G. Trinquier and J.-C. Barthelat, *J. Am. Chem. Soc.* **112**, 9121 (1990).
95. G. Trinquier and J.-P. Malrieu, *J. Phys. Chem.* **94**, 6184 (1990).
96. G. Trinquier, *J. Am. Chem. Soc.* **113**, 144 (1991).
97. S. W. Benson, "Thermochemical Kinetics." Wiley (Interscience), New York, 1976.
98. J. B. Pedley, R. D. Naylor, and S. P. Kirby, "Thermochemical Data of Organic Compounds." Chapman & Hall, New York, 1986.
99. R. Walsh, *Acc. Chem. Res.* **14**, 246 (1981).
100. R. Walsh, *J. Chem. Soc., Faraday Trans. 1* **79**, 2233 (1983).
101. R. Walsh, *in* "The Chemistry of Organic Silicon Compounds" (S. Patai and Z. Rappoport, eds.), p. 371. Wiley, New York, 1989.

102. J. M. Kanabus-Kaminska, J. A. Hawari, D. Griller, and C. Chatgilialoglu, *J. Am. Chem. Soc.* **109**, 5267 (1987).
103. B. H. Boo and P. B. Armentrout, *J. Am. Chem. Soc.* **109**, 3549 (1987).
104. J. Berkowitz, J. P. Greene, H. Cho, and B. Ruscic, *J. Chem. Phys.* **86**, 1235 (1987).
105. J. A. Pople, M. Head-Gordon, D. J. Fox, K. Raghavachari, and L. A. Curtiss, *J. Chem. Phys.* **90**, 5622 (**1989**); L. A. Curtiss, C. Jones, G. W. Trucks, K. Raghavachari, and J. A. Pople, *J. Chem. Phys.* **93**, *2537* (1990).
106. P. Ho, M. E. Coltrin, J. S. Binkley, and C. F. Melius, *J. Phys. Chem.* **89**, 4647 (1985).
107. P. Ho, M. E. Coltrin, J. S. Binkley, and C. F. Melius, *J. Phys. Chem.* **90**, 3399 (1986).
108. E. W. Ignacio and H. B. Schlegel, *J. Chem. Phys.* **92**, 5404 (1990).
109. M. B. Coolidge and W. T. Borden, *J. Am. Chem. Soc.* **110**, 2298 (1988).
110. M. J. Almond, A. M. Doncaster, P. N. Noble, and R. Walsh, *J. Am. Chem. Soc.* **104**, 4717 (1982).
111. P. N. Noble and R. Walsh, *Int. J. Chem. Kinet.* **15**, 547 (1983).
112. B. Ruscic, M. Schwarz, and J. Berkowitz, *J. Chem. Phys.* **92**, 1865 (1990); B. Ruscic, M. Schwarz, and J. Berkowitz, *J. Chem. Phys.* **92**, 6338 (1990).
113. R. C. Binning and L. A. Curtiss, *J. Chem. Phys.* **92**, 1860 (1990).
114. K. Balasubramanian and A. D. McLean, *J. Chem. Phys.* **85**, 5117 (1986).
115. W. D. Allen and H. F. Schaefer, *Chem. Phys.* **108**, 243 (1986).
116. C. W. Bauschlicher, Jr., and P. R. Taylor, *J. Chem. Phys.* **86**, 1420 (1987).
117. C. W. Bauschlicher, Jr., S. R. Langhoff, and P. R. Taylor, *J. Chem. Phys.* **87**, 387 (1987).
118. M. E. Colvin, R. S. Grev, H. F. Schaefer, and J. Bicerano, *Chem. Phys. Lett.* **99**, 399 (1983).
119. M. E. Colvin, J. Breulet, and H. F. Schaefer, *Tetrahedron* **41**, 1429 (1985).
120. M. E. Colvin, H. F. Schaefer, and J. Bicerano, *J. Chem. Phys.* **83**, 4581 (1985).
121. M. S. Gordon and M. W. Schmidt, *Chem. Phys. Lett.* **132**, 294 (1986).
122. M. S. Gordon, *Chem. Phys. Lett.* **114**, 348 (1985).
123. J. E. Rice and N. C. Handy, *Chem. Phys. Lett.* **107**, 365 (1984).
124. R. S. Grev and H. F. Schaefer, *J. Am. Chem. Soc.* **108**, 5804 (1986).
125. M. S. Gordon and D. Bartol, *J. Am. Chem. Soc.* **109**, 5948 (1987).
126. S. K. Shin, W. A. Goddard, and J. L. Beauchamp, *J. Chem. Phys.* **93**, 4986 (1990).
127. K. Balasubramanian, *J. Chem. Phys.* **89**, 5731 (1988).
128. L. G. M. Petterson and J. Schule, *J. Mol. Struct. (THEOCHEM)* **208**, 137 (1990).
129. F. K. Cartledge and R. V. Piccione, *Organometallics* **3**, 299 (1984).
130. J. Kalcher, *Chem. Phys.* **118**, 273 (1987).
131. A. C. Hopkinson, C. F. Rodriguez, and M. H. Lien, *Can. J. Chem.* **68**, 1309 (1990).
132. F. A. Cotton and G. Wilkinson, "Advanced Inorganic Chemistry," 5th Ed., p. 267. Wiley, New York, 1988.
133. J. E. Huheey, "Inorganic Chemistry," 3rd Ed, p. 843. Harper & Row, New York, 1983.
134. K. M. Ervin, S. Gronert, S. E. Barlow, M. K. Gilles, A. G. Harrison, V. M. Bierbaum, C. H. DePuy, W. C. Lineberger, and G. B. Ellison, *J. Am. Chem. Soc.* **112**, 5750 (1990).
135. S. W. Benson, *Int. J. Chem. Kinet.* **21**, 233 (1989).
136. C. J. Wu and E. A. Carter, *J. Am. Chem. Soc.* **112**, 5893 (1990).
137. M. M. Gallo, T. P. Hamilton, and H. F. Schaefer, *J. Am. Chem. Soc.* **112**, 8714 (1990).
138. C. W. Bauschlicher, Jr., and S. R. Langhoff, *Chem. Phys. Lett.* **177**, 133 (1991).
139. S. G. Lias, J. E. Bartmess, J. F. Liebman, J. L. Holmes, R. D. Levin, and W. G. Mallard, *J. Phys. Chem. Ref. Data* **17**(Suppl. 1) (1988).
140. R. A. Bair and T. H. Dunning, *J. Chem. Phys.* **82**, 2280 (1985).

141. M. W. Schmidt, P. N. Truong, and M. S. Gordon, *J. Am. Chem. Soc.* **109**, 5217 (1987).
142. K. D. Dobbs and W. J. Hehre, *Organometallics* **5**, 2057 (1986).
143. P. v. R. Schleyer and D. Kost, *J. Am. Chem. Soc.* **110**, 2105 (1988).
144. J. Moc, Z. Latajka, and H. Ratajczak, *J. Mol Struct. (THEOCHEM)* **150**, 189 (1987).
145. W. Cherry, N. Epiotis, and W. T. Borden, *Acc. Chem. Res.* **10**, 167 (1977).
146. H. A. Bent, *Chem. Rev.* **61**, 275 (1961).
147. D. C. Comeau, I. Shavitt, P. Jensen, P. R. Bunker, *J. Chem. Phys.* **90**, 6491 (1989).
148. D. G. Leopold, K. K. Murray, A. E. Stevens Miller, and W. C. Lineberger, *J. Chem. Phys.* **83**, 4849 (1985).
149. W. Kutzelnigg, *Angew. Chem., Int. Ed. Engl.* **23**, 272 (1984).
150. R. Walsh, *J. Phys. Chem.* **90**, 389 (1986).
151. R. Becerra and R. Walsh, *J. Phys. Chem.* **91**, 5765 (1987).
152. R. T. White, R. L. Espino-Rios, D. S. Rogers, M. A. Ring, and H. E. O'Neal, *Int. J. Chem. Kinet.* **17**, 1029 (1985).
153. J. Sandstrom, *Topics Stereochem.* **14**, 83 (1983).
154. R. S. Grev and H. F. Schaefer, to be submitted.
155. T. Fjeldberg, A. Haaland, B. E. R. Schilling, M. F. Lappert, and A. J. Thorne, *J. Chem. Soc., Dalton Trans.*, 1551 (1986).
156. T. Fjeldberg, A. Haaland, B. E. R. Schilling, H. V. Volden, M. F. Lappert, and A. J. Thorne, *J. Organomet. Chem.* **280**, C43 (1985).
157. H. Takeo and R. F. Curl, *J. Mol. Spectrosc.* **43**, 21 (1972).
158. G. Schultz, J. Tremmel, I. Hargittai, I. Berecz, S. Bohatka, N. D. Kagramanov, A. K. Maltsev, and O. M. Nefedov, *J. Mol. Struct.* **55**, 207 (1979).
159. M. F. Lappert, M. J. Slade, J. L. Atwood, and M. J. Zaworotko, *J. Chem. Soc., Chem. Commun.*, 621 (1980).
160. B. Cetinkaya, I. Gumrukcu, M. F. Lappert, J. L. Atwood, R. D. Rogers, and M. J. Zaworotko, *J. Am. Chem. Soc.* **102**, 2088 (1980).
161. T. Fjeldberg, P. B. Hitchcock, M. F. Lappert, S. J. Smith, and A. J. Thorne, *J. Chem. Soc., Chem. Commun.*, 939 (1985).
162. T. Fjeldberg, H. Hope, M. F. Lappert, P. P. Power, and A. J. Thorne, *J. Chem. Soc. Chem. Commun.*, 639 (1983).
163. E. A. Carter and W. A. Goddard, *J. Phys. Chem.* **90**, 998 (1986).
164. E. A. Carter and W. A. Goddard, *J. Am. Chem. Soc.* **110**, 4077 (1988).
165. E. A. Carter and W. A. Goddard, *J. Chem. Phys.* **88**, 1752 (1988).
166. J. A. Paulino and R. R. Squires, *J. Am. Chem. Soc.* **113**, 5573 (1991).
167. A. Sekiguchi, S. S. Zigler, R. West, and J. Michl, *J. Am. Chem. Soc.* **108**, 4241 (1986).
168. A. Sekiguchi, G. R. Gillette, and R. West, *Organometallics* **7**, 1226 (1988).
169. A. Sekiguchi, S. S. Zigler, K. J. Haller, and R. West, *Recl. Trav. Chim. Pays-Bas.* **107**, 197 (1988).
170. M. Bogey, H. Bolvin, C. Demuynck, and J. L. Destombes, *Phys. Rev. Lett.* **66**, 413 (1991).

ADVANCES IN ORGANOMETALLIC CHEMISTRY, VOL. 33

Organotin Heterocycles

KIERAN C. MOLLOY

School of Chemistry
University of Bath
Bath BA2 7AY, England

I

INTRODUCTION

Cyclic compounds offer a chemical diversity, in terms of both structure and reactivity, not shown by their acyclic counterparts. As ring size increases the stability of the system changes from low, owing to internal ring strain, to high, where the internal angles of the polygon match the valence angles of the participating atoms, finally diminishing again as acyclic arrays compete favorably with the formation of larger rings. The introduction of heteroatoms into the ring induces polarity, and hence chemical reactivity, and if these atoms have expandable coordination spheres, then structural variations in the whole heterocycle are enhanced accordingly. Organotin heterocycles are a class of such compounds so enriched with chemical diversity, and they have been the center of a considerable research effort over the last 30 years. It is now appropriate to draw together the themes that have emerged from that collective effort, and such is the intention of this article.

It is, however, incumbent on the reviewer both to delineate the major themes in our current state of knowledge and to do so with a brevity that is consistent with the specialist nature of the topic. Thus, of the many heterocyclic rings which occur in organotin compounds, those which arise from interactions with bidentate, chelating ligands have been omitted, save in circumstances of exceptional significance. Also omitted are the stannacarboranes (1) and compounds whose only heterocyclic ring is either the ubiquitous stannoxane system [1, X = O (2)] or a halogen-bridged equivalent [1, X = halogen (3–5)]. No attempt has been made to review the

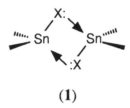

(1)

literature exhaustively. Emphasis has been placed on reports made after about 1975, although benchmark results predating this have been included. Texts covering organotin chemistry are available to readers wishing to place the reviewed topic into the wider context of organotin chemistry (6–8).

II

HETEROCYCLES CONTAINING TIN–GROUP 14 ELEMENT BONDS

The work on stannacycloalkanes and -cycloalkenes up to 1972 has been reviewed (9), and again, though in less detail, in 1982 (10). Most of the early studies concern the formation of five- and six-membered rings, and claims for smaller systems in particular should be treated with caution. More recent efforts have been directed to the synthesis of both strained and expanded rings. The synthetic methodology, however, remains dominated by the use of difunctional carbanion sources (Grignard and lithium reagents) or the hydrostannation reactions of tin dihydrides.

A. Saturated Heterocycles

1. Rings Containing One Tin Atom

1-Stannacyclopropanes are unknown, though the homocyclic 1,2,3-tristannacyclopropane has been synthesized (see Section II,D). The smallest stannacycloalkane contains a four-membered ring, but the volatility of the parent (2, R = H) has prevented isolation. The ring-substituted analog (2, R = Me), which can be distilled from high boiling ethers used as reaction solvent, has been characterized, though the yield is low (5%) (11). Ring-expanded oligomers of the parent (3–5) dominate in this and many analogous reactions [Eq. (1)]. For example, both 6 and 7 are formed in the

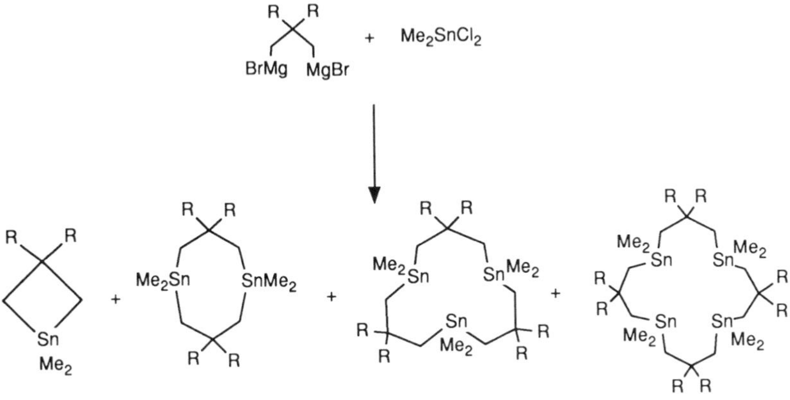

(1)

(2)   (3)   (4)   (5)

reaction of Ph_2SnCl_2 and $BrMg(CH_2)_4MgBr$, and both have been characterized crystallographically (12,13). The five-membered ring is "twisted" and of C_2 symmetry (half-chair), with an endocyclic angle at tin of approximately 95°. 1,6-Distannacyclodecane (7) has a "boat–chair–boat" arrangement like cyclodecane itself. Cycloalkanes containing only one tin

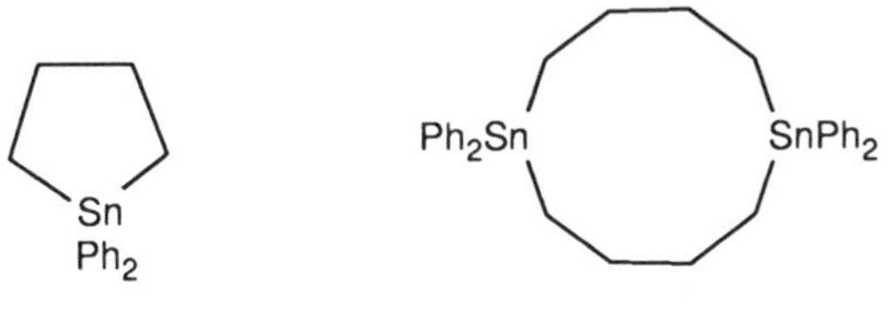

(6)   (7)

have also been prepared by pyrolysis of α,ω-bis(trimethylstannyl)alkanes [Eq. (2)] (14).

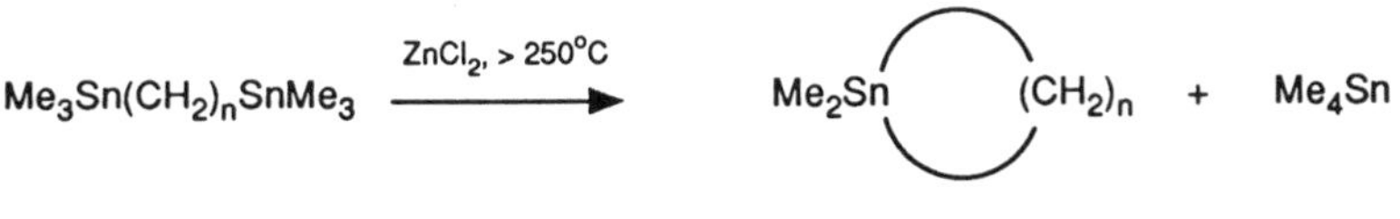

$$Me_3Sn(CH_2)_nSnMe_3 \xrightarrow{ZnCl_2, > 250°C} Me_2Sn(CH_2)_n + Me_4Sn \quad (2)$$

Rings of different sizes are readily distinguished by NMR methods, as both the ^{119}Sn chemical shift and $^2J(^{119}$Sn$-$C$-^1$H) values (to both endo- and exocyclic substituents) are sensitive to the bond angles at tin $(15,16)$. Typically, **6** and **7** have $\delta\ ^{119}$Sn values of 0 and -74 ppm, respectively (12). Further NMR data can be found in the compendia of data now available $(17-19)$.

In general, the endocyclic Sn$-$C bonds are more reactive than the exocyclic ones toward both electrophiles and nucleophiles, though as might be expected the situation is more competitive when the exocyclic bond is Sn$-$Ph (20). Ring expansion reactions provide a convenient route to oxa- and thiastannolanes [Eq. (3)] (21), whereas thermolysis in metha-

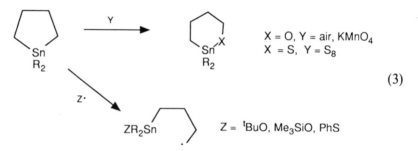

$$X = O, \ Y = \text{air, KMnO}_4$$
$$X = S, \ Y = S_8$$

$$Z = \ ^t\text{BuO, Me}_3\text{SiO, PhS}$$

$$(3)$$

nol yields higher oligomers containing up to about 7 tin atoms (22). Radicals react at the tin to yield new, ring-opened radicals, rather than with the hydrogens of α- or β-methylene groups as is the case with acyclic stannaalkanes (23).

2. Rings Containing More Than One Tin Atom

1,3-Distannacyclobutane (**8**), stabilized by bulky Me$_3$Si substitution in the 2,4-positions, has been synthesized and characterized in part by its ring cleavage reactions with nucleophiles [Eq. (4)] $(24,25)$. Surprisingly, given

$$(\text{Me}_3\text{Si})_2\text{CBr}_2 \xrightarrow{\text{BuLi}} (\text{Me}_3\text{Si})_2\text{CBrLi} \xrightarrow{\text{Me}_2\text{SnCl}_2}$$

$$(4)$$

$$(8)$$

the ring strain inherent in the molecule, brominolysis in methanol cleaves the exocyclic Sn$-$Me bond, presumably because the bulky ring substitu- ents prevent attack of the solvated bromide ion at the endocyclic bonds.

Cyclic oligomers of general formula $(Me_2SnCH_2)_n$ ($n = 3, 4$) have been synthesized using $CH_2(MgBr)_2$ and Me_2SnCl_2 (26), or by coupling $Me_2(I)SnCH_2I$ using magnesium (27). More unusually (28),

$$ (5) $$

Elegant synthetic routes to larger polystannacycloalkanes are now available and are shown in Eq. (6) (29,30) and Scheme 1 (31,32):

$$ Me_2(Na)Sn(CH_2)_nSn(Na)Me_2 \ + \ Me_2Sn[(CH_2)_nCl]_2 \xrightarrow[-78°C]{THF, NH_3(l)} \qquad (6) $$

The structure of $PhSn[(CH_2)_8]_3SnPh$ has been determined, and the Sn–Sn separation is 845 pm (33). Lewis acidity can be induced into these rings and cages by introducing halogen atoms onto the metal. For example, 3 reacts with 3 or 6 equivalents of $HgCl_2$ to form cis-1,5,9-trichloro- or 1,1,5,5,9,9-hexachlorotristannacyclodecane, respectively (29). More extensive work by Newcombe et al. has shown that stannacycloalkanes can act as anion hosts, with δ ^{119}Sn values reflecting the change in coordination number at tin with increasing halide ion concentration. Complexation is also selective depending on the size of the macrocycle cavity (34,35). Structural evidence for the environment of the complexed anions comes from two crystal structures, showing that F$^-$ lies between two tin atoms in $ClSn[(CH_2)_6]_3SnCl$ (Sn–F 212, 228 pm), whereas Cl$^-$ is similarly, but more asymmetrically, linked in $ClSn[(CH_2)_8]_3SnCl$ [Sn–Cl$^-$ 274.5, 338.8 pm] (Fig. 1) (36). In solution, Cl$^-$ jumps between tin atoms, for which the ΔE_{act} is 5.3 kcal/mol (36).

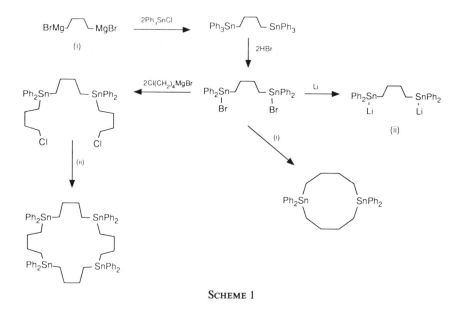

Several 1,2,4,5-tetrastannacyclohexanes are now known [Eq. (7)] (*37,38*), and the available crystallographic data (*39–41*) suggest that it is the substituents on tin, rather than carbon, which determine the conformation of the heterocycle. Methyl groups on tin induce a boat form, whereas the octaphenyl derivative has a chair structure. Interestingly, the solid-state ^{119}Sn-NMR spectrum of **10** (R = R' = Me) consists of four

$$\underset{\textbf{(10)}}{\begin{array}{c} R'\ R' \\ \diagdown\!\!\diagup \\ R_2Sn \qquad SnR_2 \\ |\qquad\qquad | \\ Br \qquad Br \end{array} \xrightarrow{\text{Na, NH}_3(l)} \begin{array}{c} R'\ R' \\ \diagdown\!\!\diagup \\ R_2Sn \qquad SnR_2 \\ |\qquad\qquad | \\ R_2Sn \qquad SnR_2 \\ \diagup\!\!\diagdown \\ R'\ R' \end{array}} \qquad (7)$$

distinct tin environments and two different ^1J couplings arising from two different Sn—Sn bonds in the molecule (278.0, 279.1 pm) (*39*). Compound **10** reacts with alkynes and allenes to generate a variety of new heterocycles [Eq. (8)] (*42*).

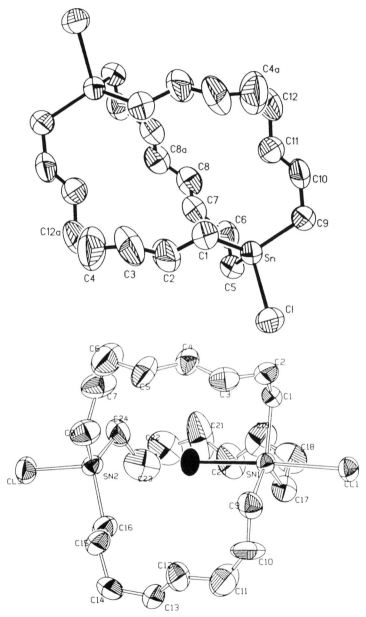

FIG. 1. Structures of (top) ClSn[(CH$_2$)$_8$]$_3$SnCl and (bottom) its complex with Cl$^-$. The coordinated anion has been highlighted. (Reproduced from Ref. *36* with permission of the American Chemical Society.)

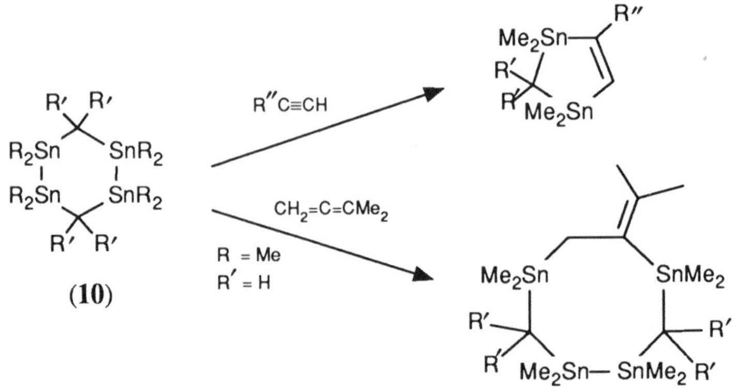

$$(8)$$

B. Unsaturated Heterocycles

The reaction between stannylenes and carbon–carbon multiple bonds generates unsaturated heterocycles containing tin and between two and four other atoms (Scheme 2). Formation of five-membered rings from [4 + 2] addition reactions are by far the most common (43). With the isolated, strained alkynes 11 and 12, both 1-stannacyclopropene (13) and

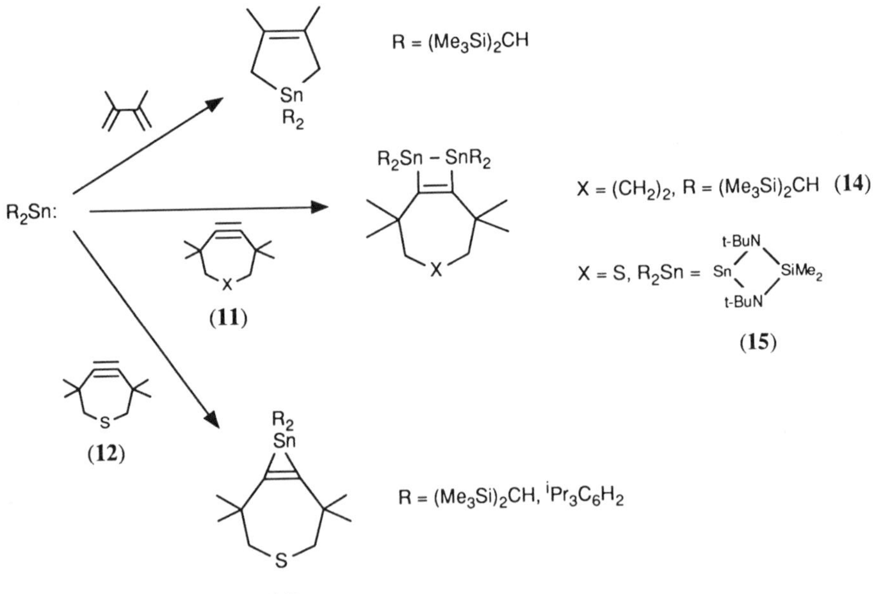

SCHEME 2

1,2-distannacyclobut-3-enes (**14, 15**) have been isolated, though in solution an equilibrium exists between product and reactants. Compound **13** is characterized by a δ [119]Sn value of -536.8 ppm, some 100 ppm upfield from the corresponding tristannapropane (Section II,D), and in the solid state by Sn—C bond lengths of 213.4 and 213.6 pm. The internal angle at tin is 36.6° (*44*). The mechanism of formation of **14** and **15** is believed to be by further stannylene insertion into the Sn—C bond of a stannacyclopropene (the kinetically favored product) to produce a less strained four-membered ring. However, a [2 + 2] reaction of stannylene dimer with alkyne is also possible for **14** (but not for the monomeric diaminostannylene used to synthesize **15**), and this is the proposed mechanism of formation of a phosphadistannacyclobutene (see Section III,B). Both **14** and **15** exhibit positive [119]Sn chemical shifts and large $^1J(^{119}\text{Sn}-^{117}\text{Sn})$ couplings [56.5 ppm, 2436 Hz (*45*) and 155 ppm, 3723 Hz (*46*), respectively]. The four-membered rings are essentially planar, and their reactivity is dominated by the Sn–Sn feature.

1-Stannacyclopentadienes (**16, 17**) can be prepared either by organoboration of bis(alkynyl)diorganostannanes [Eq. (9)] (*47,48*) or from 1,4-dilithiobutadienes and R_2SnX_2 (X = halogen) [Eq. (10)], though the latter is

$$R_2Sn(C\equiv CR')_2 \;+\; BR''_3 \;\longrightarrow\; \text{(16)} \tag{9}$$

(16)

$$R_2SnX_2 \;+\; \text{(butadiene)} \;\longrightarrow\; \text{(17)} \tag{10}$$

(17)

more common with 1,2,3,4-substituted butadienes (*49–51*). The chemical reactivity of these stannoles is summarized in Scheme 3 (*52–58*). Of particular note is their usefulness as precursors for other 1-metallacyclopentadienes (*52–54*) and, via 7-stannanorbornanes, reactive stannylenes R_2Sn (*55,56*).

Hydrostannation of bis-alkynes is most commonly used to prepare 1-stannacyclohexa-2,5-dienes (**18**) (*59–63*), though both 1-stannacyclopenta-2,4-dienes (*64*) and 1-stannacyclohepta-2,4,6-trienes (*65*) have been

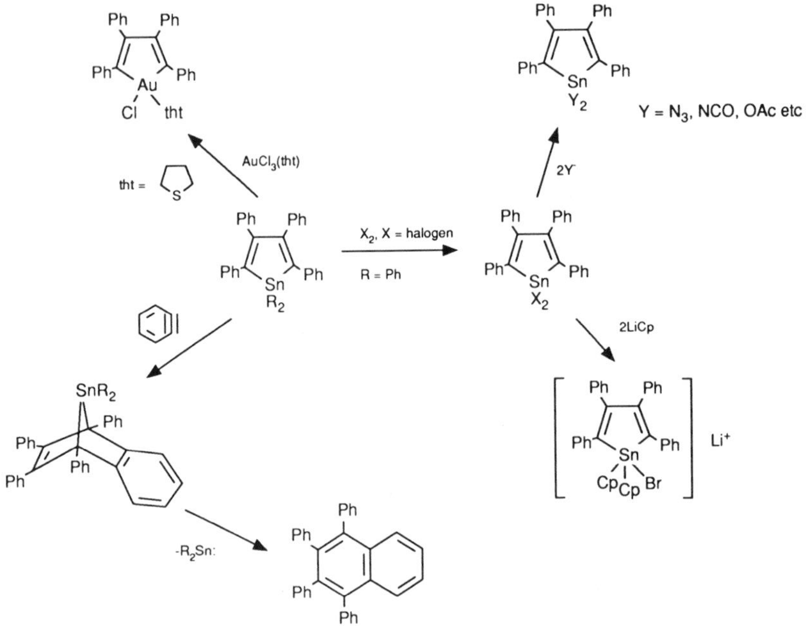

SCHEME 3

synthesized by this method. A wide variety of ring substituents have been
included in the final products [Eq. (11)]. The C-4 protons of **18** are quite

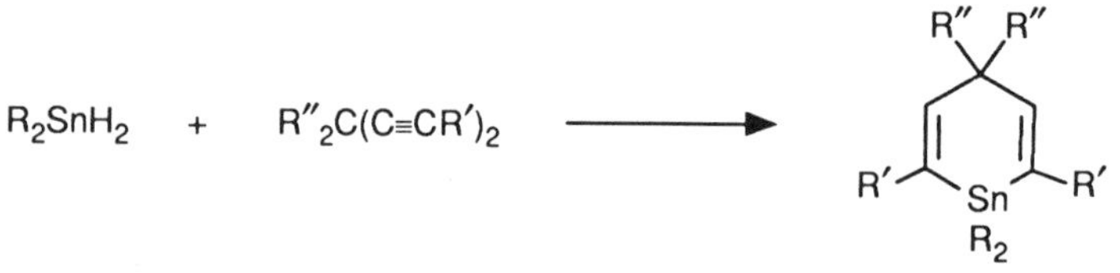

$$R_2SnH_2 \quad + \quad R''_2C(C\equiv CR')_2 \quad \longrightarrow \quad \text{(18)} \quad (11)$$

(18)

acidic and can be removed with base [typically lithium diisopropylamide
(LDA)]. This allows modification of the substituents at C-4 (*66,67*),
though in some cases the double bonds migrate to the 2,4-positions
(*67,68*). Transmetallation with BuLi affords 1,5-dilithio-1,4-dienes and
Bu$_4$Sn (*69*), from which other 1-metallacyclohexa-2,5-dienes can be prepared (*70*). Alternatively, transmetallation with BBr$_3$ (*68*), MeBBr$_2$ (*65*),
PX$_3$ (*59,62*), AsX$_3$ (*59–66,71*), and PhSbCl$_2$ (*64*) introduces B, P, As, and
Sb directly into the heterocycle in place of tin. By way of contrast, **19**
photochemically extrudes tin to give the metal-free carbacycle perylene
(**20**) in 60% yield [Eq. (12)] (*72*).

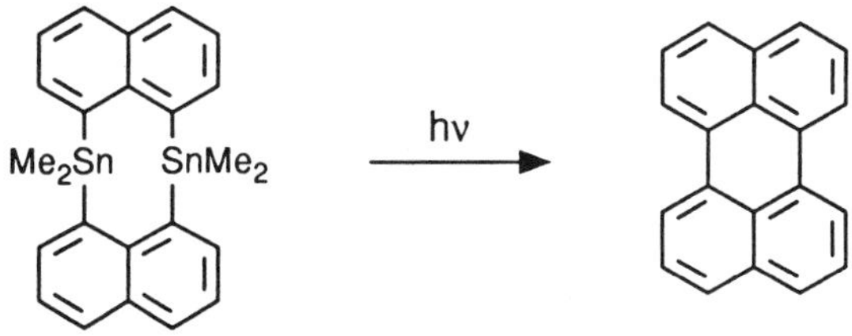

(19)   (20)

    One other unsaturated tin heterocycle which deserves mention is shown in Eq. (13). Of interest is the fact that the reaction fails to generate a compound containing a four-membered ring (21) as is found in the corresponding reaction for Si and Ge (73).

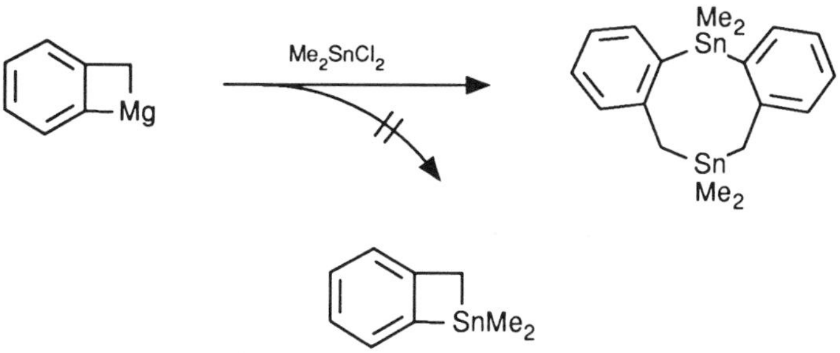

(21)

C. Rings Containing Other Heteroatoms

    Variations on the methodology already described can be used to introduce additional heteroatoms into the ring, namely, the use of dilithio (74) or di-Grignard reagents (75,76), organoboration (77,78), and hydrostannation (79–81), some examples of which are shown in Eqs. (14)–(17).

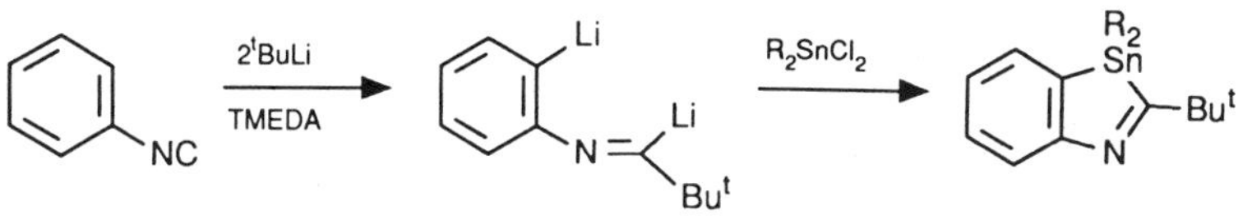

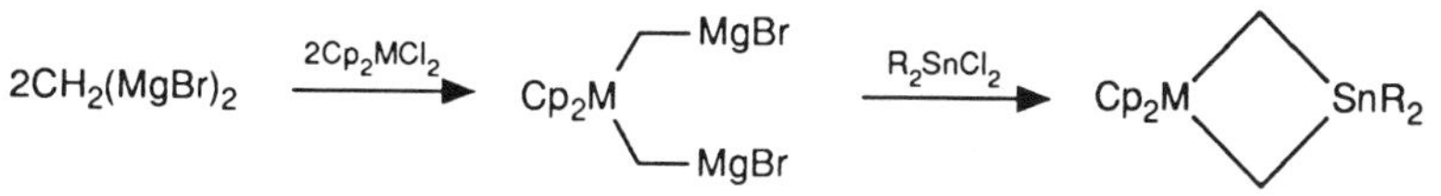

M = Ti, Zr, Hf

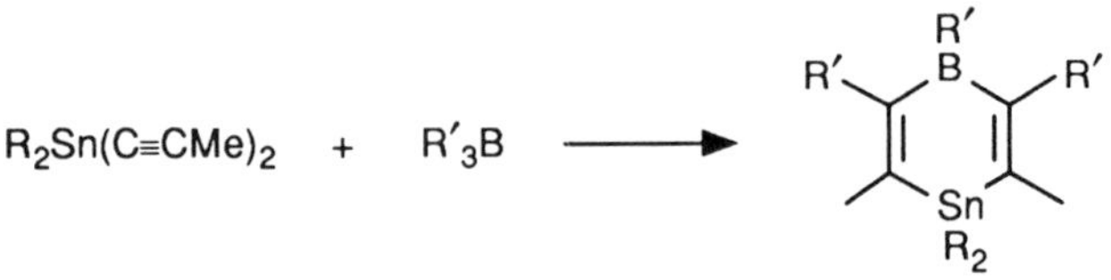

$$R_2Sn(C\equiv CMe)_2 \quad + \quad R'_3B \quad \longrightarrow \quad \text{(structure)} \tag{16}$$

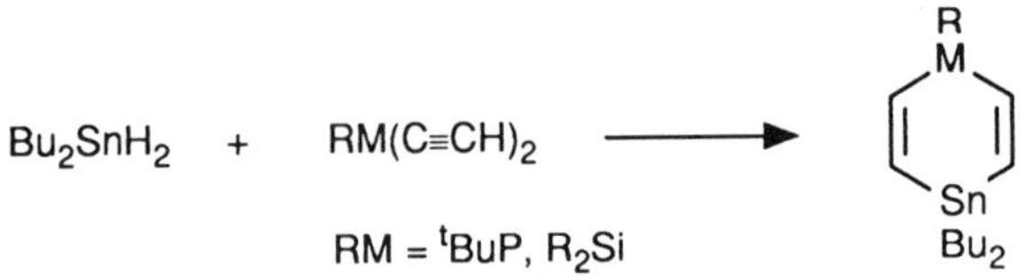

$$Bu_2SnH_2 \quad + \quad RM(C\equiv CH)_2 \quad \longrightarrow \quad \text{(structure)} \tag{17}$$

$$RM = {}^tBuP, R_2Si$$

1-Stanna-4-sila-2,5-hexadienes can alternatively be synthesised via an assumed 1-stanna-2-sila-cyclobut-3-ene intermediate (22) which can be trapped by phenylacetylene [Eq. (18)] (82).

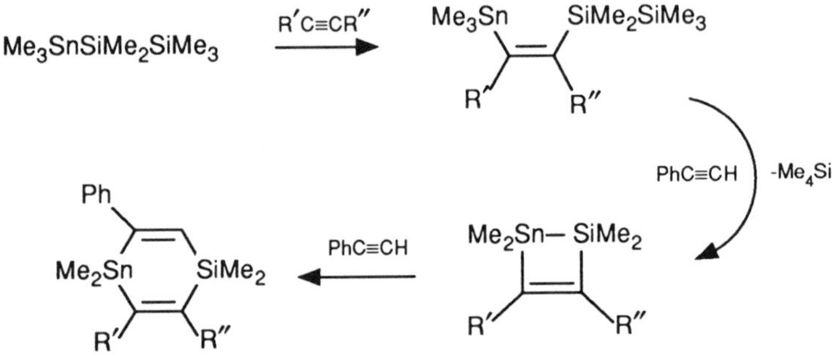

$$Me_3SnSiMe_2SiMe_3 \quad \xrightarrow{R'C\equiv CR''} \quad \text{(structure)} \tag{18}$$

(22)

1-Stanna-2-boraalkenes (23) are intermediates in the formation of unusual boron-containing heterocycles (Scheme 4) (83,84). Compound 24 contains an annelated 6-membered ring formed via P: → B donation (83). Finally, several 9-substituted 10-stanna-9,10-dihydroanthracenes (25) have been prepared, including X = O (85,86), S (87), NR (88,89), and Me$_2$Si (90).

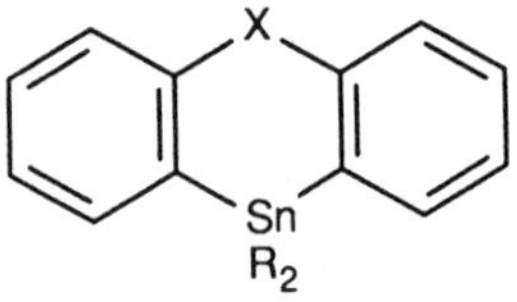

$$\text{(structure)} \qquad X = O, S, NR, SiMe_2$$

(25)

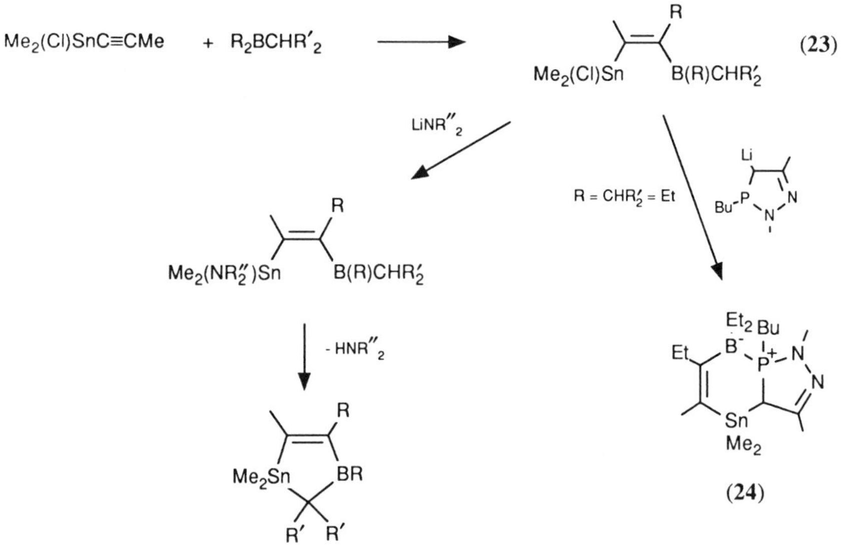

$$Me_2(Cl)SnC\equiv CMe \quad + \quad R_2BCHR'_2 \quad \longrightarrow \quad (23)$$

SCHEME 4

D. Homocyclic Polytin Compounds

Studies in the field of tin-substituted cycloalkanes stem in part from a desire to understand the modifications, in terms of conformation and reactivity, brought to the carbacycle by the inclusion of increasing numbers of heteroatoms. Although not strictly heterocycles, cyclic polystannanes allow this line of investigation to be taken to its extreme conclusion, and as such they are included in this article for comparison. The synthetic routes to cyclostannanes are summarized in Eqs. (19)–(23), with details in Table I (91–103).

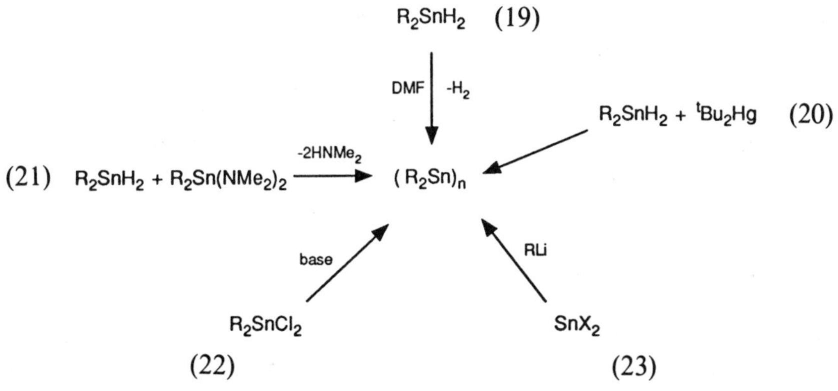

$$R_2SnH_2 \quad (19)$$

$$R_2SnH_2 + {}^tBu_2Hg \quad (20)$$

$$(21) \quad R_2SnH_2 + R_2Sn(NMe_2)_2 \xrightarrow{-2HNMe_2} (R_2Sn)_n$$

$$(22) \quad R_2SnCl_2$$

$$SnX_2 \quad (23)$$

TABLE I

SYNTHETIC ROUTES TO CYCLOSTANNANES, $(R_2Sn)_n$

Reaction	Base	R	n	Ref.
Eq. (19)	—	Ph	5, 6	*91*
		$PhCH_2$	4	*92*
		$PhCH_2$	6	*93*
Eq. (20)	—	Cy	5	*94*
		$Et,^{n,i}Bu,Ph$	6	*94*
		Et	7	*94*
Eq. (21)	—	CH_3, CD_3	6	*95*
		$(Me_3Si)_2CH$	4	*96*
Eq. (22)	$NaC_{10}H_8$	9-Phenanthryl	3	*97*
	$LiC_{10}H_8$	$2,6\text{-}Et_2C_6H_3$	3	*98*
	$LiC_{10}H_8$	$2,4,6\text{-}{}^iPr_3C_6H_2$	3	*99*
	${}^tBuMgCl$	${}^tBu$	4	*100,101*
	${}^tBuMgCl$	${}^tAm^a$	4	*101*
	Li	Ph	6	*93*
Eq. (23)^b	—	$2,6\text{-}Et_2C_6H_3$	6^c	*103*
		$2,6\text{-}Et_2C_6H_3$	3	*103*
Eq. (23)^d	—	$1,2\text{-}L_2C_6H_4{}^e$	4	*102*

^{a} ${}^tAm = C_2H_5(CH_3)_2C.$
^{b} $X = Cl.$
^{c} Bicyclo[2.2.0].
^{d} $X = 4\text{-}Me\text{-}2,6\text{-}{}^tBu_2C_6H_2.$
^{e} $L = CH_2SiMe_3$; product is RSn.

1,2,3-Tristannapropanes are air stable in the solid but are converted to the corresponding oxides $(R_2SnO)_3$ on standing in solution (*98*). Low temperature photolysis leads cleanly to the distannanes $R_2Sn=SnR_2$. Whereas the cyclotristannane is thermodynamically stable at room temperature, it is in rapid equilibrium with the distannane at elevated temperatures, with the distannane occurring in approximately 2-fold excess at 90°C (*99*). The related bicyclo[2.2.0]hexastannane (Fig. 2) has remarkable thermochromic properties, being pale yellow at −196°C and orange-red at room temperature. Moreover, the UV spectra of this compound (309, 360 nm) can be attributed to σ delocalization over the polytin framework (*103*). Given the important and commercially utilizable electronic properties of polysilanes which are now being harnessed, there is considerable potential for new polystannanes to also show interesting electronic phenomena. The chemical properties of the higher oligomers are dominated by the reactivity of the Sn—Sn bonds, and they can be the starting point for the synthesis of other heterocycles (see Section IV,A) or acyclic polystannanes by ring opening with molecular halogens (*104*).

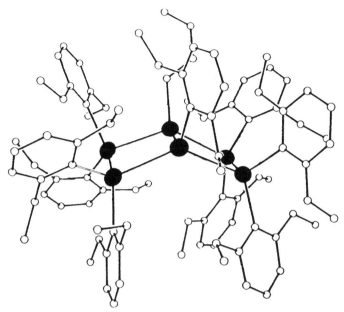

FIG. 2. The structure of bicyclo[2.2.0]hexastannane. Tin atoms are highlighted. (Reproduced from Ref. *103* with permission of the American Chemical Society.)

Selected spectroscopic and crystallographic data for a range of cyclostannanes are given in Table II (*105,106*). It is clear that the ^{119}Sn chemical shifts and couplings for the three-membered ring systems are unique. Sn—Sn bond lengths and molecular conformations of these cyclostannanes are dependent on the nature of the peripheral organic groups, which are often bulky to enhance the stability of the system. Four-membered rings are generally planar (*96,101,102*); however, both (^{t}Am$_2$Sn)$_4$ (*101*) and the fused, bicyclic hexastannane (*103*) (Fig. 2) contain puckered rings. Several cyclohexastannanes have been examined crystallographically (*93,104,105*), usually differing in the degree of solvation by crystallization solvents. The structures are always that of a chair conformation, though the torsion angles within the chair are surprisingly dependent on the nature of the occluded solvent molecules (*93*).

III

HETEROCYCLES CONTAINING TIN–GROUP 15 ELEMENT BONDS

The reader is encouraged to consult the excellent review by Veith concerning cyclic tin(II)–nitrogen compounds (*107*), which provides a complement to heterocycles containing organotin(IV).

TABLE II

Structural Data for Cyclostannanes, $(R_2Sn)_n$

R	n	δSn (ppm)	J(Sn-Sn) (Hz)	Sn-Sn (pm)	<SnSnSn (°)	Conformation	Ref.
2,6-Et$_2$C$_6$H$_3$	3	-416	2285	287.0, 285.6, 285.4	59.8, 59.7, 60.3	Planar	98
2,4,6-^{i}Pr$_3$C$_6$H$_2$	3	-379	3017				99
^{t}Bu	4	89		288.7	89.87, 90.13	Planar	101
^{t}Am	4	101	1339	292.3, 291.8, 297.0, 291.4	89.3, 88.8, 89.2, 89.0	Puckered	101
1,2-L$_2$C$_6$H$_4$[a]	4	-77[c]	624	285.2	88.21	$\bar{4}$[b]	102
(Me$_3$Si)$_2$CH	4	-231[c]	941				
Me	6	-208[c]	1067	282.9, 283.4	89.96, 90.04	Planar	96
Ph	6			277.3-279.1[d], 277, 278	108.30-121.61, 109, 114	Chair	93,105
PhCH$_2$	6	-140[c]	769	279.2-281.1[e]	105.54-112.89	Chair	106
2,6-Et$_2$C$_6$H$_3$	6[f]			281.8-293.1	87.5, 92.9	Fused, puckered rings[g]	103

[a] L = CHSiMe$_3$.
[b] Fold angle 20.8°.
[c] Data from Ref. 101.
[d] Range covers data for Ph$_{12}$Sn$_6$, Ph$_{12}$Sn$_6$·2Tol, and Ph$_{12}$Sn$_6$·2Xyl (Tol = toluene; Xyl = xylene).
[e] (PhCH$_2$)$_{12}$Sn$_6$·DMF.
[f] Bicyclo[2.2.0].
[g] Fusion angle = 131.9°.

A. *Tin–Nitrogen Heterocycles*

The early syntheses of cyclic tin–nitrogen compounds were reviewed by Jones and Lappert in 1966 (*108*). The two most common preparative routes are transamination reactions (*109*) and reactions between R_2SnCl_2 and an alkali metal amide (*110*). Some examples are shown in Eqs. (24)–(26). Dilithio reagents analogous to **28** can react with other metal dihalides

$$^tBu_2Sn(NMe_2)_2 \ + \ 2RNH_2 \ \xrightarrow[R = Me,\ PhCH_2]{-2HNMe_2} \ ^tBu_2Sn(NHR)_2 \ \xrightarrow[100\text{-}130°C]{-2RNH_2} \ ^tBu_2Sn \overset{\overset{\displaystyle R}{\underset{\displaystyle N}{|}}}{\underset{\overset{\displaystyle N}{\underset{\displaystyle R}{|}}}{}} SnBu_2{}^t \quad (24)$$

$$(26)$$

$$3\,^tBu_2SnCl_2 \ + \ KNH_2 \ \longrightarrow \ \begin{array}{c} ^tBu_2Sn \\ HN \qquad NH \\ | \qquad\qquad | \\ ^tBu_2Sn \qquad SnBu_2{}^t \\ N \\ H \end{array} \ + \ 3KCl \ + \ 3HCl \quad (25)$$

$$(27)$$

$$^tBu_2SnCl_2 \ + \ ^tBu_2Sn[N(Li)Bu^t]_2 \ \longrightarrow \ ^tBu_2Sn \overset{\overset{\displaystyle ^tBu}{\underset{\displaystyle N}{}}}{\underset{\overset{\displaystyle N}{\underset{\displaystyle ^tBu}{}}}{}} SnBu_2{}^t \ + \ 2LiCl \quad (26)$$

$$\qquad\qquad (28) \qquad\qquad\qquad\qquad\qquad (29)$$

to produce a variety of 1,3,2,4-diazametallastannetidines, incorporating Ge(IV), Se(IV), Ti(IV), Zr(IV) (*111*), Si(IV) (*112*), and Sn(II) (*113*). Known synthetic routes to more asymmetric tin–nitrogen heterocycles, that is, including additional heteroatoms in the ring, are more diverse, and in general they do not have the versatility of the synthetic methods already described. Four examples are given in Eqs. (27)–(30) (*114–117*).

$$\left. \begin{array}{c} R_2Sn(NMe_2)_2 \\ \text{or} \\ R_2Sn(OMe)_2 \end{array} \right\} \ + \ HO\frown N(H)Me \ \xrightarrow[\text{or -2MeOH}]{-2HNMe_2} \ \overset{O\frown NMe}{\underset{\underset{\displaystyle R_2}{Sn}}{\diagdown\diagup}} \quad (27)$$

Redistribution of groups between $(Bu_2SnS)_3$ and $(Bu_2SnNEt)_3$ yields six-membered rings containing Sn, S, and N [Eq. (31)], which can be distinguished by ^{119}Sn NMR (*118*). $Me_2SnSNSN$ has been synthesized by two routes [Eq. (32)] (*119,120*). The compound is dimeric through bridg-

$$(28)$$

$$[(Me_3Si)RN]_2Sn \xrightarrow{\Delta} Me_2Si \underset{N}{\overset{R}{\bigsqcup}} Sn \underset{N}{\overset{}{\bigsqcup}} SiMe_2 \;+\; (Me_3Si)RNH \;+\; Sn \quad (29)$$

$$Ph_3Sn(CH_2)_3CN \;+\; Bu_3SnN_3 \longrightarrow \quad \xrightarrow{-Bu_3SnPh} \quad (30)$$

$$(Bu_2SnS)_3 \;+\; (Bu_2SnNEt)_3 \longrightarrow \quad + \quad (31)$$

$$2(Me_3Sn)_3N \;+\; S_4N_4 \longrightarrow S \;+\; Me_3SnN{=}S{=}NSnMe_3$$

$$\Delta \;\big|\; {-}\,Me_4Sn$$

$$2(Me_3Sn)_2NSiMe_2NCS \;+\; S_4N_4 \xrightarrow[{-4Me_3SnNSNSiMe_2NCS}]{-2Me_4Sn} \left[\begin{array}{c} S{=}N \\ \| \quad \diagdown \\ N{-}\!\!\underset{Me_2}{Sn}{-}S \end{array} \right]_2 \quad (32)$$

ing Sn–N interactions (231.6 pm) (*119,121*) but reacts with Ph_4AsCl to give the monomeric anion $Me_2(Cl)SnS_2N_2^-$ (*122*), with both compounds having a trigonal bipyramidal geometry at tin. The structure of the former is shown in Fig. 3.

Structural data for other Sn–N heterocycles show that the size of the

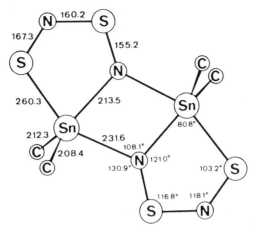

FIG. 3. Schematic representation of dimeric $Me_2SnS_2N_2$, showing structural parameters. [Reproduced from H. W. Roesky, *Adv. Inorg. Chem. Radiochem.* **22**, 254 (1978), with permission of Academic Press.]

heterocyclic ring, that is, the degree of molecular association, is determined by the combined bulk of the substituents on tin and nitrogen. The structures of **26** (R = ^{t}Bu, SO_2CH_3) and **27** all contain planar rings, even though as the ring size expands the nitrogen goes from trigonal pyramidal to planar. Sn—N bonds are in the range 203.2–209.2 pm (*110*). The five-membered heterocycle in **30**, synthesized from $MeB[N(Li)Bu^t]_2$ and $Me_2BrSnCH_2Br$, is also planar with an Sn—N bond length of 206.6 pm (*123*).

Two reaction types govern the chemistry of Sn–N heterocycles: substitution of the ring elements and addition reactions with unsaturated species to give ring-expanded products. In the former case, relatively few examples are known where N is substituted [Eq. (33)] (*124,125*); replacement of tin

$$ ^tBu_2Sn \overset{\displaystyle \underset{Me}{N}}{\underset{\underset{Me}{N}}{}} SnBu_2{}^t \quad \begin{array}{c} \xrightarrow{\ CS_2,\ -2MeNCS\ } \\[2mm] \xrightarrow{\ 2PhPH_2\ \atop -2MeNH_2\ } \end{array} \quad \begin{array}{c} (^tBu_2SnS)_2 \\[4mm] (^tBu_2SnPPh)_2 \end{array} \tag{33}$$

$$2[RNSnMe_2]_3 + 6RN=S=O \xrightarrow{\ SO_2\ } \begin{array}{c} Me_2 \\ Sn \\ RN \diagup \quad \diagdown NR \\ | \qquad\qquad | \\ RN=S \diagdown \quad \diagup SnMe_2 \\ N \\ R \end{array} \tag{34}$$

$$R = CF_3SO_2$$

is more common [Eq. (34)]. Both four- (31) and five-membered B—N rings (32) can be synthesized by reacting either 29 or 30 with MeBBr$_2$

$$(30) \qquad\qquad (31) \qquad\qquad (32)$$

(123,126), whereas Me$_2$SnS$_2$N$_2$ has proved to be a versatile reagent for the synthesis of other MS$_2$N$_2$ rings [M = Pt (127,128) and Pd (127,129)]. Insertion reactions with a range of 1,2-dipoles, for example, RC≡N, RC≡CR, PhN=C=NPh, and TosN=S=NTos (Tos = MeC$_6$H$_4$SO$_2$),

$$(35)$$

are known [Eq. (35)], which afford a facile method of expanding the ring size to six or eight atoms (124,130).

Finally, and most remarkably, the tin(II) heterocycle 1,3,2,4-diazasilas-tannetidine (33) reacts with Ph$_3$P=CH$_2$ to give a novel pentacyclic compound, Sn$_2$(CH)$_2$P$_2$(C$_6$H$_4$)$_2$Ph$_2$ (36), which can be described as containing stannate(II) anions (from SnC$_3$ coordination) and cationic, four-coordinated phosphorus. The reaction sequence, depicted in a collection of crystal structures (Fig. 4), shows that after coordination of the phosphorane to tin (34), the molecule can isomerize (35) to allow delivery of both a methylene proton as well as one from an o-phenyl group, to eliminate Me$_2$Si(NHBu^t)$_2$, and leave 36. Kinetic studies along with deuterium labeling elegantly corroborate the mechanism shown by the crystallography (131).

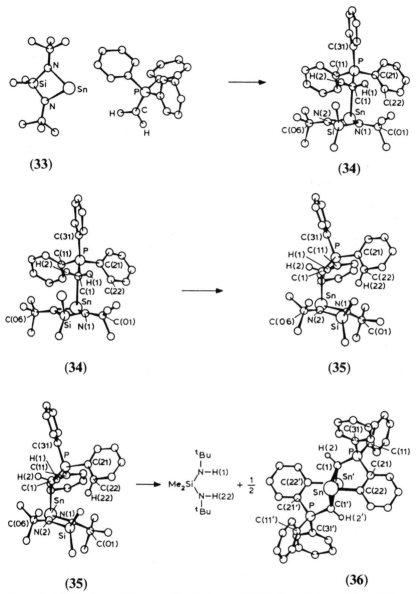

Fig. 4. Mechanism of the reaction between 1,3,2,4-diazasilastannetidine (33) and Ph$_3$P=CH$_2$, shown as a series of crystal structures. (Reproduced from Ref. *131* with permission of Elsevier Sequoia.)

B. Tin–Phosphorus Heterocycles

Much of the earlier work in the field of organotin phosphines has been reviewed by Schumann (132), including several reports on heterocyclic systems emanating from that author's group. Indeed, these studies were the first to show the diverse structural chemistry that such compounds might exhibit, although detailed structural assignments should be treated with some caution since they are based largely on infrared and cryoscopic data. Reaction of Bu_2SnCl_2 with $PhPH_2$ in the presence of base yields the cyclic trimer $(Bu_2SnPPh)_3$ (133), whereas cage structures emerge when trifunctional reagents, for example, $RSnCl_3$ and PH_3, are used (134,135) [Eq. (36)]. Neither the trigonal bipyramid (37) nor the cubane (38) structure has

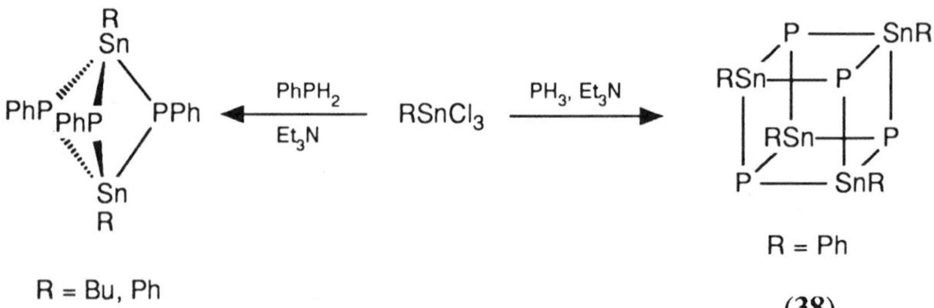

$$
\begin{array}{ccc}
\textbf{(37)} & \xleftarrow[\text{Et}_3\text{N}]{\text{PhPH}_2} \quad \text{RSnCl}_3 \quad \xrightarrow{\text{PH}_3,\ \text{Et}_3\text{N}} & \textbf{(38)}
\end{array} \tag{36}
$$

R = Bu, Ph (for **37**)   R = Ph (for **38**)

been confirmed crystallographically, which is surprising as these polyhedra are otherwise unknown in organotin chemistry (287).

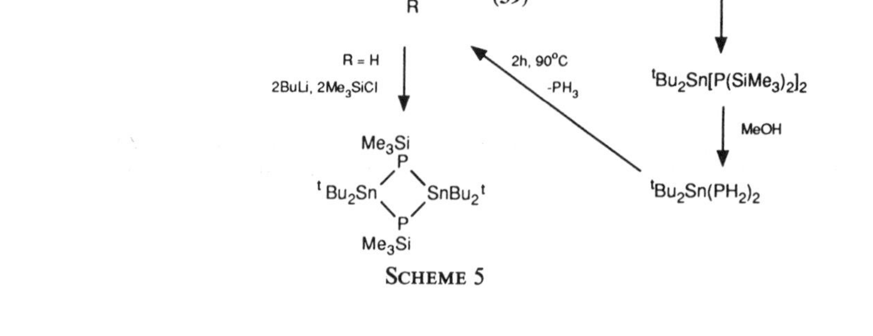

SCHEME 5

TABLE III

NMR DATA FOR HETEROCYCLIC Sn—P COMPOUNDS

Compound	$\delta\ ^{31}P$ (ppm)	$\delta\ ^{119}Sn$ (ppm)	$^1J(P-Sn)$ (Hz)	$^2J(P-Sn)$ (Hz)	Ref.
39, R = H	−261.0	46.6t	586.6		138
39, R = Me	−136.6	0.6t	741.9		137
39, R = ^{t}Bu		52.9t	954		136
39, R = SiMe$_3$	−239.6	98.9t	824.5		138
40	−299		749	93	140
41	−291		800, 931	126	141
42	−73.3	−240.1t	935		142
43	11.0		768.6	9.4	142
44	18.8				142
45	−58.3		1117.3	46.4	142
46^a	459.0	58.6d, 160.4d	796	278.3	143
48	−23.4	52.5t	896^b		145

^{a} $^1J(Sn-Sn) = 892.2$ Hz.
^{b} At 30°C.

Nearly 20 years elapsed after Schumann's early work before definitive crystallographic and NMR data were able to advance our understanding of simple cyclic stannaphosphines. The synthetic methodology used is shown in Scheme 5 (*136–138*). Both ^{31}P- and ^{119}Sn-NMR data (Table III), particularly signal multiplicity, confirm the composition of the compounds. Two examples of the Sn$_2$P$_2$ ring have been studied crystallographically, and both confirm the planarity of the heterocycle. The Sn—P bond length and internal angles at Sn or P are 254.1 and 254.6 pm and 98.0° and 82.0° [39, R = H (*138*)] or 255.6 pm and 91.9° and 88.1° [39, R = ^{t}Bu (*136*)], respectively.

Heterocycles rich in either phosphorus and/or tin can be obtained using reagents rich in these elements. Thus, Me$_4$Sn$_2$H$_2$ reduces elemental phosphorus *in the dark* to yield Me$_{12}$Sn$_6$P$_2$ (40), which loses Me$_2$Sn: on exposure to light to give Me$_{10}$Sn$_5$P$_2$ (41) (*139–141*). The structures of both compounds are shown in Fig. 5. Compound 40 has D_3 symmetry and can be viewed as arising from three Sn$_4$P$_2$ rings of boat conformation which share three bonds with each other. Compound 41 has a norbornane structure. On the other hand, ^{t}BuPP^{t}Bu^{2-} reacts with either ^{t}Bu$_2$SnCl$_2$ or Et$_2$SnCl$_2$ to give a bewildering array of heterocycles, some in very low yield [Eqs. (37) and (38)]. NMR data are given in Table III. Most striking is the ^{119}Sn chemical shift (−240 ppm) of the 1,2,3-diphosphastannirane (42), which is similar to those in the Sn$_3$ homocycles (Table II) (*142*).

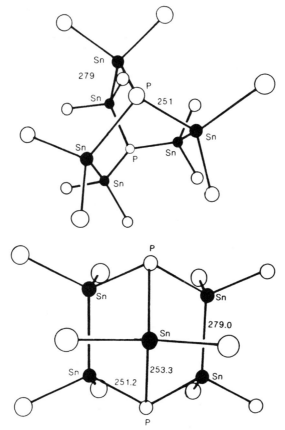

FIG. 5.   Structures of (top) $Me_{12}Sn_6P_2$ **(40)** and (bottom) $Me_{10}Sn_5P_2$ **(41)**. (Reproduced from Ref. 7 with permission of Blackie.)

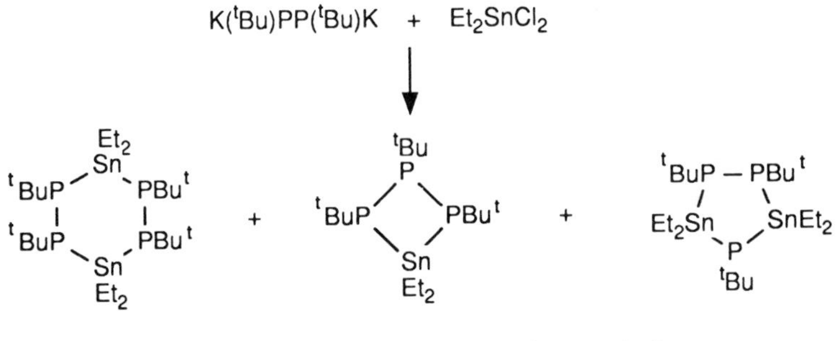

$$K(^tBu)PP(^tBu)K \quad + \quad Et_2SnCl_2$$

(45)   minor products

Several heterocycles containing Sn–P–C rings have been synthesized. The smallest of these is **46**, formed from a [2 + 2] cycloaddition between $R_2Sn{=}SnR_2$ and $^tBuC{\equiv}P$ (143). Mechanistic considerations do not favor the three-membered **47** as an intermediate (i.e., stepwise interaction of

Sn-Sn	: 287.8pm
Sn-P	: 254.4pm
Sn-C	: 220.3pm
C=P	: 171.2pm
α	: 76.1°
β	: 74.8°

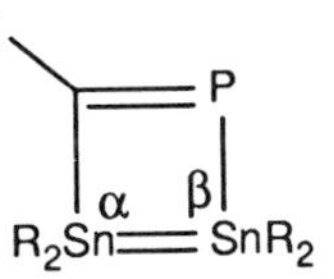

(46) 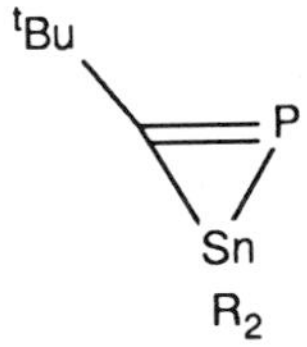   (47)

stannylene monomers), unlike the reaction between stannylenes and isolated alkynes (see Section II,B). Notably, $[(Me_3Si)_2N]_2Sn$, which is not dimeric, does not react with the phosphaalkyne. Both saturated (149) and unsaturated heterocycles (145) with the same five-membered framework can be synthesized by the method of Eq. (39). Hindered organotins [R =

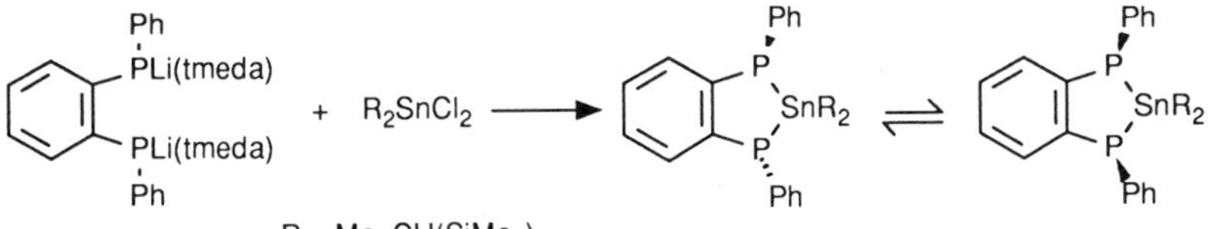

R = Me, CH(SiMe₃)₂

(48)   (49)

$HC(SiMe_3)_2$] exist as conformer **48** in solution at room temperature, whereas both conformers can be observed under the same conditions when R is Me. Compound **48** [R = $HC(SiMe_3)_2$] crystallizes with Sn—P bonds of 253.3 and 252.7 pm (145). 2,5-Distannaphospholanes (**50**) can be prepared from either the corresponding silicon compounds or RPLi$_2$ [Eq. (40)] (*144*). Additional heteroatoms can be incorporated by trapping the unstable intermediate [Me$_2$Sn=PPh] with strained rings [Eq. (41)] (*146*).

$$\text{Me}_2(\text{Cl})\text{Sn}(\text{CH}_2)_2\text{Sn}(\text{Cl})\text{Me}_2 \xrightarrow[\substack{^t\text{Bu}_2\text{Si} \quad \text{SiBu}_2{}^t \\ \text{P} \\ \text{R}}]{\text{RPLi}_2 \text{ or}} \quad {}^t\text{Bu}_2\text{Sn} \diagup \diagdown \text{SnBu}_2{}^t \qquad (40)$$

(50)

$$\text{Me}_2\text{SnCl}_2 \; + \; \text{Me}_2\text{Si} \diagup \diagdown \text{SiMe}_2 \xrightarrow{\text{THF, -40}^\circ\text{C}} [\text{Me}_2\text{Sn=P}] \; + \; \text{Me}_2(\text{Cl})\text{Si} \diagup \diagdown \text{Si(Cl)Me}_2$$

with S ring, downward arrow

$$\text{PhP} \diagup \diagdown \text{S} \quad \text{Sn} \quad \text{Me}_2 \qquad (41)$$

IV

HETEROCYCLES CONTAINING TIN–GROUP 16 ELEMENT BONDS

Heterocycles containing tin and either O, S, or, to a lesser extent, Se and Te are probably the most thoroughly studied of all the tin heterocycles. In recent years a wealth of new cluster chemistry has emerged in this area, predominantly built around aggregations of stannoxane rings (**1**, X = O), although, like zeolites, the resultant Sn–O framework also contains rings of other sizes. Two examples of clusters are shown in Fig. 6 (*147,148*). No further comment will be made on compounds of this type, since the intrinsic heterocyclic building block is well known in organotin chemistry. The major contributions to this field can be found in a review by Holmes detailing work from that author's own laboratory (*147*) and in the work of Puff and Reuter (*148*). More general aspects of organotin compounds which contain bonds to oxygen (*149*) or the heavier chalcogenides (*150*), up to 1982, have been reviewed, including several heterocycles.

A. Heterocycles with No Additional Heteroatoms

Diorganotin oxides are the ultimate hydrolysis products of diorganotin halides, and they are generally amorphous solids of unknown structure.

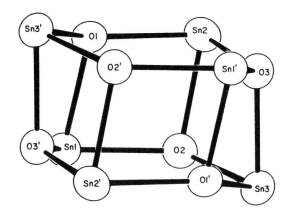

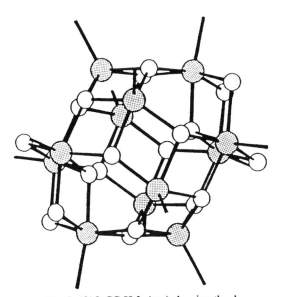

FIG. 6.   Structures of $[BuSn(O)O_2CC_5H_9]_6$ (top) showing the drum arrangement of stannoxane rings and $[(^iPrSn)_{12}O_{14}(OH)_6]^{2+}$ (bottom). Reproduced from *Inorg. Chem.* **26,** 1050 (1987) (cited in Ref. *147*) and Ref. *148* with the permission of the American Chemical Society and Elsevier Sequoia, respectively.)

That these compounds are polymeric there can be little doubt, and spectroscopic evidence favors a cross-linked, reticular lattice rather than a ladder structure (*151*, and references therein). However, if bulky hydrocarbon groups are present on tin then soluble oligomers are formed, which to date have all been trimers [Eq. (42)] (*152–154*). The $(SnO)_3$ ring (**51**) is planar

$$R_2SnCl_2 \xrightarrow[\text{- NaCl, H}_2\text{O}]{\text{NaOH}} \quad \begin{array}{c} R_2 \\ Sn \\ O \diagup \diagdown O \\ | \qquad | \\ R_2Sn \diagdown \diagup SnR_2 \\ O \end{array} \quad \xleftarrow[\text{NaHCO}_3]{\text{RMgBr}} Cl_2Sn(acac)_2 \qquad (42)$$

R = ^{t}Bu, ^{t}Am   R = 2,6-$Et_2C_6H_3$

(51)

in all three cases; geometric data are given in Table IV. The synthetic approach to the corresponding sulfides, selenides, and tellurides is similar to that of the oxides, using Na_2S, $(Me_3Si)_2S$, or NaHX (X = Se and Te) as chalcogenide sources (155–157). However, in contrast to the oxides, these compounds containing the heavier group 16 elements are rarely polymeric (158); thus, greater control over the extent of oligomerization can be exercised. With small hydrocarbon groups on tin, for example, Me, the heterocycles invariably contain six atoms in a twist–boat conformation. Structures are available for the complete series $(Me_2SnX)_3$ [X = S, Se (159), and Te (160)], including two crystal modifications of the sulfide with virtually identical structures (161,162), along with that of (Ph_2SnS_3) (163). Though the structures of all these compounds are similar, it is

TABLE IV

SELECTED STRUCTURAL DATA FOR $(R_2SnX)_n$

R	X	n	Sn—X (pm)	<XSnX (°)	<SnXSn (°)	Ref.
^{t}Bu	O	3	196.5	106.9	133.1	152, 153
^{t}Am	O	3	195.2–197.8	105.9, 106.3	133.7, 134.5	153
$Et_2C_6H_3$	O	3	192.9–196.1		135.6–137.1	154
Me[a]	S	3	241	107.7	103.0	161
Me[b]	S	3	241.1[c]	106.1–108.7	102.0–104.4	162
Ph	S	3	239.9	109.6–112.4	103.8–106.1	163
Me	Se	3	253.1[c]	107.1–110.6	100.6–101.1	159
Me	Te	3	275[c]		96[c]	160
^{t}Bu	S	2	248, 240[c]	94.0, 94.3	84.0, 87.6	164
^{t}Bu	Se	2	255	97.5	82.5	164
$^iPr_3C_6H_2$	O, S[d]	2	202.6, 204.2[e] 243.5, 243.9[f]	88.6, 89.1	101.7[e], 80.6[f]	179

[a] Tetragonal modification.
[b] Monoclinic modification.
[c] Mean value.
[d] See 55.
[e] X = O.
[f] X = S.

notable that the Sn—Te—Sn angle is markedly more acute than analogous angles in any of the other compounds (Table IV). More bulky organic groups, for example, ^tBu (164) and mesityl (155), form only four-membered rings, that is, dimers. In both ($^tBu_2SnX)_2$ (X = S and Se), the ring is planar, but whereas in the selenium derivative the ring is rhombic (all Sn—Se distances equal) the sulfide is distorted, with a short and a long bond between sulfur and tin (Table IV).

Control over the degree of oligomerization also seems possible using coordinating ligands. Thus, the estertin sulfide [(MeOCOCH$_2$CH$_2$)$_2$SnS]$_3$ is trimeric as might be anticipated from the limited bulk of the ester group. Alternatively, the related dithiocarbamate [(MeOCOCH$_2$CH$_2$)(Me$_2$NCS$_2$)SnS]$_2$ is dimeric, in which tin achieves pentacoordination through a bidentate dithiocarbamate ligand (156). [MeN-(CH$_2$CH$_2$CH$_2$)$_2$SnS]$_2$ (165) and {[CH$_2$N(Et)CH$_2$CH$_2$CH$_2$]$_2$SnS}$_2$ (166) are also dimeric, with an intramolecular N: → Sn bond expanding the coordination sphere about the metal.

Hydrolysis of RSnCl$_3$ leads to the organostannoic acids, RSn(O)OH, of unknown structure. When R is (Me$_3$Si)$_3$C, the product appears to be trimeric with a heterocyclic core akin to 51, though definitive structural data are unavailable (167). The corresponding sesquisulfides, RSnS$_{1.5}$, are better characterized (155,168). Compounds with R = Me (168) and C$_6$F$_5$ (155) are both tetramers with an adamantane cage structure (52, X = S,

(52)

Y = RSn) (Sn—S 238.1–239.5 and 239 pm, respectively). Other compounds in this series adopt the same structure, since the intensity of the ^{117}Sn satellites in the ^{119}Sn-NMR spectra of approximately 20% the total integrated spectral intensity specify three identical $^2J(^{119}$Sn–S–^{117}Sn) couplings (155). Structurally related to these sesquisulfides are 7-alkyl-1,3,5-tristanna-2,4,6-trithiaadamantanes (52, X = CH$_2$, Y = MeC and HC) (169,170). The adamantane cage in these compounds is flattened at C-7, leading to strain not present in adamantane itself (171). Hydrogen at C-7 can be used to reduce alkyl halides to hydrocarbons (169), but the

C-7—CH$_3$ bond is not as reactive as suggested by the bond strength of 40 kcal/mol calculated by MNDO methods (*170,172*).

Both the (R$_2$SnX)$_n$ and R$_4$Sn$_4$X$_6$ series have been thoroughly investigated spectroscopically. Typical v(Sn–S) values are in the range 300–330 cm^{-1}, whereas mass spectra for the methyltin compounds confirm the degree of molecular association through the isotope distribution patterns for the high mass fragments, which may be the parent ion (P) or P − Me (*173,174*). NMR data (Table V) indicate that the ^{119}Sn chemical shift is largely independent of the alkyl group on tin, or the ring size (save in the case of X = Te); indeed, the data are very similar to analogous acyclic compounds incorporating the X—Sn—X moiety [e.g., δ ^{119}Sn: Me$_2$Sn(SMe)$_2$, 137 ppm; Me$_2$Sn(SeMe)$_2$, 57 ppm (*175*)]. Within the series X = S, Se, and Te, δ ^{119}Sn values become more negative as the electronegativity of X decreases and tin becomes more shielded. UPES data also suggest strong σ delocalization in (Me$_2$SnS)$_3$ which may also contribute to

TABLE V

SELECTED NMR DATA FOR (R$_2$SnX)$_n$

(51)   (53)

Compound, R	X	X'	δ ^{119}Sn (ppm)	δ X^a (ppm)	^1J(Sn–X)^a (Hz)	^2J(Sn–X–Sn) (Hz)	Ref.
51							
^{t}Bu	O	O	−84.3			369	*152, 153*
^{t}Am	O	O	−72.9			394	*153*
Et$_2$C$_6$H$_3$	O	O	−125.0				*154*
Me	S	S	131			195	*175*
Me	Se	Se	42	−350	1228	237	*175, 176*
Me	Te	Te	−195	−860	3098	250	*157*
Me	S	Se	84, 137^b		1238^c	204, 220	*175*
53							
^{t}Bu	S	S	126			114	*157*
^{t}Bu	Se	Se	54.6		915		*157*
^{t}Bu	Te	Te	−121.0	−1099	2117		*157*
^{i}Pr$_3$C$_6$H$_2$	S	S	−48.2			177	*179*
^{i}Pr$_3$C$_6$H$_2$	O	S	11.4			137	*179*

^{a} ^{77}Se or ^{125}Te.
^{b} SeSnSe and SSnS, respectively.
^{c} J(Sn–Se).

the deshielding at tin (*176*). The Sn–O compounds clearly do not fit into this trend, a fact which most likely stems from the differing ring conformation (planar versus twist–boat) in the latter case.

Although both cyclic and acyclic compounds show similar [119]Sn chemical shifts, significant differences are seen in the shifts of X nuclei ([77]Se, [125]Te). For example, compare the values of Table V with the following chemical shifts which also incorporate the Sn—X—Sn unit: $(Me_3Sn)_2Se$, −547 ppm (*176*); $(Me_3Sn)_2Te$, −1226 ppm (*157*). The marked angular changes at X in the heterocycle (133°–96°, Table IV) on going from X = S to X = Te can no doubt be correlated with this trend. Both $^1J(Sn-X)$ and $^2J(Sn-X-Sn)$ decrease in magnitude with decreasing ring size (Table V). In contrast, however, are the trends in Mössbauer quadrupole splitting values (Δ) which are available for both [119]Sn and [125]Te in the $(R_2SnTe)_n$ series. In this case, it is the tin parameter which shows most variation, increasing as n decreases (1.64 to 1.87 mm/second), whereas Δ for the [125]Te nucleus is largely constant. It would appear that the total electron asymmetry brought about by ring constraints is more manifest at four-coordinate tin than two-coordinate tellurium (*157*).

Solid-state ^{13}C-NMR data are available for $(Me_2SnS)_3$ (*177,178*). The spectrum shows resonances arising from the three pairs of unique methyl groups in the twist–boat structure of C_2 symmetry, with a $^1J(^{13}C-^{117,119}Sn)$ value of 420 Hz (cf. 398 Hz in solution). The complexity of the spectrum also indicates that at least three crystalline modifications of this compound are present, although only two have so far been confirmed crystallographically. Also noted in the same study is the corresponding J value for polymeric $(Me_2SnO)_n$ (640 Hz), which is in keeping with a higher coordination number at tin (see above).

Although several mixed chalcogenides containing S, Se, and Te based on the trimeric ring have been observed in solution by NMR methods (*19,175*) (Table V), none have so far been isolated. Moreover, no analogous mixed derivatives based on the dimeric structure have been reported. One example of a 1-oxa-3-thia-2,4-distannetane (**55**) is known (*179*), synthesized by the sequence shown in Eq. (43). The 1-thia-2,3-distannirane

$$R_4Sn_2Br_2 \longrightarrow (54) \xrightarrow{O_2} (55) \quad (43)$$

$$R = 2,4,6\text{-}^iPr_3C_6H_2$$

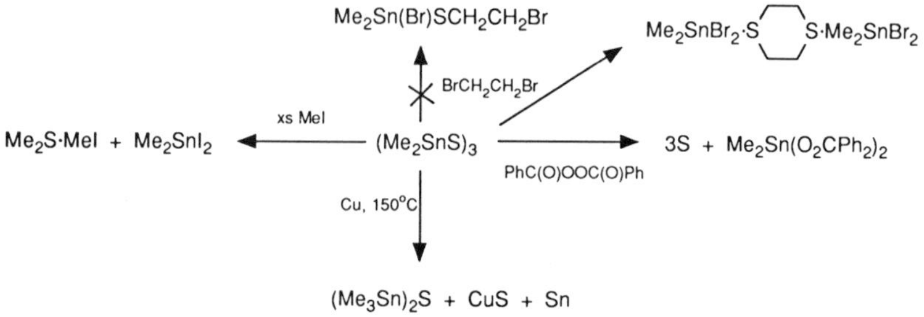

$$Me_2Sn(Br)SCH_2CH_2Br$$

$$Me_2SnBr_2 \cdot S \overset{\frown}{\underset{\smile}{}} S \cdot Me_2SnBr_2$$

$$Me_2S \cdot MeI + Me_2SnI_2 \xleftarrow{\text{xs MeI}} (Me_2SnS)_3 \xrightarrow{PhC(O)OOC(O)Ph} 3S + Me_2Sn(O_2CPh_2)_2$$

(with $\text{BrCH}_2\text{CH}_2\text{Br}$ crossed-out arrow upward, and Cu, 150°C arrow downward)

$$(Me_3Sn)_2S + CuS + Sn$$

SCHEME 6

(54) has not been isolated, but its role as a reaction intermediate is suggested by the fact that under anaerobic conditions the reaction product is the dimer $(R_2SnS)_2$. Thus, ring strain in 54 is relieved by insertion into the Sn—Sn bond, by either O under aerobic conditions or S if O_2 is absent. The structure of the planar, four-membered SnOSnS ring differs from either the rhombic or distorted Sn_2X_2 (X = S and Se) arrangements previously described. It adopts an "arrowhead" motif (see 55), arising from dissimilar Sn—O and Sn—S bond lengths and markedly different angles at either O or S. The smaller angle at the larger sulfur atom is in keeping with the trend apparent in Table IV for related compounds.

The reaction chemistry of these simple cyclic compounds has been largely neglected. Derivatives of the heavier chalcogenides are more stable hydrolytically than the oxides, presumably owing to a Sn—X bond of lower polarity. In general, Sn—S, Sn—Se, and Sn—Te bonds do not undergo reactions across the Sn—X bond to anything like the same extent as is known for Sn—O compounds (149). One cited addition reaction, latter shown to have been incorrectly formulated (180), is shown in Scheme 6 along with more common desulfurization reactions (181). Interestingly, a sulfur incorporation reaction is known [Eq. (44)], which under

$$6Me_3Sn^+ + 3S^{2-} \longrightarrow 3(Me_3Sn)_2S \xrightarrow{h\nu} (Me_2SnS)_3 + Me_4Sn \quad (44)$$

environmental conditions generates $(Me_2SnS)_3$ from $(Me_3Sn)_2S$. This has important consequences for the environmental mobility of organotins, such that in sedimentary conditions (where S^{2-} is abundant) immobile organotins can be converted to tetraorganotins of high volatility (182).

More complex tin-rich rings containing S, Se, and Te are known, as are systems in which heterocycles are linked by a spiro center. In the latter case, three compounds of the same composition, $R_8Sn_5X_6$ (X = S, R = ^iPr and ^tBu; X = Se, R = ^iPr), have been synthesized by the routes shown in Eqs. (45) and (46). The compounds are conformationally distinct in the solid state (Fig. 7), but these forms equilibrate in solution (183).

$$5\ ^tBu_2SnS\ +\ 2S \xrightarrow{\ -^tBuS\ } \ ^tBu_8Sn_5S_6 \tag{45}$$

$$4^tBu_2SnCl_2\ +\ 2Na_4SnS_4 \nearrow\ {\scriptstyle -\ SnS_2,\ -\ 8NaCl}$$

$$2(^iPr_2SnCl)_2X\ +\ Na_4SnX_4 \longrightarrow\ ^iPr_8Sn_5X_6 \tag{46}$$

$$X = S,\ Se$$

A six-membered ring containing Sn—Sn bonds (**56**) is formed from reaction of $Me_4Sn_2Cl_2$ with either Na_2Se (*184*) or NaHTe (*185*). Vibrational spectroscopic data suggest the ring does not have a chair conforma-

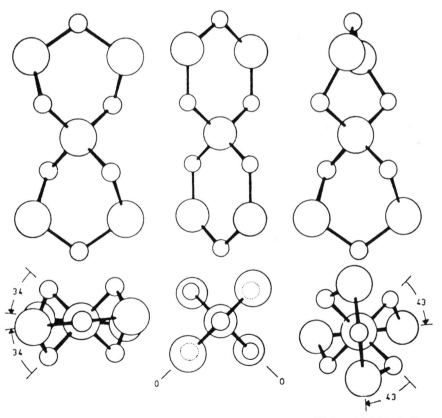

FIG. 7. Conformational isomerism in the solid-state structures of (left to right) $^tBu_8Sn_5S_6$, $^iPr_8Sn_5S_6$, and $^iPr_8Sn_5Se_6$. (Reproduced from Ref. *183* with permission of Elsevier Sequoia.)

tion, but this awaits crystallographic confirmation. Compound **56** photolytically eliminates dimethylstannylene to give **57**, along with **58**, which arises from trapping of Me_2Sn by the parent (**56**) followed by elimination of Me_2SnX [Eq. (47)] (*185*). Analogs to **58** can also be synthesized by the

$$\begin{array}{ccc} & \xrightarrow[\text{$-Me_2Sn$}]{hv} & & & (47) \\ X = \text{Se, Te} & & & & \\ (56) & & (57) & & (58) \end{array}$$

reaction of $I('Bu_2Sn)_4I$ with H_2X (X = S, Se, and Te) in the presence of base (*186*). The structures of all three compounds have been determined, and in each case the five-membered ring is almost planar. The Sn—Sn bonds in these molecules are, however, unusually long (287.0–289.8 pm). Other routes to **57** are the reactions of Me_2Sn with $(Me_2SnS)_3$ (*187*) and of elemental chalcogens with either Me_2SnH_2 (Se, Te) (*188*) or cyclo-$('Bu_2Sn)_4$ (S, Se, Te) (*189*). Vibrational spectra indicate that when R is Me the rings are nonplanar (*187*), a fact confirmed crystallographically for the selenium compound (*190*). The rings are of the envelope conformation and are liked into chains by intermolecular interactions between Se and the Sn—Sn unit of an adjacent molecule. In addition, these one-dimensional chains are cross-linked by the formation of Sn_2Se_2 rings between chains (Fig. 8). Following a theme with which the reader should now be familiar, the corresponding $'Bu$ compounds are discrete molecules, incorporating planar rings (X = S and Se). Only when steric relief accrues from the longer Sn—Te bonds does puckering of the ring follow (*189*).

Before dealing with heterocycles containing Sn, X, and C, it is pertinent to include one further ring type containing Sn—O and/or Sn—S bonds, since it represents an emergent group of macrocycles. $Ph_3SnO_2P(OPh)_2$ crystallizes as a cyclic hexamer (Fig. 9), in which individual molecules are held together by bridging C=O: → Sn interactions (*191*). This 24-membered ring is by no means an isolated example of its type (*192*), and rings containing 36 atoms are known in other hexamers (see below). The six-molecule ring would appear to be an energy minimum compared to pentamers and heptamers.

B. *Tin–X–Carbon Heterocycles (X = O, S, Se, or Te)*

There have been numerous studies of 1,3-dioxastannolanes and related 1,3-thia or mixed 1,3-oxathia analogs, largely stemming from their syn-

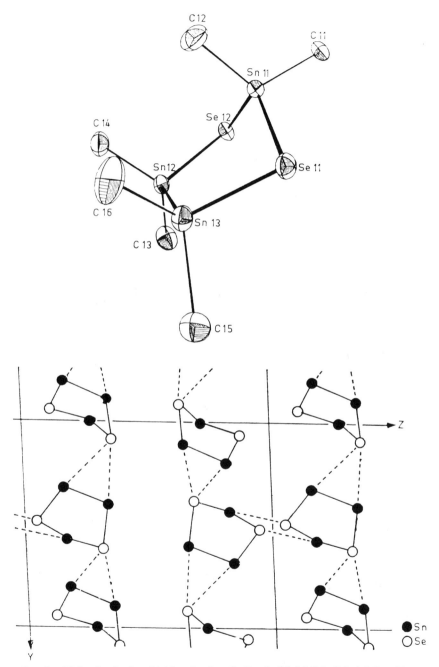

FIG. 8. Molecular (top) and lattice structure (bottom) of 2,4,5-(Me$_2$Sn)$_3$-1,3-Se$_2$. (Reproduced from Ref. *190* with permission of VEB J. A. Barth Verlag.)

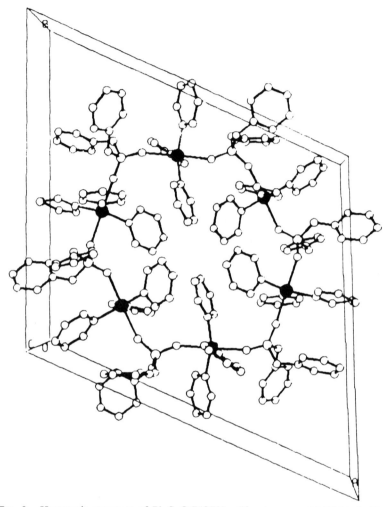

FIG. 9. Hexameric structure of $Ph_3SnO_2P(OPh)_2$. Tin atoms are highlighted. (Reproduced from Ref. *191* with permission of the American Chemical Society.)

thetic utility in various aspects of organic chemistry. In addition, the toxic action of diorganotins is thought to involve interaction with the dithiol groups of reduced lipoic acid to form a 1,3,2-dithiastanna heterocyle. Synthetic routes to these types of compounds are diverse and include reactions of diorganotin oxides (*193,194*), alkoxides (*193,195*), stannylenes (*196–198*), and intramolecular hydrostannation (*199*), some examples of which are given in Eqs. (48)–(51). The use of R_2SnO is probably the

$$R_2SnO \quad + \quad HSCH_2CH_2OH \quad \xrightarrow{-H_2O} \quad \underset{\underset{R_2}{Sn}}{S\diagdown\diagup O} \tag{48}$$

$$Bu_2Sn(OR)_2 \quad + \quad \underset{O}{O\diagdown\diagup O} \quad \longrightarrow \quad (RO)_2CO \quad + \quad \underset{\underset{Bu_2}{Sn}}{O\diagdown\diagup O} \tag{49}$$

$$\underset{\underset{R_2}{Sn}}{\overset{R' \quad R'}{O\diagdown\diagup O}} \quad \longleftarrow \quad \overset{R' \quad R'}{\underset{O \quad O}{}} \quad R_2Sn: \quad \xrightarrow{R'CHO} \quad \underset{\underset{R_2}{Sn}}{\overset{R' \quad R'}{O\diagdown\diagup O}} \tag{50}$$

$$Bu_2SnH_2 \quad + \quad Bu_2Sn(OCH_2CH=CH_2) \quad \longrightarrow \quad 2Bu_2Sn\underset{H}{OCH_2CH=CH_2} \quad \xrightarrow{40°C} \quad \underset{\underset{Bu_2}{Sn}}{\diagup\diagdown O} \tag{51}$$

most widespread, and many other Sn—O heterocycles have been made from this precursor (*200*, and references therein). The cyclic Sn—O—C products show greater thermal and hydrolytic stability than, for example, organotin alkoxides.

The structure of the 1,3-disubstituted-2-stannolanes has been the subject of some controversy. Molecular weight measurements indicate that the compounds are dimeric, and both **59** and **60** have been proposed to accommodate this fact. Proposed structure **59** stems largely from the use of these compounds to form large lactone rings (see below), though the overwhelming weight of data, both in the solid state and in solution, indicate **60** to be universally correct. The parent 2,2-dibutyl-1,3,2-dioxas-tannolane is polymeric in the solid state (**61**, R groups omitted for clarity), in which tin is in a distorted octahedral coordination (Sn—O: 204 and 251 pm for intra- and intermolecular bonds, respectively), with the butyl groups above and below the polymer plane. The five-membered ring adopts a half-chair conformation, with the carbon atoms either side of the SnO$_2$ plane (*201*). Chemically related compounds containing the same heterocyclic core can be viewed in terms of fragments of **61**. Most simply, the di-*tert*-butyl analog is dimeric ($n = 2$), in which tin is now only five-

(59)   (60)   (61)

coordinate (*202*), as is the glucose derivative (**62**) (*203*), whereas the man-

(62)   (63)

nose derivative (**63**) is pentameric with two terminal, five-coordinate tins and three internal, six-coordinate tin atoms (*204*).

As has already been mentioned, sulfur has a smaller tendency than oxygen to form polymeric compounds. The 1,3,2-dithiastannolanes are also polymeric, but the intermolecular interactions between molecules are much weaker. In the case of the 2,2-dimethyltin compound, bridging is by only one sulfur atom (Sn–S: 318 pm) while the other is nonbridging. Tin is thus five-coordinate (*205,206*). In the di-*n*-butyltin analog, both sulfur atoms bridge weakly (Sn–S: 368.8 pm) to render tin a distorted octahedral coordination sphere (*207*). Only with the bulky *tert*-butyl groups on tin are the intermolecular interactions totally removed to leave tin in a tetrahedral state (*202*). In compounds with both oxygen and sulfur in the heterocycle (1,3,2-oxathiastannolanes) it is not unexpectedly the oxygen which is involved in intermolecular interactions. Thereafter, the structure is modified by the nature of the groups bonded to tin, with the di-*tert*-butyltin compound remaining dimeric (five-coordinate tin) (*202*), whereas the di-*n*-butyltin compound is isostructural with the dimethyltin dithiastannolane, now with single Sn—O bridges forming a one-dimensional chain (*208*). In complete contrast, the related 1,3,2-oxathiastanninane (**64**) is hexameric,

(**64**)

forming an outer ring containing 36 atoms (*209*). Bridging between molecules is through the exocyclic carbonyl oxygen (218 pm), and the six-membered ring of each monomer component is very close to being flat (maximum deviation from plane 43.8 pm).

In solution, any of the stannolanes which contain oxygen are dimeric, save the 2,2-di-*tert*-butyl-1,3,2-oxathiastannolane where the bulk of the hydrocarbon groups exert their influence. The dimer is linked through oxygen bridges in each case. The dithiastannolanes are monomeric in solution. These findings summarize detailed comparisons of solid-state and solution ^{119}Sn-NMR measurements and Mössbauer quadrupole splittings (*201,202,205,208*) and are shown in Fig. 10. Upfield chemical shifts and enhanced Δ values are typical of coordination numbers greater than four at tin. Previous suggestions that 4,5-disubstituted 1,3,2-dioxastannolanes with δ ^{119}Sn values in the range -155 to -189 ppm are four-coordinate monomers are undoubtedly erroneous (*210*). The Mössbauer data for the pentameric mannose compound (**63**) can also be rationalized in terms of the two types of tin in the molecule (*201*).

The solution state behavior of the dioxastannolanes is, however, more complex than that outlined above. The components of the dimeric unit scramble rapidly between dimers, a process that can be put to effect when the dioxastannolane is synthesized from a chiral diol. In such cases, diastereoisomers containing *RR(SS)* and *SR(SR)* dimers are formed, which cause splitting of signals in the NMR spectra (most readily observed in the ^{13}C spectrum), and which enables the enantiomeric excess of chirally impure samples to be determined (*211*). In addition to the rapid scrambling of monomers between diastereoisomers which occurs at room temperature, there is a second fluxionality which inverts the configuration at tin in 4-substituted dioxastannolanes, and which can only be caused by an intramolecular process (Scheme 7). This process occurs at higher temperatures because it involves breaking of a relatively strong Sn—O$_{eq}$ bond, whereas the low temperature intermolecular scrambling requires the fis-

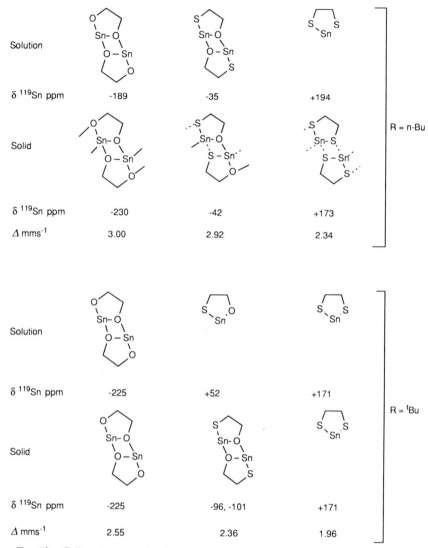

FiG. 10.   Collected structural and spectroscopic data for 1,3-chalcogen-substituted stannolanes. (Adapted from Ref. 202.)

sion of a weaker Sn—O_{ax} bond (212). That an intermediate similar to 59 is
implicated in the high temperature fluxionality also suggests that its intermediacy in certain template reactions involving dioxastannolanes cannot
be ruled out. Monomeric dithiastannolanes are also fluxional with respect
to inversion of their half-chair configurations. This process is slower than

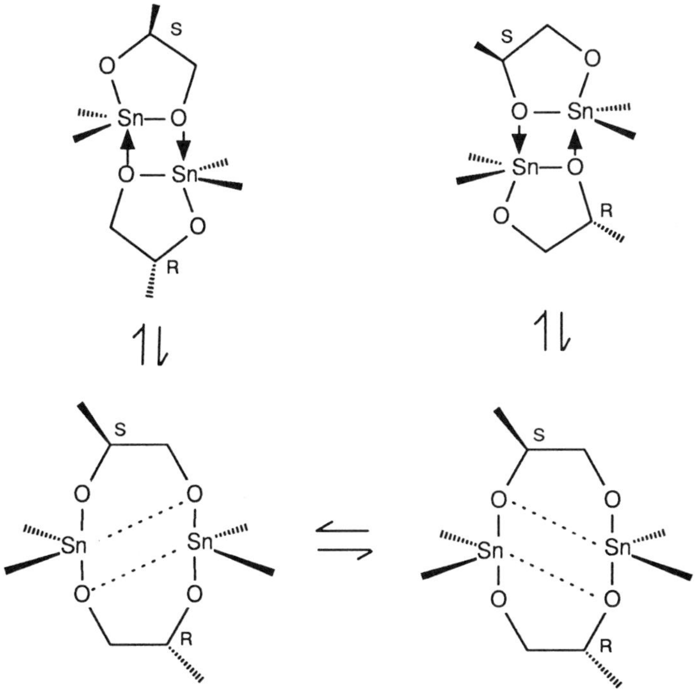

SCHEME 7

that in 1,3-dioxalanes or 1,3-dithialanes and can be studied by NMR methods. The coupling constants which are derived from such a study indicate that the half-chair is markedly puckered, such that the $-CH_2CH_2-$ component of the ring is fully staggered (213). Crystallography has confirmed that the S—C—C—S torsion angle in $Me_2SnSCH_2CH_2S$ is 61.2° (206).

The reaction chemistry of the 1,3,2-dioxastannolanes is vast and has been put to great use in various aspects of organic chemistry. Diorganotin derivatives of carbohydrates incorporate the dioxastannolane ring, and the structures of two such derivatives have already been commented on. The synthetic applications of intermediates of this type have been reviewed elsewhere (214–216) and are not discussed further. Substitution reactions have also proved useful synthetically, for example, in the formation of cyclic tetralactones [Eq. (52)] and urethanes (217–219), a subject which has also been reviewed (220). Two other substitution reactions using electrophilic carbon are shown in Eqs. (53) and (54), which typify several others of the same type (221,222).

$$(52)$$

$$(53)$$

$$(54)$$

> 60%

Addition reactions are also known, both across the polar Sn—O bond and at the metal center. With R_2SnO, a telomerization reaction takes place in which the size of the heterocycle can be enlarged in a controlled way [Eq. (55)] (223). Donor ligands, for example, pyridine (Py),

$$(55)$$

n = 2, 3, 4, 10

N,N-dimethylformamide (DMF), and dimethyl sulfoxide (DMSO), will form complexes with the tin, breaking the polymeric structure of the dioxastannolane but maintaining a coordination number of greater than four at the metal. A cis-R_2SnO_2L structure has been proposed on the basis of ^{119}Sn chemical shifts (-137 to -189 ppm) and Mössbauer data (Δ 2.11–2.82 mm/second). The donor ligand is readily lost under reduced pressure (224).

The chemistry of other five-membered heterocycles containing group 16 elements is limited in comparison with that of 1,3-disubstituted stannolanes. Oxidation (H_2O_2) of $Bu_2Sn(OMe)_2$ in the presence of aldehydes yields 1,3,5,2-trioxastannolanes (**65**) (*225*). Six- and five-membered rings

(**65**)

containing Sn, S (or O), P, and C have been synthesized by oxidation of the appropriate tin hydride with oxygen or sulfur [Eq. (56)] (*226*). The ^{1}H-

$$
\underset{\substack{| \\ Cl}}{\overset{\substack{\\ ||\ \\ O}}{Et_2Sn(CH_2)_nP(OEt)Ph}} \xrightarrow{LiAlH_4} \underset{\substack{| \\ H}}{Et_2Sn(CH_2)_nP(H)Ph} \xrightarrow[-H_2X,\ H_2]{X} Et_2Sn\overset{(CH_2)_n}{\underset{X}{\overset{}{\diagup}}}\underset{Ph}{\overset{X}{P}} \qquad (56)
$$

X = O, S
n = 2, 3

NMR spectra of the five-membered ring compounds indicate the molecules are fluxional, with inversion of configuration at phosphorus [Eq. (57)] (*227*).

$$
R_2Sn\overset{}{\underset{S}{\diagup}}\underset{Ph}{P{=}S} \rightleftharpoons R_2Sn\cdot S\overset{\text{\tiny\ldots}}{\underset{\cdot\cdot S}{P}}\diagdown Ph \rightleftharpoons R_2Sn\overset{}{\underset{S}{\diagup}}\underset{S}{\overset{\text{\tiny\ldots\ldots Ph}}{P}} \qquad (57)
$$

Unsaturated rings containing tin bonded to a group 16 element generally stem from unsaturated dithiolate ligands. Unsaturated analogs to the 1,3-dioxastannolanes have already been mentioned in passing [Eq. (50)] and, though not as widely investigated (*197,228*), do not appear to differ in behavior from the saturated stannolanes. The formation of heterocycles containing tin and S, Se, and Te by the reaction of Sn—Sn bonded compounds has been discussed at length already. The formation of Sn—O heterocycles from oxygen insertion into an Sn—Sn bond is less common [Eq. (43)] (*179*), but one such reaction has been used to form a cyclic, unsaturated, distannoxane. The 1,2-distannacyclobut-3-ene **14** is stable to oxygen in solution but can be oxidized to the stannoxane **66** by 3-chloroperoxybenzoic acid (MCPBA) or by iodination of the Sn—Sn bond followed by base hydrolysis (Scheme 8). However, excess I_2 in moist air forms

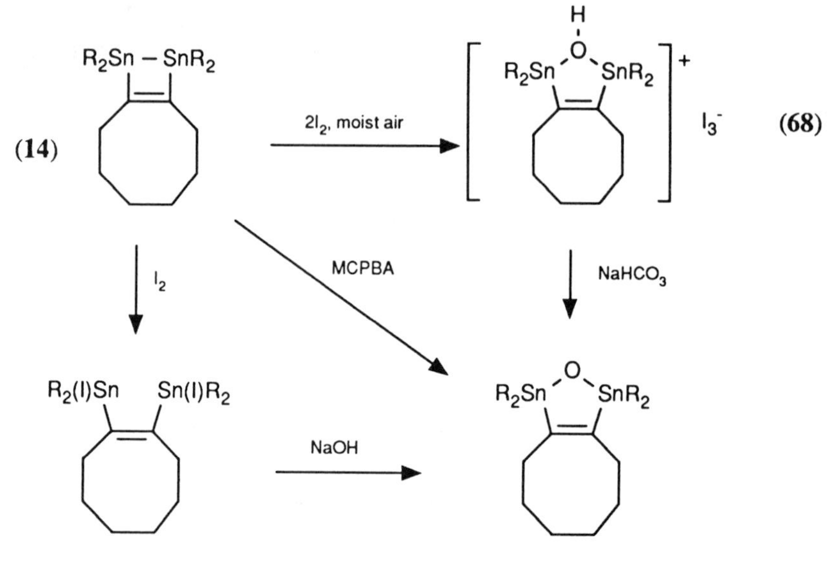

SCHEME 8

the protonated, cyclic distannoxane **68** as its triiodide salt, which can be isolated as bronze, air-stable crystals, and which can be deprotoned by NaHCO₃ to give **66** (*45*). Compound **68** contains an essentially planar five-membered heterocycle (in contrast with the half-chair arrangement of the 1,3,2-dioxastannolanes), with long Sn—O bonds (210.0, 210.8 pm) probably arising from steric congestion of the exocyclic ligands [R = CH(SiMe₃)₂].

Also unusual are orthometallation reactions involving 1-hydroxy-2,6-di-phenylbenzene, or its related lithium salt [Eqs. (58) and (59)] (*229*). Compound **69** is a stannoxane-bridged dimer in the solid state. In **70** the heterocycle is shown in schematic form, but it is identical to that in **69**.

Diorganotin derivatives of substituted aryl dithiolates have been widely studied [Eqs. (60) and (61)]. The products form three distinct structural groupings. Compounds of type **71** are tetrahedral at tin in both solid state

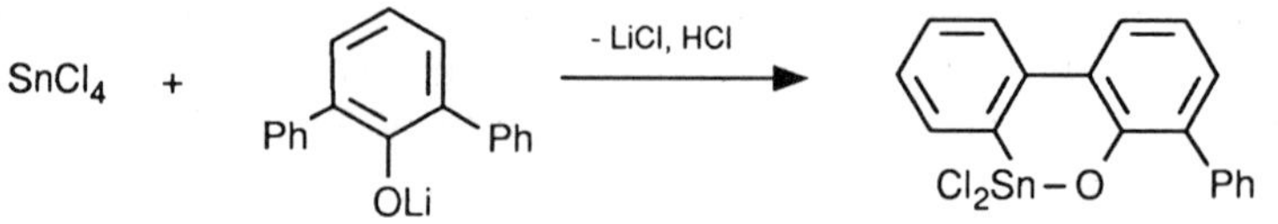

$$Sn(NMe_2)_4 \quad + \quad 2 \text{ [2,6-diphenylphenol]} \xrightarrow{-2HNMe_2} \text{(70)} \quad (59)$$

$$R_2SnO \quad + \quad \text{[dithiol]} \xrightarrow{-H_2O} \text{(71)} \quad (60)$$

$$R_2SnCl_2 \quad + \quad \text{[dithiol]} \xrightarrow[-KCl, -H_2O]{KOH} \text{(72)} \quad (61)$$

(Δ 1.27–1.68 mm/second) and solution, as are **72** in the solution state (*230*). The latter, however, aggregate in the solid state to produce a *cis*-R_2SnS_3 geometry about tin (Δ 1.93–2.62 mm/second) through a single intermolecular Sn–S interaction (*230,231*), in a proposed array similar to 2,2-dimethyl-1,3,2-dithiastannolane (*206,207*) described above. On the other hand, **73**, synthesized by the method of Eq. (61) using the 4-chloro-1,3-dithiol reagent, exists as dimers containing 12-membered rings (Sn—S: 241.9 and 242.0 pm) with no transannular Sn—S interactions

(**73**)

(232) and tetrahedral geometry about tin [v(Sn–S) 330–340 cm^{-1}; Δ 1.54–2.34 mm/second] (233).

Diorganotin dithiolates (72) are sufficiently Lewis acidic to react with quaternary ammonium salts to form five-coordinate complexes of the expected trigonal bipyramidal geometry, with only 14.3% distortion to the rectangular pyramidal form, for example, [Et$_4$N$^+$][(C$_7$H$_6$S$_2$)Ph$_2$SnCl$^-$] (234). Distortions from regular trigonal bipyramidal geometry also increase in the series 74 with increasing size of the halogen, and they are

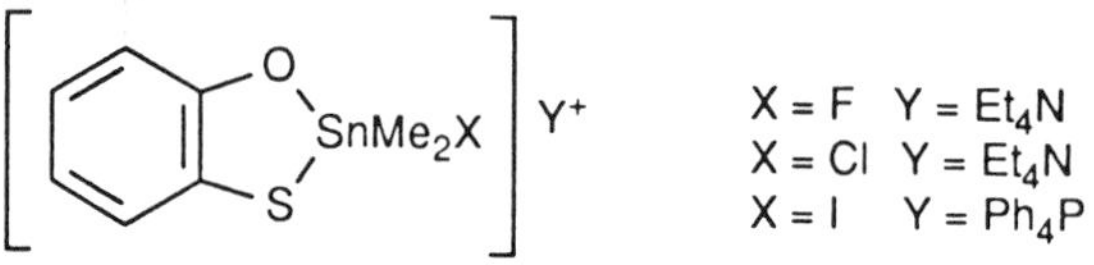

$$X = F \quad Y = Et_4N$$
$$X = Cl \quad Y = Et_4N$$
$$X = I \quad Y = Ph_4P$$

(74)

pronounced in the iodo salt, which dimerizes through weak stannoxane bridges (235). In the monoorganotin bis(dithiolates) [Eq. (62)], the structures are based on the rectangular pyramid (236), completing a sequence of structures which show collectively that tin compounds distort from trigonal bipyramids more easily than silicon or arsenic analogs. Underlining

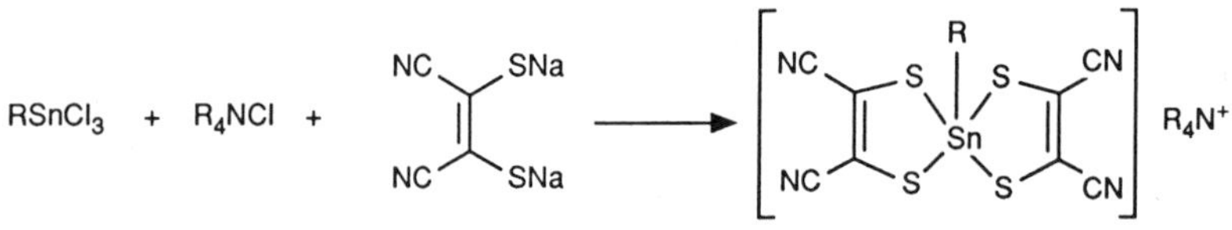

RSnCl$_3$ + R$_4$NCl + [structure] $\longrightarrow$ [structure] R$_4$N$^+$ (62)

the facile nature of the trigonal bipyramidal–rectangular pyramid interconversion, similar reactions with ethane-1,2-dithiolate [Eq. (63)] and

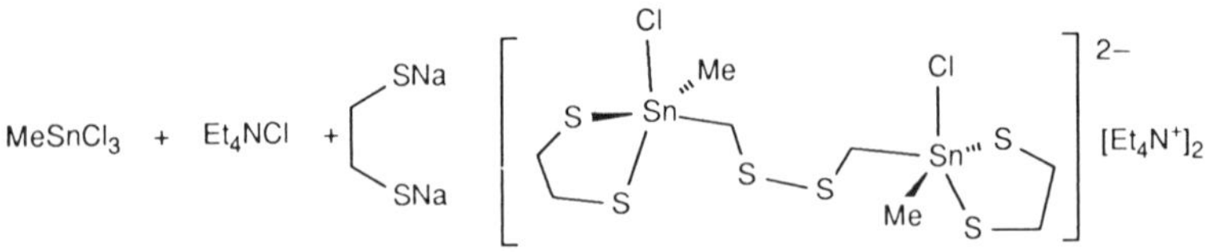

MeSnCl$_3$ + Et$_4$NCl + [structure] [structure]$^{2-}$ [Et$_4$N$^+$]$_2$ (63)

(75)

short reaction times favor the dimeric dianion 75, in which tin switches back to a trigonal bipyramidal geometry (237).

V

STANNATRANES AND RELATED SYSTEMS

This section covers heterocyclic rings which contain lone pair-bearing heteroatoms within the ring so disposed as to enable additional coordination to tin, thereby generating a collection of fused, tin-containing heterocycles. These are shown in **76** and **77**, bicyclic or diptych structures, and **78**, a tricyclic or triptych arrangement. The term "stannatrane" is also commonly used in discussions of these latter systems. These compounds

(76)   (77)   (78)

have been grouped together rather than under the headings suggested by the Sn—X bonds because of their commonality of structure. Reviews of early work in this area exist (*165,238*), as well as reviews of metallatranes in general (*239*), the latter providing a contrast between the species discussed herein and similar derivatives of other elements.

The approach to both diptych and triptych compounds in which an Sn—C bond is formed in the synthetic sequence involves use of di- and trifunctional Grignard reagents with tin halides (*75,165,240,241*). Where the molecular assembly requires the formation of Sn—X bonds (X = O and S), the X—H protons of the ligands are acidic enough to be reacted directly with organotin oxides (*242*) and halides (*243*), or to be converted to sodium salts for reaction with tin halides (*244,245*). In certain instances, the ligands are acidic enough to cause cleavage of Sn—CH$_3$ bonds (*246*). In all cases, the experimental method is enhanced by the use of high dilution techniques to minimize the formation of polymeric materials. Some examples are shown in Eqs. (64)–(67). Dichlorotin compounds can be readily converted to diorganotins by reaction with conventional carbanion sources (*75,240*).

In solution, both the diptych and triptych classes of compounds are

$$E(CH_2CH_2CH_2MgX)_2 \ + \ SnCl_4 \ \xrightarrow{- 2MgXCl} \ Cl_2Sn \longleftarrow :E \qquad (64)$$

$$E = NMe, O, S$$

$$PhSnCl_3 \ + \ S(CH_2CH_2SH)_2 \ \xrightarrow{- 2HCl} \ Ph(Cl)Sn \longleftarrow :S \qquad (65)$$

$$Ph_2SnCl_2 \quad O(CH_2CH_2SNa)_2 \ \xrightarrow{- 2NaCl} \ Ph_2Sn \longleftarrow :O \qquad (66)$$

$$Me_2SnO \ + \ N(CH_2CH_2OH)_3 \ \xrightarrow{trace\ KOH} \ MeSn \longleftarrow :N \qquad (67)$$

fluxional. The diptych compounds scramble R′ and R″ (76) through a dissociation–inversion mechanism, whereas in the more stable triptych compounds the axial donor, E, can breathe "in and out" of the tin coordination sphere, with a typical activation energy of 70 kJ/mol (165). In the solid state, the collected data in Table VI (248,249) allow the following conclusions to be drawn. The E: → Sn coordination is a function of the Lewis acidity of the metal and is weakest when R′ = R″ = X = alkyl, though even here the nature of the triptych ligand enforces five-coordination on a tetraorganotin (247). Octahedral coordination at tin is possible when two donor atoms are suitably disposed within the ring (75,166). When the Lewis acidity of tin is high owing to bonding to several electronegative atoms, and if only one donor atom E is available for bonding, then the molecules associate to expand the coordination number of tin. For example, $MeSn(OCH_2CH_2)_3N$ is a trimer through stannoxane bridges, with the central tin in a pentagonal bipyramidal coordination sphere, and the outer two tin atoms six-coordinate (246). The available structures also suggest that nitrogen is a stronger donor than either oxygen or sulfur.

TABLE VI

SELECTED STRUCTURAL DATA FOR STANNATRANES AND RELATED COMPOUNDS

Compound	X	E	R'	R''	Sn—E (pm)	<ESnR' (°)	<ESnX (°)	Ref.
76	S	O	Ph	Ph	266.0	81.8	74.6, 75.3	244
76	CH$_2$	O	Cl	Cl	238.4	91.6	71.0, 78.1	241
76	CH$_2$	NMe	S	S	256			165
76	S	NMe	Me	Me	256.6	86.9	77.9, 78.2	248
76	CH$_2$	NMe	Cl	Cl	244.1	89.4	78.3, 77.9	241
77	O	NMe	^{t}Bu	^{t}Bu	232^a	107.8–123.55	74.5–75.9	242
76	S	S	Ph	Ph	324.6	79.3	72.7, 76.1	245
76	CH$_2$	S	Cl	Cl	285.1	88.9	78.6, 79.2	241
76	S	S	Ph	Cl	280.6	89.1	83.0, 83.3	243
78	CH$_2$	N	R = Me		262.4	74.4, 74.9, 74.9		247
78	CH$_2$	N	R = Cl		237.2	80.5		249

^{a} Mean of three molecules in the asymmetric unit.

Second, the choice of geometric isomer, 76 or 77, is governed by the relative electronegativities of R, X, and E, with the most electronegative atoms going in the axial positions, allowing for the constraints imposed by the linkages within the ligands. In practice this means that 77 is rare, only occurring when X is O (242). On the other hand, several other types of ligands are known which fulfill this requirement, including dipeptides (250,251) and the anions of certain Schiff bases (252–254), and which expand the range of compounds of structure 77 considerably.

The reaction chemistry of labile halogen ligands in these compounds is not unusual. A variety of nucleophiles react with loss of Cl$^-$ (75,165,166), in addition to the reactions of carbanions already mentioned. More interestingly, the diptych tin dichlorides can be coupled with lithium metal to give Sn—Sn bonded dimers which can exist in three isomeric forms, all of which can be distinguished by their NMR spectra. The example shown in 79 is the aa isomer (both nitrogen atoms axial), other isomers being ae and ee. The parent dichloride (79, X = Cl) crystallizes with the structure shown (Sn—Sn 283.1 pm), but when X is Me all three isomers are seen, with the one shown in 79 dominant. In the former case, the electronegativities of N and Cl dictate their location exclusively in axial sites. Rotation about the Sn—Sn bond is hindered (ΔE_{act} 11.5 kcal/mol) (255). Analogous isomeric forms also exist in 80, prepared from PhSn(O)CH$_2$Sn(O)Ph and MeN(CH$_2$CH$_2$SH)$_2$ (256). The isomer shown has been investigated crystallographically and has the methylene bridge both axial and equatorial with respect to the two tin centers (ae) (256). However, in solution all three aa isomer   ae isomer

(79)   (80)

isomers interconvert, which two-dimensional NMR experiments show is in an uncorrelated manner with regard to the two tin environments (257).

VI

HETEROCYCLES CONTAINING TIN–TRANSITION ELEMENT BONDS

Metallacycles containing tin and transition elements are well known, although a surprisingly small percentage contain an organotin unit. It is therefore important to be aware of the heterocycles containing bonds between a transition metal and tin in an inorganic form in order to appreciate the totality of the advances made in recent years. Fortunately, several reviews of this area are available (258,259) to which the reader is directed. Many of the more exotic structures emanate from precursors which incorporate multiple bonds between tin and the d-block element, and, although missing much recent work, the 1986 review by Herrmann serves as a valuable entrée into this field (260).

In contrast to rings containing all p-block elements in addition to tin, the metallacycles under discussion readily form small rings, predominantly of three or four atoms in total, which in turn can fuse together to form more complex aggregates. Large rings, even of modest six-atom proportions, are scarce.

A. Heterocycles Containing Iron, Ruthenium, and Osmium

Heterocycles containing tin bonded to one of the group 8 triad are by far the most widely studied. Early work in the field has been reviewed (261).

The smallest heterocycle is that containing three atoms, SnM_2. Despite its widespread occurrence in many compounds, the isolated triangular heterocycle has been rarely observed, one example arising when a hindered stannylene, R_2Sn: [$R = CH(SiMe_3)_2$], displaces CO from $Fe_2(CO)_9$ (262) or $Fe_3(CO)_{12}$ [Eq. (68)] (263). Characterization of the product has been on

$$[(Me_3Si)_2CH]_2Sn: \quad + \quad \begin{matrix} Fe_2(CO)_9 \\ or \\ Fe_3(CO)_{12} \end{matrix} \quad \longrightarrow \quad [(Me_3Si)_2CH]_2Sn \diagdown \begin{matrix} Fe(CO)_4 \\ | \\ Fe(CO)_4 \end{matrix} \quad + \quad CO \quad (68)$$

the basis of its mass spectrum, which shows sequential loss of all CO groups from the parent and which awaits crystallographic confirmation.

More common are compounds containing the SnM_2 unit in concert with other bridging ligands to produce fused heterocycles. The simplest of the bridging groups is CO, since this is a common feature of many of the transition metal precursors. Compound 81, which comprises two fused rings and both tin and CO bridges, can be obtained by photochemical elimination of CO from acyclic $Me_2Sn[Fe(CO)_2Cp]_2$ (264). It exists in cis and trans isomeric forms in solution (81 shows the latter). Surprisingly, the $^2J(^{119}Sn-C-^1H)$ coupling is insensitive to the angular changes at tin on cyclization. The three fused-ring compound $[\mu-RR'Sn]_2Fe_2(CO)_7$ (82) (from photolysis of $[RR'SnFe(CO)_4]_2$) exists in three isomeric forms depending on the relative orientations of R and R' with respect to the bridging CO, and it shows fluxionality of R and R' between axial and equatorial sites as well as interchange of CO between bridging and terminal positions (265,266). R group exchange occurs by two processes, one at lower temperature involving "flapping" of the bimetallacycle and migration of the bridging CO (ΔG 11 kcal/mol) and a second higher temperature mode in which Sn—Fe and Fe—Fe bonds break to give a stannylene donor intermediate, $RR'Sn_2$: → Fe (ΔG 19.2 kcal/mol) (265).

Intermediate between these two forms of heterocycles is $[\mu-R_2Sn]_2Fe_2(CO)_6$ [$R = CH(SiMe_3)_2$] (83). Synthesized by reaction of R_2Sn: and $Fe_3(CO)_{11}(MeCN)$, 83 is related to 82 by loss of CO. The short

(81)   (82)   (83)

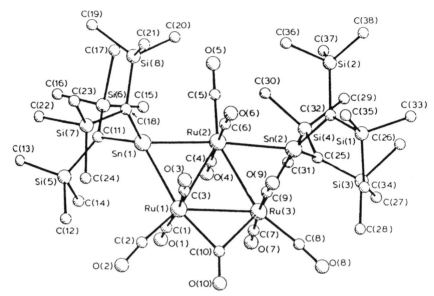

FIG. 11.   Structure of the pentametallic cluster [μ-SnR₂]₂Re₃(CO)₁₀ [R = CH(SiMe₃)₂].
The corresponding osmium compound is isostructural. (Reproduced from Ref. *263* with
permission from Elsevier Sequoia.)

iron–iron contact is presumably a formal double bond to maintain 18
electrons at iron and arises from the steric interactions between bulky R
groups which push the tin atoms apart (*263*). The now planar Fe₂Sn₂
assembly is clearly related to [Me₂SnFe(CO)₄]₂ (see below) which does not
require iron–iron bonds and, as a consequence of less hindered groups on
tin, has a less asymmetric "lozenge" shape (*267*).

    Higher aggregations of SnM₂ occur in compounds derived from the
same hindered stannylene and cluster carbonyls of the heavier group
8 elements. These reactions, taken together with those already described
for iron, show the diversity of products which can be formed from
relatively minor changes in transition metal precursor. Reaction with
Ru₃(CO)₁₂ or Ru₃(CO)₁₀(MeCN)₂ gives the pentametallic cluster [μ-
R₂Sn]₂Ru₃(CO)₁₀ [R = CH(SiMe₃)₂]. Tin atoms cap two edges of an Ru₃
triangle to give a planar structure (Fig. 11). Os₃(CO)₁₂ does not react with
R₂Sn, but Os₃(CO)₁₂(MeCN)₂ gives the isostructural pentametallic cluster
[μ-R₂Sn]₂Os₃(CO)₁₀ (Fig. 11) (*263*). On the other hand, when the osmium
source is Os₃H₂(CO)₁₀ the product (**84**) is planar and has both Sn—Os and
Os—Os bonds supported by hydride bridges (*268*). This compound is a
useful source of other compounds containing the SnOs₂ heterocycle, either
by thermal rearrangement (*268*) or by reaction with alkynes (*269*).

$$R_2Sn$$

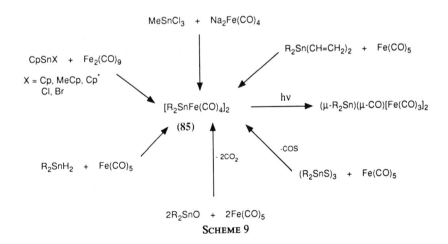

(84)

The Sn_2Fe_2 heterocycle is accessible by a number of routes (Scheme 9), and it is surprising that no analogs containing Ru or Os have been observed (267,270–275). Two different groups on tin give rise to isomeric forms analogous to **82**. Also notable is the change in hydrocarbon bonding from $(\eta^5\text{-}C_5H_5)_2Sn$ to $[(\eta^1\text{-}C_5H_5)_2SnFe(CO)_4]_2$ (272). In two structure determinations [R = Me (267) and Cp (272)] the Sn—Fe bonds (264.7 and 265.1, 267.0 pm) and internal Sn—Fe—Sn (102.04°) and Fe—Sn—Fe (77, 77.96°) are virtually identical. Addition of base cleaves the dimer to give stannylene complexes of general type base: → R_2Sn: → $Fe(CO)_4$, in which tin acts as both an acid and a base (276). The dimer is also unstable in refluxing toluene, with cyclopentadiene transfer from tin to iron occurring, along with skeletal rearrangement to give the cluster $[CpFe(CO)_2]_2Sn_2Fe_3(CO)_9$ (277).

In addition to $[Me_2SnFe(CO)_4]_2$, reaction of $Na_2Fe(CO)_4$ with $MeSnCl_3$ generates a second compound, $Me_4Sn_3Fe_4(CO)_{16}$ (**86**), in which two planar

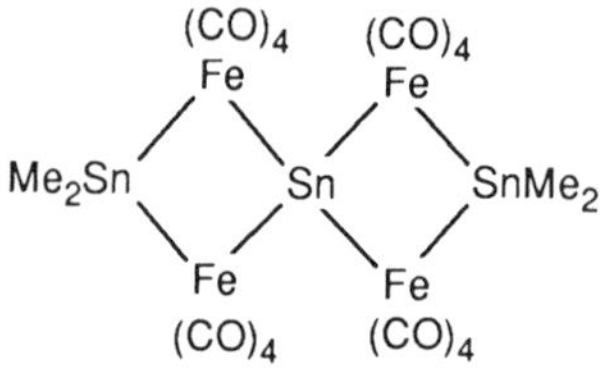

$$\begin{array}{ccc}
& (CO)_4 & (CO)_4 \\
& Fe & Fe \\
Me_2Sn & \diagdown \quad Sn \quad \diagdown & SnMe_2 \\
& Fe & Fe \\
& (CO)_4 & (CO)_4
\end{array}$$

(86)

Sn_2Fe_2 rings are linked through a spiro center (Sn). Bond lengths involving the terminal tin atoms are in keeping with those in **85** (262.5 pm) but are longer when the spiro tin is involved (274.7 pm) (278). The largest ring incorporating the Sn—Fe bond also contains four carbon atoms and is obtained from insertion of the $Fe(CO)_4$ fragment into the cyclic Sn—C bond of a stannacyclopentane [Eq. (69)] (21).

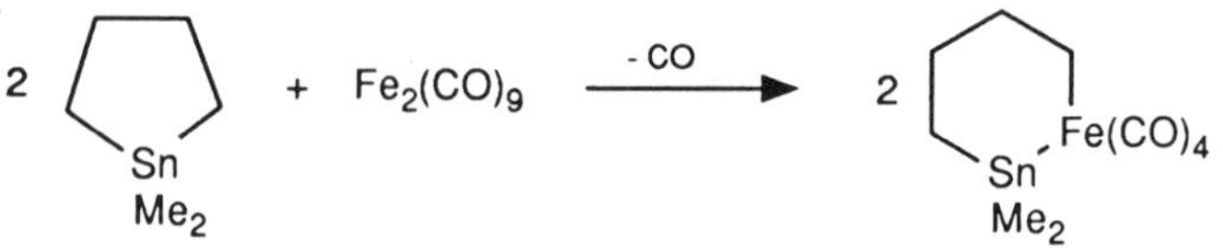

$$2 \underset{\substack{Sn \\ Me_2}}{\bigcirc} + Fe_2(CO)_9 \xrightarrow{\ -CO\ } 2 \underset{\substack{Sn \\ Me_2}}{\bigcirc}\!\!{}_{Fe(CO)_4} \qquad (69)$$

B. Heterocycles Containing Other Transition Elements

A number of the reactions and products described for group 8 elements have parallels with other d-block members. Photochemical elimination of CO from either $Me_2Sn[Co(CO)_4]_2$ or $Me_2ClSnMn(CO)_5$ gives $[Me_2SnCo(CO)_3]_2$ (**87**) and $(\mu\text{-}Me_2Sn)(\mu\text{-}CO)[Co(CO)_3]_2$ or $[Me_2SnMn(CO)_4]_2$ all of which contain triangular SnM_2 heterocycles (264). Compound **87** can also be synthesized from $Co_2(CO)_8$ and Me_2SnH_2, and it has C_{2v} symmetry akin to **82** (without the bridging CO). Also like **82**, the molecule readily exchanges axial and equatorial methyl groups (279). $MeSnCo_3(CO)_9$ (**88**), synthesized from Me_2SnCl_2 and

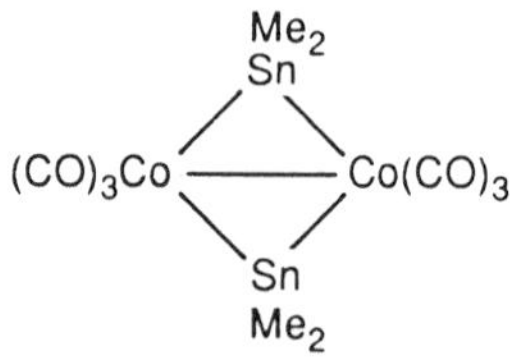

$$\underset{(87)}{\underset{\substack{Sn \\ Me_2}}{\overset{\substack{Me_2 \\ Sn}}{(CO)_3Co \diagup\!\!\!\diagdown Co(CO)_3}}} \qquad\qquad \underset{(88)}{\underset{Co(CO)3}{\overset{\substack{Me \\ Sn}}{(CO)_3Co \diagup\!\!\!\diagdown Co(CO)_3}}}$$

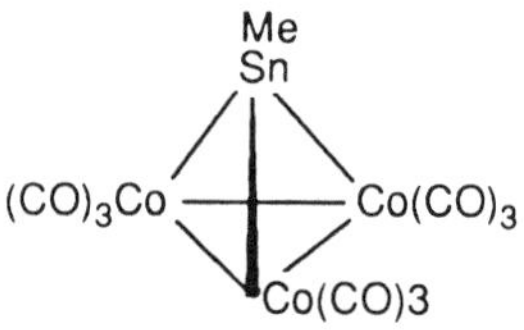

(87)   (88)

$Co_2(CO)_8$ in 34% yield, adopts a tetrahedrane structure with three fused $SnCo_2$ rings (280).

The Sn_2Co_2 ring, without a cross-ring cobalt–cobalt interaction, has been synthesized by the reaction of Me_6Sn_2 and $CpCo(CO)_2$ in a sealed tube at 160°C. The product, $[Me_2SnCo(CO)Cp]_2$, is isoelectronic and isostructural with $[Me_2SnFe(CO)_4]_2$ (85). The Sn—Co bond (254.2 pm) is slightly shorter than in the iron compound, but the internal angle at cobalt (78°) is essentially identical (281).

Both isolated and fused $SnMn_2$ rings are also known, though the former (89) also includes a hydride bridge across the Mn—Mn bond (282). Syntheses are shown in Eq. (70). Compound 90 is isoelectronic with

$[Me_2SnMn(CO)_4]_2$ (cf. 87) which has been synthesized photochemically from $ClMe_2SnMn(CO)_5$ (264). In 90, the Sn—Mn bond lengths are 260.53 and 261.78 pm, and the Sn—Mn—Sn angle 72.94°. In keeping with formal electron counting which requires an Mn—Mn bond is the Mn—Mn separation of 310.45 pm.

The $SnRe_2$ unit occurs in $Re_3(\mu\text{-}H)(\mu\text{-}SnMe_2)(CO)_{12}$, though the bridging hydride (from its inferred position) induces considerable asymmetry into the pair of Sn—Re bonds (mean values 287.1 and 267.5 pm) [Eq. (71)] (283).

Two heterocycles with a common structural unit are shown in **91** and **92**, the products of Eqs. (72) and (73), respectively (*284,285*). Though

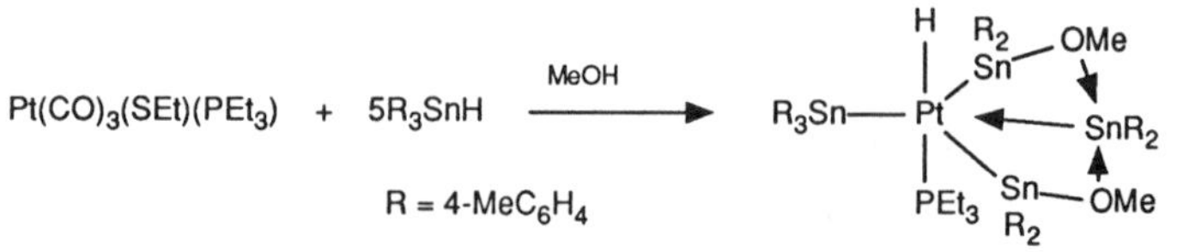

$$Pt(CO)_3(SEt)(PEt_3) \quad + \quad 5R_3SnH \quad \xrightarrow{\text{MeOH}} \quad (72)$$

R = 4-MeC$_6$H$_4$

(91)

$$W(CO)_3(PMTA) \ ^+ \ 6K + Ph_3SnCl \quad \xrightarrow{\text{EtOH}} \quad \{[(Ph_3Sn)_2[(Ph_2Sn)_2OEt]W(CO)_3]^- \quad + \quad [HW(CO)_3(SnPh_3)_3]^{2-}$$

PMTA = Me$_2$N⌒N⌒NMe$_2$

recrystallize
from ^{i}PrOH

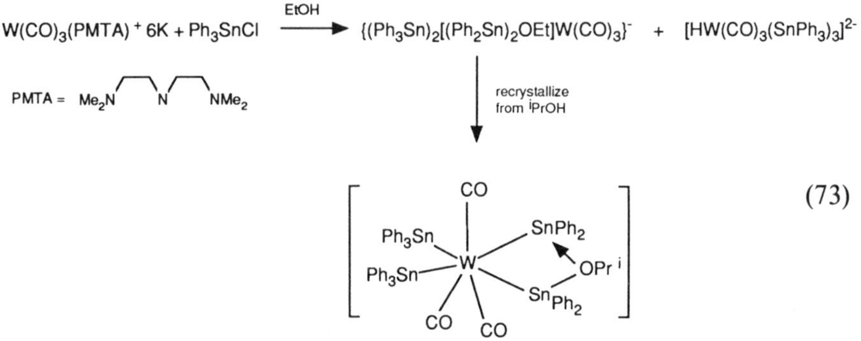

(73)

(92)

formal assignment of oxidation state is difficult and often belies the complexity of the electron distribution, both compounds can be thought of as having stannylene donors to the transition element, in which Sn(II) is complexed by either one (**92**) or two (**91**) alkoxy groups of a Sn(IV) species. An alternative electron counting of the tungsten compound would be $[(Ph_3Sn)_2(CO)_3W]^{2-}$ chelated by $[Ph_2SnO(^iPr)SnPh_2]^+$ (cf. **68**). In both cases, MSn$_2$O heterocycles are generated. Distinct Sn—Pt bond lengths to each of the three types of tin are observed in **91** (Sn^{2+}: 262.6; SnOMe: 258.5, 259.5; Ph$_3$Sn: 264.5 pm), whereas the mean Sn—W separation in **92** is 278.2 pm.

Finally, Me$_2$Sn(CH=CH$_2$)$_2$ displaces ethene from Pt(C$_2$H$_4$)$_2$(PR$_3$)$_2$ and coordinates the platinum through π-bonding vinyl groups to form what is nominally a six-membered ring (**93**) (*286*).

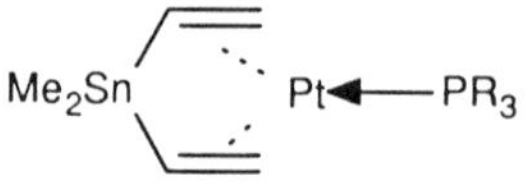

$$Me_2Sn \cdots Pt \longleftarrow PR_3$$

(93)

VII

CONCLUDING REMARKS

Major advances in both the control of ring size and heteroatom composition have been made in the last 10 years in particular. Notable omissions are rings with the heavier group 13 and 15 elements, but the synthetic methodology is available to meet these challenges. Also still largely absent are rings with functional groups, particularly halogens, on the periphery, such that further reactions of the heterocycle become possible. In particular, the possibility of fusing heterocycles through such functional groups offers the prospect of generating new generations of cage structures with all the technological implications, for example, host–guest chemistry, of these types of compounds.

REFERENCES

1. N. S. Hosmane and J. A. Maguire, *Adv. Organomet. Chem.* **30**, 99 (1990).
2. For example, S. P. Narula, S. K. Bharadwaj, H. K. Sharma, G. Mairesse, P. Barbier, and G. Nowogrocki, *J. Chem. Soc., Dalton Trans.*, 1719 (1988).
3. S.-W. Ng and J. J. Zuckerman, *J. Chem. Soc., Chem. Commun.*, 475 (1982).
4. H. Schmidbauer, T. Prost, B. Huber, O. Steigelmann, and G. Müller, *Organometallics* **8**, 1567 (1989).
5. A. G. Davies, H. J. Milledge, D. C. Puxley, and P. J. Smith, *J. Chem. Soc. (A)*, 2862 (1970).
6. A. G. Davies and P. J. Smith, *in* "Comprehensive Organometallic Chemistry" (G. Wilkinson, F. G. A. Stone, and E. W. Abel, eds.), Chap. 11, p. 519. Pergamon, Oxford, 1982.
7. K. C. Molloy, *in* "Chemistry of Tin" (P. G. Harrison, ed.), Chap. 6, p. 187. Blackie, Glasgow, 1989.
8. A. K. Sawyer, ed. "Organotin Compounds," Vols. 1–3. Dekker, New York, 1971.
9. B. C. Pant, *J. Organomet. Chem.* **66**, 321 (1974).
10. Ref. *6*, p. 542.
11. J. W. F. L. Seetz, G. Schat, O. S. Ackerman, and F. Bickelhaupt, *J. Am. Chem. Soc.* **105**, 3336 (1983).
12. A. G. Davies, M.-W. Tse, J. D. Kennedy, W. McFarlane, G. S. Pyne, M. F. C. Ladd, and D. C. Povey, *J. Chem. Soc., Chem. Commun.*, 791 (1978).
13. A. G. Davies, M.-W. Tse, J. D. Kennedy, W. McFarlane, G. S. Pyne, M. F. C. Ladd, and D. C. Povey, *J. Chem., Soc., Perkin Trans. 2*, 791 (1978).
14. E. J. Bulten and H. A. Budding, *J. Organomet. Chem.* **82**, C13 (1974).
15. M. Gielen and J. Topart, *J. Organomet. Chem.* **81**, 359 (1974).
16. M. Gielen, M. de Clerq, and J. Nasielski, *Bull. Chim. Soc. Belg.* **78**, 237 (1969).
17. P. J. Smith and A. P. Tupčiauskas, *Annu. Rep. NMR Spectrosc.* **8**, 291 (1978).
18. V. S. Petrosyan, *Prog. NMR Spectrosc.* **11**, 115 (1977).
19. B. Wrackmeyer, *Annu. Rep. NMR Spectrosc.* **16**, 73 (1985).
20. F. J. Bajer and H. W. Post, *J. Organomet. Chem.* **11**, 187 (1968).
21. E. J. Bulten and H. A. Budding, *J. Organomet. Chem.* **166**, 339 (1979).
22. E. J. Bulten and H. A. Budding, *J. Organomet. Chem.* **137**, 165 (1977).

23. A. G. Davies, B. P. Roberts, and M.-W. Tse, *J. Chem. Soc., Perkin Trans.* 2, 1499 (1977).

24. D. Seyferth and J. L. Lefferts, *J. Am. Chem. Soc.* **96,** 6237 (1974).

25. D. Seyferth and J. L. Lefferts, *J. Organomet. Chem.* **116,** 257 (1976).

26. J. W. Bruin, G. Schat, O. S. Ackerman, and F. Bickelhaupt, *J. Organomet. Chem.* **288,** 13 (1985).

27. D. Seyferth and S. C. Vick, *Synth. React. Inorg. Met.-Org. Chem.* **4,** 515 (1974).

28. A. Maercker, F. Bravers, W. Breiden, M. Jung, and H. D. Lutz, *Angew. Chem., Int. Ed. Engl.* **27,** 404 (1988).

29. K. Jurkschat, H. G. Kuivila, S. Liu, and J. A. Zubieta, *Organometallics* **8,** 2755 (1989).

30. K. Jurkschat, A. Rühlemann, and A. Tzschach, *J. Organomet. Chem.* **381,** C53 (1990).

31. M. Newcombe, Y. Azuma, and A. R. Courtney, *Organometallics* **2,** 175 (1983).

32. Y. Azuma and M. Newcombe, *Organometallics* **3,** 9 (1984).

33. M. Newcombe, M. T. Blanda, Y. Azuma, and T. J. Delord, *J. Chem. Soc., Chem. Commun.,* 1159 (1984).

34. M. Newcombe, J. H. Horner, and M. T. Blanda, *J. Am. Chem. Soc.* **109,** 7878 (1987).

35. M. Newcombe, A. M. Madonik, M. T. Blanda, and J. K. Judice, *Organometallics* **6,** 145 (1987).

36. M. Newcombe, J. H. Horner, M. T. Blanda, and P. J. Squattrito, *J. Am. Chem. Soc.* **111,** 6294 (1989).

37. K. Jurkschat and M. Gielen, *J. Organomet. Chem.* **236,** 69 (1982).

38. T. N. Mitchell, B. Fabisch, R. Wickenkamp, H. G. Kuivila, and T. J. Karol, *Si, Ge, Sn, Pb Cmpds.* **9,** 57 (1986).

39. H. Preut and T. N. Mitchell, *Acta Crystallogr.* **C45,** 35 (1989).

40. J. Meunier-Piret, M. van Meerssche, M. Gielen, and K. Jurkschat, *J. Organomet. Chem.* **252,** 289 (1983).

41. H. Preut, P. Bleckmann, T. N. Mitchell, and B. Fabisch, *Acta Crystallogr.* **C40,** 370 (1984).

42. H. Killing and T. N. Mitchell, *Organometallics* **3,** 1917 (1984).

43. J. D. Cotton, P. J. Davidson, and M. F. Lappert, *J. Chem. Soc., Dalton Trans.,* 2275 (1976).

44. L. R. Sita and R. D. Bickerstaff, *J. Am. Chem. Soc.* **110,** 5208 (1988).

45. L. R. Sita, I. Kinoshita, and S. P. Lee, *Organometallics* **9,** 1644 (1990).

46. A. Krebs, A. Jacobsen-Bauer, E. Haupt, M. Vieth, and V. Huch, *Angew. Chem., Int. Ed. Engl.* **28,** 603 (1989).

47. L. Killian and B. Wrackmeyer, *J. Organomet. Chem.* **132,** 213 (1977).

48. B. Wrackmeyer, *J. Organomet. Chem.* **364,** 331 (1989).

49. W. A. Gustavson, L. M. Principe, W. Z. M. Rhee, and J. J. Zuckerman, *J. Am. Chem. Soc.* **103,** 4126 (1981).

50. W. Z. M. Rhee and J. J. Zuckerman, *Synth. React. Inorg. Met.-Org. Chem.* **11,** 633 (1981).

51. A. J. Ashe and S. Mahmoud, *Organometallics* **7,** 1878 (1988).

52. L. Killian and B. Wrackmeyer, *J. Organomet. Chem.* **148,** 137 (1978).

53. V. R. Sandel, G. R. Buske, S. G. Maroldo, D. K. Bates, D. Whitman, and G. Sypmewski, *J. Org. Chem.* **46,** 4069 (1981).

54. R. Usón, J. Vicente, and M. T. Chicote, *J. Organomet. Chem.* **209,** 271 (1987).

55. L.-W. Gross, R. Moser, W. P. Neumann, and K.-H. Scherping, *Tetrahedron Lett.* **23,** 635 (1982).

56. C. Grugel, W. P. Neumann, and Schreiwer, *Angew. Chem., Int. Ed. Engl.* **18,** 543 (1979).

57. W. A. Gustavson, L. M. Principe, W. Z. M. Rhee, and J. J. Zuckerman, *Inorg. Chem.* **20**, 3460 (1981).
58. W. Z. M. Rhee and J. J. Zuckerman, *J. Am. Chem. Soc.* **97**, 2291 (1975).
59. G. Märkl and F. Kneidl, *Angew. Chem., Int. Ed. Engl.* **12**, 931 (1973).
60. G. Märkl and F. Kneidl, *Angew. Chem., Int. Ed. Engl.* **13**, 667 (1974).
61. G. Märkl, H. Baier, and S. Heinrich, *Angew. Chem., Int. Ed. Engl.* **14**, 710 (1975).
62. A. J. Ashe, W.-T. Chan, and E. Perozzi, *Tetrahedron Lett.,* 1083 (1975).
63. A. J. Ashe and W.-T. Chan, *J. Org. Chem.* **44**, 1409 (1979).
64. A. J. Ashe and T. R. Diephouse, *J. Organomet. Chem.* **202**, C95 (1980).
65. A. J. Ashe and F. J. Drone, *J. Am. Chem. Soc.* **109**, 1879 (1987).
66. A. J. Ashe and S.-T. Abu-Orabi, *J. Org. Chem.* **48**, 767 (1983).
67. P. Jutzi and J. Baumgärtner, *J. Organomet. Chem.* **148**, 247 (1978).
68. G. Maier, H.-J. Wolf, and R. Boese, *Chem. Ber.* **123**, 505 (1990).
69. G. Märkl, H. Baier, and R. Liebl, *Liebigs Ann. Chem.,* 919 (1987).
70. G. Märkl and D. Rudnic, *J. Organomet. Chem.* **181**, 305 (1979).
71. G. Märkl and P. Hofmeister, *Tetrahedron Lett.,* 2569 (1977).
72. J. Meinwald, S. Knapp, T. Tatsuoka, J. Finer, and J. Clardy, *Tetrahedron Lett.,* 2247 (1977).
73. H. J. R. de Boer, O. S. Ackerman, and F. Bickelhaupt, *J. Organomet. Chem.* **321**, 291 (1987).
74. J. Heinicke, *J. Organomet. Chem.* **364**, C17 (1989).
75. K. Jurkschat, J. Kalbitz, M. Dargatz, E. Kleinpeter, and A. Tzschach, *J. Organomet. Chem.* **347**, 41 (1988).
76. B. J. J. van de Heisteeg, G. Schat, O. S. Ackerman, and F. Bickelhaupt, *J. Organomet. Chem.* **308**, 1 (1986).
77. L. Killian and B. Wrackmeyer, *J. Organomet. Chem.* **153**, 153 (1978).
78. C. Bihlmayer, S.-T. Abu-Orabi, and B. Wrackmeyer, *J. Organomet. Chem.* **332**, 25 (1987).
79. H.-O. Berger, H. Nöth, G. Rub, and B. Wrackmeyer, *Chem. Ber.* **113**, 1235 (1980).
80. G. Märkl and D. Matthes, *Tetrahedron Lett.,* 2599 (1976).
81. G. Märkl, H. Baier, and R. Liebl, *Synthesis,* 842 (1977).
82. M. Murakami, Y. Morita, and Y. Ito, *J. Chem. Soc., Chem. Commun.,* 428 (1990).
83. S. Kerschl, B. Wrackmeyer, A. Willhalm, and A. Schmidpeter, *J. Organomet. Chem.* **319**, 49 (1987).
84. S. Kerschl and B. Wrackmeyer, *J. Chem. Soc., Chem. Commun.,* 1199 (1985).
85. I. Lengyel and M. J. Aaronson, *Angew. Chem., Int. Ed. Engl.* **2**, 161 (1970).
86. H. A. Meinema and J. G. Noltes, *J. Organomet. Chem.* **63**, 243 (1973).
87. W. T. Pennington, A. W. Cordles, J. C. Graham, and Y. W. Jung, *Acta Crystallogr.* **C39**, 712 (1983).
88. I. Lengyel and M. J. Aaronson, *Angew. Chem., Int. Ed. Engl.* **6**, 521 (1972).
89. V. I. Belsky, I. E. Saratov, V. O. Reikhsfeld, and A. A. Simonenko, *J. Organomet. Chem.* **258**, 283 (1983).
90. W. Z. McCarthy, J. Y. Coey, and E. R. Corey, *Organometallics* **3**, 255 (1984).
91. W. P. Neumann and K. König, *Annalen* **667**, 1 (1964).
92. W. P. Neumann and K. König, *Angew. Chem., Int. Ed. Engl.* **3**, 751 (1964).
93. H. Puff, C. Bach, H. Reuter, and W. Schuh, *J. Organomet. Chem.* **277**, 17 (1984).
94. H.-P. Ritter and W. P. Neumann, *J. Organomet. Chem.* **56**, 199 (1973).
95. B. Watta, W. P. Neumann, and J. Sauer, *Organometallics* **4**, 1954 (1985).
96. V. K. Belski, N. N. Zemlyanski, N. D. Kolosova, and I. V. Borisova, *J. Organomet. Chem.* **215**, 41 (1981).

97. J. Fu and W. P. Neumann, *J. Organomet. Chem.* **272,** C5 (1984).
98. S. Masamune and L. R. Sita, *J. Am. Chem. Soc.* **105,** 631 (1983).
99. S. Masamune and L. R. Sita, *J. Am. Chem. Soc.* **107,** 6390 (1985).
100. W. V. Farrer and H. A. Skinner, *J. Organomet. Chem.* **1,** 434 (1964).
101. H. Puff, C. Bach, W. Schuh, and R. Zimmer, *J. Organomet. Chem.* **312,** 313 (1986).
102. M. F. Lappert, W.-P. Leung, C. L. Raston, A. J. Thorne, B. W. Skelton, and A. H. White, *J. Organomet. Chem.* **233,** C28 (1982).
103. L. R. Sita and R. D. Bickerstaff, *J. Am. Chem. Soc.* **111,** 3769 (1989).
104. S. Adams and M. Dräger, *J. Organomet. Chem.* **288,** 295 (1985).
105. M. Dräger, B. Mathiasch, L. Ross, and M. Ross, *Z. Anorg. Allg. Chem.* **506,** 99 (1983).
106. D. H. Olsen and R. E. Rundle, *Inorg. Chem.* **2,** 1310 (1963).
107. M. Veith, *Chem. Rev.* **90,** 1 (1990).
108. K. Jones and M. F. Lappert, *Organomet. Chem. Rev.* **1,** 67 (1966).
109. D. Hänssgen and I. Pohl, *Angew. Chem., Int. Ed. Engl.* **13,** 607 (1974).
110. H. Puff, D. Hänssgen, N. Beckermann, A. Roloff, and W. Schuh, *J. Organomet. Chem.* **373,** 37 (1989).
111. T. Kitazume and J. M. Schreeve, *Inorg. Chem.* **16,** 2040 (1977).
112. K. E. Peterman and J. M. Schreeve, *Inorg. Chem.* **15,** 743 (1976).
113. D. Hänssgen, J. Kuna, and B. Ross, *Chem. Ber.* **109,** 1799 (1976).
114. S. Sakai, Y. Fugimura, and Y. Ishi, *J. Organomet. Chem.* **50,** 113 (1973).
115. D. Hänssgen and E. Odenhausen, *J. Organomet. Chem.* **124,** 143 (1977).
116. C. Stader and B. Wrackmeyer, *J. Organomet. Chem.* **321,** C1 (1987).
117. M. F. Mahon, K. C. Molloy, and P. C. Waterfield, *J. Organomet. Chem.* **361,** C5 (1989).
118. A. G. Davies, P. G. Harrison, J. D. Kennedy, R. J. Puddephatt, and W. McFarlane, *J. Chem. Soc. (C),* 1136 (1969).
119. H. W. Roesky and H. Weizer, *Angew. Chem., Int. Ed. Engl.* **12,** 674 (1973).
120. H. W. Roesky and H. Weizer, *Chem. Ber.* **107,** 3186 (1974).
121. H. W. Roesky, *Z. Naturforsch.* **31B,** 680 (1976).
122. T. Chivers, J. Fait, and K. J. Schmidt, *Inorg. Chem.* **28,** 3018 (1989).
123. G. Schmidt, D. Vehreschild-Yzertmann, and R. Boese, *J. Organomet. Chem.* **326,** 307 (1987).
124. D. Hänssgen, J. Kuna, and B. Ross, *J. Organomet. Chem.* **92,** C49 (1975).
125. H. Fuess, J. W. Bats, M. Diehl, L. Schönfelder, and H. W. Roesky, *Chem. Ber.* **114,** 2369 (1981).
126. W. Storch, W. J. Jacksteiss, H. Nöth, and G. Winter, *Angew. Chem., Int. Ed. Engl.* **16,** 478 (1977).
127. R. Jones, C. P. Warrens, D. J. Williams, and J. D. Woolins, *J. Chem. Soc., Dalton Trans.,* 907 (1987).
128. R. Jones, P. F. Kelly, C. P. Warrens, D. J. Williams, and J. D. Woolins, *J. Chem. Soc., Chem. Commun.,* 711 (1986).
129. R. Jones, T. G. Purcell, D. J. Williams, and J. D. Woolins, *Polyhedron* **7,** 647 (1988).
130. D. Hänssgen and I. Pohl, *Chem. Ber.* **112,** 2798 (1979).
131. M. Veith and V. Huch, *J. Organomet. Chem.* **308,** 263 (1986).
132. H. Schumann, *Angew. Chem., Int. Ed. Engl.,* **8,** 937 (1969).
133. H. Schumann and H. Benda, *Angew. Chem., Int. Ed. Engl.* **7,** 812 (1968).
134. H. Schumann and H. Benda, *Angew. Chem., Int. Ed. Engl.* **7,** 813 (1968).
135. H. Schumann and H. Benda, *Angew. Chem., Int. Ed. Engl.* **8,** 989 (1969).
136. D. Hänssgen, H. Aldenhoven, and M. Nieger, *J. Organomet. Chem.* **367,** 47 (1989).
137. D. Hänssgen and H. Aldenhoven, *Chem. Ber.* **123,** 1833 (1990).
138. D. Hänssgen, H. Aldenhoven, and M. Nieger, *Chem. Ber.* **124,** 1837 (1990).

139. B. Mathiasch and M. Dräger, *Angew. Chem., Int. Ed. Engl.* **17**, 767 (1978).
140. M. Dräger and B. Mathiasch, *Angew. Chem., Int. Ed. Engl.* **20**, 1029 (1987).
141. B. Mathiasch, *J. Organomet. Chem.* **165**, 295 (1979).
142. M. Baudler and H. Suchomel, *Z. Anorg. Allg. Chem.* **505**, 39 (1983).
143. A. H. Cowley, S. W. Hall, C. M. Nunn, and J. M. Power, *Angew. Chem., Int. Ed. Engl.* **27**, 838 (1988).
144. J. D. Andriamizaka, C. Couret, J. Escudié, and J. Satgé, *Phosphorus Sulphur* **12**, 265 (1982).
145. R. Bohra, P. B. Hitchcock, M. F. Lappert, and W.-P. Leung, *J. Chem. Soc., Chem. Commun.*, 728 (1989).
146. C. Couret, J. D. Andriamizaka, J. Escudié, and J. Satgé, *J. Organomet. Chem.* **208**, C3 (1981).
147. R. R. Holmes, *Acc. Chem. Res.* **22**, 190 (1989).
148. H. Puff and H. Reuter, *J. Organomet. Chem.* **373**, 173 (1989).
149. Ref. 6, p. 573.
150. Ref. 6, p. 604.
151. R. K. Harris and A. Sebald, *J. Organomet. Chem.* **331**, C9 (1987).
152. H. Puff, W. Schuh, R. Sievers, and R. Zimmer, *Angew. Chem., Int. Ed. Engl.* **20**, 591 (1987).
153. H. Puff, W. Schuh, R. Sievers, W. Wald, and R. Zimmer, *J. Organomet. Chem.* **260**, 271 (1984).
154. S. Masamune, L. R. Sita, and D. J. Williams, *J. Am. Chem. Soc.* **105**, 630 (1983).
155. H. Berwe and A. Haas, *Chem. Ber.* **120**, 1175 (1987).
156. O.-S. Jung, J. H. Jeong, and Y. S. Sohn, *Polyhedron* **8**, 2557 (1989).
157. C. H. W. Jones, R. D. Sharma, and S. P. Taneja, *Can. J. Chem.* **64**, 980 (1986).
158. H. Puff, A. Bongartz, R. Sievers, and R. Zimmer, *Angew. Chem., Int. Ed. Engl.* **17**, 939 (1978).
159. M. Dräger, A. Blecher, H.-J. Jacobsen, and B. Krebs, *J. Organomet. Chem.* **161**, 319 (1978).
160. A. Blecher and M. Dräger, *Angew. Chem., Int. Ed. Engl.* **18**, 677 (1979).
161. B. Menzebach and P. Bleckmann, *J. Organomet. Chem.* **91**, 291 (1975).
162. H.-J. Jacobsen and B. Krebs, *J. Organomet. Chem.* **136**, 333 (1977).
163. A. J. Edwards and B. F. Hoskins, *Acta Crystallogr.* **C46**, 1397 (1990).
164. H. Puff, R. Gattermeyer, R. Hundt, and R. Zimmer, *Angew. Chem., Int. Ed. Engl.* **16**, 547 (1977).
165. A. Tzschach and K. Jurkschat, *Pure Appl. Chem.* **58**, 639 (1986).
166. D. Schollmeyer, J. Kalbitz, H. Hartung, A. Tzschach, and K. Jurkschat, *Bull. Soc. Chim. Belg.* **97**, 1075 (1988).
167. F. Glockling and W.-K. Ng, *J. Chem. Res. (Suppl.)*, 230 (1980).
168. D. Kobelt, E. F. Paulus, and H. Scherer, *Acta Crystallogr.* **B28**, 2323 (1972).
169. Y. Ducharme, S. Latour, and J. D. Wuest, *J. Am. Chem. Soc.* **106**, 1499 (1984).
170. M. Gallant, M. Kobayashi, S. Latour, and J. D. Wuest, *Organometallics* **7**, 736 (1988).
171. A. L. Beauchamp, S. Latour, M. J. Oliver, and J. D. Wuest, *J. Am. Chem. Soc.* **105**, 7778 (1983).
172. M. J. S. Dewar and G. L. Grady, *Organometallics* **4**, 1327 (1985).
173. P. G. Harrison and S. R. Stobart, *J. Organomet. Chem.* **47**, 89 (1973).
174. A. Blecher, B. Mathiasch, and M. Dräger, *Z. Anorg. Allg. Chem.* **488**, 177 (1982).
175. A. Blecher, B. Mathiasch, and T. N. Mitchell, *J. Organomet. Chem.* **184**, 175 (1980).
176. C. Cauletti, F. Grandinetti, G. Granozzi, A. Sebald, and B. Wrackmeyer, *Organometallics* **7**, 262 (1988).

177. W. F. Manders and T. P. Lockhart, *J. Organomet. Chem.* **297**, 143 (1985).
178. T. P. Lockhart and W. F. Manders, *Inorg. Chem.* **25**, 583 (1986).
179. P. Brown, M. F. Mahon, and K. C. Molloy, *J. Chem. Soc., Chem. Commun.*, 1621 (1989).
180. V. I. Scherbakov, I. K. Grigor'eva, G. A. Razuvaev, L. N. Zakharov, R. I. Bochkova, and Y. T. Struchkov, *J. Organomet. Chem.* **319**, 41 (1987).
181. G. A. Razuvaev, V. I. Shcherbalkov, and I. K. Grigor'eva, *J. Organomet. Chem.* **264**, 245 (1984).
182. P. J. Craig and S. Rapsomanikis, *J. Chem. Soc., Chem. Commun.*, 114 (1982).
183. H. Puff, E. Friedrichs, R. Hundt, and R. Zimmer, *J. Organomet. Chem.* **259**, 79 (1983).
184. B. Mathiasch, *J. Organomet. Chem.* **141**, 189 (1977).
185. B. Mathiasch, *J. Organomet. Chem.* **194**, 37 (1980).
186. H. Puff, A. Bongartz, W. Schuh, and R. Zimmer, *J. Organomet. Chem.* **248**, 61 (1983).
187. B. Mathiasch, *J. Organomet. Chem.* **122**, 345 (1976).
188. B. Mathiasch, *Z. Anorg. Allg. Chem.* **432**, 269 (1977).
189. H. Puff, B. Breur, W. Schuh, R. Sievers, and R. Zimmer, *J. Organomet. Chem.* **332**, 279 (1987).
190. M. Dräger and B. Mathiasch, *Z. Anorg. Allg. Chem.* **470**, 45 (1980).
191. K. C. Molloy, F. A. K. Nasser, C. L. Barnes, D. van der Helm, and J. J. Zuckerman, *Inorg. Chem.* **19**, 960 (1982).
192. J. G. Masters, F. A. K. Nasser, M. B. Hossain, A. P. Hagen, D. van der Helm, and J. J. Zuckerman, *J. Organomet. Chem.* **385**, 39 (1990).
193. J.-C. Pommier, E. Mendes, and J. Valade, *J. Organomet. Chem.* **55**, C19 (1973).
194. C.-D. Hager, F. Huber, A. Silvestri, and R. Barbieri, *Inorg. Chim. Acta* **49**, 31 (1981).
195. S. Sakai, S. Furusawa, H. Matsunaga, and T. Fujinami, *J. Chem. Soc., Chem. Commun.*, 265 (1975).
196. W. P. Neumann and A. Schwarz, *Angew. Chem., Int. Ed. Engl.* **14**, 812 (1975).
197. K.-H. Scherping and W. P. Neumann, *Organometallics* **1**, 1017 (1982).
198. K. Hiller and W. P. Neumann, *Tetrahedron Lett.* **27**, 5347 (1986).
199. M. Massol, J. Barrau, J. Satgé, and B. Bouyssieres, *J. Organomet. Chem.* **80**, 47 (1974).
200. W. D. Honnick and J. J. Zuckerman, *Inorg. Chem.* **18**, 1437 (1979).
201. A. G. Davies, A. J. Price, H. M. Dawes, and M. B. Hursthouse, *J. Chem. Soc., Dalton Trans.*, 297 (1986).
202. P. A. Bates, M. B. Hursthouse, A. G. Davies, and S. D. Slater, *J. Organomet. Chem.* **363**, 45 (1989).
203. S. David, C. Pascard, and M. Cesario, *Nouv. J. Chim.* **3**, 63 (1979).
204. C. W. Hopzapfel, J. M. Kockemoer, C. M. Marais, G. J. Kruger, and J. A. Pretorius, *S. Afr. J. Chem.* **35**, 81 (1982).
205. M. Dräger, *Z. Anorg. Allg. Chem.* **477**, 154 (1981).
206. A. S. Secco and J. Trotter, *Acta Crystallogr.* **C39**, 451 (1983).
207. A. G. Davies, S. D. Slater, D. C. Povey, and G. W. Smith, *J. Organomet. Chem.* **352**, 283 (1988).
208. P. A. Bates, M. B. Hursthouse, A. G. Davies, and S. D. Slater, *J. Organomet. Chem.* **325**, 129 (1987).
209. T. P. Lockhart, *Organometallics* **7**, 1438 (1988).
210. P. J. Smith, R. F. M. White, and L. Smith, *J. Organomet. Chem.* **40**, 341 (1972).
211. C. Luchinat and S. Rolens, *J. Am. Chem. Soc.* **108**, 4873 (1986).
212. C. Luchinat and S. Rolens, *J. Org. Chem.* **52**, 4444 (1987).
213. E. W. Abel, S. K. Bhargava, K. G. Orrel, and V. Sik, *J. Chem. Soc., Dalton Trans.*, 2073 (1982).
214. S. J. Blunden, P. A. Cusack, and P. J. Smith, *J. Organomet. Chem.* **325**, 141 (1987).

215. K. C. Molloy, in "Chemistry of the Metal–Carbon Bond," Vol. 5, Chap. 11, p. 465. Wiley, Chichester, 1989.
216. M. Peyrere, J.-P. Quintard, and A. Rahm, "Tin in Organic Synthesis." Butterworth, London, 1987.
217. A. Shanzer and N. Mayer-Schocher, *J. Chem. Soc., Chem. Commun.,* 176 (1980).
218. A. Shanzer, J. Libman, and F. Frowlow, *J. Am. Chem. Soc.* **103,** 7339 (1981).
219. A. Shanzer, J. Libman, H. Gottlieb, and F. Frowlow, *J. Am. Chem. Soc.* **104,** 4220 (1982).
220. A. Shanzer, J. Libman, H. Gottlieb, and F. Frowlow, *Acc. Chem. Res.* **16,** 60 (1983).
221. A. G. Davies, P. Hua-De, and J. A.-A. Hawari, *J. Organomet. Chem.* **256,** 251 (1983).
222. T. Sato, E. Yoshida, T. Kobayashi, J. Otera, and H. Nozaki, *Tetrahedron Lett.* **29,** 3971 (1988).
223. A. G. Davies, J. A.-A. Hawari, and P. Hua-De, *J. Organomet. Chem.* **251,** 203 (1983).
224. A. G. Davies and A. J. Price. *J. Organomet. Chem.* **258,** 7 (1983).
225. R. L. Dannley, W. A. Aue, and A. K. Shrubber, *J. Organomet. Chem.* **38,** 281 (1972).
226. H. Weichmann and A. Tzschach, *J. Organomet. Chem.* **99,** 61 (1975).
227. C. Mügge, H. Weichmann, and A. Zschunke, *J. Organomet. Chem.* **192,** 41 (1980).
228. A. G. Davies and J. A.-A. Hawari, *J. Organomet. Chem.* **224,** C37 (1982).
229. G. D. Smith, P. E. Fanwick, and I. P. Rothwell, *J. Am. Chem. Soc.* **111,** 750 (1989).
230. K. Grätz, F. Huber, A. Silvestri, G. Alonzo, and R. Barbieri, *J. Organomet. Chem.* **290,** 41 (1985).
231. L. M. Epstein and D. K. Straub, *Inorg. Chem.* **4,** 1551 (1965).
232. H. Preut, K. Grätz, and F. Huber, *Acta Crystallogr.* **C40,** 941 (1984).
233. K. Grätz, F. Huber, A. Silvestri, and R. Barbieri, *J. Organomet. Chem.* **273,** 283 (1984).
234. A. C. Sau, R. O. Day, and R. R. Holmes, *J. Am. Chem. Soc.* **103,** 1264 (1981).
235. J. F. Vollano, R. O. Day, and R. R. Holmes, *Organometallics* **3,** 750 (1984).
236. R. O. Day, J. M. Holmes, S. Shafieezad, V. Chandrasekhar, and R. R. Holmes, *J. Am. Chem. Soc.* **110,** 5377 (1988).
237. R. R. Holmes, S. Shafieezad, J. M. Holmes, and R. O. Day, *Inorg. Chem.* **27,** 1232 (1988).
238. A. Tzschach and K. Jurkschat, *Comm. Inorg. Chem.* **3,** 35 (1983).
239. M. G. Voronkov and V. P. Baryshok, *J. Organomet. Chem.* **239,** 199 (1982).
240. K. Jurkschat and A. Tzschach, *J. Organomet. Chem.* **272,** C13 (1984).
241. K. Jurkschat, J. Schilling, C. Mügge, A. Tzschach, J. Meunier-Piret, M. van Meerssche, M. Gielen, and R. Willem, *Organometallics* **7,** 38, (1988).
242. R. G. Swisher and R. R. Holmes, *Organometallics* **3,** 365 (1984).
243. M. Dräger, *Z. Anorg. Allg. Chem.* **527,** 169 (1985).
244. M. Dräger, *Chem. Ber.* **114,** 2051 (1981).
245. M. Dräger and H.-J. Guttman, *J. Organomet. Chem.* **212,** 171 (1981).
246. R. G. Swischer, R. O. Day, and R. R. Holmes, *Inorg. Chem.* **22,** 3692 (1983).
247. K. Jurkschat, A. Tzschach, and J. Meunier-Piret, *J. Organomet. Chem.* **315,** 45 (1986).
248. M. Dräger, *J. Organomet. Chem.* **251,** 209 (1983).
249. K. Jurkschat, A. Tzschach, J. Meunier-Piret, and M. van Meerssche, *J. Organomet. Chem.* **290,** 285 (1985).
250. F. Huber, H.-J. Haupt, H. Preut, R. Barbieri, and M. T. Lo Guidice, *Z. Anorg. Allg. Chem.* **432,** 51 (1977).
251. H. Preut, B. Mundus, F. Huber, and R. Barbieri, *Acta Crystallogr.* **C42,** 536 (1986).
252. H. Preut, F. Huber, H.-J. Haupt, R. Cefalù, and R. Barbieri, *Z. Anorg. Allg. Chem.* **410,** 88 (1974).
253. H. Preut, H.-J. Haupt, F. Huber, R. Cefalù, and R. Barbieri, *Z. Anorg. Allg. Chem.* **407,** 257 (1974).

254. H. Preut, F. Huber, R. Barbieri, and N. Bertazzi, *Z. Anorg. Allg. Chem.* **423**, 75 (1976).
255. K. Jurkschat, A. Tzschach, C. Mügge, J. Meunier-Piret, M. van Meerssche, G. van Binst, C. Wynants, M. Gielen, and R. Willem, *Organometallics* **7**, 593 (1988).
256. R. Willem, M. Gielen, J. Meunier-Piret, M. van Meerssche, K. Jurkschat, and A. Tzschach, *J. Organomet. Chem.* **277**, 335 (1984).
257. C. Wynants, G. van Binst, C. Mügge, K. Jurkschat, A. Tzschach, H. Pepermans, M. Gielen, and R. Willem, *Organometallics* **4**, 1906 (1985).
258. F. Glockling, in Ref. *7*, p. 245.
259. M. S. Holt, W. L. Wilson, and J. H. Nelson, *Chem. Rev.* **89**, 11 (1989).
260. W. A. Herrmann, *Angew. Chem., Int. Ed. Engl.* **25**, 56 (1986).
261. E. H. Brookes and R. J. Cross, *Organomet. Rev. (A)* **6**, 227 (1970).
262. J. D. Cotton, P. J. Davidson, and M. F. Lappert, *J. Chem. Soc., Dalton Trans.*, 2275 (1976).
263. C. J. Cardin, D. J. Cardin, G. A. Lawless, J. M. Power, M. B. Power, and M. B. Hursthouse, *J. Organomet. Chem.* **325**, 203 (1987).
264. K. Triplett and M. D. Curtis, *Inorg. Chem.* **15**, 431 (1976).
265. G. W. Grynkewich and T. J. Marks, *Inorg. Chem.* **15**, 1307 (1976).
266. T. J. Marks and G. W. Grynkewich, *J. Organomet. Chem.* **91**, C9 (1975).
267. C. J. Gilmore and P. Woodward, *J. Chem. Soc., Dalton Trans*, 1387 (1972).
268. C. J. Cardin, D. J. Cardin, H. E. Parge, and J. M. Power, *J. Chem. Soc., Chem. Commun.*, 609 (1984).
269. C. J. Cardin, D. J. Cardin, J. M. Power, and M. B. Hursthouse, *J. Am. Chem. Soc.* **107**, 505 (1985).
270. R. B. King and F. G. A. Stone, *J. Am. Chem. Soc.* **82**, 3833 (1960).
271. J. D. Cotton, S. A. R. Knox, I. Paul, and F. G. A. Stone, *J. Chem. Soc. (A)*, 264 (1967).
272. P. G. Harrison, T. J. King, and J. A. Richards, *J. Chem. Soc., Dalton Trans.*, 2097 (1975).
273. A. B. Cornwell, P. G. Harrison, and J. A. Richards, *J. Organomet. Chem.* **108**, 47 (1976).
274. R. A. Burnham, M. A. Lyle, and S. R. Stobart, *J. Organomet. Chem.* **125**, 179 (1977).
275. V. Sriyunyongwat, R. Hani, T. A. Albright, and R. Geanangel, *Inorg. Chim. Acta* **122**, 91 (1986).
276. T. J. Marks and A. R. Neuman, *J. Am. Chem. Soc.* **95**, 769 (1973).
277. T. J. McNeese, S. S. Wreford, D. L. Tipton, and R. Bau, *J. Chem. Soc., Chem. Commun.*, 390 (1977).
278. R. M. Sweet, C. J. Fritchie, and R. A. Schunn, *Inorg. Chem.* **6**, 749 (1967).
279. R. D. Adams, F. A. Cotton, W. R. Cullen, D. L. Hunter, and L. Mihichuk, *Inorg. Chem.* **14**, 1395 (1975).
280. K. E. Schwarzhans, *Z. Naturforsch.* **34B**, 1456 (1979).
281. J. Weaver and P. Woodward, *J. Chem. Soc., Dalton Trans.*, 1060 (1973).
282. R. Carreño, V. Riera, M. A. Ruiz, V. Jeannin, and M. Philoche-Levisalles, *J. Chem. Soc., Chem. Commun.*, 15 (1990).
283. B. T. Huie, S. W. Kirtley, C. B. Knobler, and H. D. Kaesz, *J. Organomet. Chem.* **213**, 45 (1981).
284. J. F. Almeida, K. R. Dixon, C. Eaborn, P. B. Hitchcock, A. Pidock, and V. Vinaixa, *J. Chem. Soc., Chem. Commun.*, 1315 (1982).
285. G. L. Rochfort and J. E. Ellis, *J. Organomet. Chem.* **250**, 277 (1983).
286. A. Christofides, M. Griano, J. L. Spencer, and F. G. A. Stone, *J. Organomet. Chem.* **178**, 273 (1979).
287. The cubane structure has recently been established for $(RSn)_8$ $(R = 2,6\text{-}Et_2C_6H_3)$. See L. R. Sita and I. Kinoshita, *Organometallics*, **9**, 2865 (1990).

ADVANCES IN ORGANOMETALLIC CHEMISTRY, VOL. 33

Halogenoalkyl Complexes of Transition Metals

HOLGER B. FRIEDRICH and JOHN R. MOSS

Department of Chemistry
University of Cape Town
Rondebosch 7700, South Africa

I

INTRODUCTION

Halogenoalkyl (or haloalkyl) complexes of the main group metals have been extensively studied and shown to be important reagents in synthetic chemistry (1–3). Halogenoalkyl transition metal complexes of type $[L_yM\{(CH_2)_nX\}]$ (L_y = ligands, M = transition metal, X = Cl, Br, I, or F, $n \geq 1$) are less well known, although they have recently been recognized as a separate class of alkyl compounds (4). Complexes $[L_yM\{(CH_2)_nX\}]$ can either be considered as simple examples of functionalized alkyl compounds of transition metals or as metal-substituted alkyl halides. Either way, they can be versatile precursors for a wide range of useful compounds. Complexes of the type $[L_yMCH_2X]$ can be useful precursors to methylene complexes $[L_yM=CH_2]^+$ (5), hydroxymethyl complexes $[L_yMCH_2OH]$ (6), and methylene-bridged complexes $L_yMCH_2M'L_x$ (7), all of which have been implicated in many catalytic processes (8–10). The methylene complexes are also promising reagents for electrophilic cyclopropanations (11,12) and for alkene methathesis (13). Successive formation of formyl, formaldehyde, hydroxymethyl, carbene, and alkyl surface intermediates has been proposed for the Fischer–Tropsch reaction and related processes

235

$(8,14-16)$. In general, many metal–C_1 complexes are believed to be important and useful as models for catalytic intermediates (17).

Complexes of type $[L_yM\{(CH_2)_nX\}]$, where $n > 1$, have been shown to be useful precursors for hetero- and homobimetallic $\mu(\alpha,\omega)$ alkanediyl complexes, $[L_xM(CH_2)_nM'L_y]$ (where ML_x is not necessarily the same as $M'L_y$). Such hydrocarbon-bridged binuclear compounds have been proposed as models for intermediates in the Fischer–Tropsch reaction $(18,19)$ and other significant catalytic processes $(20-23)$. Some $[L_yM\{(CH_2)_nX\}]$ complexes are precursors to cyclic carbene complexes (Section III), whereas others have been shown to have synthetic utility in organic chemistry (24).

As the subject of transition metal halogenoalkyl compounds has not previously been reviewed, our aim is to cover this topic comprehensively, and we hope that we have included all the most important papers in this field. We have restricted this article to monohalogenoalkyl transition metal complexes; although polyhalogenoalkyl compounds such as $[CpFe(CO)_2CBr_3]$ (25), $[Re(CO)_5(CF_2CF_3)]$ (26), and $[4CNpyCo-(DH)_2(CH_2CF_3)]$ (27) are known (py = pyridine, Cp = η^5-C_5H_5, and DH = monoanion of dimethylglyoxime), we have considered these polyhalogenoalkyl compounds to be outside the scope of this article. We begin this review of monohalogenoalkyl compounds $[L_yM\{(CH_2)_nX\}]$, with the iron group for historical reasons and treat each group thereafter in order of increasing group number in the periodic table.

II

MONOHALOGENOMETHYL TRANSITION METAL COMPLEXES

A. Iron, Ruthenium, and Osmium

The first syntheses of methoxymethyl and halogenomethyl transition metal complexes were reported by Jolly and Pettit (28) in 1966 and Green et al. in 1967 (29). Both papers reported the synthesis of $[CpFe(CO)_2CH_2OCH_3]$ (1) and $[CpFe(CO)_2CH_2Cl]$ (2). Green further reported the bromomethyl analog of 2, $[CpFe(CO)_2CH_2Br]$ (3). Similar synthetic routes were used by both groups. The methoxymethyl complex was synthesized by the reaction of chloromethyl methyl ether with the iron anion [Eq. (1)]. The chloro- and bromomethyl complexes were synthesized by reacting the appropriate methoxymethyl complex from Eq. (1) with HX gas [Eq. (2)]. Similarly, the acetoxymethyl iron complex, $[CpFe(CO)_2\{CH_2OC(O)Me\}]$, reacts with HCl gas to give $[CpFe(CO)_2CH_2Cl]$ (30). Complex 2 was also obtained by reacting $[CpFe(CO)_2CH_2NMe_2]$ with CH_3COCl (31).

$$Na^+[M]^- + ClCH_2OCH_3 \rightarrow [MCH_2OCH_3] + NaCl$$

$$M = CpFe(CO)_2$$

(1)

$$[MCH_2OCH_3] + HX \rightarrow [MCH_2X] + CH_3OH$$

$$M = CpFe(CO)_2; X = Cl, Br$$

(2)

The iodomethyl complex $[CpFe(CO)_2CH_2I]$ (4) has been prepared by two routes (32). The first route involves the reaction of complex 1 with HI gas, as shown in Eq. (2). The second route involves the reaction of complex 3 with NaI [Eq. (3)]. The methoxymethyl complex (1) was isolated as an air-sensitive oil, whereas the halogenoalkyl complexes were isolated as air- and light-sensitive solids. The bromomethyl complex (3) was found to be less stable than the chloromethyl complex (2) in all respects, and the iodomethyl complex (4) was found to be even less stable.

$$[CpFe(CO)_2CH_2Br] + NaI \rightarrow [CpFe(CO)_2CH_2I] + NaBr$$

(3)

King and Braitsch prepared a number of halomethyl complexes of transition metals by reacting the appropriate transition metal anion with dihalomethanes according to Eq. (4) (33). CH_2ClI was used to prepare the chloromethyl derivatives of $[MCH_2X]$ directly. Since the C—I bond is weaker than the C—Cl bond, the C—I bond is preferentially cleaved in most cases. This method is a good synthetic route to the chloromethyl complexes of Mo, W, and Mn, but not for those of Fe.

$$Na[M] + XCH_2X' \rightarrow [MCH_2X] + NaX'$$

(4)

$$M = CpMo(CO)_3^-, CpW(CO)_3^-, CpFe(CO)_2^-, Mn(CO)_5^-$$
$$XCH_2X' = \text{dihaloalkane, e.g., } CH_2ClI$$

Facile conversion of 2 to 1 was achieved by reaction with NaOMe (29). The Cl in $[CpFe(CO)_2CH_2Cl]$ (2) proved to be highly susceptible to nucleophilic attack, and a number of complexes of type $[CpFe(CO)_2CH_2Z]$ [Z = OEt, SEt, $O(CH_2)_2NMe_2H$, $OCR_2CH=CH_2$ (R = Me or H)] have been synthesized (29). The complexes $[CpFe(CO)_2CH_2X]$ have also been shown to react with neutral nucleophiles, L (L = tertiary phosphines, amines, sulfides, and $AsPh_3$), to form complexes of the type $[CpFe(CO)_2CH_2L]^+$ or $[CpFe(L)_2(CO)]^+$, depending on the size and pK_a of the ligand (L), the halide (X), and the solvent used (32,34,35). The rate of the reaction of $[CpFe(CO)_2CH_2X]$ with L increases in the sequence Cl < Br < I. Furthermore the relative rates of reaction of complex 3 with L were found to increase with increasing pK_a and decreasing cone angle of L. Proposed pathways for the reactions of $[M(L_y)(CH_2X)]$ (M = transition metal) complexes with ligands L are shown in Scheme 1.

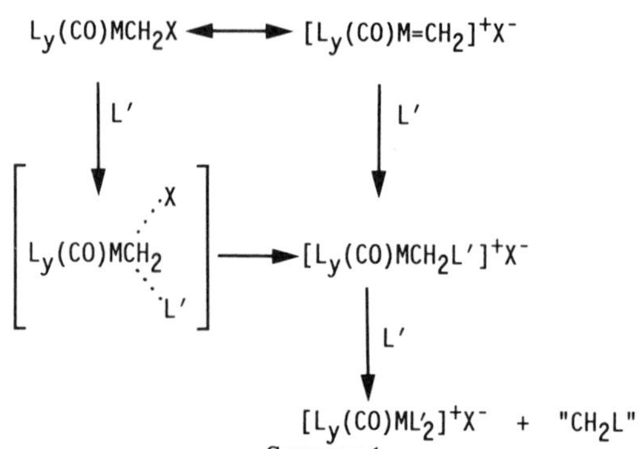
Ly(CO)MCH2X
[Ly(CO)M=CH2]+X-
L'
L'
X
Ly(CO)MCH2
L'
[Ly(CO)MCH2L']+X-
L'
[Ly(CO)ML'2]+X- + "CH2L"

SCHEME 1

The reaction of complex **3** with $PPh_2(CH_2)_2PPh_2$ (dppe) yielded a dicationic ylide complex, $[CpFe(CO)_2CH_2P(Ph)_2(CH_2)_2P(Ph)_2CH_2Fe(CO)_2Cp]^{2+}$. Although complex **2** did not react with $P(OMe)_3$ or NEt_3 under normal conditions, it reacted with the above ligands and with all other neutral nucleophiles tried, in the presence of the halide abstractor $TlBF_4$, to give $[CpFe(CO)_2CH_2L]BF_4$ (*36*). The ylide type complexes obtained from the reactions of halogenomethyl complexes with ligands, L, are in general fairly stable. This is in sharp contrast to most uncoordinated ylides, which are essentially unstable. These ylides are thus stabilized by coordination to a transition metal complex. Interest is currently being shown in these stabilized ylide complexes as versatile substrates for further synthetic manipulations in organometallic chemistry (*37*). A more common route to transition metal ylide type complexes involves halide metathesis using preformed ylides (*38,39*).

Jolly and Pettit found that the addition of $AgBF_4$ to $[CpFe(CO)_2CH_2Cl]$ (**2**) gave a species which could convert cyclohexene to norcarane in 80% yield (*28*). The intermediacy of $[CpFe(CO)_2(=CH_2)]^+$ (**5**) was proposed (Scheme 2). Workup of the reaction mixture (in the absence of cyclohexene), however, gave mainly the ethylene complex $[CpFe(CO)_2(H_2C=CH_2)]^+$, as well as $[CpFe(CO)_2CH_3]$ and $[CpFe(CO)_3]^+$. Brookhart has

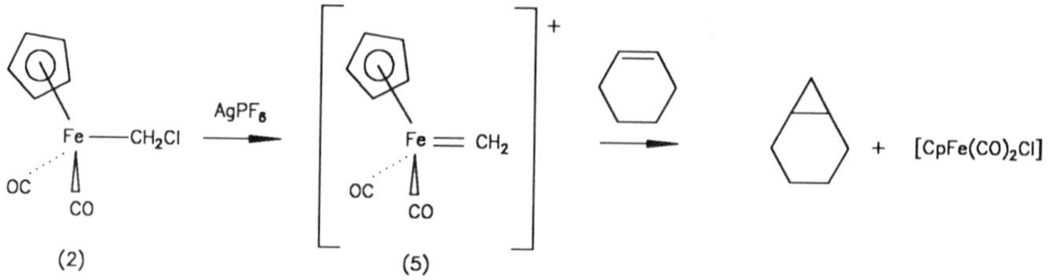
Fe
CH2Cl
OC
CO
AgPF6
Fe
CH2
OC
CO
+
[CpFe(CO)2Cl]

(2) (5)

SCHEME 2

since reported that **5** is too unstable for observation by ^{1}H NMR at $-80\,^\circ$C
(*40*).

The existence of **5** as an intermediate was also proposed by Green *et al.*
(*29*), whereas Steven and Beauchamp have shown that **5** occurs in the
electron-impact mass spectrum of [CpFe(CO)$_2$CH$_2$OCH$_3$] (**1**) if protonation agents are present (*41*). Complex **5** was also observed in the electron-
impact mass spectra of the complexes [CpFe(CO)$_2$CH$_2$X] (X = Cl, Br, or
I). The relative abundance of this carbene species under mass spectral
conditions depends on the relative abilities of the halogens as leaving
groups (I > Br > Cl) (*42*). Complex **5** was also obtained from the reaction
of **2** with AgPF$_6$, where it was successfully used as an alkylating agent
toward coordinated ligands (*43*). Bodnar and Cutler reported that **2**, after
reaction with AgPF$_6$, can react with exogenous CO to give a stable ($\eta^2 - $C,
C) ketene complex (**6**), which transforms to its carbomethoxymethyl analog (**7**) in the presence of methanol (Scheme 3) (*44*).

The triphenylphosphine-containing complexes [CpFe(PPh$_3$) (CO)CH$_2$-
OCH$_3$] (**8**) and [CpFe(PPh$_3$)(CO)CH$_2$Cl] (**9**) as well as [CpFe-
(PPh$_3$)(CO)CH$_2$I] (**11**) were briefly reported by Davison *et al.* (*45*). Flood
et al. later reported the full characterization of **8** and **9** (*46*). Complex **8** was
synthesized by the irradiation of [CpFe(CO)$_2$CH$_2$OCH$_3$] (**1**) in the presence of PPh$_3$. Product **8** was then reacted with HCl gas to yield **9** as shown
in Eq. (2) (*46*). Optically pure $(-)$[CpFe(PPh$_3$)(CO)CH$_2$Cl] (**9**) was prepared by reacting $(-)$[CpFe(PPh$_3$) (CO)CH$_2$OMen] (Men = menthyl) with
HCl at 0 $^\circ$C. Flood also prepared the bromo- (**10**) and iodomethyl (**11**)
analogs of **9** by reacting [CpFe(PPh$_3$)(CO)CH$_2$OMen] with HBr or HI gas
by a method similar to that shown in Eq. (2). Only **10** was reported to be
sufficiently stable to characterize. Again, the carbon of the chloromethyl
complex (**9**) was reported to be susceptible to nucleophilic attack by nucleophiles such as OMe$^-$, H$^-$, and CN$^-$. Complex **9** could also be converted to the alkyl compounds, [CpFe(PPh$_3$)(CO)R], using Grignard reagents or RLi [R = CH$_3$, CH$_2$CH$_3$, CH(CH$_3$)$_2$, C$_6$H$_5$, or C$_3$H$_5$].
Furthermore complex **9** reacts with NaOC$_{10}$H$_{19}$ to give [CpFe-
(CO)$_2$CH$_2$OMen] (*45*). As with [CpFe(CO)$_2$CH$_2$Cl] (*28*), **9**, **10**, and **11**
were also found to be sources of [CH$_2$] (*46*). Hence, for example,

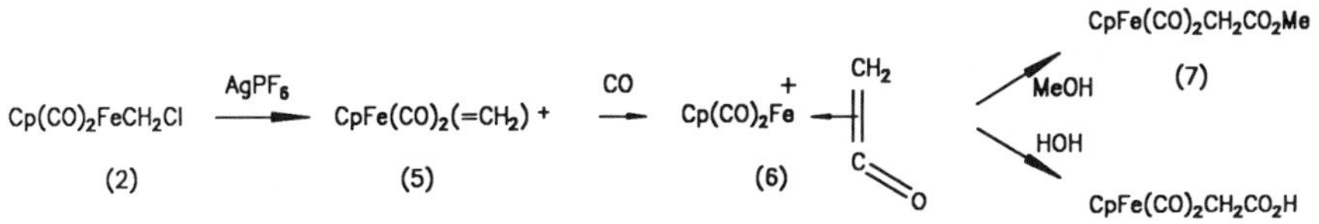

SCHEME 3

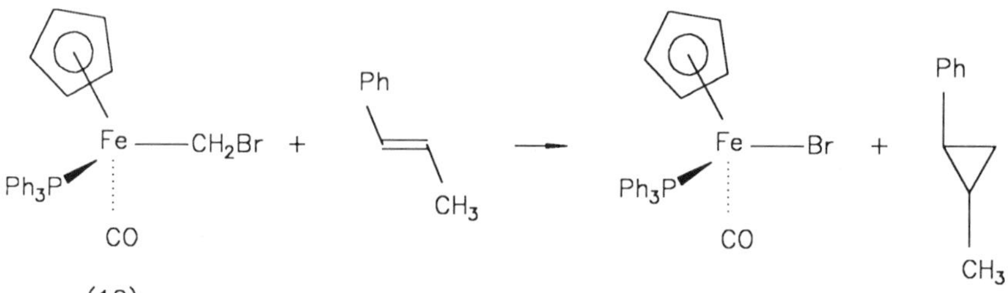

(10)

SCHEME 4

[CpFe(PPh$_3$)(CO)CH$_2$Br] reacted with [PhCH=CHCH$_3$] to yield a substituted cyclopropane (Scheme 4).

Complex **9** and [CpFe(o-BP)(CO)CH$_2$Cl] [o-BP = tri(o-biphenyl)-phosphite] were also prepared by the reaction of HCl with [CpFe(L)-(CO)CH$_2$OEt] (L = PPh$_3$, o-BP) (47). Alkylation of the two complexes with sodium tert-butyl acetoacetate and pyrroline cyclohexanone enamine yielded six of the eight possible alkylation products. The two products where L was o-BP and the nucleophile was tert-butyl acetoacetate did not form, presumably because of excessive steric hindrance. The excess of one diastereomer over the other ranged from 10 to 64%.

The iodomethyl complex [Fe{P(OCHMe$_2$)$_3$}$_2$(CO)$_2$(I)CH$_2$I], whose molecular structure is shown in Fig. 1, was prepared by the addition of HX to the formaldehyde complex [Fe{P(OCHMe$_2$)$_3$}$_2$(CO)$_2$(η^2-CH$_2$O)] (48). This reaction proceeds via the intermediacy of [Fe{P(OCHMe$_2$)$_3$}$_2$(CO)$_2$(I)-CH$_2$OH]. The geometry around the Fe atom, in the iodomethyl complex, is approximately octahedral. Compared to the length of organic C—I bonds, the length of the C—I bond in [Pt(PPh$_3$)$_2$(I)CH$_2$I](see Section II,E), and also the sum of the covalent radii (210 pm), the length of the FeCH$_2$—I bond (221 pm) is unusually long. Correspondingly, the Fe—CH$_2$ bond is somewhat short. This effect may be due to a significant contribution from the ionic resonance structure [Fe=CH$_2$]$^+$ I$^-$. The weakening of the C—I bond is associated with a larger than usual Fe—CH$_2$—I angle of 120°.

The ruthenium complex [CpRu(CO)$_2$CH$_2$Cl] was reported by Moss and co-workers (34,49). Unlike [CpFe(CO)$_2$CH$_2$Cl], [CpRu(CO)$_2$CH$_2$Cl] did not react with PPh$_3$ in methanol. The reaction of [Cp*Ru(CO)$_2$CH$_2$OCH$_3$] with HCl gives [Cp*Ru(CO)$_2$CH$_2$Cl] (50) (Cp* = η^5-C$_5$Me$_3$).

Roper and co-workers reported that [Os(PPh$_3$)$_2$(CO)$_2$(η^2-CH$_2$O)] proved to be a useful synthetic precursor for stable hydroxymethyl, methoxymethyl, and halomethyl complexes (51,52). Thus [Os(PPh$_3$)$_2$(CO)$_2$(η^2-CH$_2$O)] reacts with excess HX (X = Cl, Br, or I) to yield the halomethyl

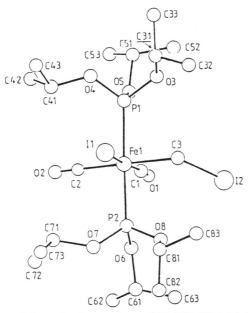

FIG. 1. Diagram of the molecular structure of [Fe{P(OCHMe$_2$)$_3$}$_2$(CO)$_2$(I)CH$_2$I]. [By permission from the Verlag der Zeitschrift der Naturforschung; from H. Berke, R. Birk, G. Huttner, and L. Zsolnai, *Z. Naturforsch.* **39B**, 1380 (1984).]

complexes [Os(PPh$_3$)$_2$(CO)$_2$(X)CH$_2$X]. The reaction proceeds with the hydroxymethyl complex [Os(PPh$_3$)$_2$(CO)$_2$(X)CH$_2$OH] acting as an intermediate. The proposed mechanism is shown in Scheme 5. The reactivity order observed for the hydrogen halides was HI > HBr > HCl. It is interesting to note that this trend is directly opposite to that observed for the reaction of the Fe complex **1** with HX gas to form the complexes [CpFe(CO)$_2$CH$_2$X] (*53*). The postulated intermediate osmium methylene complex could not be isolated.

[Os(PPh$_3$)$_2$(CO)$_2$(I)CH$_2$I] (*51*) reacts with SeH$^-$ to initially form [Os(PPh$_3$)$_2$(CO)$_2$I(CH$_2$SeH)] (*54*). On recrystallization, HI is rapidly lost and [Os(PPh$_3$)$_2$(CO)$_2$(η^2-CH$_2$Se)] is obtained. The reaction of

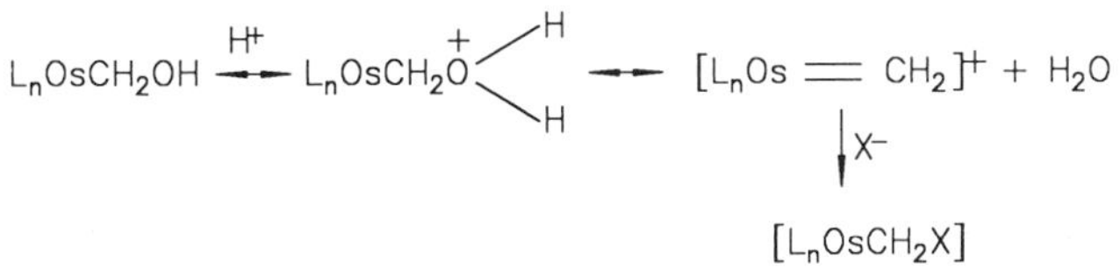

$$L_nOsCH_2OH \xrightarrow{H^+} L_nOsCH_2O\overset{+}{\diagup}\diagdown\overset{H}{\underset{H}{}} \longrightarrow [L_nOs = CH_2]^+ + H_2O$$

$$\downarrow X^-$$

$$[L_nOsCH_2X]$$

SCHEME 5

[Os(PPh$_3$)$_2$(CO)$_2$(Cl)CH$_2$Cl] with methanol affords [Os(PPh$_3$)$_2$(CO)$_2$-(Cl)CH$_2$OCH$_3$] (55). The chloromethyl complex [Os(PPh$_3$)$_2$(NO)(Cl)$_2$-CH$_2$Cl] (12) was prepared by the reaction of [Os(PPh$_3$)$_2$(NO)Cl(=CH$_2$)] with Cl$_2$ (56). Complex 12 undergoes a rearrangement reaction to give [Os(PPh$_3$)(NO)(Cl)$_3$CH$_2$PPh$_3$]. This rearrangement is thought to proceed via the cationic methylene complex [Os(PPh$_3$)$_2$(NO)Cl$_2$(=CH$_2$)]$^+$.

B. Chromium, Molybdenum, and Tungsten

The reactions of CH$_2$N$_2$ with [CpCr(NO)$_2$X] (X = Cl or Br), in the presence of Cu powder, gave the stable green complexes [CpCr-(NO)$_2$CH$_2$X] (13) in high yield (57). The complexes [(η^5-C$_5$H$_4$-CH$_3$)Cr(NO)$_2$CH$_2$X] (14) and [Cp*Cr(NO)$_2$CH$_2$X] (15) (X = Cl or Br) were prepared in a similar fashion. The iodomethyl complexes [(η-C$_5$R$_5$)Cr(NO)$_2$CH$_2$I] (R = H or Me) were more conveniently prepared by the reaction of the corresponding chloromethyl complexes with NaI.

The reaction of [CpCr(NO)$_2$CH$_2$Cl] with AgBF$_4$ results in migration of the methylene moiety into a C—H bond of the Cp ring to give the cation [(η^5-C$_5$H$_4$CH$_3$)Cr(NO)$_2$]$^+$ and not [CpCr(NO)$_2$CH$_2$]$^+$, as might have been expected. The cause of the intramolecular reactivity is believed to be due to the extremely strong π-withdrawing power of the two nitrosyl ligands working together, making the Cr–methylene species generated by the halide abstraction very electrophilic. Unlike the analogous iron complex [CpFe(CO)$_2$CH$_2$Cl] (see Section II,A), none of the complexes 13 or 14 transfer methylene to cyclohexene, thus showing the importance of the metal and its associated ligands on the reactivity of L$_n$MCH$_2$X. Thus, for example, treatment of complex 14 (X = Cl) with AgBF$_4$, followed by addition of PPN$^+$Cl$^-$ (PPN = Ph$_3$PNPPh$_3^+$), leads to a 9:1 mixture of the 1,2- and 1,3-Me$_2$Cp products [(η-C$_5$H$_3$Me$_2$)Cr(NO)$_2$Cl] and [(η-C$_5$H$_4$Me)Cr(NO)$_2$Cl]. Evidence for intramolecular methylene migration in the gas phase is observed in the mass spectra of the (η-C$_5$H$_4$R) (R = H or Me) complexes. Intermolecular CH$_2$ transfer is observed when complexes 15 are treated with Ag$^+$ in the presence of cyclohexene, although the yields of norcarane are very low (≤ 5%).

The [(η-C$_5$R$_5$)Cr(NO)$_2$CH$_2$X] complexes are more conventional in their reactions with nucleophiles, giving stable complexes of the type [(η-C$_5$R$_5$)Cr(NO)$_2$CH$_2$Z] (Z = CN, OTs, OMe, and OEt). The reaction of [CpCr(NO)$_2$CH$_2$I] with PPh$_3$ gives the stable ylide complex [CpCr(NO)$_2$CH$_2$PPh$_3$]$^+$ I$^-$. The molecular structure of [Cp*Cr(NO)$_2$CH$_2$I] has been determined. The molecule adopts a piano stool geometry with a Cr—C bond distance of 209.3(11) pm, which is believed to be short for a

Cr—C single bond. Thus, as observed for the iodomethyl iron complex $[Fe\{P(OCHMe_2)_3\}_2(CO)_2(I)CH_2I]$ (see Section II,A), the CH_2I carbon atom may have some sp^2 character (57).

Dodd and Johnson prepared the halomethyl complexes $[Cr(H_2O)_5CH_2X]^{2+}$ (X = Cl, Br, or I) via the reaction shown in Eq. 5 (58). These authors noted that the $[Cr(H_2O)_5CH_2X]^{2+}$ complexes are also obtained by the reaction of a large excess of Cr^{2+} with CHX_3 (X = Cl, Br, or I). This they attributed to the conversion of $[Cr(H_2O)_5CHX_2]^{2+}$, which forms initially, by reaction with excess Cr^{2+} [Eq. (6)]. The presence of these $[Cr(H_2O)_5CH_2X]^{2+}$ complexes, as intermediates, was earlier proposed in the stepwise reduction of haloforms and methylene halides by $CrSO_4$ in N,N-dimethylformamide (DMF)–H_2O (59). A kinetic and mechanistic study on the formation of $[Cr(H_2O)_5CH_2Cl]^{2+}$ has been reported (60). This study implied that the rate-determining step is a bimolecular halogen atom abstraction reaction of $CrCHCl_2^{2+}$ and Cr^{2+} to form $CrCl^{2+}$ and $CrCHCl^{2+}$. The latter intermediate rapidly exchanges with Cr^{2+} to form the intermediate Cr_2CHCl^{4+}, which in turn gives Cr^{3+} and $CrCH_2Cl^{2+}$ in a protonolysis step.

$$2\ Cr^{2+}(aq) + CH_2X_2 \rightarrow [Cr(H_2O)_5X]^{2+} + [Cr(H_2O)_5CH_2X]^{2+}$$

$$X = Cl, Br, I \tag{5}$$

$$[Cr(H_2O)_5CHCl_2]^{2+} + 2\ Cr^{2+}(aq) + H^+ \rightarrow$$
$$[Cr(H_2O)_5CH_2Cl]^{2+} + [Cr(H_2O)_6]^{3+} + [Cr(H_2O)_5Cl]^{2+} \quad (6)$$

The complex $[Cr(H_2O)_5CH_2Cl]^{2+}$ was found to react with IBr with electrophilic cleavage of the Cr—C bond to give CH_2ClI as the only volatile product [Eq. (7)] (61). Similarly $[Cr(H_2O)_5CH_2X]^{2+}$ (X = Cl, Br, or I) reacted with Y_2 (Y = Br or I) to afford CH_2XY, $[Cr(H_2O)_6]^{3+}$, and Y^- (62,63). The rate of the reaction was found to follow the order $k(IBr) \cong k(Br_2) \gg k(I_2)$. Physical and kinetic evidence suggests that the reactions proceed via a S_E2 mechanism with an "open" transition state.

$$[Cr(H_2O)_5CH_2Cl]^{2+} + IBr + H_2O \rightarrow CH_2ClI + [Cr(H_2O)_6]^{3+} + Br^- \tag{7}$$

Aquation of the complexes $[Cr(H_2O)_5CH_2X]^{2+}$ gives $[Cr(H_2O)_6]^{3+}$, MeOH, and X^- for X = Cl and Br. The mechanism proposed for the reactions involves the formation of a chromium–methanol intermediate formed by elimination of HX after an S_N2 attack of H_2O at the carbon. In contrast, the aquation of $[Cr(H_2O)_5CH_2I]^{2+}$ gives $[Cr(H_2O)_6]^{3+}$ and CH_3I. The mechanism for this reaction is proposed to involve a carbanion type transition state. Surprisingly, the rate constants for these aquation reactions at 25°C follow the order $CH_2Br \gg CH_2Cl > CH_2I$ (64).

The complexes $[Cr(H_2O)_5CH_2X]^{2+}$ (X = Cl, Br, or OCH_3) were found to react with mercuric nitrate [Eq. (8)] (58,65). This reaction is believed to involve a binuclear displacement of Cr by attack of Hg on the carbon atom. In contrast, the reaction of $[Cr(H_2O)_5CH_2I]^{2+}$ with Hg^{2+} involves an abstraction of I^- by Hg^{2+} [Eq. (9)]. The reaction of $[Cr(H_2O)_5CH_2I]^{2+}$ with Cr^{2+} as shown in Eq. (10) has also been reported (66). This reaction is believed to proceed via a carbon-bridged dinuclear chromium intermediate.

$$[Cr(H_2O)_5CH_2X]^{2+} + Hg^{2+} \rightarrow HgCH_2X^+ + Cr^{3+} \tag{8}$$

$$[Cr(H_2O)_5CH_2I]^{2+} + Hg^{2+} \rightarrow HgI^+ + CH_2 + Cr^{3+} + 5H_2O \tag{9}$$

$$[Cr(H_2O)_5CH_2I]^{2+} + 2\ Cr^{2+}(aq) + 2\ H^+ \rightarrow 2\ [Cr(H_2O)_6]^{3+} + [Cr(H_2O)_5I]^{2+} + CH_4 \tag{10}$$

The position trans to the CH_2Cl group has been found to be several orders of magnitude kinetically more labile than the same position in the corresponding inorganic complex of Cr^{3+} (i.e., $[Cr(H_2O)_5Cl]^{2+}$) [Eq. (11)] (67,68). Relatively rapid reactions occur between $[Cr(H_2O)_5CH_2Cl]^{2+}$ and two acac$^-$ ligands to give trans-$[Cr(acac)_2(H_2O)Ch_2Cl]$. The coordination of the oxygen atom of the acac$^-$ ligand presumably occurs at the position trans to the Cr—C bond, followed by chelation with migration of the acetyl acetonate (acac$^-$) to sites cis to the Cr—C bond. The water molecule trans to the CH_2Cl group of $[Cr(acac)_2(H_2O)CH_2Cl]$ can easily be replaced by ligands (L) such as CH_3OH or pyridine to give $[Cr(acac)_2(L)CH_2Cl]$. The molecular structure of $[Cr(acac)_2(py)CH_2Cl]$ has been determined. Both the Cr—C and C—Cl bond lengths are within normal parameters, although the Cr—C—Cl bond angle is a little larger than average (see Table I in Section IV). Notable is the Cr—N (pyridine) bond length (220 pm), which is longer than the average length of such bonds (210 pm), indicating a weakening of the Cr–ligand bond trans to the Cr—CH_2Cl bond. This may well account for the lability of the trans ligand (69).

$$[Cr(H_2O)_5CH_2Cl]^{2+} + SCN^- \rightarrow trans-[Cr(H_2O)_4(NCS)CH_2Cl]^+ + H_2O \tag{11}$$

Green et al. reported the synthesis of $[CpMo(CO)_3CH_2X]$ [X = OMe (16), Cl (17), or Br (18)] (29). Compounds 16–18 were prepared according to a method similar to that shown in Eqs. (1) and (2). Complex 17 was also prepared in very high yield by the reaction of $[CpMo(CO)_3\{CH_2OC(O)Me\}]$ with HCl (30). An alternative route to complex 17 involved reacting the molybdenum anion $[CpMo(CO)_3]^-$ with $ClCH_2I$ [Eq. (4)] (33). Reaction of $Na[CpMo(CO)_3]$ with CH_2I_2 was reported to give $[CpMo(CO)_3CH_2I]$ (19) in low yield (33). Complex 19 has only been partially characterized and is reported as being unstable in air. Complex 16 was isolated as an air-sensitive oil; complexes 17 and 18 are air- and light-sensitive solids.

Facile conversion of **17** to **16** was achieved by reaction with NaOMe (*29*). The carbene complex [CpMo(CO)$_3$(CH$_2$)]$^+$ was generated by the reaction of **17** with AgPF$_6$ (*43*). The reactivity of [CpMo(CO)$_3$CH$_2$Cl] (**17**) with PPh$_3$ differs from that of its tungsten analog [CpW(CO)$_3$CH$_2$Cl] (see later). Compound **17** reacts with PPh$_3$ in methanol to give [CpMo(PPh$_3$)(CO)$_2${C(O)CH$_2$OCH$_3$}], [CpMo(PPh$_3$)(CO)$_2$Cl], and [Cp-Mo(PPh$_3$)$_2$(CO)Cl] depending on the reaction time (*35*). Unlike [CpMo(CO)$_3$CH$_3$] which underwent "CO insertion," complex **17** did not react with C$_6$H$_{11}$CN (*70*). Proposed pathways for the reactions of the [M(L)CH$_2$X] complex with ligands, L, are shown in Scheme 1. In a similar way to iron analog **2**, complex **17** reacts with L (L = py, SMe$_2$, or SMePh), using TlBF$_4$ as an electrophilic substitution promoter, to give ylide type complexes [CpMo(CO)$_3$CH$_2$L]BF$_4$ (*36*).

The pentamethylcyclopentadienyl analogs of compounds **16–19**, [Cp*Mo(CO)$_3$CH$_2$X] (X = OCH$_3$, Cl, Br, or I), were recently reported (*71*). In contrast to [Cp*W(CO)$_3$CH$_2$Cl], [Cp*Mo(CO)$_3$CH$_2$Cl] reacts with PPh$_3$ to give [Cp*Mo(PPh$_3$)(CO)$_2$Cl] and not the ylide type product obtained from [Cp*W(CO)$_3$CH$_2$Cl] (see later in this section). As with the Cp complexes, the Cp* molybdenum complex [Cp*Mo(CO)$_3$CH$_2$Cl] is reported to be less stable than its tungsten analog [Cp*W(CO)$_3$CH$_2$Cl].

The complexes *trans*-[CpMo{P(OPh)$_3$}(CO)$_2$CH$_2$X] (X = Cl, Br, or I) were prepared by the novel route of utilizing the ability of the formyl complex [CpMo{P(OPh)$_3$}(CO)$_2$CHO] to transfer hydride in the presence of electrophilic reagents (*17*) as shown in Scheme 6. The direct conversion of the formyl complex to the halomethyl complex occurs by the action of HX. The stereochemistry of the initial formyl complex was preserved in the halomethyl products.

The complexes [CpW(CO)$_3$CH$_2$X] [X = OCH$_3$ (**20**), Cl (**21**), or Br (**22**)] were first reported by Green *et al.* (*29*). They were prepared according to the method shown in Eqs. (1) and (2). Facile conversion of **21** to **20** was achieved by reaction with NaOMe, though the yield was low. King and

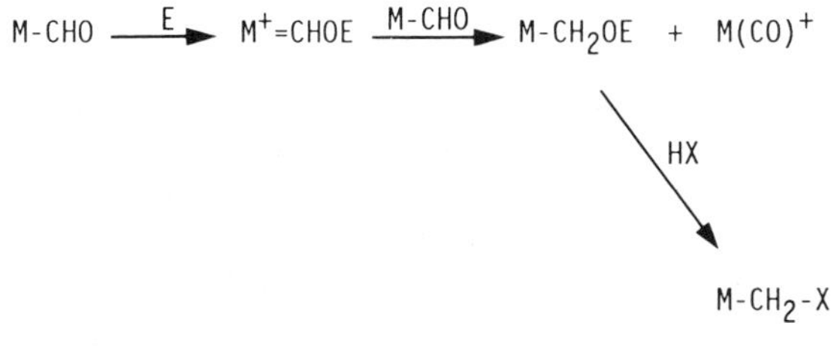

$$M\text{-CHO} \xrightarrow{\text{E}} M^+\text{=CHOE} \xrightarrow{\text{M-CHO}} M\text{-CH}_2\text{OE} + M(CO)^+$$

$$\downarrow \text{HX}$$

$$M\text{-CH}_2\text{-X}$$

E = H or CH$_3$;   X = Cl,Br,I

SCHEME 6

Braitsch prepared **21** by the alternative route of reacting the tungsten anion with $ClCH_2I$ [see Eq. (4)] (*33*). The reaction of the tungsten anion with CH_2I_2 gave the new (but only partially characterized) compound $[CpW(CO)_3CH_2I]$ (**23**), in poor yield. This iodomethyl complex was reported to be highly unstable. Recent evidence suggests that compound **23** is more stable than previously thought, and it can be prepared in high yield by the reaction of $[CpW(CO)_3CH_2Br]$ with NaI. Like the analogous iron complexes **2–4** these halogenomethyl tungsten complexes form carbene (methylene) species under mass spectral conditions, with the relative abundance of these species following the order I > Br > Cl (*42,72*).

These halogenomethyl complexes react with neutral nucleophiles, L (e.g., tertiary phosphines, SMe_2, $AsPh_3$, NMe_3, and py), to give either the ylide type products $[CpW(CO)_3CH_2L]^+$ or the disubstituted complexes $[CpWL_2(CO)_2]^+$, depending on the halide, the pK_a and cone angle of the ligand, and the solvent used. Kinetic evidence showed that the reaction rates followed the order I > Br > Cl and indicated that the ylide type product is formed via a concerted (S_N2) mechanism. The rate of ylide formation was significantly faster for phosphines with small cone angles and high pK_a values. The reactions of the tungsten halogenoalkyl complexes were significantly slower than the corresponding reactions of the analogous iron complexes, $[CpFe(CO)_2CH_2X]$. Thus, the metal has a significant effect on the reactivity of $[L_yMCH_2X]$ (*42,72*).

The pentamethylcyclopentadienyl analogs of compounds **20–23**, $[Cp^*W(CO)_3CH_2X]$ (X = OCH_3, Cl, Br, or I), were recently reported (*71*). $[Cp^*W(CO)_3CH_2X]$ reacted with PPh_3 to form the ylide complex $[Cp^*W(CO)_3CH_2PPh_3]X$ (X = Cl or I) (*71*). Similarly, $[CpW(CO)_3CH_2Cl]$ reacts with PPh_3 in acetonitrile or methanol to give $[CpW(CO)_3CH_2PPh_3]Cl$ (*35,49*). The Cp* tungsten haloalkyl complexes are reported to be more stable than their Cp analogs and have a stability order of $CH_2Cl > CH_2Br > CH_2I$, as is found for the Cp analogs.

C. *Manganese and Rhenium*

King and Braitsch found $[Mn_2(CO)_{10}]$ to be the only product of the reaction between $Na[Mn(CO)_5]$ and CH_2ClI at room temperature (*33*). Moss and Pelling, however, found that the same reaction at $-20°C$ gave $[Mn(CO)_5CH_2Cl]$ (**24**) in 50% yield (*34*). Compound **24** had previously been referred to by Jolly and Pettit (*28*). Moss and co-workers also reported the synthesis of, and full characterization data for, *cis*-$[Mn(PPh_3)(CO)_4CH_2Cl]$ (*34,49*). Complex (**24**) reacts with PPh_3 in CH_3CN at room temperature to give $[Mn(PPh_3)_2(CO)_3Cl]$ (*35*).

[Mn(CO)$_5$CH$_2$I] (**25**) has been prepared by Gladysz and co-workers by the reaction of [Mn(CO)$_5$CH$_2$OCH$_3$] with (CH$_3$)$_3$SiI (*73*). The fluoromethyl complex [Mn(CO)$_5$CH$_2$F] was prepared by the reaction of Na[Mn(CO)$_5$] with Cl(CO)CH$_2$F to give [Mn(CO)$_5$C(O)CH$_2$F] and subsequent decarbonylation of the acyl complex (*74*). Complex **24** can be prepared similarly by the decarbonylation of the acyl complex [Mn(CO)$_5$C(O)CH$_2$Cl] (*75,76*).

A novel method for the synthesis of the complexes [Mn(CO)$_5$CH$_2$X] (X = Cl, Br, or I) involves the facile electrophilic halogen exchange between BX$_3$ (X = Cl, Br, or I) and [Mn(CO)$_5$CH$_2$F] as shown in Eq. (12). The formation of the strong B—F bond is believed to be the driving force for the overall reaction (*25*). The bromoalkyl complex [Mn(CO)$_5$CH$_2$Br] was converted to [Mn(CO)$_5$CH$_3$] by reaction with HSn(^{n}Bu$_3$) (*77*). Complex **25** did not appear to react with PPh$_3$ in CD$_3$CN but decomposed to form [Mn(CO)$_5$I] (*78*). If a CH$_2$Cl$_2$ solution of complex **25** is allowed to stand for several days, [Mn(CO)$_5$I] as well as [Mn(CO)$_4$I]$_2$ are isolated. Attempts to carbonylate **25** under CO pressure (up to 100 atm) always gave [Mn(CO)$_5$I], with at most trace amounts of carbonylated product present (*73*). This contrasts with the ready carbonylation of [Mn(CO)$_5$CH$_3$] and agrees with theoretical considerations, which state that the migratory ability of a methyl group that is electronegatively substituted should be considerably lower than that of a CH$_3$ unit (*79*). As expected, complex **25** was observed to act as a methylene transfer reagent, reacting with cyclohexene to form norcarane (*78*).

$$[Mn(CO)_5CH_2F] + BX_3 \rightarrow [Mn(CO)_5CH_2X] + BX_2F \qquad (12)$$

As was observed for [CpMo{P(OPh)$_3$}(CO)$_2$CHO], the formyl complexes *mer,trans*-[Re{P(OPh)$_3$}$_2$(CO)$_3$CHO] and *cis*-[Mn(PPh$_3$)(CO)$_4$CHO] react with HX, utilizing the ability of some formyls to transfer hydride in the presence of electrophilic reagents, to give *mer,trans*-[Re{P(OPh)$_3$}$_2$-(CO)$_3$CH$_2$X] and *cis*-[Mn(PPh$_3$)(CO)$_4$CH$_2$X] (X = Cl, Br, or I), respectively, with retention of stereochemistry (as shown in Scheme 6). The complexes *mer,trans*-[Mn(PPh$_3$)$_2$(CO)$_3$(CHO)] and *mer,trans*-[Mn{P-(OPh)$_3$}$_2$(CO)$_3$(CHO)] react with methyl triflate as the initial electrophile to give the corresponding methoxymethyl species (Scheme 6), which then react with HX to give the halomethyl complexes *mer,trans*-[Mn-(PPh$_3$)$_2$(CO)$_3$CH$_2$X] and *mer,trans*-[Mn{P(OPh)$_3$}$_2$(CO)$_3$CH$_2$X] (X = Cl, Br, or I) (*17*).

The complex [LRe(CO)(NO)CH$_2$I]BF$_4$ (L = 1,4,7-triazacyclononane) was prepared by oxidation of [LRe(CO)(NO)(CH$_3$)]BF$_4$ with I$_2$. The complex [LRe(CO)(NO)CH$_2$I]BF$_4$ reacts in aqueous solution to give the complex [{LRe(CO)(NO)}$_2$(μ-CH$_2$OCH$_2$)]I$_2$, a complex with a previously unknown 2-oxapropane-1,3-diide bridge. This reaction is believed to be due to the hydrolysis of Re—CH$_2$I with H$_2$O, followed by dimerization to give the —CH$_2$—O—CH$_2$— bridged product (80). The Cp* complex [Cp*Re(PPh$_3$)(NO)(Cl)CH$_2$Cl]BF$_4$ was obtained in the decomposition of the dichloromethane complex [Cp*Re(PPh$_3$)(NO)(ClCH$_2$Cl)]BF$_4$ (81).

D. Cobalt, Rhodium, and Iridium

Several halogenomethyl cobalt complexes are known. [CpCo-(PMe$_3$)(CO)] or [CpCo(PMe$_3$)(C$_2$H$_4$)] reacted with CH$_2$ClI at low temperature ($-50°$C) to give a mixture of [CpCo(PMe$_3$)(I)CH$_2$Cl] (26) and [CpCo(PMe$_3$)(Cl)I] (27) (82,83). Complex 26 is stable in the solid state; however, in solution at temperatures exceeding $-30°$C it eliminates CH$_2$ to form 27. The reaction of 26 with KOMe gives the methoxymethyl cobalt(III) complex [CpCo(PMe$_3$)(I)CH$_2$OCH$_3$] (28). This methoxymethyl complex can also be prepared in a one-pot synthesis with [CpCo(PMe$_3$)(CO)], CH$_2$ClI, and methoxide (Scheme 7).

Addition of an equimolar quantity of a two-electron donor, L [where L = PMe$_3$, P(OMe)$_3$, or CNMe], to 26 produces the corresponding [CpCo(PMe$_3$)(L)CH$_2$Cl]$^+$ cations. These cations can also be prepared by the reaction of [CpCo(PMe$_3$)(CO)] with CH$_2$ClI in the presence of L. The cations react at room temperature with PMe$_3$ to produce the ylidic [CpCo(PMe$_3$)(L)CH$_2$PMe$_3$]$^{2+}$ dications. Similarly, Klein and Hammer obtained cobalt-coordinated ylidic alkylidene–trimethylphosphorane ligands by the oxidative additions of 1,1-dichloroalkanes to Co(PMe$_3$)$_4$ (84). They proposed that the formation of these complexes involved chloroalkyl cobalt intermediates.

The halomethyl complexes [CpCo(PMe$_3$)(X)CH$_2$X] (X = Br or I) have been proposed as labile intermediates in the formation of [CpCo(PMe$_3$)(η-CH$_2$E)] (E = S or Se) from [CpCo(PMe$_3$)(CO)] according to Scheme 8. In contrast, [CpCo(PMe$_3$)(I)CH$_2$Cl], reacting with NaSH or NaSeH, gave only [CpCo(PMe$_3$)(Cl)I] (85). Similarly, halomethyl intermediates are

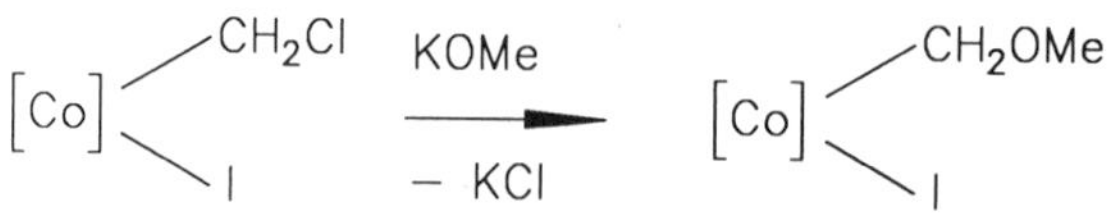

$$[\text{Co}] \overset{\text{CH}_2\text{Cl}}{\underset{\text{I}}{<}} \quad \xrightarrow[-\text{KCl}]{\text{KOMe}} \quad [\text{Co}] \overset{\text{CH}_2\text{OMe}}{\underset{\text{I}}{<}}$$

(26)   (28)

$$[\text{Co}] = \text{CpCo(PMe}_3)$$

SCHEME 7

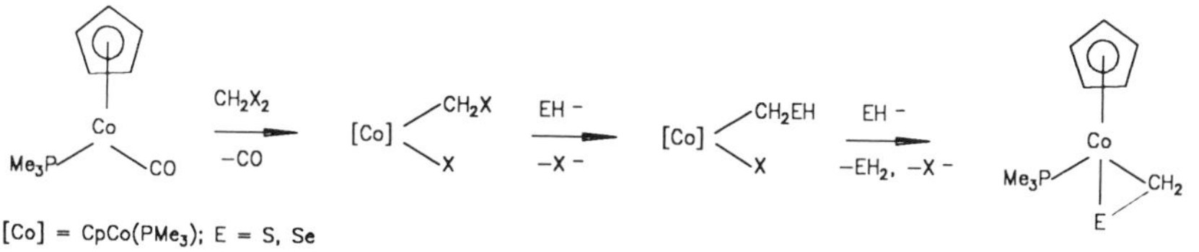

$[Co] = CpCo(PMe_3); E = S, Se$

SCHEME 8

proposed in the reactions of $[CpCo(\mu\text{-}PMe_2)]_2$ with CH_2X_2 (X = Br or I) to give $[\{CpCoX\}_2(\mu\text{-}PMe_2)_2]$ and $[\{CpCoX\}_2(\mu\text{-}PMe_2)(\mu\text{-}CH_2PMe_2)]$ (Scheme 9) (86).

The cobalt halomethyl compound $[Cp^*Co(CO)(Cl)CH_2Cl]$ (29) was recently prepared by Dahl and co-workers (87). This complex was synthesized by the photolytic reaction of $[Cp^*Co(CO)_2]$ with CH_2Cl_2 and represents the first example of an oxidative addition product from a nonporphyrin cobalt(I) species. Compound 29 is reported to be relatively air stable in the solid state and is stable at room temperature in noncoordinating solvents. The molecular configuration of complex 29 conforms to the three-legged piano stool geometry, with normal $Co\text{-}CH_2Cl$ (199 pm) and $CH_2\text{-}Cl$ (180 pm) bond lengths and a normal $Co\text{-}CH_2\text{-}Cl$ bond angle of 116.3° (Fig. 2). In a room temperature photolytic reaction, 29 reacts with PMe_3 to give two products, $[Cp^*Co(PMe_3)(Cl)CH_2Cl]$ and $[Cp^*Co(PMe_3)Cl_2]$.

The remarkable chloromethyl complex $[Co(CO)_4CH_2Cl]$ was prepared by the decarbonylation of the chloroacetyl complex $[Co(CO)_4\{C(O)\text-}CH_2Cl\}]$. This reaction is reversible and thus represents the first example of a carbonylation/decarbonylation reaction couple of a MCH_2X versus $MC(O)CH_2X$ pair. A mixture of the decarbonylated and carbonylated species reacted with PPh_3 to give $[Co(PPh_3)(CO)_3\{C(O)CH_2Cl\}]$ (30). This

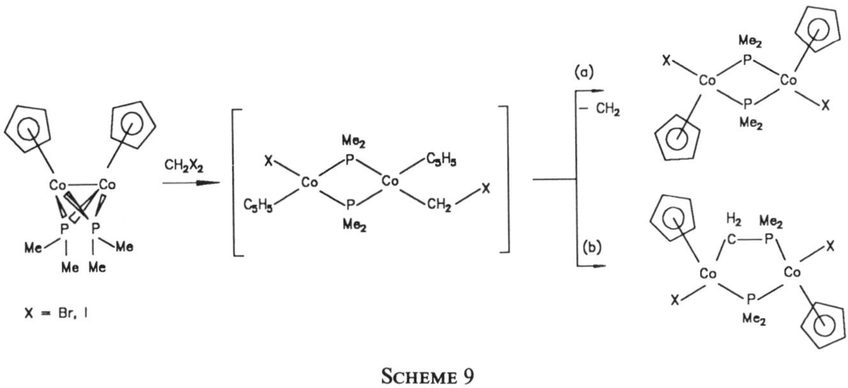

SCHEME 9

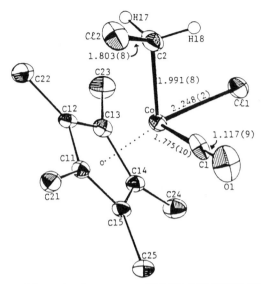

FIG. 2.   Diagram of the molecular structure of [Cp*Co(CO)(Cl)CH₂Cl] (**29**). [Reprinted with permission from W. L. Olson, D. A. Nagaki, and L. F. Dahl, *Organometallics* **5**, 630 (1986). Copyright (1986) American Chemical Society.]

is the first example where a chloromethyl transition metal complex reacts with PPh₃ to yield a CO substitution product, and it is also the first example of a ligand-induced migratory insertion reaction in this class of compounds. The chloroacetyl complex (**30**) could be decarbonylated at 40°C to give the corresponding chloromethyl derivative [Co(PPh₃)(CO)₃CH₂Cl] (**31**). The latter complex (**31**) can in turn be carbonylated to give **30** again. An alternative route to complex **31** involves the reaction of Na[CoPPh₃(CO)₃] and ClCH₂I (*88*). The molecular structure of complex **31** showed no unusual features, with Co—CH₂Cl and CH₂—Cl bond lengths of 202 and 179 pm, respectively, and Co—CH₂—Cl bond angle of 114.9°.

The complexes [Co([14]aneN₄)(H₂O)CH₂X]²⁺ (X = Cl, Br, or I; [14]aneN₄ = the tetraazacyclotetradecane ligand) were prepared by Bakac and Espenson (*89*). Irradiation of these complexes with visible light led to the homolysis of the Co—C bond. The unusual bromomethyl complex, [(*Co*)CH₂Br], where (*Co*) represents vitamin B₁₂ₛ, was prepared by Smith *et al.* according to Eq. (13) (*90*).

$$[(Co)] + BrCH_2Br \rightarrow [(Co)CH_2Br]$$

$$(Co) = \text{vitamin } B_{12s}$$

(13)

The halomethyl cobaloxime complexes $[Co(DH)_2\{P(OCH_3)_3\}CH_2X]$ (**32**) (X = Cl, Br, or I; DH = dimethylglyoximato ligand) have also been reported (*91,92*), as have a wide range of cobaloxime type halogenoalkyl complexes, for example, $[Co(DH)_2LCH_2X]$ (X = Cl, Br, or I, L = py or 4-CN-py; X = Br or I, L = ^tBupy; X = I or Cl, L = H_2O; X = Br, L = PPh_3, P^tBu_3, $PEtPh_2$, 2-NH_2-py, or 1-*N*-Me-imidazole; X = Cl, L = NH_3). The properties of these complexes have been comprehensively reviewed fairly recently, along with other substituted cobaloxime complexes, with special emphasis on the effect of the axial ligand (L) on structures and coordination chemistry (*93*). The CH_2X group has been observed to affect the chemistry of these systems. Thus, the rate of dissociation of $P(OCH_3)_3$ from complexes **32** follows the order $CH_2Cl > CH_2Br$ (*91*). The related complex $[Co(bae)CH_2Cl]$ [bae = *N,N'*-bis(acetylacetonato)ethylenediimine] has also been reported (*94*).

Treatment of the cobalt porphyrins $[Co(TPP)X]$ (X = Cl, Br, or I) and $[Co(OEP)Br]$ (TPP = *meso*-tetraphenylporphine and OEP = octaethylporphine) with diazomethane gives the corresponding halomethyl cobalt(II) porphyrins $[Co(TPP)CH_2X]$ (X = Cl, Br, or I) and $[Co(OEP)CH_2Br]$, respectively. These reactions are proposed to proceed via the insertion of the carbene moity into a Co—N bond, and only subsequent nucleophilic attack by the halide results in the formation of the CH_2X group (*95*).

Rhodium complexes of the general formula $[CpRh(L)(L')CH_2I]PF_6$ and $[CpRh(L)_2CH_2I]PF_6$ were prepared according to the method shown in Eqs. (14), (15), and (16) (*82,96–98*). Rhodium halogenomethyl complexes display interesting and varied chemistry. The compounds $[CpRh(PMe_3)\{P(OMe)_3\}CH_2I]PF_6$ and $[CpRh\{P(OEt)_3\}_2CH_2I]PF_6$ react with NaI to give the neutral dialkylphosphonate complexes $[CpRh(PMe_3)\{P(O)(OMe)_2\}CH_2I]$ and $[CpRh\{P(OEt)_3\}\{P(O)(OEt)_2\}CH_2I]$ in a Michaelis–Arbuzov type reaction (*96*). The complexes $[CpRh(PMe_3)_2CH_2I]^+$ (**34**) and $[CpRh(PR_3)(L)CH_2I]PF_6$ [L = $P(OMe)_3$, PR_3 = PMe_3, PMe_2Ph, or $PMePh_2$; L = $P(OEt)_3$, PR_3 = PMe_3] isomerize thermally, or on addition of NEt_3 (as catalyst), to form ylide rhodium(III) complexes $[CpRh(PMe_3)(I)CH_2PMe_3]PF_6$ and $[CpRh(L)(I)CH_2PR_3]PF_6$, respectively (*83,96–98*). In all cases the ligand with the strongest donor properties migrates from the metal to the carbon. The reaction of $[CpRh(PMe_3)_2CH_2I]^+$ with the softer base SMe^- gives a mixture of $[CpRh(PMe_3)(SMe)CH_2PMe_3]I$ (**35**) and $[CpRh(PMe_3)_2CH_2SMe]I$. Complex **35** has been shown to form rapidly via the intermediate $[CpRhI(PMe_3)CH_2PMe_3]I$ (*97*). In none of the observed reactions of the rhodium haloalkyl complexes has the breaking of a Rh—PMe_3 bond been observed. It has therefore been proposed that the isomerism proceeds via an intramolecular mechanism involving a four-center transition state

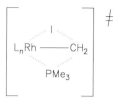

FIG. 3.   Proposed transition state in the isomerization of complex **35**.

formed with the aid of the catalyst (Fig. 3). Substitution of the carbon-bonded iodine in $[CpRh(PMe_3)_2CH_2I]I$ with PMe_3 and that in $[CpRh(PMe_3)\{P(O)(OMe)_2\}CH_2I]$ with L [L = PMe_3, PMe_2Ph, $PMePh_2$, PMeEtPh, $P(OMe)_3$, or SMe_2] leads to the formation of the ylide complexes $[CpRh(PMe_3)_2CH_2PMe_3]I_2$ and $[CpRh(PMe_3)\{P(O)(OMe)_2\}CH_2L]I$, respectively.

$$[CpRhLL'] + CH_2I_2 \rightarrow [CpRhLL'CH_2I]I \xrightarrow[-I^-]{PF_6^-} [CpRhLL'CH_2I]PF_6 \qquad (14)$$

L	L'	L	L'
PMe_3	PMe_2Ph	PMe_3	P^iPr_3
PMe_3	$PMePh_2$	PMe_3	CNMe
PMe_3	$P(OMe)_3$	PEt_3	$P(OMe)_3$
PMe_3	$P(OEt)_3$	PMe_2Ph	$P(OMe)_3$
		$PMePh_2$	$P(OMe)_3$

$$[CpRhL_2] + CH_2I_2 \rightarrow [CpRhL_2CH_2I]I \xrightarrow[-I^-]{PF_6^-} [CpRhL_2CH_2I]PF_6 \qquad (15)$$

$$L = CNMe,\ ^tBuCN,\ P(OEt)_3$$

$$[CpRh(PMe_3)_2] + CH_2I_2 \rightarrow [CpRh(PMe_3)_2CH_2I]I \xrightarrow{BF_4^-} [CpRh(PMe_3)_2CH_2I]BF_4 \qquad (16)$$

(33)

The chloromethyl rhodium complexes $[CpRh(PMe_3)_2CH_2Cl]PF_6$ were prepared according to Eq. (17), whereas $[CpRh(PMe_3)\{P(OMe)_3\}CH_2Br]^+$ and $[CpRh(PMe_3)\{P(OMe)_3\}CH_2Cl]^+$ were prepared by the reaction of $[CpRh(PMe_3)\{P(OMe)_3\}]$ with CH_2Br_2 and CH_2ClI, respectively (96). The reaction of complex **33** with CH_2Br_2 gave $[CpRh(PMe_3)_2CH_2Br]Br$, but not in pure form (96). Similarly, $[CpRh(R_2PCH_2CH_2PR_2)]$ (R = Me or Ph) reacted with CH_2I_2 in the presence of NH_4PF_6 to give $[CpRh(R_2PCH_2CH_2PR_2)CH_2I]PF_6$. The latter complex (where R =

$$[CpRh(PMe_3)_2] + CH_2ClX \rightarrow [CpRh(PMe_3)_2CH_2Cl]X \xrightarrow[-X^-]{PF_6^-} [CpRh(PMe_3)_2CH_2Cl]PF_6$$
$$X = Cl,\ I \qquad (17)$$

Me) isomerized in the presence of NEt_3 to form the cation [Cp-$(I)RhCH_2PMe_2CH_2CH_2PMe_2]^+$, which contains an α-C,ω-P-bonded cyclic ylide ligand (98).

The neutral complex $[CpRh\{P(OMe)_3\}(I)CH_2I]$ can be synthesized by the reaction of $[CpRh\{P(OMe)_3\}(C_2H_4)]$ or $[CpRh\{P(OMe)_3\}_2]$ with CH_2I_2 (96). The ethylene complex $[CpRh(PMe_3)(C_2H_4)]$ undergoes an oxidative substitution reaction with CH_2I_2, CH_2Br_2, and CH_2ICl to give $[CpRh(PMe_3)(X')CH_2X]$ (99–101) [see Eq. (18)]. Thus, displacement of the ethylene group and the formation of a Rh—C and a Rh—X bond cause the oxidation of Rh(I) to Rh(III). Complex 36 can also be prepared by the reaction of $[CpRh(L)(L')]$ $[L = PMe_3, L' = CO, PPh_3, or P(OPh)_3]$ with CH_2I_2 (96, 100).

$$[CpRh(PMe_3)(C_2H_4)] + CH_2XX' \rightarrow [CpRh(PMe_3)(X')CH_2X] + C_2H_4 \qquad (18)$$

	X	X'
(36)	I	I
	Br	Br
	Cl	I

Because the carbon-bonded iodide in 36 is very labile, complex 36 reacts with a wide range of anionic nucleophiles, as shown in Scheme 10

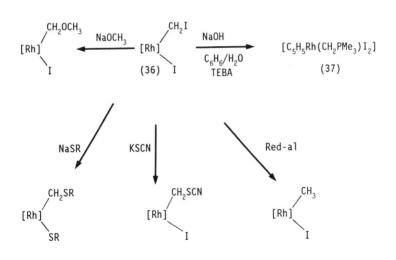

$[Rh] = CpRh(PMe_3)$; $R = CH_3, C_6H_5$; red-al $= Na[NaAlH_2(OC_2H_4OCH_3)_2]$

SCHEME 10

(101,102). Notable is the isomerization of **36** in the presence of [NEt$_3$(CH$_2$Ph)]Cl (TEBA) to [CpRh(I)$_2$CH$_2$PMe$_3$] (**37**). Complex **37** is also formed, together with the expected ylide product [CpRh(PMe$_3$)-(I)CH$_2$NEt$_3$]I, when **36** is reacted with NEt$_3$ *(100)*. Furthermore, complex **36** reacts with neutral nucleophiles with substitution of the carbon-bonded iodide as shown in Scheme 11. The reactions of **36** with py, AsPh$_3$, SMe$_2$, and SC$_4$H$_8$ are considerably slower than those with PPh$_3$ and P^{i}Pr$_3$. It is notable that the ylides CH$_2$AsPh$_3$ and CH$_2$SMe$_2$ are stabilized when coor-

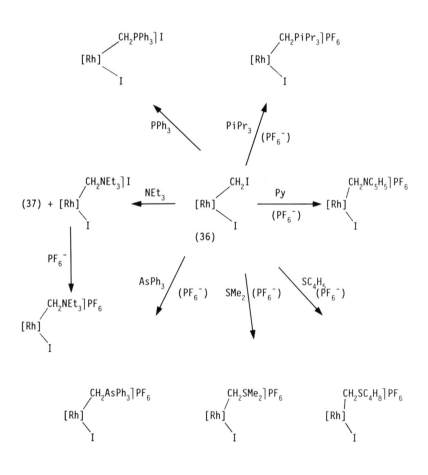

[Rh] = CpRh(PMe$_3$)

SCHEME 11

2 LiEHPh
− 2 LiI
− PhEH2
Rh
OC
CH2I
I
E = As, P
Rh
OC
CH2
E
Ph

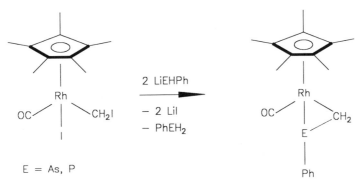

SCHEME 12

dinated to the metal. Neither of the ylides is stable in their uncoordinated form (*101,103,104*).

Werner and Paul reported that the highly unstable molecule CH_2Se can be fixed by the reaction of [CpRh(PMe_3)(I)CH_2I] with NaSeH, to give [CpRh(PMe_3)(η^2-CH_2Se)] (*105*). A similar reaction was carried out by Headford and Roper, using [Os(PPh_3)$_2$(CO)$_2$(I)CH_2I], to obtain [Os(PPh_3)$_2$(CO)$_2$(η^2-CH_2Se)] (*54*). [Cp*Rh(CO)(I)CH_2I] has been shown to react similarly (*106*). The complex [Cp*Rh(CO)(I)CH_2I] also reacts with LiEHPh (E = As or P) to give [Cp*Rh(CO)CH_2EPh] (Scheme 12). Ph—As=CH_2 is not known in its "free" (uncoordinated) form (*107*). Similar complexes to those above are obtained in fairly good yield by reacting [CpRh(PMe_3)(I)CH_2I] with NaSH and NaTeH. The reaction has been proposed to proceed as shown in Scheme 13 (*108*).

Whereas [CpRh$(CO)_2$] (**38**) does not react with CH_2Br_2 or CH_2I_2, the reaction of [Cp*Rh$(CO)_2$] with CH_2I_2 gives [Cp*Rh(CO)(I)CH_2I] (**39**) (*109*), presumably because of the increased electron density on the rhodium caused by the Cp* group. In general, the reaction of [Cp*Rh(CO)L] [L = PMe_3, PMe_2Ph, $P(OMe)_3$, and CO] with CH_2IX (X = I or Cl) gives complexes of the type [Cp*Rh(L)(I)CH_2X] [Eq. (19)]. Complex **40** can also be prepared by reacting [Cp*Rh(PMe_3)C_2H_4] with CH_2I_2. Complexes **39** and **40** show a surprisingly different reactivity. Although **40** reacts with $NaOCH_3$, as does the analogous Cp complex **36**, to form

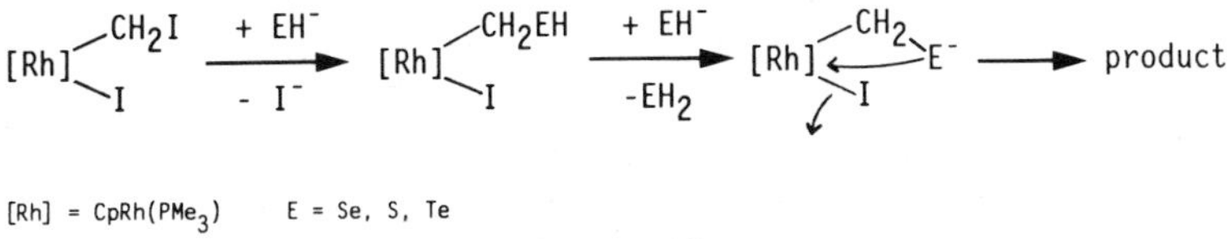

[Rh]
CH2I
I
+ EH⁻
- I⁻
[Rh]
CH2EH
I
+ EH⁻
-EH2
[Rh]
CH2
E⁻
I
product
[Rh] = CpRh(PMe3)
E = Se, S, Te

SCHEME 13

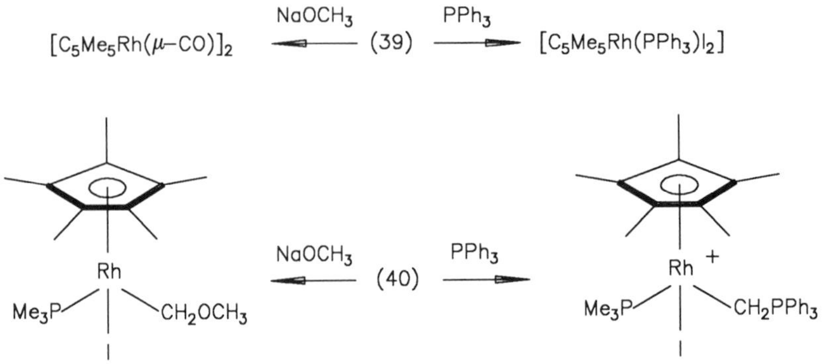

$$[C_5Me_5Rh(\mu\text{-CO})]_2 \xrightarrow[\text{NaOCH}_3]{} (39) \xrightarrow{\text{PPh}_3} [C_5Me_5Rh(PPh_3)I_2]$$

SCHEME 14

$[C_5Me_5Rh(PMe_3)(I)CH_2OCH_3]$ (41), complex 39 reacts to form the dimer $[Cp^*Rh(\mu\text{-CO})]_2$ (Scheme 14). The reaction of 39 with PMe_3 leads to a ligand rearrangement and formation of 40. The reaction of 39 with PPh_3 results in the formation of $[Cp^*Rh(PPh_3)I_2]$ (42), whereas the reaction of 40 with PPh_3 gives the expected ylide complex $[Cp^*Rh(PMe_3)\text{-}(I)CH_2PPh_3]I$ (43). Complex 39 reacts with $PPh_2(CH_2)PPh_2$ (dppm), with displacement of a CO group *and* substitution of the carbon-bonded iodide, to form the complex $[Cp^*Rh(I)CH_2PPh_2CH_2PPh_2]I$ (44). This complex is similar to the complex $[CpRh(I)CH_2PMe_2(CH_2)_2PMe_2]PF_6$, discussed earlier. This reaction is in contrast to the reaction of $[CpFe(CO)_2CH_2Br]$ with dppe, which forms $[\{CpFe(CO)_2CH_2\}_2\{\mu\text{-dppe}\}]^{2+}$, namely, a dimeric product.

$$[Cp^*RhL(CO)] \xrightarrow[-CO]{CH_2IX} [Cp^*RhL(I)CH_2X] \qquad (19)$$

	L	X
(40)	PMe$_3$	I
	PMe$_3$	Cl
	PMe$_2$Ph	I
	P(OMe)$_3$	I

Several macrocyclic rhodium halogenomethyl complexes have been prepared. The halomethyl porphyrin complexes $[(TPP)RhCH_2X]$ (X = Cl, Br, or I; TPP = the dianion of tetraphenylporphyrin) were prepared by the reaction of electrochemically generated $[(TPP)Rh]$ and CH_2X_2 (110). The complexes where X = Cl (111) and I (112) had previously been reported, the latter by the reaction of $[(TPP)RhI]$ with CH_2N_2. The macrocyclic rhodium(I) complexes $[RhL]X$ {(a): L = [14]aneS$_4$, X = Cl; (b): L = Me$_4$[14]aneS$_4$, X = Cl; (c): L = Me$_4$[14]aneS$_4$, X = BPh$_4$} react with CH_2Cl_2 in an oxidative addition reaction to give *trans*-$[RhL(Cl)CH_2Cl]X$ ([14]aneS$_4$ = tetrathiacyclotetradecane). The conformation of the macrocyclic ligand was found to affect the σ bascity of the rhodium(I) ion. Thus, the rate of oxidative addition of CH_2Cl_2 to (b) was 7 times faster than that to (c) (*113*). The oxidative addition of CH_2Cl_2 to [Rh(DO)(DOH)pn] [(DO)(DOH)pn = 1(diacetylmonoximato-imino-3-(diacetyl-monoxime-imino)propane] affords *trans*-[Rh(DO)(DOH)pn(Cl)CH$_2$Cl] (*114*).

Marder *et al.* showed that the square-planar cationic complexes [M(dmpe)$_2$]Cl (M = Rh or Ir) react with CH_2Cl_2 to give *trans*-[M(dmpe)$_2$(Cl)CH$_2$Cl]Cl·CH$_2$Cl$_2$ in high yields (*115*) [dmpe = PMe$_2$(CH$_2$)$_2$PMe$_2$]. Similarly, the neutral complex [Rh(PMe$_3$)$_3$Cl] reacts with CH_2Cl_2 to give two products, namely *mer,cis*-[Rh(PMe$_3$)$_3$(Cl)$_2$CH$_2$Cl] (**45**) and what is believed to be [Rh(PMe$_3$)$_2$(Cl)$_3$CH$_2$PMe$_3$] (**46**). Preliminary results indicate that complex **45** reacts slowly with PMe$_3$ via Cl$^-$ displacement from Rh, rather than RhCH$_2$Cl, to give *trans*-[Rh(PMe$_3$)$_4$(Cl)CH$_2$Cl]Cl. Complex **46** is believed to form from the isomerization of **45**. This chloride/phosphine exchange reaction appears to be promoted by the PMe$_3$ ligand trans to the chloromethyl group. The molecular structure of [Rh(dmpe)$_2$(Cl)CH$_2$Cl]Cl·CH$_2$Cl$_2$ has been reported and is shown in Fig. 4.

The reaction of [Ir(PPh$_3$)$_2$(CO)Cl] with CH_2N_2 was investigated by Mango and Dvoretzky and gave what was assumed from its infrared spectrum and reactivity to be the chloromethyl compound [Ir(PPh$_3$)$_2$(CO)CH$_2$Cl] (*116*). It appeared that a CH_2 group had been inserted into a metal–chlorine bond. The authors proposed that this reaction proceeded via a carbene intermediate (Scheme 15). The chloromethyl complex catalyzes the decomposition of excess CH_2N_2 to ethylene. Deuteration experiments have shown, however, that the Ir—CH$_2$Cl group does not participate in the dimerization reaction. Assuming Scheme 15 to be a valid equilibrium (i.e., left to right being a migratory insertion and right to left an α-elimination), Roper and co-workers concluded that the position

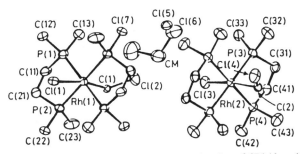

FIG. 4. Diagram of a unit cell containing two molecules of [Rh(dmpe)$_2$(Cl)CH$_2$Cl]-Cl·CH$_2$Cl$_2$. CM is the carbon atom of the CH$_2$Cl$_2$ solvent molecule. [By permission from the Royal Society of Chemistry; from T. B. Marder, W. C. Fultz, J. C. Calabrese, R. L. Harlow, and D. Milstein, *J. Chem. Soc., Chem. Commun.*, 1543 (1987).]

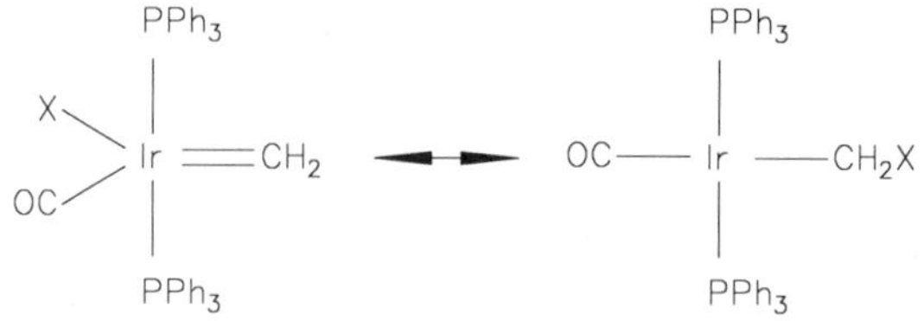

SCHEME 15

of the equilibrium would depend on X (the halogen), and a weak C—I bond would therefore favor the methylene form (117). Reaction of [Ir(PPh$_3$)$_2$(CO)I] with CH$_2$N$_2$ in tetrahydrofuran (THF) at $-50°$C proved this to be correct, yielding [Ir(PPh$_3$)$_2$(CO)I(=CH$_2$)] in 93% yield.

Harrison and Stobart reported that the reaction of [Ir(CO)$_2$(μ-pz)]$_2$ (pzH = pyrazole) with CH$_2$I$_2$ gave a mixture (70:20:10 ratio) of [Ir$_2$(CO)$_4$(μ-pz)$_2$(I)CH$_2$I] (47), [Ir$_2$(CO)$_4$(μ-pz)$_2$I$_2$(μ-CH$_2$)] (48), and [Ir(CO)$_2$(μ-pz)I]$_2$ (49) (118). Heating the mixture in benzene resulted in compounds 48 and 49 in a 70:30 ratio (Scheme 16). The rearrangement of the iodomethyldiiridium species to form the isomeric methylene complex (Scheme 16) is the first authenticated example of this type of transformation. The reaction of [Ir(cod)(μ-pz)]$_2$ (cod = 1,5-cyclooctadiene) (50) and [Ir$_2$(PPh$_3$)(CO)(μ-pz)$_2$] with CH$_2$I$_2$ afforded stable binuclear iodomethyl derivatives, namely, [Ir$_2$(cod)$_2$(μ-pz)$_2$(I)CH$_2$I] (51) and [Ir$_2$(PPh$_3$)$_2$(CO)$_2$(μ-pz)$_2$(I)CH$_2$I] (52), respectively (118). Although 51 does not undergo "oxidative isomerization" to the diiridium(III) methylene complex, complex 52 does, thus forming [Ir$_2$(PPh$_3$)$_2$(CO)$_2$(μ-pz)$_2$(I)$_2$(μ-CH$_2$)] (119). Mass spectral analysis indicates that the isomerization process is intramolecular. The chloromethyl analog of 51 was formed by a photochemically induced net two-electron reduction of [Ir(cod)(μ-pz)]$_2$(50) in CH$_2$Cl$_2$, leading to the two-center oxidative addition product [Ir$_2$(cod)$_2$(μ-pz)$_2$(Cl)CH$_2$Cl] (120).

Similarly, the complex [Ir$_2$(PPh$_3$)$_2$(CO)$_2$(μ-1,8-(NH)$_2$C$_{10}$H$_6$)(I)CH$_2$I] (53), prepared by the oxidative addition of CH$_2$I$_2$ to [Ir$_2$(PPh$_3$)$_2$(CO)$_2$(μ-1,8-(NH)$_2$C$_{10}$H$_6$)], undergoes thermal oxidative isomerization to give

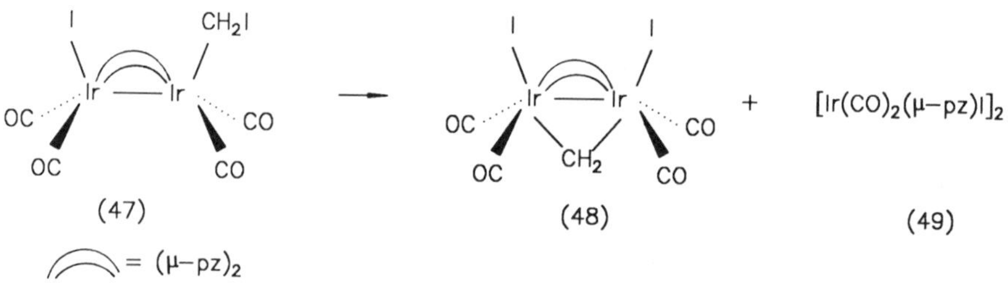

(47)   (48)   (49)

⌒ = (μ-pz)$_2$

SCHEME 16

the methylene-bridged complex $[Ir_2(PPh_3)_2(CO)_2\{\mu\text{-}1,8\text{-}(NH)_2C_{10}H_6\}$-$(I)_2(\mu\text{-}CH_2)]$ (**54**), as well as the diiodo complex $[Ir_2(PPh_3)_2(CO)_2\{\mu\text{-}1,8\text{-}$(NH)_2C_{10}H_6\}(I)_2]$ (*121*). Of interest is the structural change produced in this oxidative isomerization process. Thus, the CO and PPh$_3$ ligands are arranged in trans positions in **53** but cis positions in **54**. This transformation supports the proposal of Brost and Stobart (*119*) that the PPh$_3$ ligand dissociates from the Ir center, relieving conformational rigidity at the metal center and thus allowing intramolecular reorganization of **53** to **54**.

The dinuclear complex $[Ir_2(CO)_4(\mu\text{-}C_5H_4NS)_2]$ reacts with CH$_2$I$_2$ to give the dinuclear iodomethyl complex $[Ir_2(CO)_4(\mu\text{-}C_5H_4NS)_2(I)CH_2I]$ (*122*). Unlike previous observations (Scheme 16) (*118,119*) the complex $[Ir_2(CO)_4(\mu\text{-}C_5H_4NS)_2(I)CH_2I]$ does not form $[\{Ir(CO)_2(\mu\text{-}C_5H_4NS)I\}_2(\mu\text{-}CH_2$)] on heating.

The complexes $[Ir(PMe_3)_2(CO)(Cl)_2CH_2Cl]$, $[Ir(PMe_3)_2(CO)(Cl)(Br)$-CH$_2$Br], $[Ir(PMe_3)_2(CO)(Cl)(I)CH_2I]$, and $[Ir(PMe_3)_2(CO)(Cl)_2CH_2$-OCH$_3$] were prepared by the oxidative addition of CH$_2$Cl$_2$, CH$_2$Br$_2$, CH$_2$I$_2$, and ClCH$_2$OCH$_3$, respectively, to trans-$[Ir(PMe_3)_2(CO)(Cl)]$ (*123*). Evidence suggests that the chloromethyl methyl ether reacts via a S$_N$2 type reaction pathway, whereas the dihalomethanes appear to react via a radical pathway (*123*). Complexes very similar to those above were prepared by the addition of CH$_2$IX (X = Cl or I) to $[IrL_2(CO)(Cl)]$ (L = PMe$_3$, PMe$_2$Ph, or PMePh$_2$) (*124*). The *trans*-phosphine complexes $[IrL_2$-(CO)(Cl)(I)CH$_2$X] were obtained when L = PMe$_3$ or PMePh$_2$, whereas a mixture of both *cis*- and *trans*-phosphine isomers were obtained when L = PMe$_2$Ph. Addition of methanol to a solution of these isomers resulted in the isomerization of the cis complex to the trans complex. In contrast, heating $[Ir(PMe_3)_2(CO)(Cl)(I)CH_2I]$ in methanol gave the methoxymethyl derivative $[Ir(PMe_3)_2(CO)(Cl)(I)CH_2OCH_3]$. Another chloromethyl complex of Ir, $[Ir(PPh_3)_2(Cl)_2CH_2Cl]$, was obtained from the spontaneous rearrangement of the coordinatively unsaturated acyl complex $[Ir(PPh_3)_2Cl_2\{C(O)CH_2Cl\}]$ (*125*).

E. Palladium and Platinum

The first halomethylpalladium(II) complex, *cis*-$[Pd(PPh_3)_2(I)CH_2I]$ (**55**) was obtained from $[Pd(PPh_3)_4]$ and CH$_2$I$_2$ (*126*). Complex **55** failed to react with PPh$_3$, although it reacted with the more basic PEt$_3$ to yield *cis*-$[Pd(PPh_3)_2(I)CH_2PEt_3]I$.

Reaction of dichloro(2,2,*N*,*N*-tetramethyl-3-buten-l-amine)palladium-(II) with CH$_2$N$_2$ gives the carbene insertion product, *a*-chloro-*b*-(chloromethyl)-*d*,*c*-(2,2,*N*,*N*-tetramethyl-3-buten-1-amine(palladium(II)) (**56**), together with analogous ethoxymethyl and methyl complexes (*127*).

Similarly, **56** reacts with sodium methoxide to give small quantities of the methoxymethyl analog of **56**. In the same way, treatment of dichloro(2,2-dimethyl-3-buten-1-yl methyl sulfide)palladium(II) with CH_2N_2 gives the expected chloromethyl complex and a cyclopropanation product.

In general, reaction of CH_2N_2 with a range of palladium dichloride and dibromide complexes containing bidentate chelating ligands has been found to lead to the formation of mono(halomethyl) products (*128*) (Fig. 5). Methylpalladium derivatives are obtained as by-products in these reac-

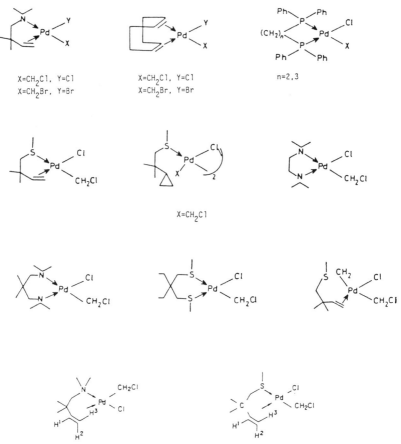

FIG. 5. Diagrams of the halogenomethyl palladium complexes prepared by McCrindle *et al.* [Adapted with permission from the American Chemical Society, Elsevier Sequoia S. A., and the Royal Society of Chemistry; from R. McCrindle, G. Ferguson, G. J. Arsenault, A. J. McAlees, B. L. Ruhl, and D. W. Sneddon, *Organometallics* **5**, 1171 (1986); R. McCrindle and D. W. Sneddon, *J. Organomet. Chem.* **282**, 413 (1985); R. McCrindle, G. J. Arsenault, R. Farwaha, A. J. McAlees, and D. W. Sneddon, *J. Chem. Soc., Dalton Trans.*, 761 (1989).]

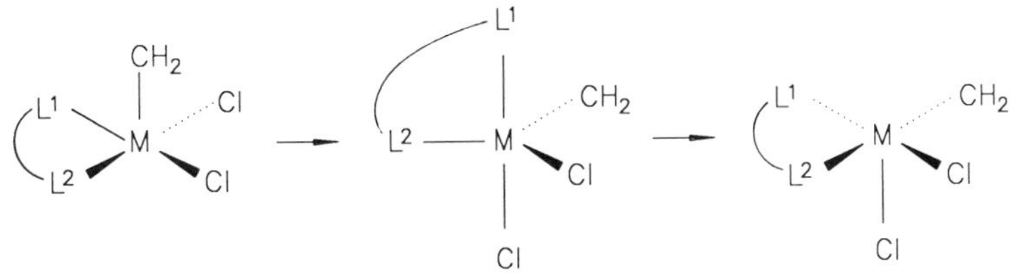

SCHEME 17

tions if the chelating ligand is an olefin. Performing the reactions at $-65\,°C$ reduces the proportion of the methyl product. The insertion of CH_2 into a Pd—Br bond occurs less readily than CH_2 insertion into the corresponding Pd—Cl bond. The insertion process has been proposed to be preceded by a series of steps (Scheme 17). The bromomethyl derivatives are less stable than their chloromethyl analogs. All the chloromethyl complexes showed some tendency to revert to the starting dichloro complexes. Chloromethyl palladium compounds are also formed by the reaction of dichloropalladium complexes with bis(chloromethyl)mercury, as well as with a preformed chloromethyl complex by ligand exchange. Although reaction times are longer, yields are higher for the transmetallation reactions.

The strongly nucleophilic complexes $[Pd(np_3)]$ and $[Pt(PPh_3)np_3]$ [np_3 = tris(2-diphenylphosphinoethyl)amine] react with CH_2Cl_2 at room temperature to give the five-coordinate chloromethyl complexes $[M(np_3)CH_2Cl]^+$ in very high yield. The complexes were isolated as their BPh_4^- salts (129). During the investigation of the thermal decomposition of 1,4-tetramethylene-bis(tri-n-butylphosphine)platinum(II) in CH_2Cl_2, Young and Whitesides postulated the formation of the complex $[Pt(L_2)(Cl)CH_2Cl]$ on the basis of its ^{31}P- and ^1H-NMR spectra, as well as its chemical behavior (130). Young and Whitesides also synthesized $[Pt(bipy)(CH_2)_4(Cl)CH_2Cl]$(bipy = bipyridine) (Scheme 18).

The chloromethyl complexes cis- and trans-$[Pt(PPh_3)_2(Cl)CH_2Cl]$ were synthesized by the photoinduced oxidative addition of CH_2Cl_2 to

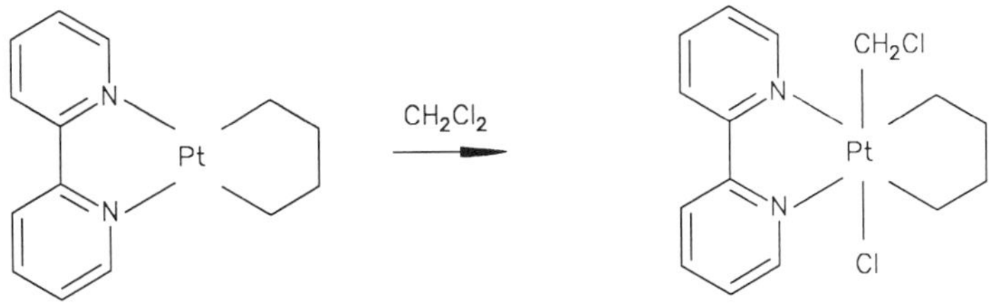

SCHEME 18

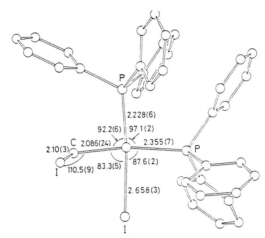

FIG. 6. Diagram of the molecular structure of [Pt(PPh$_3$)$_2$(I)CH$_2$I]. [By permission from the Royal Society of Chemistry; from N. J. Kermode, M. F. Lappert, D. W. Skelton, A. H. White, and J. Holton, *J. Chem. Soc., Chem. Commun.*, 698 (1981).]

[Pt(PPh$_3$)$_2$(C$_2$H$_4$)] (*131*). The reaction appears to involve free radicals and is inhibited by duroquinone. No reaction occurs in the dark. When exposed to light, *cis-/trans*-[Pt(PPh$_3$)$_2$(Cl)CH$_2$Cl] decomposes in solution with loss of CH$_2$, to give [Pt(PPh$_3$)$_2$(Cl)$_2$].

The reaction of [Pt(PPh$_3$)$_2$(η-C$_2$H$_4$)] with CH$_2$XY (CH$_2$I$_2$, CH$_2$Br$_2$, CH$_2$ClBr, CH$_2$BrI, or CH$_2$ClI) gave *trans*- (and usually *cis*-) [Pt(PPh$_3$)$_2$(Y)CH$_2$X] (*132*). Production of the complexes where X = Cl, Y = Br and where X = Cl, Y = I also gave halogen-redistributed products in smaller yields. Thus, complexes where X = Br, Y = Cl, X = Y = Br, and X = Y = Cl were also obtained from these reactions (*132–134*). Heating a mixture of the cis and trans isomers in CH$_2$Cl$_2$ caused complete conversion to the trans isomers. For CH$_2$Br$_2$ or CH$_2$Cl$_2$ additions, there was a preponderance of the trans products at 25° C (Cl > Br > I). The molecular structure of the iodomethyl complex, [Pt(PPh$_3$)$_2$(I)CH$_2$I], has been reported (*134*), with Pt—CH$_2$I and CH$_2$—I bond lengths of 209 and 210 pm, respectively. These values are within the expected range (unlike the complex [Fe{P(OCHMe$_2$)$_3$}$_2$(CO)$_2$(I)CH$_2$I] reported earlier), as is the value of the Pt—CH$_2$—I bond angle of 110.5°. The structure is shown in Fig. 6.

Both the cis and trans forms of [Pt(PPh$_3$)$_2$(I)CH$_2$Cl] (**57**) react with PPh$_3$ to give the ylide complex, *cis*-[Pt(PPh$_3$)$_2$(Cl)CH$_2$PPh$_3$]$^+$ (**58**) (Scheme 19) (*134,135*). The novelty of the reaction in Scheme 19 is the specific formation of **58**, which involves the migration of Cl from C to Pt with loss of I$^-$

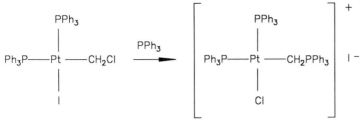

SCHEME 19

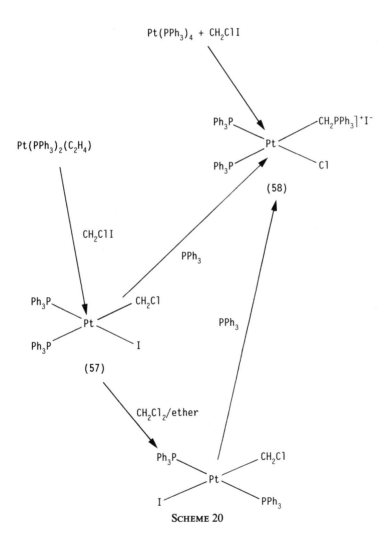

SCHEME 20

from the coordination sphere of the metal. This result implies that complex 57 is an intermediate in the reaction of $[Pt(PPh_3)_4]$ with CH_2ClI to give $[Pt(PPh_3)_2(Cl)CH_2PPh_3]^+$ (Scheme 20) (135).

Lappert and co-workers reported that the reaction of $[Pt(PEt_3)_3]$ with CH_2I_2 gave trans-$[Pt(PEt_3)_2(I)CH_2PEt_3]$ (59), which they believed formed via the α-functionalized alkyl, cis-$[Pt(PEt_3)_2(I)CH_2I]$ (126), and which has since been isolated (136). Similar ylides were obtained by reacting cis- or trans-$[Pt(PPh_3)_2(X)CH_2X]$ (X = Cl, Br, or I) with PEt_3, P^nBu_3, $PMePh_2$, $PEtPh_2$, or PPh_3.

The trans isomers of $[Pt(PPh_3)_2(Y)CH_2X]$ [X = Y = I (60); X = Cl, Y = Br (61); or X = Cl, Y = I (62)] were treated with 2 equivalents of $AgBF_4$ in the presence of 1 equivalent of L to give the cationic metallocycles $[Pt\{PPh_2(2\text{-}C_6H_4CH_2)\}(PPh_3)L]BF_4$ [L = PPh_3 and $P(C_6H_4Me\text{-}4)_3$] (63 and 64) (Fig. 7) (132). The formation of the metallocycle (63) involves attack by the carbon center of the halomethyl substituent on the ortho position of a phosphine phenyl group. This is in stark contrast to previous reports of attack on phosphorus atoms. The reaction of trans-60, trans-61, and trans-62 with only 2 equivalents of $AgBF_4$ gave $[Pt\{PPh_2(2\text{-}C_6H_4CH_2)\}(PPh_3)(S)]BF_4$ (65) (S = solvent molecule) as the sole product (Fig. 7). The cyclic products were obtained irrespective of variations in X or Y (132).

The iodoalkanes CH_2I_2 and CH_2ClI oxidatively add to $[Pt(phen)Me_2]$ (phen = 1,10-phenanthroline), via a free radical mechanism, to give a mixture of complexes 66, 67a, and 68a and 66, 67b, 68b, 67c, and 68c, respectively, with the cis adducts (68) slowly isomerizing to the trans adducts (67) (137,138) (Fig. 8). In a similar way, CH_2Br_2, CH_2BrCl, and CH_2Cl_2 add to $[Pt(phen)Me_2]$ to form the complexes 67d, 68d, 67e, 68e, 67c, and 68c. The reactions of CH_2I_2 and CH_2ClI were very fast, whereas

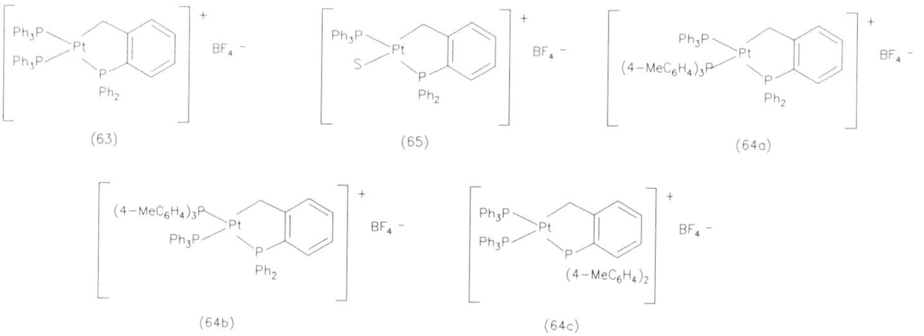

FIG. 7. Cationic metallocycles (63–65) formed from the reactions of $[Pt(PPh_3)_2(Y)CH_2X]$ with $AgBF_4$ in the presence of L.

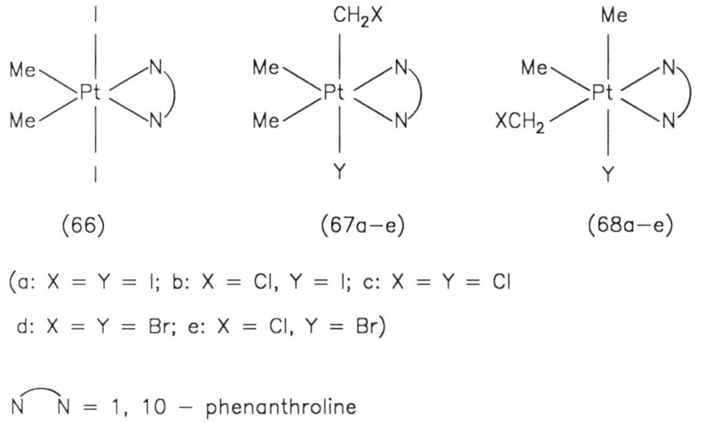

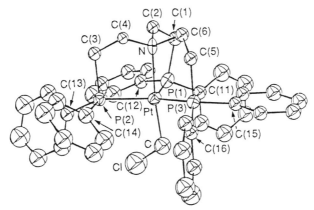

(66)   (67a−e)   (68a−e)

(a: X = Y = I; b: X = Cl, Y = I; c: X = Y = Cl

d: X = Y = Br; e: X = Cl, Y = Br)

N N = 1, 10 − phenanthroline

FIG. 8.   Molecular formulas of complexes 66–68.

those of CH_2ClBr, CH_2Br_2, and CH_2Cl_2 were slow. In all cases a mixture of cis and trans isomers was obtained (139).

The oxidative addition reaction of CH_2Cl_2 to [Pt(PPh₃)(tdpea)] [tdpea = tris(2-diphenylphosphinoethyl)amine] affords [Pt(tdpea)CH₂-Cl]⁺, which was isolated as its BPh₄⁻ salt. The structure of [Pt(tdpea)CH₂Cl]⁺ is shown in Fig. 9. The bond angles and distances within the $PtCH_2Cl$ group are in the expected range, with the CH_2Cl group at the axial site of the slightly distorted trigonal bipyramidal geometry of the metal center (140).

FIG. 9.   Diagram of the molecular structure of [Pt(tdpea)CH₂Cl]⁺. [By permission from the Royal Society of Chemistry; from C. A. Ghilardi, S. Midollini, S. Monetti, A. Orlandini, G. Scapacci, and A. Traversi, *J. Chem. Soc., Dalton Trans.*, 2293 (1990).]

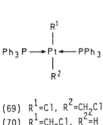

Ph₃P → Pt ← PPh₃ structure with R^1, R^2

(69) R^1=Cl, R^2=CH₂Cl
(70) R^1=CH₂Cl, R^2=H
(71) R^1=Cl, R^2=Me
    R^1=CH₂Cl, R^2=Me
    R^1=CH₂I, R^2=CF₃

(72) R^1=Cl, R^2=CH₂Cl
(73) R^1=R^2=CH₂Cl
    R^1=R^2=CH₂I
    R^1=CH₂Cl, R^2=CH₂OMe
    R^1=R^2=CH₂OMe
    R^1=Cl, R^2=CH₂OMe

X=Cl, Y=Cl
X=Cl, Y=CH₂Cl
X=CH₂Cl, Y=Cl

(74) R^1=Cl, R^2=CH₂Cl, R^3=Ph; n=3
(75) R^1=R^2=CH₂Cl, R^3=Ph; n=3
    R^1=R^2=CH₂Cl, R^3=C₆H₁₁; n=2

(76) R^1=R^2=CH₂Cl, R^3=Ph
    R^1=R^2=CH₂Cl, R^3=Et
    R^1=Cl, R^2=CH₂Cl, R^3=Ph
    R^1=Cl, R^2=CH₂Cl, R^3=Et

R^1=Cl, R^2=CH₂Cl, R^3=C₆H₁₁
R^1=H, R^2=CH₂Cl, R^3=Et
R^1=Me, R^2=CH₂Cl, R^3=Et

R^1=R^2=CH₂Cl, L^1=PEt₃, L^2=BuᵗNC
R^1=Cl, R^2=CH₂Cl, L=PEt₃, L^2=BuᵗNC

X=CH₂Cl

X=CH₂Cl, Y=Cl
X=Cl, Y=CH₂Cl

X=C, Y=CH₂Cl
X=Si, Y=CH₂Cl

R^1=CH₂Cl, R^2=Cl
R^1=Cl, R^2=CH₂Cl

FIG. 10. Diagrams of the platinum halogenomethyl complexes prepared by McCrindle *et al.* [Adapted with permission from the American Chemical Society and the Royal Society of Chemistry; from R. McCrindle, G. Ferguson, G. J. Arsenault, A. J. McAlees, B. L. Ruhl, and D. W. Sneedon, *Organometallics* **5**, 1171 (1986); R. McCrindle, G. J. Arsenault, R. Farwaha, M. J. Hampden-Smith, R. E. Rice, and A. McAlees, *J. Chem. Soc., Dalton Trans.*, 1773 (1988).]

McCrindle *et al.* have prepared a range of both mono and bis halomethyl and methoxy methyl platinum complexes (Fig. 10) (*141,142*). The halomethyl complexes were prepared by treating platinum(II) halide complexes with CH_2N_2. Methoxy methyl complexes were obtained when the reaction was carried out in the presence of methanol. With only one exception, the methylene insertion appears to occur only when the metal–halogen bond is trans to a group of high trans influence (olefin, phosphine, alkyl, or hydride). A proposed mechanism for the insertion of the methylene moiety into the Pt—Cl bond is shown in Scheme 21. The one exception is the formation of *trans*-[Pt(PPh₃)₂(Cl)CH₂Cl] (**69**) from *trans*-[Pt(PPh₃)₂Cl₂], which appears to be the only case where direct insertion into a Pt—Cl bond trans to a ligand (Cl) of low trans effect may have taken place. Two structures of these chloromethyl complexes have been reported. The values of the Pt—CH_2Cl (~ 202 pm) and CH_2—Cl (~ 175 pm) bond lengths and the Pt—CH_2—Cl bond angles (~ 115°) were found to be within the expected range (*142*). The complex *trans*-[Pt(PPh₃)₂(H)CH₂Cl] (**70**) was observed to rearrange to the *trans*-chloro(methyl) complex (**71**). The proposed mechanism for this rearrangement is shown in Scheme 22 (*141*).

Displacement of cod from complexes **72** and **73** occurs by treatment with 1,3-bis(diphenylphosphino)propane to give complexes **74** and **75**. Similarly, **73** reacts with 2 molar equivalents of PPh_3 to give **76**. In neither example is the C of the CH_2Cl group attacked by the nucleophile. Com-

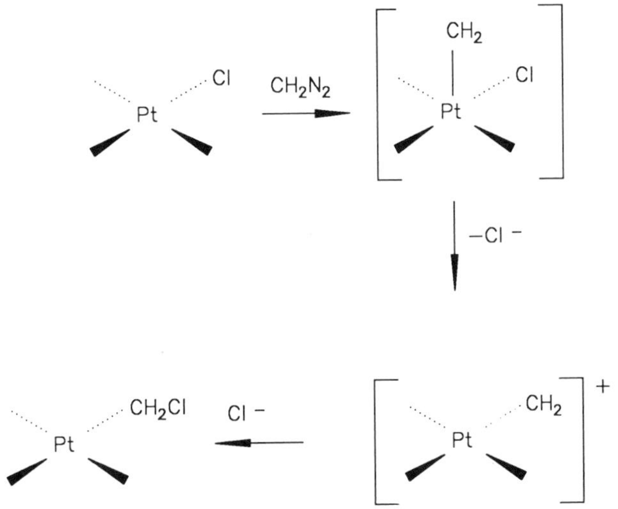

SCHEME 21

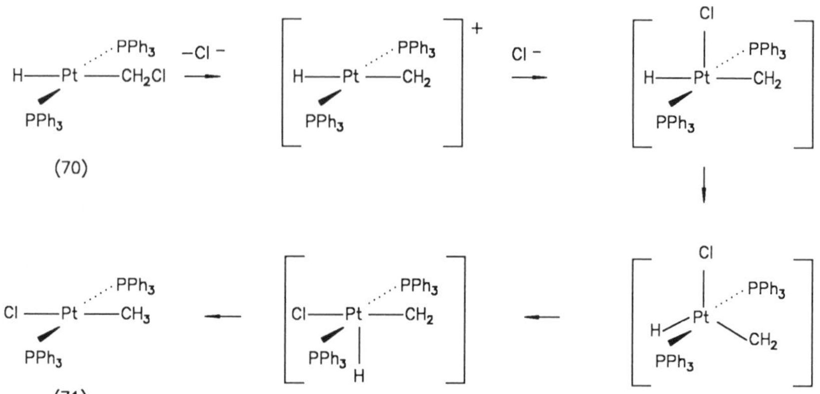

$$\left[\begin{array}{c} \text{H} \overset{\cdots PPh_3}{\underset{PPh_3}{\longrightarrow Pt \longrightarrow CH_2Cl}} \end{array} \right] \overset{-Cl^-}{\longrightarrow} \left[\begin{array}{c} \text{H} \overset{\cdots PPh_3}{\underset{PPh_3}{\longrightarrow Pt \longrightarrow CH_2}} \end{array} \right]^{+} \overset{Cl^-}{\longrightarrow} \left[\begin{array}{c} \overset{Cl}{\underset{PPh_3}{\text{H} \longrightarrow Pt \longrightarrow CH_2}} \overset{\cdots PPh_3}{} \end{array} \right]$$

(70)

$$\left[\begin{array}{c} \text{Cl} \overset{\cdots PPh_3}{\underset{PPh_3}{\longrightarrow Pt \longrightarrow CH_3}} \end{array} \right] \longleftarrow \left[\begin{array}{c} \text{Cl} \overset{\cdots PPh_3}{\underset{PPh_3 \, | \, H}{\longrightarrow Pt \longrightarrow CH_2}} \end{array} \right] \longleftarrow \left[\begin{array}{c} \overset{Cl}{\underset{PPh_3}{\text{H} \longrightarrow Pt \searrow CH_2}} \overset{\cdots PPh_3}{} \end{array} \right]$$

(71)

SCHEME 22

plex **73** reacts with 2,2-diethyl-1,3-bis(methylthio)propane to give a product which is postulated to be **77** (Fig. 11) (*143*).

Treatment of [Pt(cod)(CH$_2$I)$_2$] with 3 or more equivalents of PPh$_3$ surprisingly gave the zwitterionic bis(phosphonium ylide) complex [Pt(PPh$_3$)I(CH$_2$PPh$_3$)$_2$]I (**78**). Complex **78** was also obtained by treatment of [Pt(PPh$_3$)$_2$(CH$_2$Cl)$_2$] (**79**) with 1 equivalent of PPh$_3$ and an excess of NaI. The failure to isolate the expected [Pt(PPh$_3$)$_2$(CH$_2$I)$_2$] from the ligand displacement reaction is rationalized by assuming that the nucleophilic displacement of iodide becomes competitive with cyclooctadiene replacement in the bis(iodomethyl) complex. The complex [Pt(PPh$_3$)$_2$(CH$_2$I)$_2$] can, however, be prepared by the reaction of a large excess of NaI with **79**.

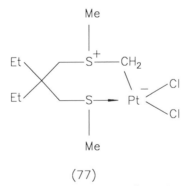

(77)

FIG. 11.   Proposed structure of complex **77**.

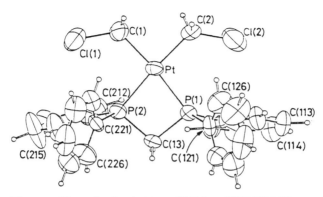

FIG. 12.   Diagram of the molecular structure of [Pt(dppm)(CH₂Cl)₂]. [By permission from the Royal Society of Chemistry; from N. W. Alcock, P. G. Pringle, P. Bergamini, S. Sostero, and O. Traverso, *J. Chem. Soc., Dalton Trans.*, 1553 (1990).]

The reaction of excess PMe₃ with [Pt(cod)(CH₂I)₂] gave a high yield of the bis(ylide) complex, [Pt(PMe₃)₂(CH₂PMe₃)₂]I₂ (*144*).

   Insertion of CH₂ into the platinum–halogen bond of [Pt(dppm)(X)₂] (X = Cl, Br, or I) to give the complexes [Pt(dppm)(CH₂X)₂] has also been reported (*145*). The stability of these complexes in solution follows the order Cl > Br > I. The structure of the dichloromethyl complex [Pt(dppm)(CH₂Cl)₂] has been determined (Fig. 12). The Pt—CH₂Cl and CH₂—Cl bond lengths are within the expected range. The angle between the two CH₂Cl groups is 84.2°. The repulsion between the P and C atoms is believed to be responsible for this small angle. Both Cl atoms lie in the PtP₂ plane. The complex [Pt(dppm)(CH₂Cl)₂] (**80**) was found to react with HCl to give [Pt(dppm)(Cl)CH₂Cl]. Ethylene was obtained as a major product on decomposition of [Pt(dppm)(CH₂I)₂] on irradiation. Treatment of **80** with neutral nucleophiles, L, resulted in a ring expansion to give [Pt(CH₂PPh₂CH₂PPh₂)(L)CH₂Cl]Cl (L = NC₅H₅, PPh₃, or PPh₂H); this ring expansion is unique to dppm chelates (*145*).

   McCrindle *et al.* reported that [Pt(cod)(CH₃)Cl] reacted with CH₂N₂ to give [Pt(cod)(CH₃)CH₂Cl] (**81**), which is stable under ambient conditions, in good yield (*146*). Complex **81** can also be prepared from [Pt(cod)(Cl)CH₂Cl] and MeLi (*146*). The carbene insertion reaction is again proposed to proceed via a platinum carbene intermediate. Complex **81** rapidly transforms to [Pt(cod)Cl(CH₂CH₃)] (**82**) when dissolved in CDCl₃ containing (CF₃)₂CHOH. Reacting **82** with CH₂N₂ gave largely [Pt(cod)(CH₂CH₃)CH₂Cl] (**83**), as well as small quantities of [Pt(cod)(CH₃)CH₂Cl] and [Pt(cod)(CH₂Cl)₂]. Complex **83**, when dis-

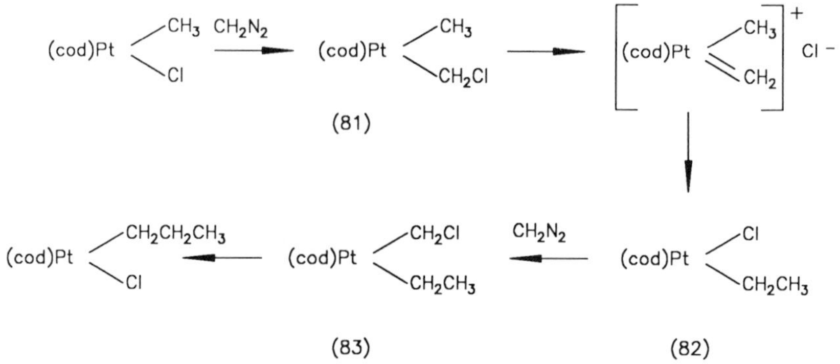

SCHEME 23

solved in $CDCl_3/(CF_3)_2CHOH$ (9:1) gave $[Pt(cod)Cl(CH_2CH_2CH_3)]$. since insertion of a methylene group into a metal–alkyl bond has been proposed as a key step in the polymerization of CH_2N_2 by transition metal complexes (116,147,148), these authors propose Scheme 23 as a model for chain growth.

The reaction of cis-bis(tricyclohexylphosphino)platinum(II)dichloride with a large excess of CH_2N_2 gave immediate evolution of ethylene gas. Once all the CH_2N_2 had been consumed, the starting complex was recovered in essentially quantitative yield. The proposed mechanism involves chloromethyl complexes (Scheme 24).

Attempts to prepare $[Pt(PCy_3)_2(CH_2Cl)_2]$ by reacting $[Pt(cod)(CH_2Cl)_2]$ with 2 molar equivalents of PCy_3 gave cis-$[Pt(PCy_3)(Cl)_2(CH_2CH_2PCy_3)]^+$ (84) in $CDCl_3$(Cy = cyclohexyl): Adding MeOH to this solution gave a mixture of 84 and $[Pt(PCy_3)(CH_2CH_2PCy_3)(\mu\text{-}Cl)]_2^{2+}$, that is, complexes containing the novel phosphinoethyl moiety. Addition of NH_4PF_6 to this mixture gave $[Pt(PCy_3)(CH_2CH_2PCy_3)(\mu\text{-}Cl)]_2(PF_6)_2$ (149).

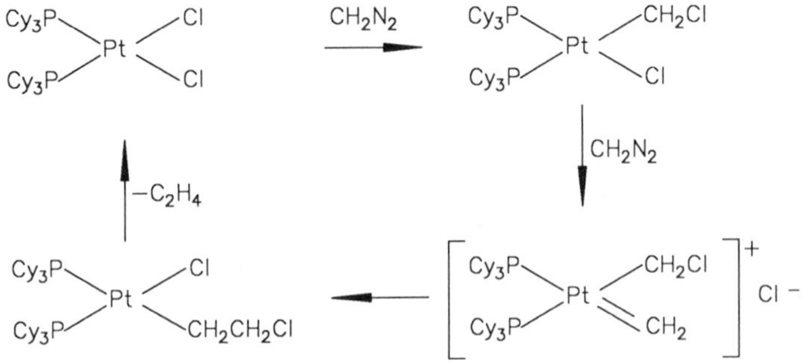

SCHEME 24

F. *Gold*

The complexes $[Au(L)CH_2Cl]$ **(85)** $[L = PPh_3, PEt_3,$ and $P(OPh)_3]$ were first prepared by Nesmeyanov *et al.* from $[Au(PPh_3)Cl]$ and CH_2N_2 (*150*). They were later shown to react via a trans oxidative addition reaction with $[Pt(bipy)Me_2]$ to give the heterobimetallic complexes $[Cl(Me)_2(bipy)-PtCH_2AuL]$ (*151*). Kinetic evidence showed that the C—Cl bond of complex **85** is strongly activated by the metal (by at least a factor of 10^3) to oxidative addition relative to CH_3Cl. Complex **85** decomposed back to $[Au(PPh_3)Cl]$ with loss of CH_2. Complex **85** was shown to act as a methylene transfer reagent, reacting with 3-cyclohexenol to give bicyclo[4.1.0]heptan-2-ol. The reaction of **85** with KI afforded $[Au(PPh_3)I]$; the latter complex is believed to be the decomposition product of $[Au(PPh_3)CH_2I]$, which is very unstable. Complex **85** reacts with aqueous NaCN to give $[CH_3PPh_3]^+ [Au(CN)_2]^-$ (*150*).

Fackler and co-workers reported that the reaction of $[Au(CH_2)_2PPh_2]_2$ in neat CH_2XX' $(X = Cl, X' = Br$ or $I)$ initially gave the dinuclear Au(II) alkyl halides $[\{Au(CH_2)_2PPh_2\}_2(X')CH_2X]$ $[X' = Br$ **(86)**, $X' = I$ **(87)**] (Scheme 25) (*152*). The crystal structure of $[\{Au(CH_2)_2PPh_2\}_2(Br)CH_2Cl]$ **(86)** (Fig. 13) shows evidence of a close contact between one Au center and the Cl of the coordinated CH_2Cl (Au–Cl 289.5 pm). Another unusual feature of this structure is the very small Au—C—Cl bond angle of 96.2°. This is about 13° smaller than the 109.5° angle for an sp^3 carbon center. Remarkably, the Au—C—Cl bond angle in the very similar complex $[\{Au(CH_2)_2PPh_2\}_2(I)CH_2Cl]$ **(87)** (they differ only in the halogen bonded to the Au unit not containing the CH_2Cl group) is 115.6°, that is, within the expected range (See Fig. 14.) The Au–Cl "bond" distance of the latter complex is 318.5 pm, significantly longer than that of complex **86**. Notable, too, is a significant difference in the Au—C—Cl bond length of the two complexes **86** and **87** (see Table I in Section IV).

Allowing the reaction solution of complex **86** to stand for 45 minutes results in the formation of $[\{Au(CH_2)_2PPh_2\}_2(\mu\text{-}CH_2)ClBr]$ (Scheme 26).

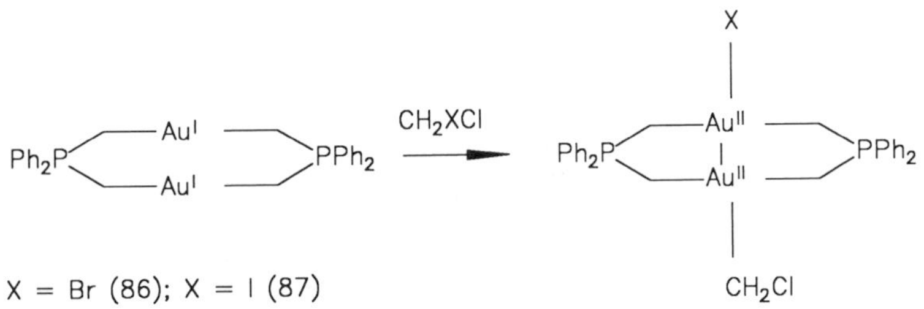

X = Br (86); X = I (87)

SCHEME 25

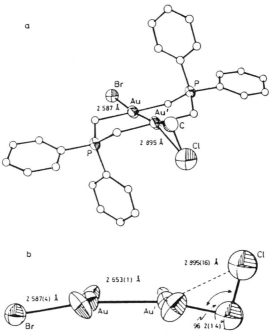

FIG. 13. (a) Diagram of the molecular structure of [{Au(CH$_2$)$_2$PPh$_2$}$_2$(Br)CH$_2$Cl] (86). (b) The Br—Au—Au′—CH$_2$Cl backbone of 86. [By permission from the Royal Society of Chemistry; from H. H. Murray III, J. P. Fackler, and D. A. Tocher, *J. Chem. Soc., Chem. Commun.*, 1278 (1985).]

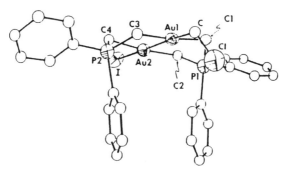

FIG. 14. Diagram of the molecular structure of [{Au(CH$_2$)$_2$PPh$_2$}$_2$(I)CH$_2$Cl] (87). [Reproduced with permission from H. H. Murray, J. P. Fackler, and A. M. Mazany, *Organometallics* 3, 1310 (1984). Copyright (1984) American Chemical Society.]

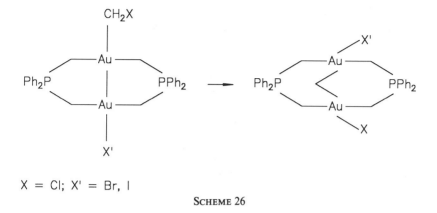

X = Cl; X' = Br, I

SCHEME 26

This rearrangement is postulated to involve the formation of a gold(I)–gold(III) carbene species, along with the Au—Cl bond at the Au(III) center, prior to the rearrangement to the final A-frame product (*153*). Allowing the [{Au(CH$_2$)$_2$PPh$_2$}$_2$(I)CH$_2$Cl] reaction solution to stand also gives the bridging (methylene)gold(III) ylide dimer [(μ-CH$_2$){Au-(CH$_2$)$_2$PPh$_2$}$_2$ICl] (Scheme 26).

III

ω-MONOHALOGENOALKYL TRANSITION METAL COMPLEXES

In 1967, King and Bisnette reported that the reaction of Na[CpM(CO)$_3$] (M = Mo or W) with 1,3-dibromopropane and 1,4-dibromobutane yielded [CpM(CO)$_3${(CH$_2$)$_n$Br}] (n = 3 or 4, M = Mo or W) (*154*). The physical properties of these compounds resemble those of other [CpM(CO)$_3$R] (R = alkyl) complexes. The tungsten complexes were observed to be more stable than their molybdenum analogs. Attempts to synthesize compounds of the type [L$_x$M—(CH$_2$)$_n$—M'L$_y$] containing two different metal atoms by the reactions of [CpM(CO)$_3${(CH$_2$)$_n$Br}] (n = 3 or 4) with Na-[CpFe(CO)$_2$] and Na[Mn(CO)$_5$], yielded only [{CpFe(CO)$_2$}$_2$(μ-CH$_2$)$_n$] (n = 3 or 4) and what was later shown to be a dimanganese cyclic carbene complex [(CO)$_5$MnMn(CO)$_4${$\overline{\text{CO(CH}_2\text{)}_2\text{CH}_2}$}] (*155,156*). This was presumably due to the high nucleophilicity of the iron and manganese anions.

The ω-haloalkyl manganese compounds [Mn(CO)$_5${(CH$_2$)$_n$Cl}] were prepared by the reaction of the [Mn(CO)$_5$]$^-$ anion with Cl(CO)(CH$_2$)$_n$Cl, to give [Mn(CO)$_5${C(O)(CH$_2$)$_n$Cl}] (n = 3 or 4), followed by decarbonylation of this acyl product. The reaction of the anion $[Mn(CO)_5]^-$ with $[Mn(CO)_5\{(CH_2)_3Cl\}]$ (88) gave the carbene complex $[(CO)_5MnMn(CO)_4\{\overline{CO(CH_2)_2CH_2}\}]$. The complexes $[Mn(CO)_5\{(CH_2)_nCl\}]$ readily reacted with phosphines, L, to give the carbonyl-inserted products $[MnL(CO)_4\{C(O)(CH_2)_nCl\}]$. These products could in turn be decarbonylated to give $[MnL(CO)_4\{(CH_2)_nCl\}]$ ($n = 3$ or 4). The complexes $[MnL(CO)_4\{(CH_2)_nCl\}]$ reacted with I^- to give the cyclic carbene complexes $[MnL(I)(CO)_3\{\overline{CO(CH_2)_{n-1}CH_2}\}]$ [L = CO, $n = 3$ or 4; L = $P(OCH_2)_3CCH_3$; $n = 3$] (157).

Recently $[Mn(CO)_5\{(CH_2)_nBr\}]$ ($n = 3$ or 4) have been prepared by decarbonylation of the corresponding acyl complexes (76). The haloethyl complex $[Mn(CO)_5(CH_2CH_2Br)]$ is believed to be a very short-lived intermediate in the decarbonylation of $[Mn(CO)_5\{C(O)CH_2CH_2Br\}]$, but it was not observed. $[CpMo(CO)_3\{(CH_2)_3Br\}]$ (89) reacted with PPh_3 in acetonitrile to give $[CpMo(PPh_3)(CO)_2\{\overline{CO(CH_2)_2CH_2}\}]Br$ (90) (Scheme 27) (158). This cyclic carbene species is presumably formed by initial attack of PPh_3 on the Mo, resulting in CO insertion to form the acyl derivative $[CpMo(PPh_3)(CO)_2\{C(CO)(CH_2)_3Br\}]$. The latter species spontaneously undergoes internal nucleophilic attack of the acyl oxygen atom on the γ carbon atom to displace Br^- and form the cyclic carbene cation (90) (Scheme 28) (158).

The reaction of $[CpMo(CO)_3]^-$ with $I(CH_2)_3I$ in THF led to $[CpMo(CO)_3\{(CH_2)_3I\}]$ (91), which reacted further under reflux to give $[CpMo(CO)_2I\{\overline{CO(CH_2)_2CH_2}\}]$ (92), a complex very similar to 90 (159). Complex 92 can also be obtained by the reaction of 89 with LiI; however, the analogous bromide complex is not obtained from a similar reaction with LiBr. It is proposed that the formation of 92 involves the attack of the nucleophile I^- on $[CpMo(CO)_3\{(CH_2)_3X\}]$ (X = Br or I) to generate the intermediate acyl species $[CpMo(CO)_2I\{C(O)(CH_2)_3X\}]^-$, which undergoes

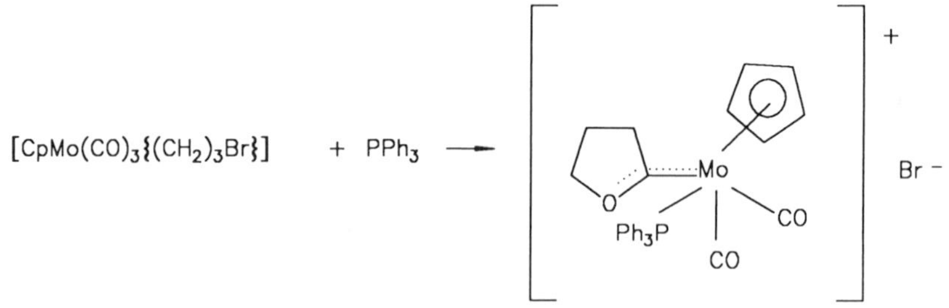

$[CpMo(CO)_3\{(CH_2)_3Br\}]$ + PPh_3 —→

(90)

SCHEME 27

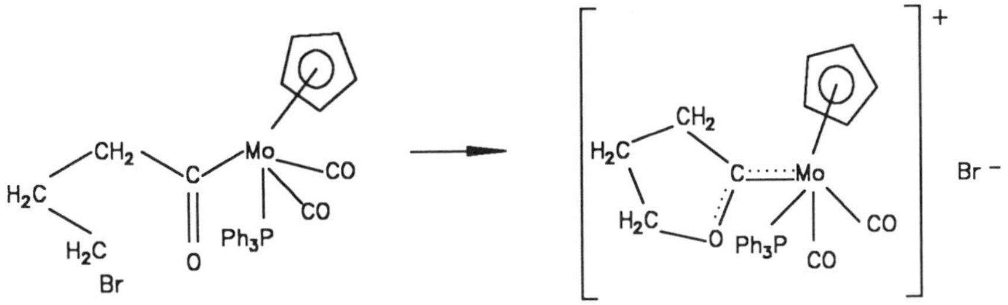

(90)

SCHEME 28

rapid, spontaneous cyclization to form **92** (*159*). Interestingly, [CpMo(CO)₃{(CH₂)₄I}] reacts with LiI to form only [CpMo(CO)₃I] and not the expected 2-oxacyclohexylidene complex (*159*). As in the reactions shown in Schemes 27 and 28, **89** reacts with other nucleophiles, such as SPh⁻ and CN⁻, to give [CpMo(CO)₂X{$\overline{\text{CO(CH}_2)_2\text{CH}_2}$}] **(93)** (X = SPh or CN) (*160*). The cyclization was shown to be independent of the Cp ring, and substitution of a proton of the ring by Me, C(O)Me, or SiMe₃ yielded products similar to **93**.

The tungsten analog of **89**, [CpW(CO)₃{(CH₂)₃Br}] **(94)**, undergoes carbene formation with I⁻ and CN⁻ to form [CpW(CO)₂X{$\overline{\text{CO(CH}_2)_2\text{CH}_2}$}] (X = I or CN), though the reaction proceeds more slowly than that for Mo. The proposed mechanism for the carbene formation is depicted in Scheme 29. The key steps in Scheme 29 are the migration of the alkyl chain to an adjacent carbonyl group and coordination of the nucleophile to the metal atom. However, the formation of cyclic carbenes appears to be dependent on the halide of the haloalkyl group involved (*161*). Hence, [CpMo(CO)₃{(CH₂)₃Cl}] **(95)** reacts with PPh₃ to give *trans*-[CpMo(PPh₃)(CO)₂{C(O)(CH₂)₃Cl}], that is, a halogenoacyl complex. A similar reaction with [CpMo(CO)₃{(CH₂)₄I}] and PPh₃ gave the acyl complex [CpMo(PPh₃)(CO)₂{C(O)(CH₂)₄I}]. Acyl complexes were also obtained in the reaction of [CpMo(CO)₃{(CH₂)ₙBr}] (n = 4 or 5) with PPh₃ [Eq. (20)] (*161*). These results are thought to be due to competition between two possible reaction pathways after nucleophilic attack by PPh₃: either rapid intramolecular cyclization by the cis isomer formed initially or intramolecular geometrical isomerization to form the *trans*-haloacyl complex. Hence, **95** forms the haloacyl complex because Cl is a poorer leaving group than Br (slowing the attack of the acyl oxygen), and consequently the rate of cyclization is decreased relative to isomerization. Similarly, when the halo-

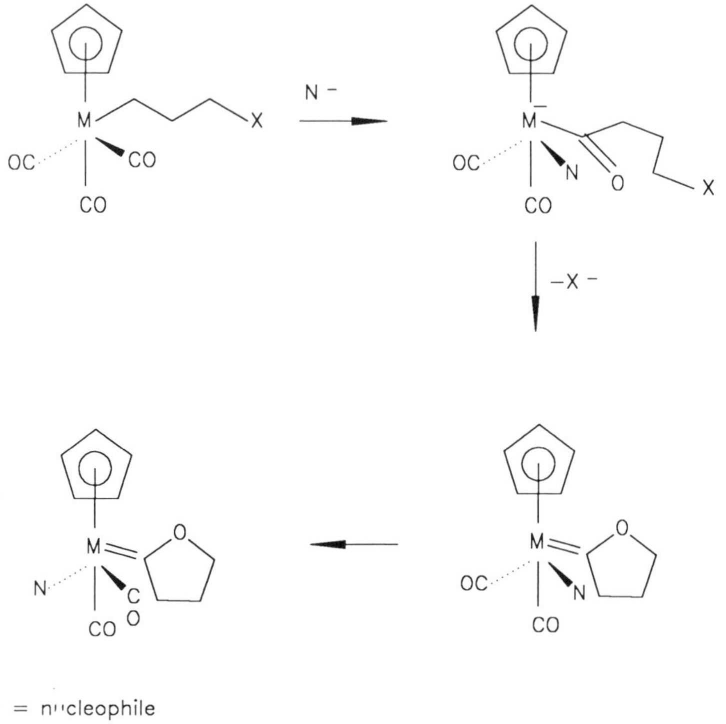

N^- = nucleophile

SCHEME 29

gen is further removed from the metal atom, formation of the *trans*-haloacyl becomes favored owing to the retardation of the rate of ring closure for rings containing more than five members (*161*).

$$[CpMo(CO)_3\{(CH_2)_nBr\}] + PPh_3 \rightarrow trans\text{-}[CpMo(PPh_3)(CO)_2\{C(O)(CH_2)_nBr\}] \quad (20)$$

The complexes *trans*-[CpMoL(CO)$_2$\{(CH$_2$)$_3$Br\}] [L = PPh$_3$ or P(OMe)$_3$] were prepared from Na[CpMoL(CO)$_2$] and Br(CH$_2$)$_3$Br (*160*). Neither complex reacts with I$^-$ to produce carbene complexes.

The iodide complexes, [CpMo(CO)$_3$\{(CH$_2$)$_3$I\}] [M = Mo (**96**) or W (**97**)], were originally prepared by Winter and co-workers (*162*). Both **96** and **97** react with excess iodide to form [CpM(CO)$_2$I\{$\overline{CO(CH_2)_3CH_2}$\}]. Winter and co-workers found that transition metal anions may induce similar cyclization reactions. Hence, reaction of excess Na[CpMo(CO)$_3$] with Br(CH$_2$)$_3$Br yields **89**, which reacts with further Na[CpMo(CO)$_3$] to form an intermediate anionic acyl complex, which in turn undergoes elimination of Br$^-$ to form the carbene ligand in a manner similar to that depicted in Scheme 29. $Na[CpW(CO)_3]$ reacts with **89** to give $[Cp(CO)_3MoW(CO)_2\{\overline{CO(CH_2)_2CH_2}]Cp]$. However, the reaction of **89** with $[Mn(CO)_5]^-$ leads exclusively to the dimanganese species $[(CO)_5MnMn\{\overline{CO(CH_2)_2CH_2}\}(CO)_4]$, whereas reaction with $[CpFe(CO)_2]^-$ produces $[\{CpFe(CO)_2\}_2\{\mu\text{-}(CH_2)_3\}]$. Both reactions eliminate $[CpMo\text{-}(CO)_3]^-$. $[CpW(CO)_3]^-$ does not react with $[CpW(CO)_3\{(CH_2)_2Br\}]$ at the tungsten center; instead, the bromide is displaced to give the dimetal-lopropane $[\{CpW(CO)_3\}_2\{\mu\text{-}(CH_2)_3\}]$ in moderate yield. Similarly, $[CpW(CO)_3]^-$ reacts with $I(CH_2)_4I$ to yield $[\{CpW(CO)_3\}_2\{\mu\text{-}(CH_2)_4\}]$ via the intermediate $[CpW(CO)_3\{(CH_2)_4I\}]$. The complex $[\{CpMo(CO)_3\}_2\{\mu\text{-}(CH_2)_4\}]$ is prepared similarly, by the reaction of $[CpMo(CO)_3]^-$ with $I(CH_2)_4I$, involving $[CpMo(CO)_3\{(CH_2)_4I\}]$ as an intermediate (*162*).

The reaction of $[CpMo(CO)_2(CNMe)]^-$ with $I(CH_2)_3I$ in THF gives *cis*-$[CpMo(CO)_2I\{\overline{C(NMe)(CH_2)_2CH_2}\}]$ (**98**). It is proposed that complex **98** forms via the intermediate $[CpMo(CO)_2(CNMe)\{(CH_2)_3I\}]$ (Scheme 30) (*163*). In reactions similar to those in Scheme 30, $[CpM(CO)_2(CNMe)]^-$ reacts with $I(CH_2)_nI$ to give the cyclic carbene species, *cis*-$[CpM\text{-}(CO)_2\{\overline{CN(Me)(CH_2)_{n-1}CH_2}\}]$ (M = Mo, $n = 2$ or 3; M = W, $n = 2$) (*164*). Again, halogenoalkyl intermediates are implicated.

The reactions of $[Cp^*M(CO)_3]^-$ (M = Mo or W) with $Br(CH_2)_3Br$ give the halogenopropane complexes $[Cp^*M(CO)_3\{(CH_2)_3Br\}]$ (**99**). These reactions are reported as being cleaner than those of the Cp analogs, presumably owing to the greater nucleophilicity of the $[Cp^*M(CO)_3]^-$ anions, imparted by the Cp* group, relative to the $[CpM(CO)_3]^-$ anions.

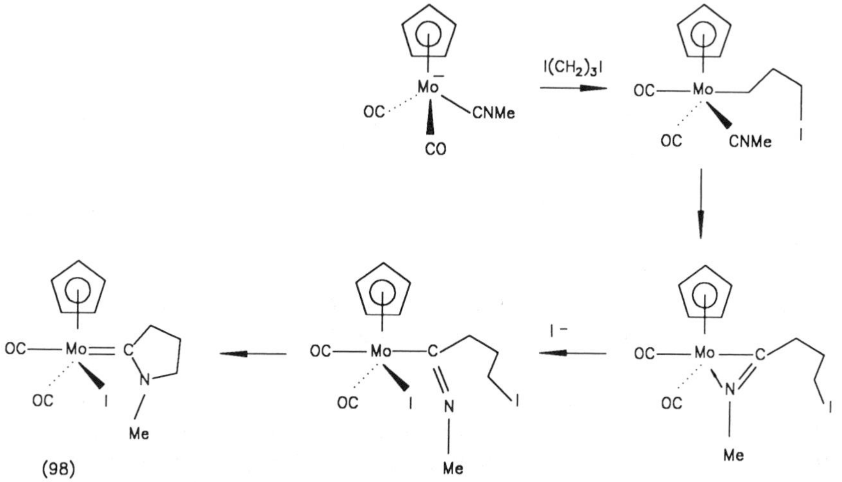

SCHEME 30

The reaction of complex **99** (M = Mo) with LiI gives the *cis*- and *trans*-carbenes, [Cp*MoI(CO)$_2$$\{$=$\overline{\text{C}}$(CH$_2$)$_3$$\dot{\text{O}}\}$]. In a similar way to [CpMo(CO)$_3$$\{$(CH$_2$)$_3Br\}$], **99** (M = Mo) reacts with CN$^-$ to give *trans*-[Cp*Mo(CO)$_2$(CN)$\{$$\overline{\text{C}}$(CH$_2$)$_3$$\dot{\text{O}}\}$]. Although [CpMo(CO)$_3$$\{$(CH$_2$)$_3Br\}$] does not react with Br$^-$ or SCN$^-$ to form a bromocarbene complex, the [Cp*Mo(CO)$_3$$\{$(CH$_2$)$_3Br\}$] complex does react with Br$^-$ and SCN$^-$, forming the respective cyclic carbene complexes, [Cp*Mo(CO)$_2$-Z$\{$$\overline{\text{C}}$(CH$_2$)$_3$$\dot{\text{O}}\}$] (Z = Br or SCN) (*165*).

A further difference in the reactivity of the Cp*Mo complex **99** relative to the Cp analog **89** is illustrated by its reaction with [CpFe(CO)$_2$]$^-$. Whereas [CpMo(CO)$_3$$\{$(CH$_2$)$_3Br\}$] reacts with [CpFe(CO)$_2$]$^-$ with elimination of both Br$^-$ and [CpMo(CO)$_3$]$^-$ to form [$\{$CpFe(CO)$_2$$\}_2$$\{\mu$-(CH$_2$)$_3$$\}$], the Cp*Mo complex **99** reacts with [CpFe(CO)$_2$]$^-$ to form [Cp*Mo-(CO)$_3$(CH$_2$)$_3$Fe(CO)$_2$Cp] (*165*). As with its Cp analog, [Cp*W(CO)$_3$-$\{$(CH$_2$)$_3$Br$\}$] reacts with I$^-$ to give [Cp*WI(CO)$_2$$\{$=$\overline{\text{C}}$(CH$_2$)$_3$$\dot{\text{O}}\}$].

The complexes [CpFe(CO)$_2$$\{$(CH$_2$)$_nX\}$] (n = 3–10, X = Br; n = 3, X = Cl) and [Cp*Fe(CO)$_2$$\{$(CH$_2$)$_nBr\}$] (n = 3–5) were synthesized by reacting Na[CpFe(CO)$_2$] or Na[Cp*Fe(CO)$_2$] with an excess of the α,ω-dihaloalkanes [Eqs. (21) and (22)] (*19, 166*). The reaction of Na[CpFe(CO)$_2$] with Br(CH$_2$)$_2$Br was found to yield only [CpFe(CO)$_2$]$_2$ and ethylene (*167*).

$$\text{Na[CpFe(CO)}_2] + \text{X(CH}_2)_n\text{X} \xrightarrow{-20°C,\ THF} \text{[CpFe(CO)}_2\{(\text{CH}_2)_n\text{X}\}] + \text{NaX} \qquad (21)$$

$$n = 3,\ X = Cl;\ n = 3-10,\ X = Br$$

$$\text{Na[Cp*Fe(CO)}_2] + \text{Br(CH}_2)_n\text{Br} \xrightarrow{-50°C,\ THF} \text{[Cp*Fe(CO)}_2\{(\text{CH}_2)_n\text{Br}\}] + \text{NaBr} \qquad (22)$$

$$n = 3-5$$

The complexes [CpFe(CO)$_2$$\{$(CH$_2$)$_3X\}$] (X = Cl or Br) and [Cp-Fe(CO)$_2$$\{CH_2CH_2$CH(CH$_3$)Br$\}$] have recently been shown to have good organic synthetic utility as cyclopropane precursors when reacted with AgBF$_4$ (*24*). In contrast, the formation of cyclobutane and cyclopentane, from the corresponding [CpFe(CO)$_2$$\{$(CH$_2$)$_nBr\}$] (n = 4 or 5) complexes, was slow and inefficient. The reaction of the γ-haloalkyl iron complexes to produce cyclopropanes is rationalized as being due to participation of iron in the cleavage of the γ-carbon–halogen bond (*24*).

[CpFe(CO)$_2$$\{$(CH$_2$)$_3Br\}$] was shown to react with Na[CpFe(CO)$_2$] to yield [$\{$CpFe(CO)$_2$$\}_2$$\{\mu$-(CH$_2$)$_n$$\}$] (n = 3–5). The synthesis of [CpFe(CO)$_2$$\{\mu$-(CH$_2$)$_3$$\}$Mo(CO)$_3$Cp], attempted unsuccessfully by King and Bisnette using [CpMo(CO)$_3$$\{$(CH$_2$)$_3Br\}$] and Na[CpFe(CO)$_2$] (*154*), was achieved by Moss and co-workers by reacting [CpFe(CO)$_2$$\{$(CH$_2$)$_3Br\}$] (**100**) with

Na[CpMo(CO)$_3$] (19) [Eq. 23)]. The reaction of **100** with PPh$_3$ resulted in the formation of two complexes: a carbene complex **101** and an acyl complex **102,** as shown in Eq. (24) (19). Knox and co-workers synthesized the iodo analog of **100**, [CpFe(CO)$_2$\{(CH$_2$)$_3$I\}] (**103**) (in low yield and not fully characterized), by reacting Na[CpFe(CO)$_2$] with excess I(CH$_2$)$_3$I (168). Complex **103** was then reacted with Na[CpRu(CO)$_2$] to give the mixed metal polymethylene-bridged complex [Cp(CO)$_2$Fe(CH$_2$)$_3$Ru-(CO)$_2$Cp].

[CpFe(CO)$_2$\{(CH$_2$)$_3$Br\}] + Na[CpMo(CO)$_3$] →

$$[CpFe(CO)_2(CH_2)_3Mo(CO)_3Cp] + NaBr \quad (23)$$

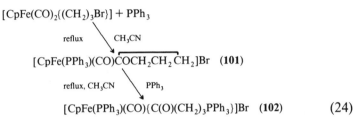

[CpFe(CO)$_2$\{(CH$_2$)$_3$Br\}] + PPh$_3$

reflux \ CH$_3$CN

[CpFe(PPh$_3$)(CO)$\overline{\text{COCH}_2\text{CH}_2\text{CH}_2}$]Br (**101**)

reflux, CH$_3$CN \ PPh$_3$

[CpFe(PPh$_3$)(CO)\{C(O)(CH$_2$)$_3$PPh$_3$\}]Br (**102**)   (24)

The complexes [CpFe(CO)$_2$\{(CH$_2$)$_n$I\}] ($n = 3$–10) and [Cp*Fe(CO)$_2$-\{(CH$_2$)$_n$X\}] ($n = 3$–5) have since been prepared in good yield and fully characterized (166) by the following route [Eq. (25)]. The iodoalkyl complexes have been shown to be good precursors to heterobimetallic alkane-diyl complexes of the type [(η^5-C$_5$R$_5$)(CO)$_2$Fe(CH$_2$)$_n$ML$_y$] (R = H, ML$_y$ = Ru(CO)$_2$Cp, Mo(CO)$_3$Cp, W(CO)$_3$Cp, or Re(CO)$_5$, $n = 3$–6; R = CH$_3$, ML$_y$ = Ru(CO)$_2$Cp, $n = 3$–5, ML$_y$ = Re(CO)$_5$, $n = 4$) ($169,170$). In contrast to their reactions with NaI, both [CpFe(CO)$_2$\{(CH$_2$)$_3$Br\}] and [Cp*Fe(CO)$_2$\{(CH$_2$)$_3$Br\}] react with LiI to form the carbene complexes [CpFe(CO)I\{$=\overline{\text{C(CH}_2)_3\text{O}}$\}] and [Cp*Fe(CO)I\{$=\overline{\text{C(CH}_2)_3\text{O}}$\}], respectively ($171$). The ruthenium haloalkyl complexes [CpRu(CO)$_2$-\{(CH$_2$)$_n$X\}] [X = Cl, $n = 3$ (172); X = Br or I, $n = 3$ or 4 ($42,173$)] have also been prepared, by routes similar to that shown in Eq. (21) and (25).

$$[(\eta^5\text{-C}_5\text{R}_5)\text{Fe(CO)}_2\{(CH_2)_n\text{Br}\}] + \text{NaI} \rightarrow [(\eta^5\text{-C}_5\text{R}_5)\text{Fe(CO)}_2\{(CH_2)_n\text{I}\}] + \text{NaBr} \quad (25)$$

R = H, $n = 3$–10; R = CH$_3$, $n = 3$–5

A very rare haloethyl complex, [Ir(PMe$_2$Ph)$_2$(CO)(Br)$_2$(CH$_2$CH$_2$Br)] (**104**), was prepared by Deeming and Shaw by reacting [Ir(PMe$_2$Ph)$_2$-(CO)(C$_2$H$_4$)]BPh$_4$ with Br$_2$ (174). The above haloethyl complex was found to react with MeOH to form [Ir(PMe$_2$Ph)$_2$(CO)(Br)$_2$-(CH$_2$CH$_2$OCH$_3$)]. Pyrolysis of complex **104** gave mainly ethylene and [Ir(PMe$_2$Ph)$_2$(CO)(Br)$_3$]. The reaction of **104** with PMe$_2$Ph gave what was believed to be [Ir(PMe$_2$Ph)$_2$(CO)(Br)$_2$(CH$_2$CH$_2$PMe$_2$Ph)]. Harrison

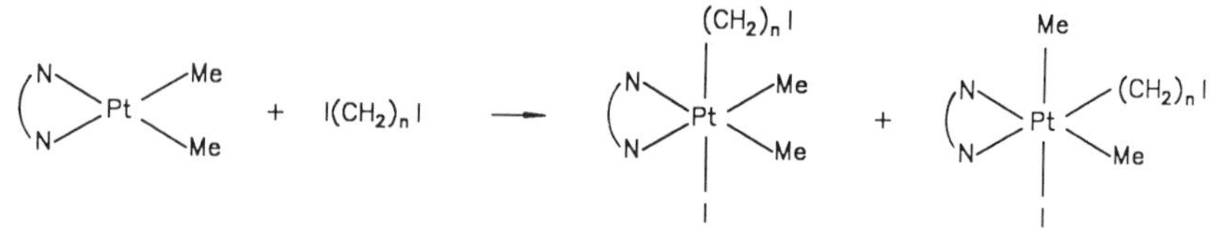

(106)   (107)   (108)

a: n = 0   n = 1
b: n = 1
c: n = 2
d: n = 3
e: n = 4
f: n = 5

SCHEME 31

and Stobart reported that the reaction of $[Ir(CO)_2(\mu\text{-pz})]_2$ (**105**) with 1,3-diiodopropane led to the formation of the stable μ-iodoalkyl complex $[Ir_2(CO)_4(\mu\text{-pz})_2I\{(CH_2)_3I\}]$ (*118*). Reaction of **105** with $I(CH_2)_2I$ yielded only $[Ir(CO)_2(\mu\text{-pz})I]_2$.

Monaghan and Puddephatt investigated the oxidative addition of $I(CH_2)_nI$ ($n = 1-5$) to (dimethyl-1,10-phenanthroline)platinum(II) (**106**), to give both $M(CH_2)_nX$ and $M-(CH_2)_n-M$ (M = transition metal, X = halogen, $n \geq 1$) complexes as represented in Schemes 31 and 32 (*138*). For $n = 3-5$, the reaction of **106** with $I(CH_2)_nI$ gave a mixture of complexes **107** and **109**, which were easily separated. With a large excess of $I(CH_2)_nI$, only **107** was formed, by a S_N2 mechanism (*138*). Where $n = 2$, the products were **109c** and ethylene rather than **107c**. Similarly, no haloalkyl products were found in the reactions of $Br(CH_2)_2Br$ with **106**. Complexes **109** could be prepared by reacting pure **107** and **106** (Scheme 32). Overall rates of reaction for both Schemes 31 and 32 follow the sequence $n =$

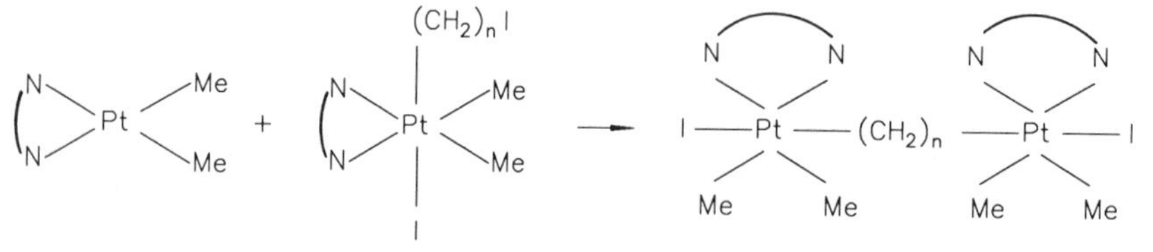

(106)   (107)   (109)

c: n = 2; d: n = 3; e: n = 4; f: n = 5

SCHEME 32

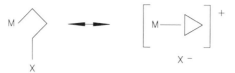

FIG. 15. Proposed resonance forms for [M{CH₂)₃X}] type complexes.

$2 \gg 3 > 4 \cong 5 \gg 1$. The low reactivity of **107b** and **108** is thought to be due to steric hindrance to attack by **106**. In general, though, there is good evidence for metal activation of the C—I bond (*137*). Monaghan and Puddephatt propose that the commonly observed activation of C—X bonds in complexes of the form MCH₂X and MCH₂CH₂X may be due to the lowering of energy in the transition state by contributions from the resonance forms [M=CH₂]⁺ X⁻ or [MC₂H₄]⁺ X⁻, respectively (*137*). The smaller effect for [M{(CH₂)₃X}] may then be due to the resonance forms shown in Fig. 15, thus making displacement of the halide easier.

Further haloalkyl platinum complexes were synthesized by reacting [PtMe₂(bipym)] (**110**) (bipym = 2,2′-bipyrimidine) with excess α,ω-diiodoalkanes to give the adducts **111a–d** (Fig. 16) (*175*). Complexes **111a–d** react with [Pt₂Me₄(μ-SMe₂)₂] to give **112a–d**, respectively, and not **113** as might be expected (Fig. 17). The lack of reactivity of the Pt(II) center to intramolecular oxidative addition is believed to be due to steric hindrance and/or ring strain effects in the transition state (*175*).

Intermolecular oxidative addition between **111** and complex **110** gave the (μ-hydrocarbyl)diplatinum(IV) complex (**114**) (Fig. 18) (*175*). The oxidative addition of I(CH₂)ₙI (*n* = 3–6) to [Pt₂Me₄(μ-pyen)] (**115**) [pyen = bis(2-pyridyl)ethylenediimine] gave the iodoalkyl complexes [Pt₂Me₄(μ-pyen)(I₂){(CH₂)ₙI}₂] (**116**) (*176*). Monitoring the reaction of **115** with I(CH₂)₅I showed that the complex [Pt₂Me₄(μ-pyen)(I₂)I{(CH₂)ₙI}] (**117**) formed at an intermediate stage and reacted

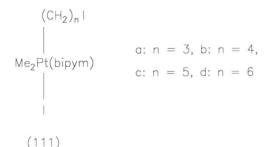

a: n = 3, b: n = 4,

c: n = 5, d: n = 6

(111)

FIG. 16. Molecular formulas of complexes **111**.

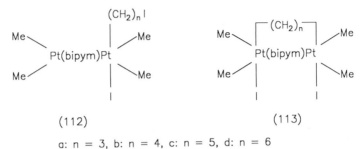

a: n = 3, b: n = 4, c: n = 5, d: n = 6

FIG. 17.   Molecular formulas of complexes **112** and **113**.

further to give a complex of type **116**. The failure of **117** to undergo intramolecular oxidative addition to form a μ-polymethylene complex is believed to be due to unfavorable conformational effects.

The vitamin B_{12s} haloalkyl complexes $[(Co)\{(CH_2)_nX\}]$ were prepared according to Eq. (26) (*90*). When $n = 4$, the above reaction gives $[(Co)(CH_2)_4(Co)]$ as the major and $[(Co)\{(CH_2)_4Br\}]$ as the minor products. The complex $[(Co)\{(CH_2)_4Br\}]$ reacts with (Co) to give $[(Co)(CH_2)_4(Co)]$ (*90*). The related complexes $[Co(DH)_2py\{(CH_2)_nX\}]$ (X = Br, $n = 3-7$; X = I, $n = 3$; DH = monoanion of dimethylglyoxime) were prepared according to Eq. (27) (*177*). These haloalkyl complexes, where $n = 4-6$, reacted with $[Co(DH)_2pyCl]$ in the presence of sodium borohydride to give the dinuclear polymethylene-bridged cobaloximes [py(DH)₂Co-$(CH_2)_n$Co(DH)₂py], as shown in Eq. (28). Neither the vitamin B_{12s}

$$(Co) + Br(CH_2)_nBr \rightarrow [(Co)\{(CH_2)_nX\}]$$

$$(Co) = \text{vitamin } B_{12s}, \; n = 3, \; X = Br, Cl \tag{26}$$

$$[Co(DH)_2pyCl] + X(CH_2)_nX \xrightarrow{\text{MeOH, NaBH}_4} [Co(DH)_2py\{(CH_2)_nX\}]$$

$$X = Br, \; n = 3-7; \; X = I, \; n = 3 \tag{27}$$

$$[Co(DH)_2py\{(CH_2)_3X\}] + [Co(DH)_2pyCl] \rightarrow [py(DH)_2Co(CH_2)_nCo(DH)_2py] \tag{28}$$

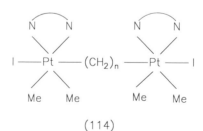

(114)

FIG. 18.   Molecular formulas of complexes **114**.

complex $[(Co)\{(CH_2)_3X\}]$ (X = Br or Cl) nor the cobaloxime complex $[Co(DH)_2py\{(CH_2)_3X\}]$ (X = Br or I) reacted to form a 1,3-propanediyl compound, presumably owing to steric factors (90,177). The bromopropylcobaloxime complex gave cyclopropane and the cobaloxime bromide on thermolysis (178).

Complexes of the type $[Rh(TPP)(RX)]$ $[RX = C_nH_{2n}X$ ($n = 3-5$, X = Cl or Br; $n = 3-6$, X = I); TPP = dianion of tetraphenylporphyrin] were prepared by Anderson et al. (179). The nature of RX was found to determine the overall electrochemical behavior for the reduction of $[Rh(TTP)(RX)]$. For some complexes, specifically those where X = Br and I, the bound alkyl halide could be reduced without cleavage of the metal–carbon bond. This resulted in the electrochemically initiated conversion of $[Rh(TPP)(RX)]$ to a $[Rh(TPP)(R)]$ complex. The E_{pc} value for this reduction was dependent on the chain length and halide of the RX group and followed the trend predicted for alkyl halides. The reduction of the bound RX occured at $E_{\frac{1}{2}}$ values significantly less negative that those for reduction of free RX under the same solution conditions.

Collman et al. prepared halopropyl rhodium complexes by the reaction of dihalopropanes with a macrocyclic rhodium(I) complex [Eq. (29)] (180). The ratio of mononuclear to binuclear product was found to be solvent dependent, with the variations in product ratios in different solvents being related to the solubility of the Rh(I) macrocycle.

$$2\ Rh(I) + X(CH_2)_nX \rightarrow XRh\{(CH_2)_nX\}$$

(29)

Rh = $[C_2(DO)(DOBF_2)_{pn}]$
= [difluoro{N,N'-bis(3-pentanon-2-ylidene)-1,3-diaminopropane dioximato}borate]

IV

CONCLUDING REMARKS

Two notable points from the aforegoing discussion are as follows. (1) By far the majority of the known monohalogenoalkyl compounds are of the group VIII transition elements; there are very few early transition metal halogenoalkyl compounds known. (2) Very few monofluoroalkyl metal complexes have been prepared. The lack of early transition metal halogenoalkyl compounds may be in part due to the high electropositivity of these metals, which facilitates α- and β-elimination reactions. Related hydride elimination reactions almost certainly occur more easily for early transition metal alkyl compounds than for later transition metal compounds. In this regard it is particularly noteworthy that one of the only early transition metal haloalkyl compounds mentioned is the fluoroethyl scandium compound [Cp*$_2$Sc(CH$_2$CH$_2$F)], which was proposed as a very short-lived intermediate which underwent β-fluoride elimination even more readily than its ethyl analog [Cp*$_2$Sc(CH$_2$CH$_3$)] (*181*). Remarkable, too, is the fact that the complex [Ir(PMe$_2$Ph)$_2$(CO)(Br)$_2$(CH$_2$CH$_2$Br)] is the only haloethyl complex to have been fully characterized. All other attempts to prepare such complexes were unsuccessful. This supports the belief that transition metal ethyl complexes are very susceptible to β-elimination reactions.

We have shown, however, that a significant number of monohalogenoalkyl transition metal compounds have now been prepared, many of them in the last decade. A variety of routes have been used to synthesize these compounds, but the most common ones are (1) the reaction of an alkoxyalkyl compound with anhydrous HX (X = Cl, Br, or I), (2) the reaction of a nucleophilic transition metal compound with a dihaloalkane, and (3) the insertion of a methylene group into a metal–halogen bond. The most common route to the ω-halogenoalkyl complexes is by reaction of an anionic metal species with the α,ω-dihaloalkane. These monohalo-

TABLE I

M—C AND C—X BOND LENGTHS AND M—C—X BOND ANGLES OF SOME HALOGENOALKYL TRANSITION METAL COMPLEXES

Complex	M—C bond length (pm)	C—X bond length (pm)	M—C—X bond angle (°)	Ref.
[Fe{P(OCHMe$_2$)$_3$}$_2$(CO)$_2$(I)CH$_2$I]	205(2)	221(2)	120(1)	*48*
[Cp*Cr(NO)$_2$CH$_2$I]	209.3(11)	212.7(11)	115.1(5)	*57*
[Cr(acac)$_2$(H$_2$O)CH$_2$Cl]	207.7(6)	178.5(6)	115.6(3)	*69*
[Cp*Co(CO)(Cl)CH$_2$Cl]	199.1(8)	180.3(8)	116.3(4)	*87*
[Co(PPh$_3$)(CO)$_3$CH$_2$Cl]	202.2(3)	179.4(3)	114.9(1)	*88*
[Rh(dmpe)$_2$(Cl)CH$_2$Cl]Cl · CH$_2$Cl$_2$	216.1(2)	175.4(11)	119.9(7)	*115*
[Ir$_2$(CO)$_4$(μ-C$_5$H$_4$NS)$_2$(I)CH$_2$I]	217.3(15)	213.3(15)	115.0(7)	*122*
[Pt(PPh$_3$)$_2$(I)CH$_2$I]	208.6(24)	210(3)	110.5(9)	*134*
[Pt(tdpea)CH$_2$Cl]BPh$_4$	206(4)	176(5)	118.3(21)	*140*
[Pt{CH$_2$CHC(CH$_3$)$_2$CH$_2$N(CH$_3$)$_2$}(Cl)CH$_2$Cl]^a	205.7(13)	170.2(15)	114.6(7)	*142*
[Pt{CH$_2$CHC(CH$_3$)$_2$CH$_2$N(CH$_3$)$_2$}(Cl)CH$_2$Cl]^{a,b}	203(2)	175(2)	115.4(6)	*142*
[Pt{CH$_2$CHC(CH$_3$)$_2$CH$_2$N(CH$_3$)$_2$}(Cl)CH$_2$Cl]^b	201.6(11)	175.3(11)	114.8(6)	*142*
[Pt{CH$_2$CHC(CH$_3$)$_2$CH$_2$S(CH$_3$)}(Cl)CH$_2$Cl]	203.2(7)	177.8(8)	114.0(4)	*142*
[Pt(dppm)(CH$_2$Cl)$_2$]	206.9(8)	176.5(10)	120.4(5)^c	*145*
	208.2(9)	173.3(9)	120.4(5)^c	
[{Au(CH$_2$)$_2$PPh$_2$}$_2$(I)CH$_2$Cl]	200(4)	175(4)	115.6(2.1)	*152*
[{Au(CH$_2$)$_2$PPh$_2$}$_2$(Br)CH$_2$Cl]	210(3)	178(4)	96.2(14)	*153*

^{a} These two molecules differ in the orientation of Cl and CH$_2$Cl.
^{b} Two forms exist for this structure.
^{c} Data from the Cambridge Crystallographic Data Centre.

genoalkyl compounds are generally sufficiently stable to be isolated and fully characterized by standard physical techniques. A number of X-ray structural studies have been carried out on the halogenomethyl compounds, and the results are summarized in Table I. Some of the structures show novel features. As yet, no molecular structure has been reported for a halogenoalkyl compound of the type $[L_yM\{(CH_2)_nX\}]$, where $n > 1$.

Since the monohalogenoalkyl complexes have a functionality on carbon, they have additional reactivity potential compared to the alkyl analogs. Thus, the halogenoalkyl complexes show interesting and novel chemistry and can be used, for example, as precursors for many important classes of compounds including acyclic and cyclic carbene complexes, homo- and heterobimetallic hydrocarbon-bridged compounds, as well as a range of other important functionalized alkyl compounds. The subject of halogenoalkyl transition metal compounds has grown from a few poorly characterized compounds in the 1960s to an ever increasing number of fully characterized compounds showing novel reactivity.

REFERENCES

1. D. Seyferth and H. Shih, *J. Org. Chem.* **39,** 2329 (1974).
2. J. Barluenga, P. J. Campers, J. C. Garcia-Martin, M. A. Roy, and G. Asensio, *Synthesis,* 893 (1979), and references therein.
3. D. Seyferth and S. B. Andrews, *J. Organomet. Chem.* **30,** 151 (1971), and references therein.
4. F. A. Cotton and G. Wilkinson, "Advanced Inorganic Chemistry," 5th Ed., p. 1128. Wiley, New York, (1988).
5. D. L. Thorn, *Organometallics* **1,** 879 (1982), and references therein.
6. V. Guerchais and C. Lapinte, *J. Chem. Soc., Chem. Commun.,* 663 (1986), and references therein.
7. Y. C. Lin, J. C. Calabrese, and S. S. Wreford, *J. Am. Chem. Soc.* **105,** 1679 (1983).
8. G. Henrici-Olivé and S. Olivé, *Angew. Chem., Int. Ed. Engl.* **15,** 136 (1976).
9. C. P. Casey, M. A. Andrews, D. R. McAllister, and J. E. Rinz, *J. Am. Chem. Soc.* **102,** 1927 (1980).
10. J. C. Calabrese, D. C. Roe, D. L. Thorn, and T. H. Tulip, *Organometallics* **3,** 1223 (1984).
11. C. P. Casey, W. H. Miles, H. Tukada, and J. M. O'Connor, *J. Am. Chem. Soc.* **104,** 3761 (1982).
12. C. P. Casey, "Reactive Intermediates" (M. Jones and R. A. Moss, eds.), Vol. 2, Chap. 4. Wiley, New York, 1981.
13. E. J. O'Connor and P. Helquist, *J. Am. Chem. Soc.* **104,** 1869 (1982).
14. J. P. Collman, L. S. Hegedus, J. R. Norton, and R. G. Finke, "Principles and Applications of Organotransition Metal Chemistry," 2nd Ed. University Science Books, Mill Valley, California, 1987.
15. F. Fischer and H. Tropsch, *Brennst.-Chem.* **7,** 97 (1926).
16. A. R. Cutler, *J. Am. Chem. Soc.* **101,** 604 (1979), and references therein.
17. D. H. Gibson, S. K. Mandal, K. Owens, W. E. Sattich, and J. O. Franco, *Organometallics* **8,** 1114 (1989).

18. L. Pope, P. Sommerville, M. Laing, K. R. Hindson, and J. R. Moss, *J. Organomet. Chem.* **112**, 309 (1976).
19. J. R. Moss, *J. Organomet. Chem.* **231**, 229 (1982).
20. H. Sinn and W. Kaminsky, *Adv. Organomet. Chem.* **18**, 99 (1980).
21. G. A. Somorjai and S. M. Davies, *Platinum Met. Rev.* **27**, 54 (1983).
22. F. Garnier, P. Krausz, and J. E. Dubois, *J. Organomet. Chem.* **170**, 195 (1979).
23. P. Petrici and G. Vitulli, *Tetrahedron Lett.* **21**, 1897 (1979).
24. C. P. Casey and L. J. Smith, *Organometallics* **7**, 2419 (1988).
25. T. G. Richmond and D. F. Shriver, *Organometallics* **2**, 1061 (1983); T. G. Richmond and D. F. Shriver, *Organometallics* **3**, 305 (1984).
26. H. D. Kaesz, R. B. King, and F. G. A. Stone, *Z. Naturforsch. B: Anorg. Chem., Org. Chem., Biophys., Biol.* **15B**, 763 (1960).
27. M. F. Summers, P. J. Toscana, N. Bresciani-Pahor, G. Nardin, L. Randaccio, and L. G. Marzilli, *J. Am. Chem. Soc.* **105**, 6259 (1983).
28. P. W. Jolly and R. Pettit, *J. Am. Chem. Soc.* **88**, 5044 (1966).
29. M. L. H. Green, M. Ishaq, and R. N. Whiteley, *J. Chem. Soc. (A),* 1508 (1967).
30. S. E. Himmel, G. B. Young, D. C. M. Fung, and C. Hollingshead, *Polyhedron* **4**, 348 (1985).
31. E. K. Barefield and D. J. Sepelak, *J. Am. Chem. Soc.* **101**, 6542 (1979).
32. G. C. A. Bellinger, H. B. Friedrich, and J. R. Moss, *J. Organomet. Chem.* **366**, 175 (1989).
33. R. B. King and D. M. Braitsch, *J. Organomet. Chem.* **54**, 9 (1973).
34. J. R. Moss and S. Pelling, *J. Organomet. Chem.* **236**, 221 (1982).
35. S. Pelling, C. Botha, and J. R. Moss, *J. Chem. Soc., Dalton Trans.,* 1495 (1983).
36. E. K. Barefield, P. McCarten, and M. C. Hillhouse, *Organometallics* **4**, 1682 (1985).
37. See references cited in J. F. Hoover and J. M. Stryker, *Organometallics* **7**, 2028 (1988).
38. H. Schmidbaur, *Angew. Chem., Int. Ed. Engl.* **22**, 907 (1983).
39. W. C. Kaska, *Coord. Chem. Rev.* **48**, 1 (1983).
40. M. Brookhart and G. O. Nelson, *J. Am. Chem. Soc.* **99**, 6099 (1977).
41. A. E. Steven and J. L. Beauchamp, *J. Am. Chem. Soc.* **100**, 2584 (1978).
42. H. B. Friedrich, Ph.D. Thesis, University of Cape Town, Cape Town, South Africa (1990).
43. T. W. Bodnar and A. R. Cutler, *Organometallics* **4**, 1558 (1985).
44. T. W. Bodnar and A. R. Cutler, *J. Am. Chem. Soc.* **105**, 5926 (1983).
45. A. Davison, W. C. Krussel, and R. C. Michaelson, *J. Organomet. Chem.* **72**, C7 (1974).
46. T. C. Flood, F. J. Disanti, and D. L. Miles, *Inorg. Chem.* **15**, 1910 (1976).
47. J. E. Jensen, L. L. Campbell, S. Nakanishi, and T. C. Flood, *J. Organomet. Chem.* **244**, 61 (1983).
48. H. Berke, R. Birk, G. Huttner, and L. Zsolnai, *Z. Naturforsch., B: Anorg. Chem., Org. Chem.* **39B**, 1380 (1984).
49. C. Botha, J. R. Moss, and S. Pelling, *J. Organomet. Chem.* **220**, C21 (1981).
50. G. O. Nelson and C. E. Sumner, *Organometallics* **5**, 1983 (1986).
51. G. R. Clark, C. E. L. Headford, K. Marsden, and W. R. Roper, *J. Organomet. Chem.* **231**, 335 (1982).
52. C. E. L. Headford and W. R. Roper, *J. Organomet. Chem.* **198**, C7 (1980).
53. H. B. Friedrich and J. R. Moss, unpublished observations (1984).
54. C. E. L. Headford and W. R. Roper, *J. Organomet. Chem.* **244**, C53 (1983).
55. K. L. Brown, G. R. Clark, C. E. L. Headford, K. Marsden, and W. R. Roper, *J. Am. Chem. Soc.* **101**, 503 (1979).
56. A. F. Hill, W. R. Roper, J. M. Waters, and A. H. Wright, *J. Am. Chem. Soc.* **105**, 5939 (1983).

57. J. L. Hubbard and W. K. McVicar, *J. Am. Chem. Soc.* **108**, 6422 (1986); J. L. Hubbard and W. K. McVicar, *Organometallics* **9**, 2683 (1990).
58. D. Dodd and M. D. Johnson, *J. Chem. Soc. (A),* 34 (1968).
59. C. E. Castro and W. C. Kray, *J. Am. Chem. Soc.* **88**, 4447 (1966).
60. J. H. Espenson and J. P. Leslie II, *Inorg. Chem.* **15**, 1886 (1976).
61. J. H. Espenson and G. J. Samuels, *J. Organomet. Chem.* **113**, 143 (1976).
62. J. H. Espenson and D. A. Williams, *J. Am. Chem. Soc.* **96**, 1008 (1974).
63. J. C. Chang and J. H. Espenson, *J. Chem. Soc., Chem. Commun.,* 233 (1974).
64. J. I. Byington, R. D. Peters, and L. O. Spreer, *Inorg. Chem.* **18**, 3324 (1979).
65. J. H. Espenson and A. Bakac, *J. Am. Chem. Soc.* **103**, 2728 (1981).
66. R. S. Nohr and L. O. Spreer, *Inorg. Chem.* **13**, 1239 (1974); R. S. Nohr and L. O. Spreer, *J. Am. Chem. Soc.* **96**, 2618 (1974).
67. W. R. Bushey and J. H. Espenson, *Inorg. Chem.* **16**, 2722 (1977).
68. A. Bakac, J. H. Espenson, and L. P. Miller, *Inorg. Chem.* **21**, 1557 (1982).
69. H. Ogino, M. Shoji, Y. Abe, M. Shimura, and M. Shimoi, *Inorg. Chem.* **26**, 2542 (1987).
70. Y. Yamamoto and H. Yamazaki, *J. Organomet. Chem.* **24**, 717 (1970).
71. J. R. Moss, M. L. Niven, and P. M. Stretch, *Inorg. Chim. Acta* **119**, 177 (1986).
72. H. B. Friedrich and J. R. Moss, manuscript in preparation.
73. K. C. Brinkman, G. D. Vaughn, and J. A. Gladysz, *Organometallics* **1**, 1056 (1982).
74. K. Noack, U. Schaerer, and F. Calderazzo, *J. Organomet. Chem.* **8**, 517 (1967).
75. F. Calderazzo, K. Noack, and U. Schaerer, *J. Organomet. Chem.* **6**, 265 (1966).
76. A. P. Masters and T. S. Sorensen, *Can. J. Chem.* **68**, 502 (1990).
77. T. G. Richmond, A. M. Crespi, and D. F. Shriver, *J. Organomet. Chem.* **3**, 314 (1984).
78. K. C. Brinkman, Ph.D Thesis, University of California, Los Angeles, California (1984).
79. H. Berke and R. Hoffman, *J. Am. Chem. Soc.* **100**, 7224 (1978).
80. C. Pomp, H. Duddeck, K. Wieghardt, B. Nuber, and J. Weiss, *Angew. Chem., Int. Ed. Engl.* **26**, 924 (1987).
81. C. H. Winter and J. A. Gladysz, *J. Organomet. Chem.* **354**, C33 (1988).
82. H. Werner and L. Hofman, *J. Organomet. Chem.* **289**, 141 (1985).
83. H. Werner, *Angew. Chem., Int. Ed. Engl.* **22**, 927 (1983).
84. H. F. Klein and R. Hammer, *Angew. Chem., Int. Ed. Engl.* **15**, 42 (1976).
85. L. Hofman and H. Werner, *Chem. Ber.* **118**, 4229 (1985).
86. R. Zolk and H. Werner, *J. Organomet. Chem.* **303**, 233 (1986).
87. W. L. Olson, D. A. Nagaki, and L. F. Dahl, *Organometallics* **5**, 630 (1986).
88. V. Galamb, G. Palyi, R. Boese, and G. Schmid, *Organometallics* **6**, 861 (1987).
89. A. Bakac and J. H. Espenson, *Inorg. Chem.* **28**, 3901 (1989).
90. E. L. Smith, L. Mervyn, P. E. Huggleton, D. W. Johnson, and N. Straw, *Ann. N.Y. Acad. Sci.* **112**, 565 (1964).
91. R. J. Guschl, R. S. Stewart, and T. L. Brown, *Inorg. Chem.* **13**, 417 (1974).
92. R. C. Stewart and L. G. Marzilli, *J. Am. Chem. Soc.* **100**, 817 (1978).
93. N. Bresciani-Pahor, M. Forcolin, L. G. Marzilli, L. Randaccio, M. F. Summers, and P. J. Toscano, *Coord. Chem. Rev.* **63**, 1 (1985).
94. W. D. Hemphill and D. G. Brown, *Inorg. Chem.* **16**, 766 (1977).
95. H. J. Callot and E. Schaeffer, *J. Organomet. Chem.* **145**, 91 (1978).
96. H. Werner, L. Hofman, R. Feser, and W. Paul, *J. Organomet. Chem.* **281**, 317 (1985).
97. R. Feser and H. Werner, *Angew. Chem.* **92**, 960 (1980).
98. H. Werner, L. Hofman, and W. Paul, *J. Organomet. Chem.* **236**, C65 (1982).
99. H. Werner and W. Paul, *J. Organomet. Chem.* **236**, C71 (1982).
100. H. Werner, R. Feser, W. Paul, and L. Hofman, *J. Organomet. Chem.* **219**, C29 (1981).
101. H. Werner, W. Paul, R. Feser, R. Zolk, and P. Thometzek, *Chem. Ber.* **118**, 261 (1985).

102. H. Werner, *Pure Appl. Chem.* **54,** 177 (1982).
103. M. C. Henry and G. Wittig, *J. Am. Chem. Soc.* **82,** 563 (1960).
104. E. J. Corey and M. Chaykovsky, *J. Am. Chem. Soc.* **87,** 1353 (1965).
105. H. Werner and W. Paul, *Angew. Chem., Int. Ed. Engl.* **22,** 316 (1983).
106. H. Werner and W. Paul, *Angew. Chem.* **96,** 68 (1984).
107. H. Werner, W. Paul, and R. Zolk, *Angew. Chem.* **96,** 617 (1984).
108. W. Paul and H. Werner, *Angew. Chem.* **95,** 333 (1983).
109. W. Paul and H. Werner, *Chem. Ber.* **118,** 3032 (1985).
110. J. E. Anderson, C.-L. Yao, and K. M. Kadish, *J. Am. Chem. Soc.* **109,** 1106 (1987).
111. J. E. Anderson, C.-L. Yao, and K. M. Kadish, *Inorg. Chem.* **25,** 718 (1986).
112. H. J. Callot and E. Schaeffer, *Nouv. J. Chim.* **4,** 311 (1980).
113. T. Yoshida, T. Heda, T. Adachi, K. Yamamoto, and T. Higuchi, *J. Chem. Soc., Chem. Commun.,* 1137 (1985).
114. J. P. Collman, D. W. Murphy, and G. Dolcetti, *J. Am. Chem. Soc.* **95,** 2687 (1973).
115. T. B. Marder, W. C. Fultz, J. C. Calabrese, R. L. Harlow, and D. Milstein, *J. Chem. Soc., Chem. Commun.,* 1543 (1987).
116. D. Mango and I. Dvoretzky, *J. Am. Chem. Soc.* **88,** 1654 (1966).
117. G. R. Clark, W. R. Roper, and A. H. Wright, *J. Organomet. Chem.* **273,** C17 (1984).
118. D. G. Harrison and S. R. Stobart, *J. Chem. Soc., Chem. Commun.,* 285 (1986).
119. R. D. Brost and S. R. Stobart, *J. Chem. Soc., Chem. Commun.,* 498 (1989).
120. J. V. Caspar and H. B. Gray, *J. Am. Chem. Soc.* **106,** 3029 (1984).
121. M. J. Fernandez, J. Modrego, F. J. Lahoz, J. A. López, and L. A. Oro, *J. Chem. Soc., Dalton Trans.,* 2587 (1990).
122. M. A. Ciriano, F. Viguri, L. A. Oro, A. Tiripicchio, and M. Tiripicchio-Camellini, *Angew. Chem., Int. Ed. Engl.* **26,** 444 (1987).
123. J. A. Labinger, J. A. Osborn, and N. J. Coville, *Inorg. Chem.* **19,** 3236 (1980).
124. M. A. Bennett and G. T. Crisp, *Aust. J. Chem.* **39,** 1363 (1986).
125. D. M. Blake, A. Vinson, and R. Dye, *J. Organomet. Chem.* **204,** 257 (1981).
126. N. J. Kermode, M. F. Lappert, B. W. Skelton, A. H. White, and J. Holton, *J. Organomet. Chem.* **228,** C71 (1982).
127. R. McCrindle and D. W. Sneddon, *J. Organomet. Chem.* **282,** 413 (1985).
128. R. McCrindle, G. J. Arsenault, R. Farwaha, A. J. McAlees, and D. W. Sneddon, *J. Chem. Soc., Dalton Trans.,* 761 (1989).
129. C. A. Ghilardi, S. Midollini, S. Moneti, A. Orlandini, and J. A. Ramirez, *J. Chem. Soc., Chem. Commun.,* 304 (1989).
130. G. B. Young and G. M. Whitesides, *J. Am. Chem. Soc.* **100,** 5808 (1978).
131. O. J. Scherer and H. Jungmann, *J. Organomet. Chem.* **208,** 153 (1981).
132. Z.-Y. Yang and G. B. Young, *J. Chem. Soc., Dalton Trans.,* 2019 (1984).
133. C. Engelter, J. R. Moss, M. L. Niven, L. R. Nassimbeni, and G. Reid, *J. Organomet. Chem.* **232,** C78 (1982).
134. N. J. Kermode, M. F. Lappert, D. W. Skelton, A. H. White, and J. Holton, *J. Chem. Soc., Chem. Commun.,* 698 (1981).
135. C. Engelter, J. R. Moss, L. R. Nassimbeni, M. L. Niven, G. Reid, and J. C. Spiers, *J. Organomet. Chem.* **315,** 255 (1986).
136. R. A. Head, *J. Chem. Soc., Dalton Trans.,* 1637 (1982).
137. P. K. Monaghan and R. J. Puddephatt, *Inorg. Chim. Acta* **76,** L237 (1983).
138. P. K. Monaghan and R. J. Puddephatt, *J. Chem. Soc., Dalton Trans.,* 595 (1988).
139. P. K. Monaghan and R. J. Puddephatt, *Organometallics* **4,** 1406 (1985).
140. C. A. Ghilardi, S. Midollini, S. Monetti, A. Orlandini, G. Scapacci, and A. Traversi, *J. Chem. Soc., Dalton Trans.,* 2293 (1990).

141. R. McCrindle, G. J. Arsenault, R. Farwaha, M. J. Hampden-Smith, R. E. Rice, and A. McAlees, *J. Chem. Soc., Dalton Trans.*, 1773 (1988).
142. R. McCrindle, G. Ferguson, G. J. Arsenault, A. J. McAlees, B. L. Ruhl, and D. W. Sneddon, *Organometallics* **5**, 1171 (1986).
143. R. McCrindle, G. J. Arsenault, and R. Farwaha, *J. Organomet. Chem.* **296**, C51 (1985).
144. J. F. Hoover and J. M. Stryker, *Organometallics* **7**, 2082 (1988).
145. N. W. Alcock, P. G. Pringle, P. Bergamini, S. Sostero, and O. Traverso, *J. Chem. Soc., Dalton Trans.*, 1553 (1990).
146. R. McCrindle, G. J. Arsenault, R. Farwaha, M. J. Hampden-Smith, and A. J. McAlees, *J. Chem. Soc., Chem. Commun.*, 943 (1986).
147. H. Werner and J. H. Richards, *J. Am. Chem. Soc.* **90**, 4976 (1968).
148. R. C. Brady and R. Pettit, *J. Am. Chem. Soc.* **102**, 6181 (1980).
149. R. McCrindle, G. Ferguson, G. J. Arsenault, M. J. Hampden-Smith, A. J. McAlees, and B. L. Ruhl, *J. Organomet. Chem.* **390**, 121 (1990).
150. A. N. Nesmeyanov, É. G. Perevalova, E. I. Smyslova, V. P. Dyadchenko, and K. I. Grandberg, *Izv. Akad. Nauk SSSR, Ser. Khim.*, 2610 (1977).
151. G. J. Arsenault, M. Crespo, and R. J. Puddephatt, *Organometallics* **6**, 2255 (1987).
152. H. H. Murray, J. P. Fackler, and A. M. Mazany, *Organometallics* **3**, 1310 (1984).
153. H. H. Murray III, J. P. Fackler, and D. A. Tocher, *J. Chem. Soc., Chem. Commun.*, 1278 (1985).
154. R. B. King and M. B. Bisnette, *J. Organomet. Chem.* **7**, 311 (1967).
155. C. P. Casey, *J. Chem. Soc., Chem. Commun.*, 1220 (1970); C. P. Casey and R. L. Anderson, *J. Am. Chem. Soc.* **93**, 3554 (1971).
156. A. Irving, J. M. Garner, and J. R. Moss, *Organometallics* **9**, 2836 (1990).
157. C. H. Game, M. Green, J. R. Moss, and F. G. A. Stone, *J. Chem. Soc., Dalton Trans.*, 351 (1974).
158. F. A. Cotton and C. M. Lukehart, *J. Am. Chem. Soc.* **93**, 2672 (1971).
159. N. A. Bailey, P. L. Chell, A. Mukhopadhyay, H. E. Tabborn, and M. J. Winter, *J. Chem. Soc., Chem. Commun.*, 215 (1982).
160. N. A. Bailey, P. L. Chell, A. Mukhopadhyay, D. Rogers, H. E. Tabborn, and M. J. Winter, *J. Chem. Soc., Dalton Trans.*, 2397 (1983).
161. F. A. Cotton and C. M. Lukehart, *J. Am. Chem. Soc.* **95**, 3552 (1973).
162. H. Adams, N. A. Bailey, and M. J. Winter, *J. Chem. Soc., Dalton Trans.*, 273 (1984).
163. H. Adams, N. A. Bailey, V. A. Osborn, and M. J. Winter, *J. Organomet. Chem.* **284**, C1 (1984).
164. V. A. Osborn and M. J. Winter, *Polyhedron* **5**, 435 (1986).
165. N. A. Bailey, D. A. Dunn, C. N. Foxcroft, G. R. Harrison, M. J. Winter, and S. Woodward, *J. Chem. Soc., Dalton Trans.*, 1449 (1988).
166. H. B. Friedrich, P. A. Makhesha, J. R. Moss, and B. K. Williamson, *J. Organomet. Chem.* **384**, 325 (1990).
167. L. G. Scott, M.Sc. Thesis, University of Cape Town, Cape Town, South Africa (1984).
168. M. Cook, N. J. Forrow, and S. A. R. Knox, *J. Chem. Soc., Dalton Trans.*, 2435 (1983).
169. H. B. Friedrich, J. R. Moss, and B. K. Williamson, *J. Organomet. Chem.* **394**, 313 (1990).
170. S. J. Archer, K. P. Finch, H. B. Friedrich, J. R. Moss, and A. M. Crouch, *Inorg. Chim. Acta* **182**, 145 (1991).
171. H. Adams, N. A. Bailey, M. Grayson, C. Ridgway, A. J. Smith, P. Taylor, M. J. Winter, and C. E. Housecroft, *Organometallics* **9**, 2621 (1990).
172. K. P. Finch, M.Sc. Thesis, University of Cape Town, Cape Town, South Africa (1988).
173. H. B. Friedrich, M. Gafoor, and J. R. Moss, unpublished results.

174. A. J. Deeming and B. L. Shaw, *J. Chem. Soc. (A),* 376 (1971).
175. J. D. Scott and R. J. Puddephatt, *Organometallics* **5,** 1538 (1986).
176. J. D. Scott, M. Crespo, C. M. Anderson, and R. J. Puddephatt, *Organometallics* **6,** 1772 (1987).
177. K. P. Finch and J. R. Moss, *J. Organomet. Chem.* **346,** 253 (1988).
178. D. G. Brown, *Prog. Inorg. Chem.* **18,** 172 (1973).
179. J. E. Anderson, Y. H. Liu, and K. M. Kadish, *Inorg. Chem.* **26,** 4174 (1987).
180. J. P. Collman, J. I. Brauman, and A. M. Madonik, *Organometallics* **5,** 218 (1986).
181. B. J. Burger, Ph.D. Thesis, California Institute of Technology, Pasadena, California (1987).

ADVANCES IN ORGANOMETALLIC CHEMISTRY, VOL. 33

Bulky or Supracyclopentadienyl Derivatives in Organometallic Chemistry[1]

CHRISTOPH JANIAK*,† and HERBERT SCHUMANN*

* Institut für Anorganische und Analytische Chemie
Technische Universität Berlin
W-1000 Berlin 12, Germany
† BASF AG
Zentralbereich Kunststofflaboratorium — Polyolefine
6700 Ludwigshafen, Germany

I

INTRODUCTION

Cyclopentadiene (1), a simple organic compound first found in the volatile parts of coal tar, has become one of the most important ligands used in organometallic chemistry. More than 80% of all known organometallic complexes of the transition metals contain the cyclopentadienyl fragment or a derivative thereof.

[1] In commemoration of the late Prof. Jerold J. Zuckerman (deceased December 4, 1987).

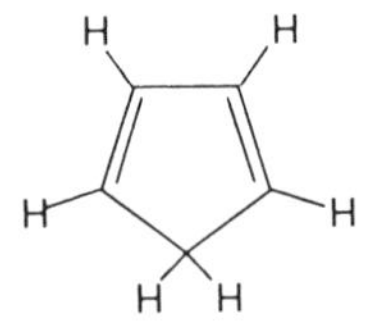

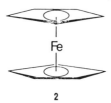

1

Organometallic chemistry was greatly stimulated by the fortuitous discovery of ferrocene, $(\eta^5\text{-}H_5C_5)_2Fe$ (2) (1). The recognition and explanation
of the remarkable structural and bond theoretical properties of ferrocene
by Fischer (2), Wilkinson and Woodward (3a,b,4), and Dunitz (5) established the study of direct metal–carbon bonds as an independent disci-

Fe

2

pline, as we know it today. Ever since, the cyclopentadienyl ligand (Cp)[2]
has played a key role in the development of this field (6). Cp complexes (π
or σ bonded to the central atom) are now known for all main group and
transition metals and semimetals, as well as for the lanthanoid and the
accessible part of the actinoid series.

From its first appearance in 1967 (7), the pentamethylcyclopentadienyl
analog Me_5C_5 also found extensive application. Remarkable differences are
observed between analogous normal- and pentamethylcyclopentadienyl
complexes (8,9). Replacement of all hydrogen atoms by methyl substituents alters the steric and electronic influence of the Cp ring and gives rise to
increased steric bulk, solubility, stability, and electron donor character,
resulting in turn in differing reactivities and spectral properties of Me_5C_5–
metal complexes relative to their H_5C_5–metal counterparts. In general, the

---

[2] In this article, Cp or cyclopentadiene/yl denotes not only $H_5C_5(H)$ but all substituted
cyclic analogs of the general formula $R^IR^{II}R^{III}R^{IV}R^VC_5(H)$, including the bare C_5 skeleton. If
$H_5C_5(H)$ is meant specifically, it will be explicitly denoted as n-Cp or normal cyclopenta-
diene/yl. As indicated, the abbreviation *Cp* stands for the *diene* as well as the *dienyl* form. It
will be clear what is meant from the context or by additions such as Cp *anion* or Cp *metal
complex*. Unless indicated otherwise, Cp is assumed to be η^5 pentahaptobonded in a metal
complex.

enhanced oxidative, reductive, and thermal stability makes the penta-methylcyclopentadienyl complexes easier to isolate and their chemical reactivity easier to study. It is conceivable that even bulkier ring substituents than methyl may induce chemical and physical properties different from both n- and Me_5-Cp. Beginning about 10 years ago (in 1980) and especially in the last 5 years, Cp ligands with extremely sterically demanding substituents have started to receive wider attention.

Before embarking on a discussion of the effects of such bulky Cp modifications, however, a working definition for "sterically demanding cyclopentadienyl ligands" will be established. We restrict ourselves to highly substituted Cp rings which carry at least three substituents larger than a methyl group. This deliberate exclusion of rings systems with only one or two large groups serves to limit the size of the article. Also, the underlying concepts of using bulky Cp's can be more clearly demonstrated by concentrating on the extreme cases.[3] We also refrain from including systems where a Cp moiety is part of a fused aromatic ring, as, for example, indenyl or fluorenyl. These benzo- and dibenzocyclopentadienyls, although related to Cp, exert different electronic effects (11) and represent an interesting area of research by themselves. Octahydrofluorene and related tricyclic derivatives as well as C_5-substituted tetrahydroindenyl, however, are covered by our presentation (see below and the Appendix). Furthermore, also excluded are cyclophane–metal complexes (metallocenophanes), with the rings joined by three or more bridges, even though they would fit into the scope of this article. However, reviews on this subject are already available (12).

For the purpose of discussing the remaining large variety of bulky Cp's in a more systematic manner, it is helpful to single out the "simple," more symmetric Cp's and to view the remainder as derivatives of such basic systems. The basic bulky Cp systems which are already more common in organometallic chemistry are sketched below in their anionic form as ball-and-stick models with the hydrogens included.[4]

---

[3] In the literature, the use of the term "bulky Cp" is rather unspecific. Depending on the authors' perspective, a monosubstituted Cp with, for example, a *tert*-butyl group can be considered as sterically demanding. For systems which are not included here but are also often considered as bulky, see, for example, the references given in (10).

[4] Bond lengths and angles and minimum energetic conformations are averaged from single-crystal X-ray structures where available. Hence, the sketches do not show "nicely flat" molecules as one might draw them by hand. The drawings may require a closer look in cases such as pentabenzyl- and the perisopropyl-Cp's to see the substituent arrangement. Note that viewing CH_2 in the perethyl-Cp's as S gives the isomorphic pentakis(methylthio)-Cp [$(MeS)_5C_5$], tetrakis(methylthio)-Cp [$(MeS)_4HC_5$], and tris(methylthio)-Cp [$(MeS)_3H_2C_5$], respectively.

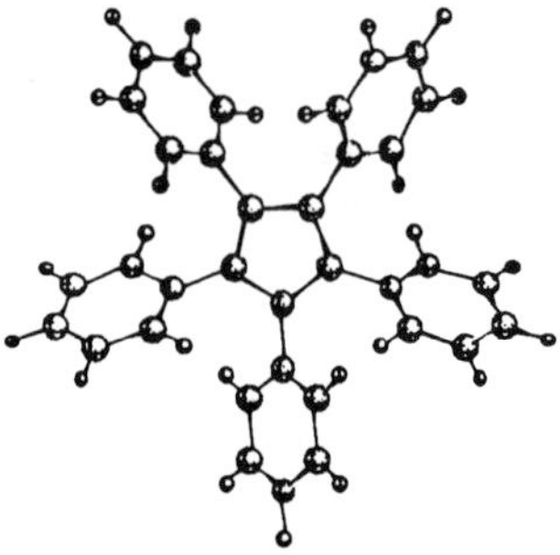

pentaphenyl-Cp, Ph_5C_5

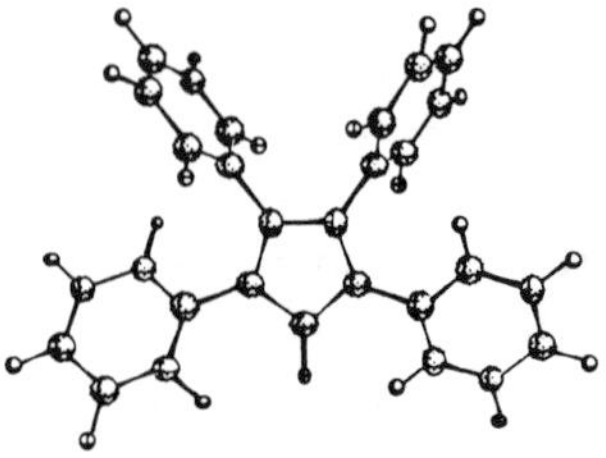

tetraphenyl-Cp, $Ph_4H\,C_5$

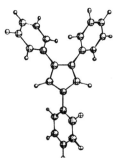

triphenyl-Cp, $Ph_3H_2C_5$

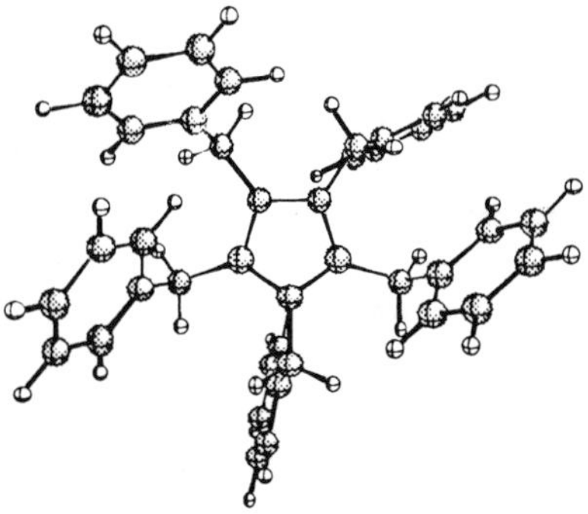

pentabenzyl-Cp, $(PhCH_2)_5C_5$

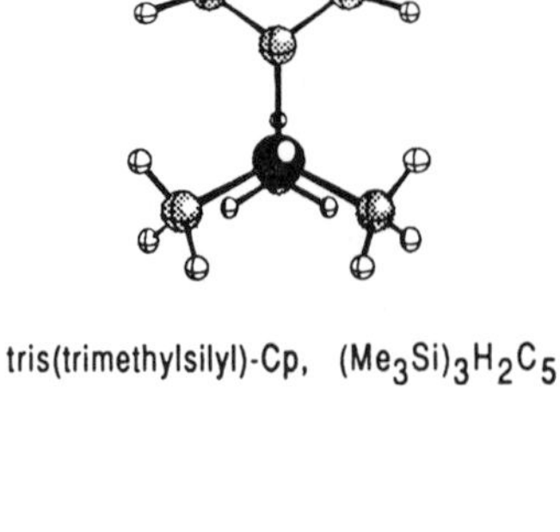

tris(trimethylsilyl)-Cp, $(Me_3Si)_3H_2C_5$ (1,2,4-isomer)

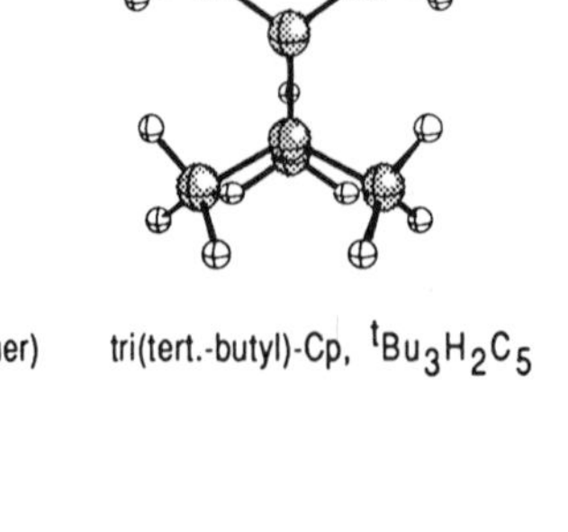

tri(tert.-butyl)-Cp, $^tBu_3H_2C_5$

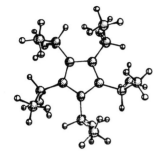

penta(isopropyl)-Cp, $^iPr_5C_5$

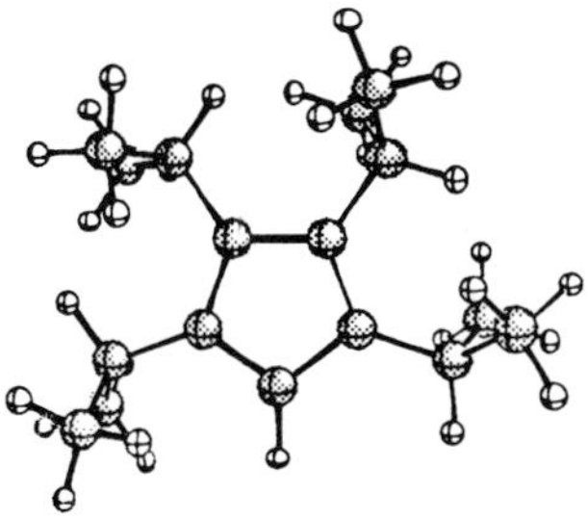

tetra(isopropyl)-Cp, $^iPr_4HC_5$

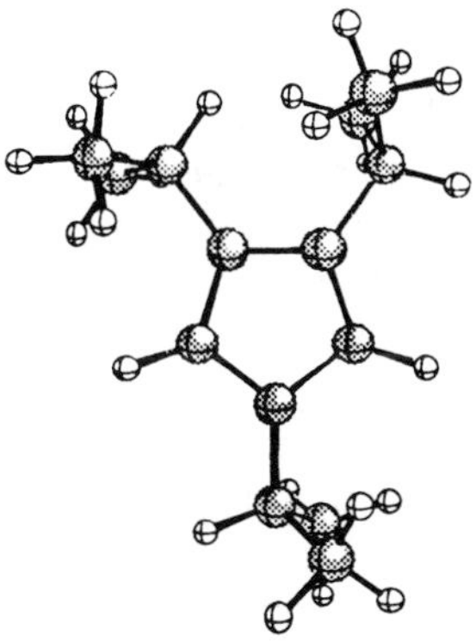

tri(isopropyl)-Cp, $^iPr_3H_2C_5$

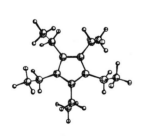

pentaethyl-Cp, Et_5C_5

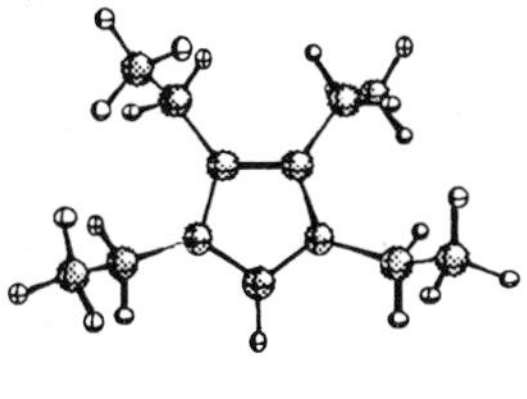

tetraethyl-Cp, Et_4HC_5

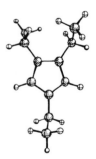

triethyl-Cp, $Et_3H_2C_5$

pentakis(methoxycarbonyl)-Cp, $(MeO_2C)_5C_5$

We then consider any additional hydrogen substitution, be it on the C_5 ring or on an R group, as a modification of the basic system. As such, the tetra- or pentasubstituted Cp's could, of course, therefore be considered as a modification of the tri- and tetrasubstituted homologs. However, the often wider application of R_4HC_5 and R_5C_5 relative to $R_3H_2C_5$ justifies their separate listing. On the other hand, a derivative like $Ph_3(XPh')HC_5$, for example, could not only be regarded as a Ph_4HC_5 but also as a $Ph_3H_2C_5$ modification, depending on its synthesis and/or connection to related complexes. (In the Appendix, where all bulky Cp complexes and their modifications are listed in a systematic fashion, such derivatives appear in two sections.) Furthermore, η^2 or η^4 metal-bound cyclopenta*dienes*, R_5C_5R'(R, R' = H, alkyl, etc.), are also viewed as a modification. For R' = H a deprotonation is often carried out, and the R_5C_5–M derivative is described as well. In this case, the *diene* complex is not listed under the heading "modifications" but in front of the deprotonated analog to better illustrate the derivatization. Finally, trisubstituted Cp's have two isomers available: the 1,2,3 and the 1,2,4 form, the latter being the preferred modification and major product in the preparation for obvious steric reasons.

In comparing the overall size of the bulky Cp's to *n*-Cp (1) we would like to coin the additional term supracyclopentadienes for these systems to describe their large molecular size. These bulky or supra-Cp's and their applications as ligands in organometallic complexes are the subject of this article. Other basic bulky Cp's described in the literature, in addition to those sketched above, are di(isobutyl)(*tert*-butyl)-Cp (*13*), penta(isopentyl)-Cp [isopentyl = $(H_5C_2)_2CH$—] (*14*), and tris(methoxycarbonyl)-Cp (*15,16*). However, these systems have had very limited application so far, with just one or two metal complexes described. Then there are some Cp systems which we would classify as basic, such as tetra-*n*-propyl-, tetra(neopentyl)-(neopentyl = ^{t}BuCH$_2$], and tetrakis(methoxycarbonyl)-Cp, which are not yet known and which have only been utilized in a modified form, namely, tetra-*n*-propyl(*tert*-butyl)-Cp (*13*), tetra(neopentyl)(*tert*-butyl)-Cp (*13*), and tetrakis(methoxycarbonyl)methyl-Cp (*17*). Of course, all metal complexes of these ligands are also listed in the Appendix, and some are even discussed in other sections. Our decision not to include depictions of these still rare supra-Cp species will allow the reader a better visualization of the more common and important bulky Cp's shown above.

Also included in this review, albeit not pictured above, are tetrasubstituted fulvalenes (**3**) (*18*), ethylene-bridged bis(tetrahydroindenyl) complexes (**4**) (*19*), and a dibornacyclopentadienyl ligand (**5**) (*20*). They, as well, have hitherto a very limited (**3, 5,** see Appendix) or very special application (**4,** see Section IV, D). We note that the ligands sketched schematically in **4** and **5** are chiral Cp's; **3** is prochiral. The ethylene-bis(tetrahydroindenyl) system (**4**) is drawn in one of its enantiomeric conformers; flipping one tetrahydroindenyl ring while keeping the other fixed would give the meso form. For more details on this subject, see Section IV, D.

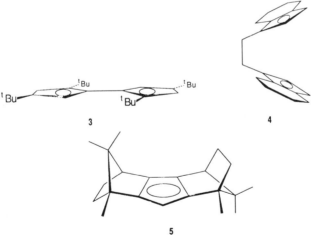

Then there are some mixed systems, where the C$_5$ moiety carries two, three, or more different bulky substituents, such that it is difficult to deduce or assign the resulting supra-Cp to one of the above basic Cp's. Examples are Ph$_2$R(MeO$_2$C)(EtO$_2$C)C$_5$ (R = H, Me) (*21,22*) and (Me$_3$Si)$_2$-

(Me$_2$NCH(Me))(Ph$_2$P)HC$_5$ (*23*). They are considered as mixed bulky Cp's and listed in a separate section at the end of the Appendix.

A borderline situation is encountered concerning the treatment of substituted fulvenes and cyclopenta*dienylides* (of the general formula **6**) as well as cyclopenta*dienones* (**8**) in this review. They will not be considered if resonance form **6** seems to be prevalent. However, if the characterization indicates a dominance of the ylidic mesomer **7**, reference will be made (see under "Fulvenes, cyclopentadienylides" in the Appendix). We note that

cyclopentadienones, including tetraphenylcyclopentadienone (Ph$_4$C$_4$-C=O), can very well function as tetrahapto, η^4-bonded complex ligands (**8**) (*24*). Such systems will not be part of this article. There is, however, a proteolytic equilibrium between the cyclopentadienone complex (**8**) and a hydroxy-Cp–metal species (**9**). Ph$_4$C$_5$OH, now a pentahapto ligand, is definitely a modification of the Ph$_4$HC$_5$ bulky Cp and as such is listed accordingly in the Appendix, provided the complex is sufficiently characterized.

In concluding the presentation of highly substituted Cp's we would also like to mention some interesting supracyclopentadienes which so far have escaped attention as ligands in organometallic complexes: tetra(*tert*-butyl)-Cp (*25*), 3,5-dimethyl- and 3,5-di-*n*-propyl-1,2,4-triphenyl-Cp (*26*), and 1,2,3,5-tetraphenyl-4-*p*-(7-cycloheptatrienyl)phenyl-Cp (*27*).

Most of the bulky ring substituents are simple hydrocarbon moities, either alkyl or aryl groups, and consequently the supra-Cp's can be divided into two classes: peralkyl and peraryl systems. The use of bulky cyclopentadienes with functional group substituents causes problems in organometallic synthesis. Although there is a sizable number of Cp–metal complexes with only *one* functional group on the C_5 ring (*28*) (see also under "Modifications" subheadings in the Appendix), examples for persubstituted systems are limited.

The ionic alkali metal or ammonium salts of $1,2,3\text{-}(OHC)_3H_2C_5$ and $1,2,3\text{-}/1,2,4\text{-}(NC)_3H_2C_5$ as well as $(NC)_4HC_5$ and $(NC)_5C_5$ are known (*28,29*). Furthermore, ionic N-bonded manganese, rhenium, and iron tricyano- and pentacyano-Cp's (*30*) as well as silver tricyanocyclopentadienides have been mentioned (*28*). However, we are not aware of π-bonded (or pentahapto) metal complexes with these systems.

The permethoxycarbonyl ligand included in the listing of the bulky Cp's illustrates the problem encountered with functional groups adjacent to the C_5 core. The Cp system will most often coordinate to the metal center via the functional groups. Only in special cases (depending on the metal) is an η^5 bonding mode observed (see Appendix). Per(methylthio)chloro-Cp's, $(MeS)_nCl_{5-n}C_5$ ($n = 3, 4, 5$), and related $(RS)_3Cl_2C_5$ derivatives ($R = {}^nBu$, Ph) are the only other highly substituted Cp's with a bulky functionality directly adjacent to the C_5 core known to us to form π-complexes (with manganese). In addition, vicinal thioether groups in the π-bonded Cp can still coordinate other metals (e.g., palladium) to give bimetallic mono- or dichelate complexes (see Appendix).

One might imagine many fascinating applications for supracyclopentadienyl metal systems with functional groups. However, this will most likely require a spacer unit to prevent charge transfer between the C_5 and functional moiety, if the C_5–metal coordination is to be preserved. Such systems exist or have recently been developed, for example, in connection with the search for a liquid crystalline organometallic complex (see Section IV,A,2,b).

Within organometallic chemistry, synthetic, analytical, physical, applied, and theoretical chemists are involved in research projects dealing with highly substituted Cp–metal complexes. For some, supra-Cp's are a major focus of their research interest, whereas for many others they are just a sideline where the bulkiness is a minor feature in the comparative study of certain phenomena. Going through the literature often revealed little unifying interest, sometimes with no cross-referencing to similar bulky Cp systems or even different metals carrying the same supra-Cp. Hence, the reader might appreciate a certain difficulty in searching the literature for a topic as structurally diverse as the one chosen for this article. A computer-based literature search assisted the authors' personal knowledge and the citation searching of key references, but it still may not have located all possible articles. While we are confident to provide the reader with over 98% coverage of the literature in this field (up to December 1990), we extend our apologies to researchers whose contributions we have missed.

II

SYNTHESES OF SUPRACYCLOPENTADIENES

The advancement of supra-Cp–metal chemistry followed along the lines of convenience concerning the ligands employed. It is no accident that the perarylated Cp coordination chemistry was developed first and most extensively. The feasible synthetic pathways to the parent dienes had already been worked out by organic chemists, and the dienes were fairly easy to prepare from inexpensive, commercially available starting meterials. Tri- and tetraphenylcyclopentadiene have been known for over 90 years (31,32). A convenient synthesis for tetra- as well as for pentaphenylcyclopentadiene was described in 1925 (33).

The peralkylated dienes followed later as many of these systems are more difficult to obtain. Organometallic reagents and techniques are often required in their syntheses, involving, for example, a sequential metallation/alkylation or a template synthesis, the latter yielding directly a bulky Cp–metal complex. Thus, it is no surprise that the most recent supra-Cp developments were carried out by organometallic chemists themselve in their (never ending) search for new ligands. Tetra- and penta(isopropyl)-Cp (34,35) as well as the perethyl-Cp's [sequential metallation/alkylation (36); template synthesis (37)] exemplify this development.

As indicated above, the in-depth exploration of the chemistry of the supra-Cp ligands and their transition metal complexes depends on economical high yields and large-scale routes to the bulky dienes (or their bromine derivatives). A large-scale synthesis is convenient because of the high molecular weight or inherent low mole/mass ratio of bulky Cp's. Recently some bulky Cp's or their immediate precursors have become commercially available, although at a high price (38). Most, however, are not. Either way, synthesis of the desired bulky Cp will most likely be required by the experimentalist before he embarks on the metal complex-forming reactions. Thus, we have compiled some typical synthetic procedures for the basic bulky cyclopentadienes, thereby demonstrating that many of these compounds can be obtained in rather simple, low-step syntheses, which in our view are even easier than the method for the often used pentamethylcyclopentadiene analog.

Three general synthetic strategies can be distinguished for the synthesis of substituted cyclopentadienes in the laboratory: (1) building up the C_5 ring system from a combination of C_3, C_2, or C_1 units, where the bulky group is already attached; (2) starting from the unsubstituted cyclopentadiene backbone, C_5H_6, and replacing the hydrogens by the desired bulky groups; and (3) derivatization of an already sterically demanding C_5 system. The modified basic Cp systems (see Appendix) were also obtained by methods similar to the ones outlined below, and small-scale template syntheses according to method 1 figure prominently among them. (For the template syntheses, see Schemes 15–22 in Section III.) This section details conventional ("organic") procedures for highly substituted Cp's.

A. Pentaphenylcyclopentadiene

See Scheme 15 for the template synthesis of pentaphenyl-Cp from palladium acetate and diphenylacetylene. Scheme 1 illustrates the preparation of pentaphenyl-Cp according to method 3, and Scheme 2 shows the synthesis by way of method 1.

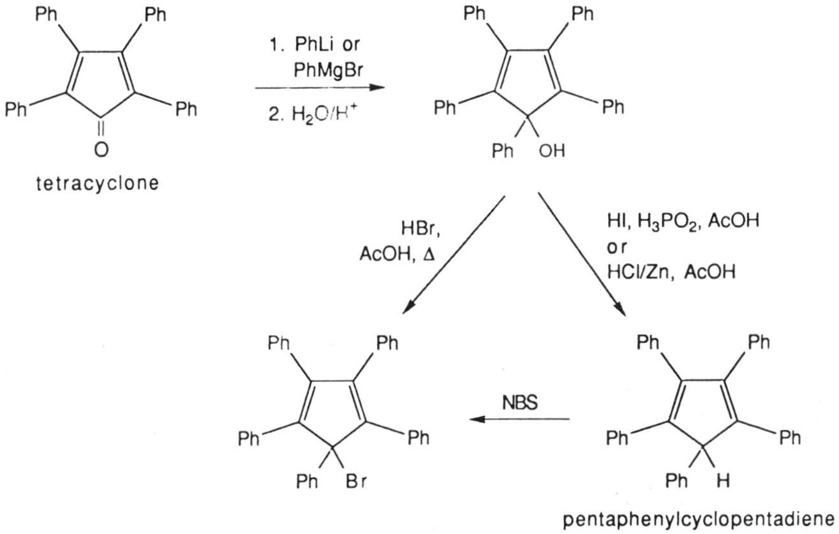

SCHEME 1. Synthesis of pentaphenyl-Cp from tetracyclone (tetraphenylcyclopentadienone). The procedure is suitable for 50 to over 100 g quantities of the starting material. The overall yield of pentaphenyl-Cp is 50–70% based on tetracyclone (33,39,40). In general, substituted cyclopentadienones of all kinds (41) can function as precursors for the attachment of an additional ligand as shown here. AcOH is acetic acid, and NBS is N-bromosuccinimide.

B. Tetraphenylcyclopentadiene

Method 1 is readily applicable for the coupling of substituted phenylacetophenones to give eventually symmetrically substituted tetraarylcyclopentadienes (Scheme 3) (45). (cf. 10). Unsymmetrically substituted 1,5-

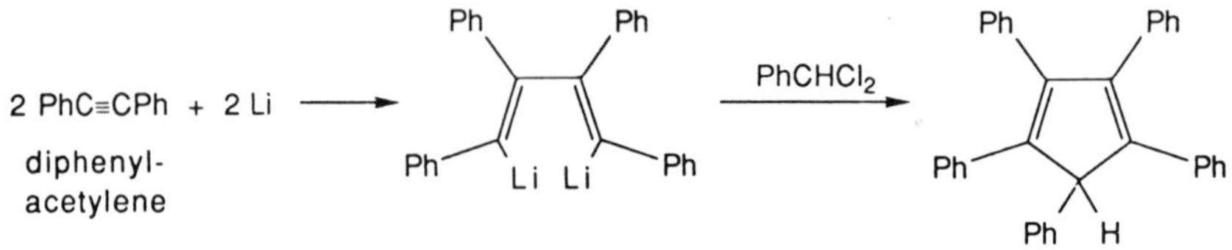

SCHEME 2. Condensation synthesis of pentaphenyl-Cp from diphenylacetylene (tolane) and benzal chloride. Yields are 24–51% (42).

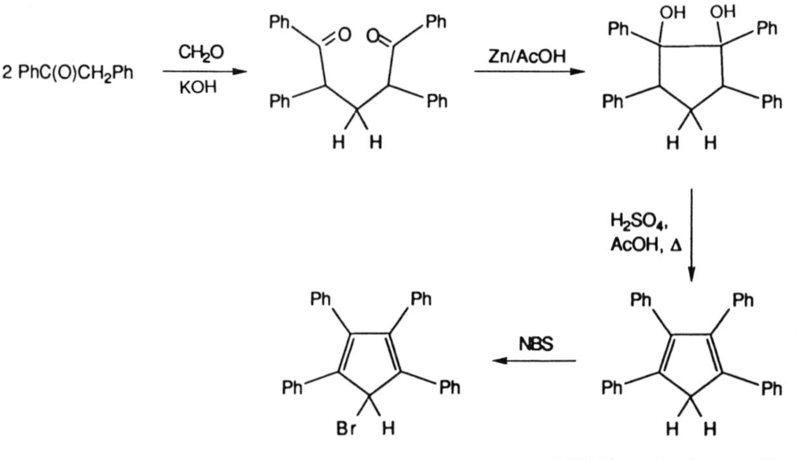

SCHEME 3. Preparation of tetraphenyl-Cp by condensation of phenylacetophenone (desoxybenzoin) and formaldehyde in a base-catalyzed reaction followed by reductive cyclization of the 1,5-diketone with zinc in acetic acid and dehydration of the 1,2-diols (32,33,43). The yield is 60%. The bromide synthesis is based on Ref. (44) (yield 70%).

diones or Cp's can be obtained via Michael addition of substituted phenylacetophenone to phenylacrylophenone [PhC(=CH$_2$)C(O)Ph] (45). In addition, base-catalyzed condensation of disubstituted benzil, R'C(O)—C(O)R', and 1,3-diarylacetone, RCH$_2$C(O)CH$_2$R, offers a general route to tetraarylcyclopentadienones with different groups at positions

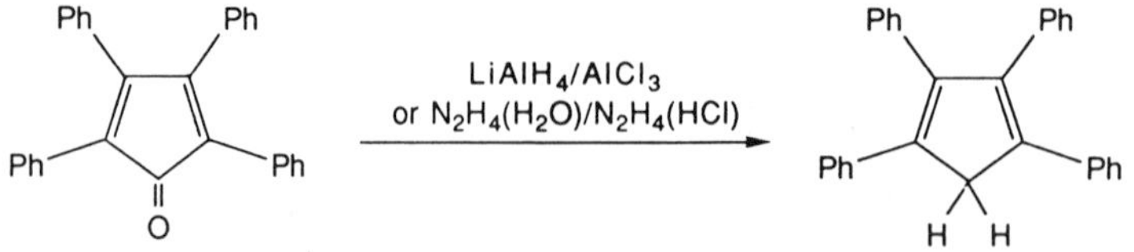

SCHEME 4. Preparation of tetraphenyl-Cp by reduction of tetraphenylcyclopentadienone. Yields range from 80 to 97% (47,48; see also 49).

2,5 and 3,4 (**10**) (*46*), which are starting materials for pentaarylcylopenta-
dienes (Scheme 1) or tetraaryl-Cp's according to Scheme 4.

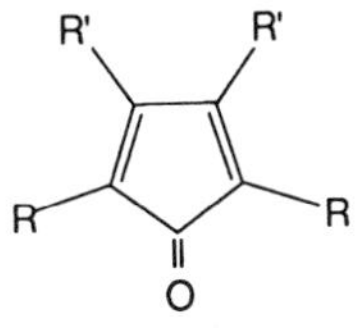

10

C. *Triphenylcyclopentadiene*

Method 1 is a convenient route to triphenyl-Cp's (Scheme 5).

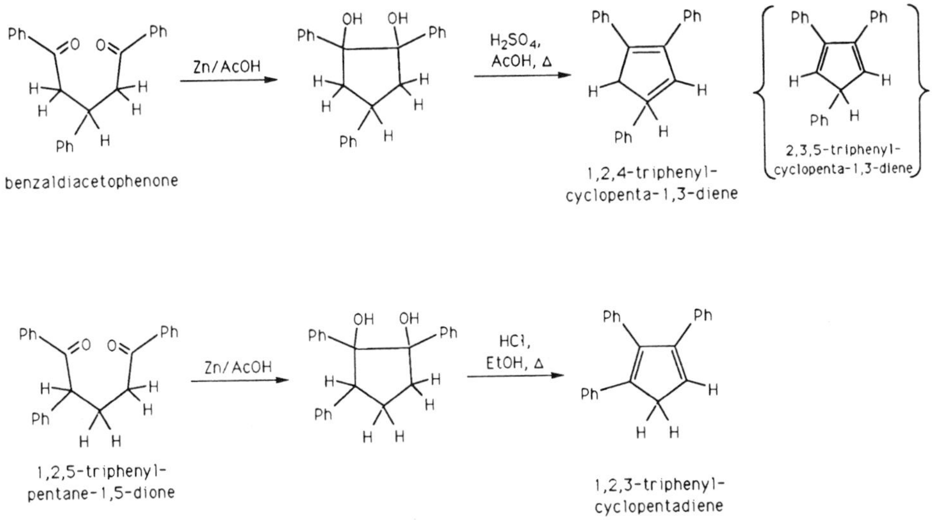

SCHEME 5.   Preparation of both triphenylcyclopentadiene isomers through the reduction
of 1,5-diones followed by dehydration (*31,50*). Note that in the reaction at top dehydration
does not yield the straightforward 2,3,5 but the thermodynamically favored 1,2,4 isomer,
which possesses both a stabilizing stilbene and a 1,4-diphenylbutadiene conjugated system
(See footnote 13 in Ref. *26*).

D. *Pentabenzylcyclopentadiene*

Pentabenzyl-Cp may be synthesized from the unsubstituted C_5H_6 backbone (method 2) according to Scheme 6.

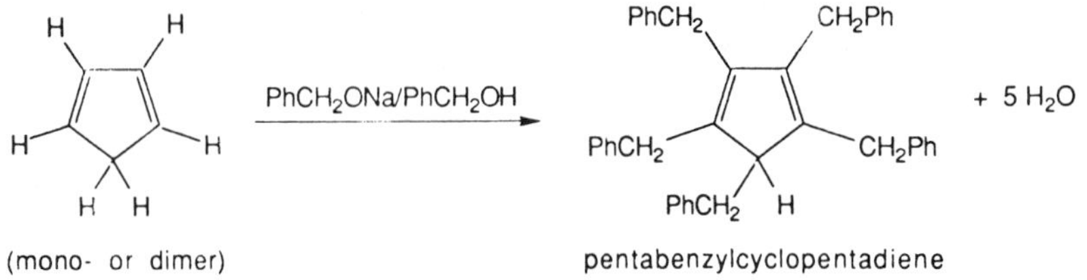

SCHEME 6.   Formation of pentabenzyl-Cp in a one-pot synthesis from normal cyclopentadiene (26,51). Yields are 15–30% based on C_5H_6.

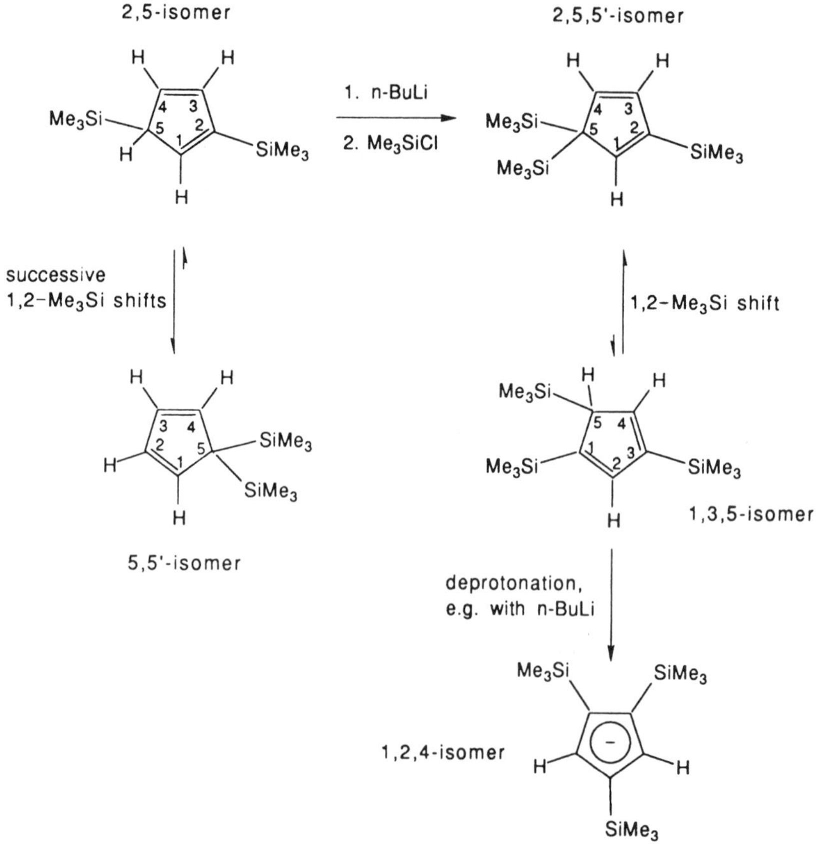

SCHEME 7.   Preparation of tris(trimethylsilyl)-Cp from the bis(trimethylsilyl)-Cp (52) homolog in about 50% yield (53,54).

E. *Tris(trimethylsilyl)cyclopentadiene*

See Scheme 18 for the synthesis of hexakis(trimethylsilyl)stannocene through the sequential deprotonation/silylation of stannocene. Method 2 is most commonly used for the preparation of tris(trimethylsilyl)-Cp (Scheme 7). Deprotonation of the bis- or tris(trimethylsilyl)cyclopentadiene proceeds via the C—H acidic 2,5 or 2,4,5 isomers, which are present in the equilibrium mixture of the 5,5′ or 2,5,5′ isomers in a small percentage (*53*). The 1,2-Me₃Si migrations are characterized as 1,5-sigmatropic (metallatropic) rearrangements of the allylic Me₃Si groups. This fluxional behavior makes trimethylsilyl-substituted Cp's or η^1-cyclopentadienyl compounds of main group elements, in general, interesting systems in themselves. Fluxional Cp systems have been studied intensively by a variety of physical methods (*54–56*).

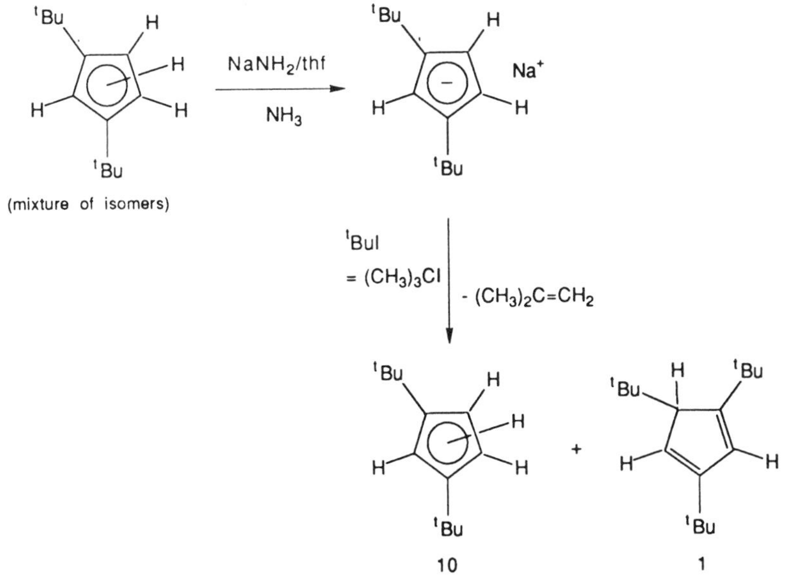

SCHEME 8.   Preparation of 1,3,5-tri(*tert*-butyl)cyclopentadiene (*34*) in low yield (9%) via a metallation/alkylation of di(*tert*-butyl)-Cp (*57*). Elimination of HI from the *tert*-butyliodide is the favored, competing reaction. The mixture of di- and tri(*tert*-butyl)-Cp can be separated by distillation. thf is tetrahydrofuran.

F. Tri(tert-butyl)cyclopentadiene

See also Scheme 17 for the template synthesis of tri(*tert*-butyl)-Cp from a dimeric tantalum alkylidyne complex and 3,3-dimethyl-1-butyne. Schemes 8 and 9 show the syntheses following method 2. The 1,3,5-substituted tri(*tert*-butyl)cyclopentadiene is obtained as the single isomer from both routes.

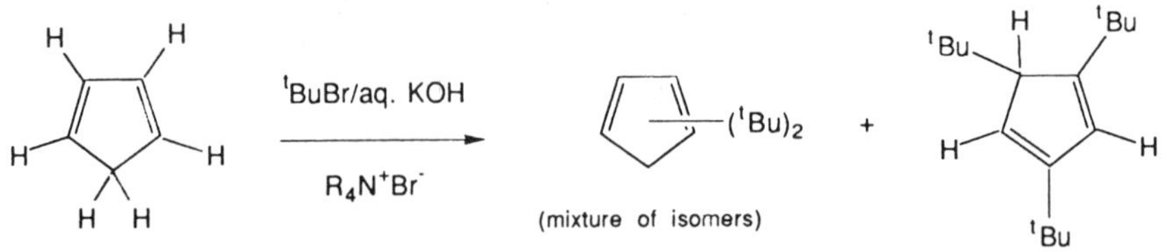

SCHEME 9. Formation of di- and tri(*tert*-butyl)cyclopentadiene in a high yield (90% combined yield) by phase transfer-catalyzed alkylation of unsubstituted cyclopentadiene. Preferably, the tri(*tert*-butyl)-Cp is prepared from the disubstituted homolog by the same method (*58*).

G. Perisopropylcyclopentadienes

For a possible template synthesis of penta(isopropyl)-Cp complexes, see Scheme 19. Scheme 10 gives the synthetic pathways leading to perisopropyl-Cp's according to method 2.

H. Perethylcyclopentadienes

Examples for template syntheses of perethyl-Cp's and their modifications are given in Schemes 16 and 20. Perethyl-Cp's may be synthesized via methods 2 and 3 (Schemes 11 and 12). The synthesis of Et_5C_5 from 3-bromo-3-hexene has been reported recently (*60b*).

I. Pentakis(methylthio)cyclopentadiene

The template synthesis for the intermediate tris- and tetrakis(methylthio)-Cp's as the chloro modification is described in Scheme 21. Synthesis of pentakis(methylthio)-Cp from unsubstituted cyclopentadiene (method 2) is shown in Scheme 13.

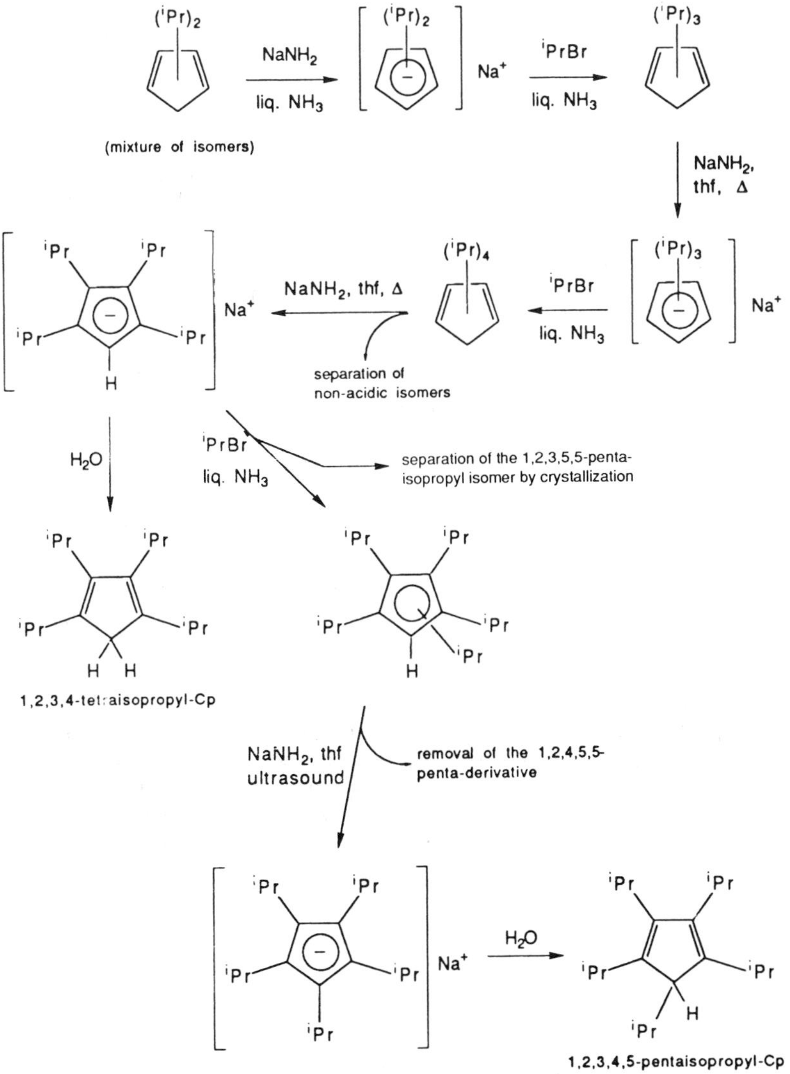

(^{i}Pr)$_2$

NaNH$_2$

liq. NH$_3$

(mixture of isomers)

$\left[\begin{array}{c} (^i\text{Pr})_2 \\ \ominus \end{array} \right]$ Na$^+$

^{i}PrBr

liq. NH$_3$

(^{i}Pr)$_3$

NaNH$_2$,
thf, Δ

$\begin{bmatrix} ^i\text{Pr} & ^i\text{Pr} \\ ^i\text{Pr} & \ominus & ^i\text{Pr} \\ & H & \end{bmatrix}$ Na$^+$

NaNH$_2$, thf, Δ

(^{i}Pr)$_4$

^{i}PrBr

liq. NH$_3$

$\left[\begin{array}{c} (^i\text{Pr})_3 \\ \ominus \end{array} \right]$ Na$^+$

separation of
non-acidic isomers

H$_2$O

^{i}PrBr
liq. NH$_3$

separation of the 1,2,3,5,5-penta-
isopropyl isomer by crystallization

^{i}Pr　^{i}Pr

^{i}Pr　　　^{i}Pr

H　H

1,2,3,4-tetraisopropyl-Cp

^{i}Pr　^{i}Pr

^{i}Pr　　　^{i}Pr

^{i}Pr

H

NaNH$_2$, thf
ultrasound

removal of the 1,2,4,5,5-
penta-derivative

$\begin{bmatrix} ^i\text{Pr} & ^i\text{Pr} \\ ^i\text{Pr} & \ominus & ^i\text{Pr} \\ & ^i\text{Pr} & \end{bmatrix}$ Na$^+$

H$_2$O

^{i}Pr　　^{i}Pr

^{i}Pr　　　　^{i}Pr

^{i}Pr　H

1,2,3,4,5-pentaisopropyl-Cp

SCHEME 10. Reaction sequence for the synthesis of tri-, tetra-, and penta(isopropyl)cyclo-pentadienes (34,35) from di(isopropyl)-Cp (59) via a series of metallations/alkylations. Tri(isopropyl)-Cp is obtained in a 4:1 mixture of the 1,2,4- and 1,2,3-substituted isomers. The C—H acidic tetra(isopropyl) isomers are separated from the 5,5'-geminal dialkylated forms by their transformation in the sodium salt and evaporation of the nonmetallated components. Subsequent hydrolysis yields 1,2,3,4-tetra(isopropyl)cyclopentadiene in an iso-merically pure form. Similarly, 1,2,3,4,5-penta(isopropyl)-Cp is purified after separation of the 1,2,3,5,5 derivative as the main alkylation product. For more details and the respective yields, see Refs. 34 and 35.

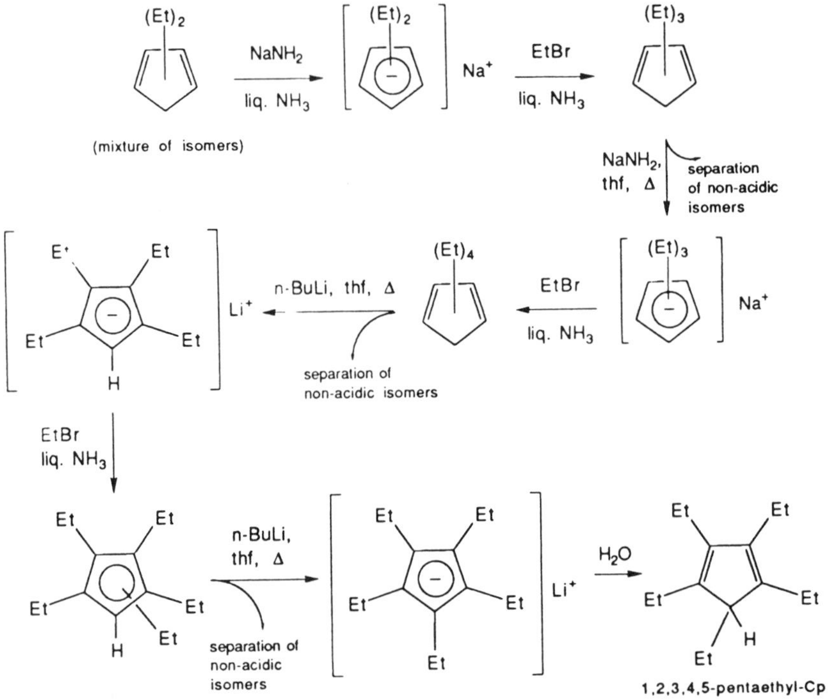

SCHEME 11. Stepwise alkylation of diethylcyclopentadiene (57,59), yielding isomeric mixtures of triethyl-, tetraethyl-, and pentaethylcyclopentadiene. In analogy to the perisopropyl-Cp's (see Scheme 10), purification of the nongeminally dialkylated forms is achieved via conversion to the alkali metal salts and further to transition metal complexes (36).

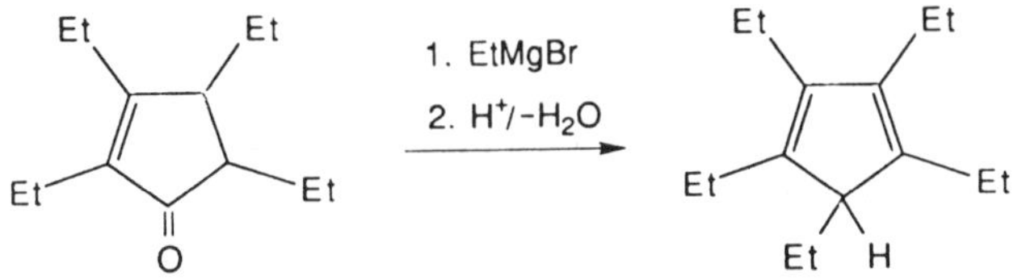

SCHEME 12. Formation of 1,2,3,4,5-pentaethyl-Cp from the reaction of 2,3,4,5-tetraethylcyclopent-2-enone (60a) with ethylmagnesium bromide (36). 1,2-addition of the Grignard reagent followed by hydrolysis may afford an allyl alcohol as an intermediate, which was not isolated, however. Acidic workup of the reaction mixture and distillation induced H_2O elimination and yielded the diene directly.

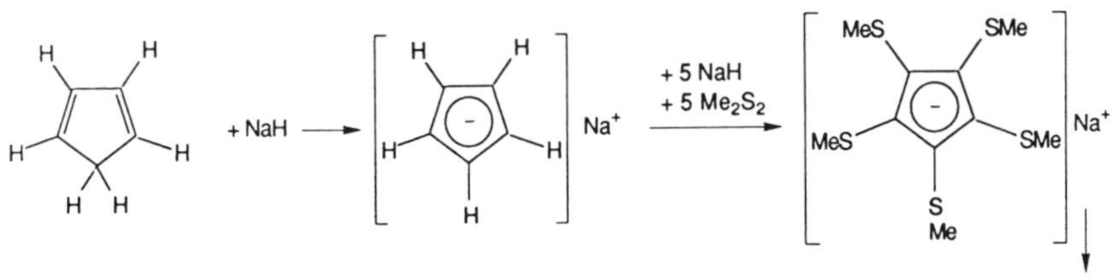

SCHEME 13. One-pot preparation of pentakis(methylthio)cyclopentadienide as the sodium salt. The yield is 51% (61).

J. Penta(methoxycarbonyl)cyclopentadiene

Synthesis of penta(methoxycarbonyl)-Cp from smaller building blocks (method 1) is depicted in Scheme 14.

This rather elaborate collection of synthetic routes leading to supra-Cp systems is also intended to provide the reader with some potential solutions for problems he or she might have in designing bulky Cp's. In addition to the substituted cyclopentadiene hydrocarbons whose syntheses are outlined above, the respective bromo-Cp's are also ligand precursors of considerable utility for Cp–metal complex-forming reactions (see Section III and

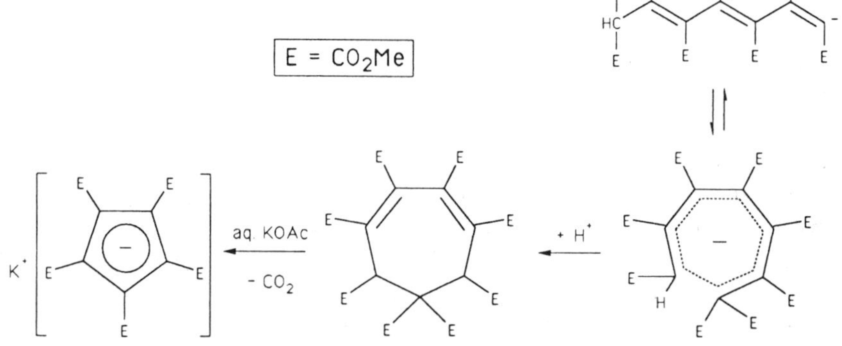

SCHEME 14. Reaction between diethyl malonate and dimethyl acethylenedicarboxylate to give octamethylcyclohepta-1,3-trien-1,2,3,4,5,6,6,7-octacarboxylate (62,63), which is hydrolyzed to the 1,2,3,4,5-penta(methoxycarbonyl)cyclopentadienyl potassium salt (62,64).

the Appendix). In some cases the preparation of the bromo derivative has been included in the sequences for the hydrocarbons (Schemes 1 and 3).

Some of the supra-Cp's, especially the perphenylated systems, are also the subject of synthetic and physical organic chemistry studies. They are, for example, products of cyclopentadienone reductions (see above and *40,47–49*), ring-opening reactions of anionic tricyclopropyl compounds (*65*), or the irradiation of phenyl-substituted vinylcyclopropenes (cf. Scheme 22) (*66*). Fragmentation of the heptaphenylcycloheptatrienyl radical leads to stilbene and the pentaphenyl-Cp anion (*67*). The gas-phase decomposition of perarylated Cp cations was investigated by mass spectrometry (*68*). The acidities of radical cations from substituted and unsubstituted cyclopentadienes were compared (*69*), and radical anions were studied by ultraviolet spectroscopy (*70*). We note accounts on the organic chemistry of tri- and tetraphenyl-substituted cyclopenta*dienylides* (*71*). The reaction of perarylated Cp's with oxygen has received some attention (*72*), and they are valued as dienes in comparative studies of the Diels–Alder reaction (*73*). Perphenylated diazocyclopentadienes were prepared (*74*) as precursors to carbenacyclopentadienes, which were reacted further with olefins (*75*). Recently, organic electroluminescent devices which emit bright blue light could be constructed from tetra- and pentaphenyl-Cp (*76*).

III

SYNTHETIC METHODS FOR BULKY CYCLOPENTADIENYL METAL COMPLEXES

In a first approximation supra-Cp metal complexes can be prepared the same way as normal or other Cp–metal and organo–metal bonds in general. The methods used most often (see Appendix) are the metathesis reaction [Eq. (1)] followed in number by oxidative additions [Eq. (2)] and metallation/deprotonation reactions [Eq. (3)]. The latter is especially important for the cyclpentadienyl alkali metal compounds. A useful variation of reaction (3) is the formation of CpTl in an acid/base reaction from cyclopentadiene and thallium ethoxide [Eq. (3b)]. This represents a convenient route to cyclopentadienylthallium compounds, which are also valued (in place of Cp alkalis) as mild Cp-transfer reagents for the synthesis of difficultly isolable cyclopentadienyl derivatives (*77*).

$$CpM + M'X \rightarrow CpM' + MX \qquad (1)$$

M = metal, especially of the alkaline group; M′ = metal or metal-containing fragment

$$CpHal + M^{m+}L_n \rightarrow CpM^{(m+2)+}L_{n-x}Hal + x\,L \tag{2}$$

Hal = halogen, especially bromine;
M^{m+} = low-, quite often zero-, valent metal; L = mostly CO

$$CpH + M \rightarrow CpM + \tfrac{1}{2}H_2 \tag{3a}$$

$$CpH + TlOC_2H_5 \rightarrow CpTl + C_2H_5OH \tag{3b}$$

Reaction (2) illustrates the elegant synthetic potential of the bromo supra-Cp's. In addition, bulky halo- and in particular the bromocyclopentadienes are also good sources of long-lived radicals which could react with low-valent metal complexes [Eq. (4)]. The radicals are generated by treatment of the substituted bromo-Cp's with zinc (46) or silver (33). So far, radicals have been only employed once, namely, in the formation of decaphenylnickelocene (78).

$$2\,CpBr + Zn\ or\ 2\,Ag \xrightarrow[\substack{-ZnBr_2/ \\ -2\,AgBr}]{} 2\,Cp\cdot \xrightarrow{2\,M^{m+}L_n} 2\,CpM^{(m+1)+}L_{n-x} + x\,L \tag{4}$$

In addition to the rather conventional procedures described above, where the already complete Cp moiety is bonded to a metal fragment, template synthesis (often discovered accidentally) is a distinct possibility for supra-Cp–metal complexes. We can distinguish three different modes of template synthesis:

1. Two or more fragments form a Cp ring with the assistance of a metal center to give a Cp–metal complex. Prominent examples for this mode include Ph_5C_5–metal compounds of molybdenum (79) and palladium (80) from diphenylacetylene (Scheme 15), perethyl-Cp (and modified)

$$[Pd(OAc)_2]_3 + 7\,PhC\equiv CPh \xrightarrow{6\ MeOH} 2\,PhC(OMe)_3 + 6\,AcOH + Pd^0$$
$$+\ [Ph_5C_5Pd]_2(\mu\text{-}PhC\equiv CPh)$$

Scheme 15. Template synthesis of pentaphenylcyclopentadienyl from palladium acetate and diphenylacetylene (tolane). Yields of the Pd complex are 44–55% (80).

tungsten complexes (37,81–84) from dialkylacetylenes and a tungstenacyclobutadiene fragment (Scheme 16; see also Appendix) as well as various supra-Cp–tantalum derivatives from a tantalum alkylidyne complex and alkynes (Scheme 17; Appendix) (13).

2. Bulky substituents are introduced on the Cp ring of a Cp–metal complex. Examples are the formation of 1,1',2,2',4,4'-hexakis(trimethylsilyl)stannocene from normal stannocene and trimethylsilylchloride in a

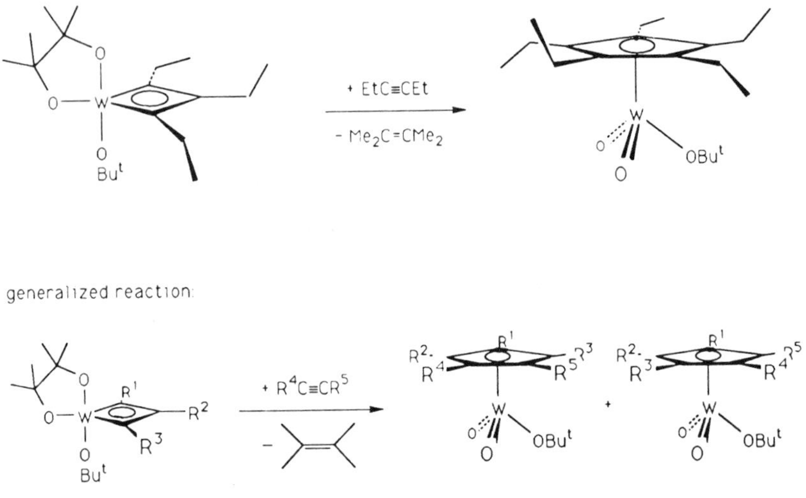

generalized reaction:

$R^1 = {}^tBu; R^2 = R^3 = Me;$ $R^4 = R^5 = Et$
$R^1 = R^2 = R^3 = Et;$ $R^4 = R^5 = Me$
$R^1 = R^2 = R^3 = R^4 = Et;$ $R^5 = CH_2CH_2(Me_4)C_5M(CO)_2$

M = Co, Rh

SCHEME 16. Formation of tungsten perethyl-Cp complexes (and modifications) from a tungstenacyclobutadiene complex and alkynes (37,81–84). The generalized reaction indicates the formation of different isomers (more than two are possible) when three or more different substituents R are present. The W(C'Bu)(dme)Cl$_3$ complex also reacts with alkynes to give bulky Cp derivatives (82) (dme = dimethoxyethane). Molecule drawings are schematic.

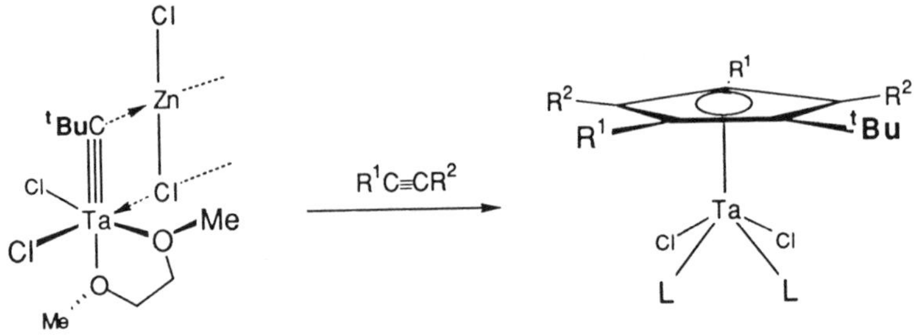

$R^1 = R^2 = Me;$ (L)$_2$ = MeC≡CMe or L = Cl
$R^1 = R^2 = Et, {}^nPr;$ L = Cl
$R^1 = H; R^2 = {}^tBu;$ (L)$_2$ = HC≡C^{t}Bu or L = Cl
$R^1 = H; R^2 = {}^iBu;$ L = Cl (two Cp isomers)

SCHEME 17. Formation of supracyclopentadienyl derivatives from the alkylidyne moiety in the dimeric tantalum complex and alkynes (13). Drawings of molecules are schematic.

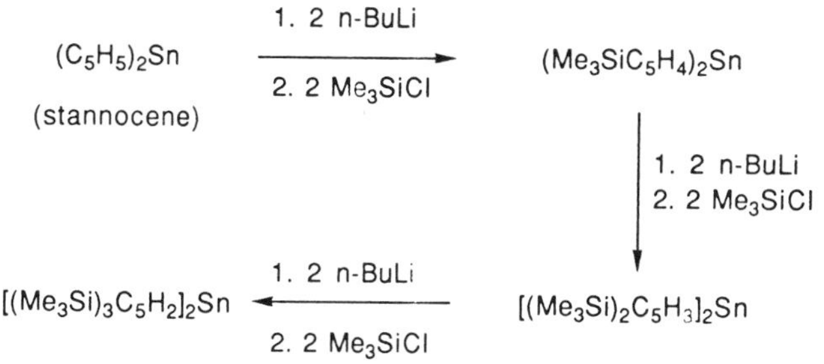

(C5H5)2Sn
(stannocene)
1. 2 n-BuLi
2. 2 Me3SiCl
(Me3SiC5H4)2Sn
1. 2 n-BuLi
2. 2 Me3SiCl
[(Me3Si)3C5H2]2Sn
1. 2 n-BuLi
2. 2 Me3SiCl
[(Me3Si)2C5H3]2Sn

SCHEME 18. Stepwise synthesis of 1,1′,2,2′,4,4′-hexakis(trimethylsilyl)stannocene through the sequential deprotonation (lithiation)/silylation of stannocene (*85*).

sequential lithiation/silylation reaction (Scheme 18) (*85*) and the one-pot synthesis of pentaisopropyl- and pentaisopentylcobalticinium salts from the reaction of pentamethylcobalticinium with base and methyl or ethyl iodide (Scheme 19) (*14*). Both syntheses were modeled after the established reactions with ferrocene and $[Me_6C_6FeC_5H_5]^+$. $(C_5H_5)_2Fe$ can be readily dilithiated (*86*), and the hexaalkylation of the hexamethylbenzene ligand in the ferrocenium cation using a base and a halide is also known (*87*).

CH3I
tBuOK
Co +
Co +
C2H5I
KOH
Co +

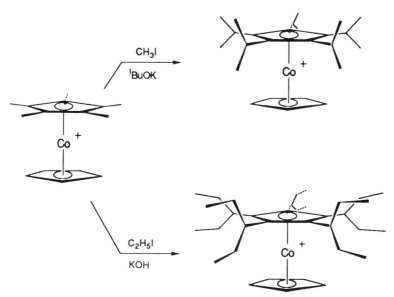

SCHEME 19. Reaction of pentamethylcobalticinium (PF_6^- salt) with excess base and alkyl iodides to give penta(isopropyl)- and penta(isopentyl)cyclopentadienylcobalticinium in 80 and 70% yield, respectively (*14*). Molecule drawings are schematic.

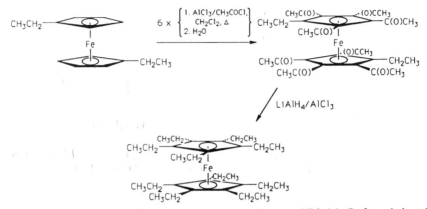

SCHEME 20. Synthesis of decaethylferrocene through repeated Friedel–Crafts acylation of 1,1'-diethylferrocene, followed by reduction (88).

The possibility of carrying out typical aromatic electrophilic substitution chemistry, such as Friedel–Crafts acylation, on ferrocene led to the synthesis of decaethylferrocene (Scheme 20) (88) as well as perethylated Cp-manganese tricarbonyls (89). A metallation/Me_2S_2 reaction sequence of a perchloro Cp ligand was also employed in the preparation of functionally

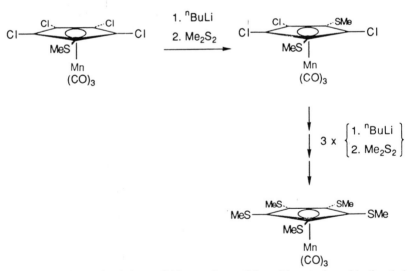

SCHEME 21. Successive halogen–lithium exchange followed by reaction with dimethyl-disulfane. The tris- and tetrakis(methylthio)chloro-Cp metal complexes were also isolated and fully characterized (90,91).

substituted permethylthio Cp–manganese tricarbonyls, as depicted in Scheme 21 (*90,91*).

3. A transition metal mediates rearrangement of a substituted (noncyclo) C_5 species. This includes ring expansion reactions of, for example, substituted 3-vinyl-1-cyclopropene (cf. Scheme 22) or cyclobutene, both bonded to a metal to give supracyclopentadiene Rh and Ru complexes (*92–94*). The rearrangement of vinylcyclopropenes to cyclopentadienes can also occur photolytically (*66*).

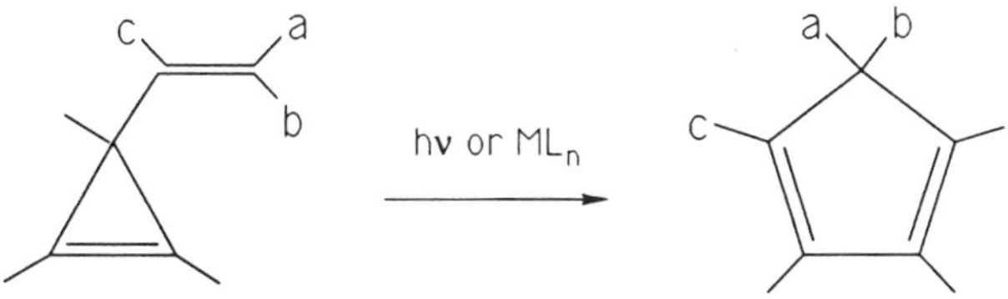

SCHEME 22.   General schematic representation of the photochemical or transition metal-induced rearrangement of substituted vinyl cyclopropenes to substituted cyclopentadienes (*66,92,93*). A transition metal-promoted C—C activation involving a proton transfer (a or b = H) can give η^5-bonded cyclopentadienyl–metal complexes (*93*).

The large steric volume of some of the bulky Cp's is, of course, the source of slightly different reactivity patterns in complex formation when compared with normal or pentamethylcyclopentadienyl reaction schemes. For example, many metathetical reactions between the pentaphenylcyclopentadienyl anion and transition metal halides fail in etheral solvents [such as tetrahydrofuran (thf) and ether] (*40,95*), although the same reaction with the tetraphenylcyclopentadienyl anion proceeds without problems (*48,96,97*). Aside from the "high steric volume," a "lower base strength" hypothesis is invoked for the pentaphenyl-Cp anion to account for the failures (*40*). The desired decaphenyl- or pentaphenylmetallocenes were eventually obtained using much less basic solvents, such as xylenes (*40*).

Similar observations were made with the pentabenzyl-Cp ligand. All efforts to prepare organolanthanoid derivatives from the reactions between SmI_2, YbI_2, or $YbCl_2$ and $(PhCH_2)_5C_5Li$ or K in thf, between Yb and $(PhCH_2)_5C_5H$ in liquid ammonia, or between Sm or Yb and $[(PhCH_2)_5C_5]_2Sn$ or $(PhCH_2)_5C_5Tl$ as ligand carriers did not afford the desired products (*98*).

It was also found that the synthesis of decabenzylferrocene proved to be quite dependent on the solvent and the reaction temperature. Reduction of $FeCl_2$ to metallic iron is the major process on reaction with $(PhCH_2)_5C_5Li$

in thf at room temperature. A combination of low temperatures and diethyl ether as the solvent was best in suppressing the reductive side reaction (*98*).

The synthesis of hexakis(trimethylsilyl)ferrocene also turned out to be much more complicated than one might have thought. Above 0°C only trace amounts of the metallocene could be isolated from the reaction of $(Me_3Si)_3H_2C_5Li$ and $FeCl_2$ in thf. An initial temperature of $-95°C$ had to be used for the combination of the reactants in thf to obtain the desired product in modest yield (*99*). Strong indications for the intermediate formation of a highly reactive half-sandwich complex of the type [CpFeX] (X = Cl, Br) were found in an investigations of this phenomenon (*100,101*).

IV

CONCEPTS FOR USING BULKY CYCLOPENTADIENES

One can envision different approaches to the discussion of Cp–metal chemistry and properties in a systematic fashion. We could follow a line of discussion based on the different bulky ligands or discuss each group in the periodic table with their bulky ligand combinations. However, we choose a third approach and focus on the areas of interest and the *concepts* which are behind the application of sterically demanding cyclopentadienyl ligands. Hence, we present only a few complexes in greater detail which serve to illustrate the cases at hand. Nevertheless, an overview of metal–supra-Cp–ligand systems can be obtained from our compilation in the Appendix. We see four conceptual areas of research interest and applications in the use of bulky cyclopentadienyl ligands (aside from the basic, general interest in novel ligand systems and their metal combinations):

1. *Imparted kinetic stabilization* is probably the most important idea for using bulky ligands, following a common and well-established concept in molecular chemistry to enable the isolation of thermodynamically unstable species. Shielding the reactive center with bulky groups slows the rate of decomposition through bimolecular homocollisions or through reactions with, for example, water and dioxygen. Of course, quite often less bulky ligands permit the isolation of a desired species if the necessary precautions are taken, for example, rigorous exclusion of impurities or working at low temperature. However, in these cases it may be desirable to improve the handlability of the compound to make it more amenable to characterization or further studies.

2. *Enforcing novel molecular structures* is facillitated by steric requirements. Bulky ligands can shift the steric versus electronic interplay in
molecular structures, and one can observe structures which are substantially different from those seen with the analogous less bulky ligands. Bulky
ligands are capable of stabilizing complexes in otherwise unaccessible local
minima on the potential surface. If lone pairs or empty low-lying orbitals
are available on the metal center, a sterically demanding ligand can be
expected to shield these orbitals from interactions with other coordination
centers.

3. *The study of ring rotation dynamics* is made easier. Bulky ligands are
more likely to lock the cyclopentadienyl ring in certain positions, thereby
slowing ring rotation and allowing a more detailed investigation of this
small phenomenon, the interest in which originated with the birth of
organometallic chemistry through the discovery of ferrocene. In addition,
the steric interplay between the bulky substituents on the C_5 ring also
affects the rotations about the C_5–substituent bonds.

4. *The induction of chirality* in Cp–metal derivatives may also be studied. There are different ways that even achiral substituents on a cyclopentadienyl ring can give chiral metal complexes. The induction of
chirality can proceed through their substitution pattern and/or a hindered
ring or substituent rotation. The isotactic polymerization of propylene by
means of metallocene catalysts is one example where such a metallocenic
chirality has already been employed in an important stereoselective
synthesis.

In the following each of the above concepts is elaborated in more detail and
illustrated with specific examples.

A. Kinetic Stabilization

Achieving the preparation of thermodynamically unstable compounds
by inducing kinetic inertness is a challenging and fascinating task for a
synthetic chemist, especially in the organometallic field. All organometallic
complexes are thermodynamically unstable in the presence of water and/or
dioxygen, and most are also labile, kinetically unstable, that is, they decompose more or less quickly to oxidation or hydrolysis products. A small
number of complexes are also thermodynamically unstable (and labile)
even under the exclusion of obvious reactive impurities such as air or
water. Intermolecular collisions among the complexes or with solvent
molecules induce a decomposition process leading to other organometallic
derivatives (often of a polymeric nature).

A brief comment is in place here, to avoid any confusion between the thermodynamic terms *stable/unstable* and the kinetic descriptors *labile/inert*. Inert complexes simply have no suitable low-energy pathway for the reaction available, or, in other words, the free enthalpy of activation ($\Delta G\ddagger$) is very high even if there are more (thermodynamically) stable products. A stable complex has a large positive free enthalpy of reaction ($\Delta G°$) for its decomposition (*102*). This is illustrated in Fig. 1. Quite often, however, the terms inert/labile are replaced by kinetically stable/unstable. For an interesting essay on the different meaning of the simple term *stable* in the chemistry and physics community, see Ref. *103*.

The reactive site in a Cp–M species is not just the metal center but also the cyclopentadienyl ring. For *p*-elements in particular, theoretical (Xα-scattered wave) molecular orbital calculations and experimental ultraviolet photoelectron spectroscopy place the HOMO (highest occupied molecular orbital) primarily on the C_5 rings in their Cp compounds (*104*). Substituting the hydrogens on a H_5C_5 cyclopentadienyl ring by bulky alkyl or aryl groups shields both the metal and Cp reaction sites by blocking the access to the coordination spheres. A space-filling model of the decabenzylmetallocenes of germanium, tin, and lead (*105,106*) (Fig. 2) illustrates this shielding effect. A similar orientation of the benzyl groups, shielding both the metal and the cyclopentadienyl ring, can also be seen in the structures of the air-stable pentabenzylcyclopentadienylthallium and -indium modifications (*107–109*) (Figs. 10 and 11).

Experimentally an enhanced inertness is clearly seen in a comparison between the normal and decabenzylmetallocenes of Ge and Sn, as an example. Although the normal or even pentamethyl-Cp metallocenes are air and water sensitive (*110–112*), the decabenzylmetallocene analogs can be stored in air for days or weeks without apparent decomposition (*105,106*).

A "shielding effect" which is the basis for the stabilization by bulky groups raises the barrier for a associative mechanism, thereby increasing

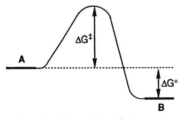

$\Delta G\ddagger$ large: **A** inert with respect to reaction to **B**
small: **A** labile with respect to reaction to **B**

$\Delta G°$: **A** is unstable with respect to **B** ($\Delta G° < 0$);
B is stable with respect to **A** ($\Delta G° > 0$)

Fɪɢ. 1. Illustration of thermodynamic stability/unstability versus kinetic lability/inertness.

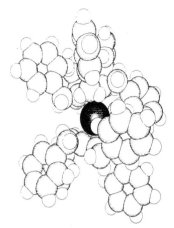

FIG. 2. Space-filling model of decabenzylgermanocene, [(PhCH$_2$)$_5$C$_5$]$_2$Ge (*105*), based on X-ray data. The stannocene and plumbocene derivatives are isostructural (*106*).

the inertness of the Cp—M bond. However, it is not true that bulky Cp complexes can be generally expected to be less reactive than their less bulky analogs. The opposite can be true if a dissociative process is operating, as evidenced by a recent kinetic study. The activation parameters ($\Delta H\ddagger$ and $\Delta S\ddagger$) for the carbonyl displacement in R$_5$C$_5$Ru(CO)$_2$Br with phosphorus donor ligands increase from R$_5$ = Ph$_5$ to Me$_4$Et to H$_5$, leading to approximate relative values of 20:14:1 for the reaction velocity (*113*). This reveals an enhanced lability of CO in the pentaphenyl derivative and implies that bulky Cp's perform much better in stabilizing the transition state, or the ligand deficient intermediate, than the H$_5$C$_5$ ligand (assuming the same bond strength for Ru—CO in the three complexes) (*113*).

Because most papers in this field deal in some way or another with the increased kinetic stability (toward an associative mechanism) of supra-Cp complexes, we discuss only some of the more spectacular and quantitative examples of this effect.

1. *Stabilization Effects from Penta- and Tetraphenylcyclopentadiene*

The perphenylated cyclopentadienyl systems are among the more interesting ligands in terms of imparted stabilization. It is often noted that the Ph$_5$C$_5$ ligand is capable of stabilizing various "unusual" oxidation states (*80,114,115*). For example, Ph$_5$C$_5$ stabilizes the 19-electron anion radicals Ph$_5$C$_5$M(CO)$_2^-$ (M = Co, Rh) compared to their unsubstituted cyclopentadienyl analogs (*116*), and the otherwise highly reactive CpRh(CO)(PR$_3$)$^+$ radical cation can be formed at room temperature with

$Cp = Ph_5C_5$. Moreover, with phenyl groups attached to the Cp ring additional effects may operate and enhance the shielding stabilization.

a. Stereorigidity. A contribution to the strong stabilization effect observed with the pentaphenyl-Cp system may stem from its stereorigidity. The rotation of an individual phenyl group is already restricted by the ensemble in tetraphenyl-Cp (*48,97*; Section IV,C,2) and probably even more so in the pentaphenyl homolog. In the solid state a rather high barrier for the rotation about the $C_5 - M$ vector can be anticipated as well (*117*). In a metallocene the ten phenyl substituents severely restrict a bending of the Cp(centroid)–metal–Cp(centroid) vector or an opening of the coordination sphere on the metal. They tend to lock the Cp rings into parallel positions. The space-filling model of a decaphenylmetallocene with tin as the central atom (Fig. 3) supports the notion that there is little room for an opening up of the metal coordination sphere (*118*).

These views might be supported by a comparison with the decabenzyl- and octaphenylmetallocene analogs. The five benzyl substituents are also quite sterically demanding, as illustrated by the space-filling models (Fig. 2), but at the same time they give a much more flexible bulky substituent ensemble, as is evident from their metal complexes. The Cp rings in a metallocene have the freedom to bend back (*105,106*). Consequently, the decaphenylmetallocenes of Ge, Sn, and Pb are even more air stable than the decabenzyl homologs. They appear to be inert "indefinitely" on exposure to air.

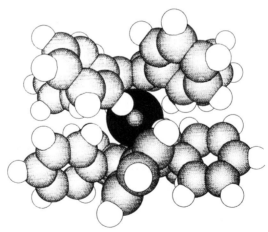

FIG. 3. Space-filling model of decaphenylstannocene, $(Ph_5C_5)_2Sn$, based on X-ray data (*118*).

Most decaphenylmetallocenes also exhibit an unsurpassed thermal stability for sandwich complexes. The decaphenyl Ge, Sn, and Pb derivatives do not decompose until above 350°C (under nitrogen) (39), in contrast with around 100°C for the decabenzyl analogs (106). For *sym*-penta- and decaphenylferrocene and -ruthenocene an extraordinary degree of thermal and oxidative stability is noted (40): they are unchanged in air (!) at 315°C and volatilize only at 250–300°C in the mass spectrometer.

The closely related octaphenylmetallocenes of group 14 (Ge, Sn, and Pb), available for comparison, are markedly less thermally [decomposition at about 200°C (43)] and kinetically stable. On contact with thf decomposition is observed within minutes, which can be followed by decolorization and the appearance of a precipitate after the initial dissolution (43). However, this reaction with thf, which also prevents the synthesis of the above octaphenylmetallocenes in this solvent, is not quite understood, especially since normal stannocene can be handled in thf and shows a base-induced polymerization with formation of a Sn(IV) species only after standing for a couple of days (119).

Octaphenyl–transition metal complexes, on the other hand, exhibit great resistance toward ligand addition and show very slow redox processes. We can only speculate that an opening of the metal coordination sphere (two "missing" phenyl rings, ring–metal–ring bending) together with a possible electronic effect of the phenyl groups (see below) increases the reactivity and Lewis acidity of the main group metal center (cf. thf ring-opening reactions by strong Lewis acids, such as BF_3, zirconocene cations, etc.)

The molecular structure of $(Ph_4HC_5)_2TiCl$ (97) shows that a bent metallocene arrangement is possible with a Cp–M–Cp bending angle very similar to the unsubstituted analogs. The missing phenyl substituent on the opposite Cp ring, together with slightly elongated and varying Ti—C bond lengths, apparently allows for a bent structure (compared to decaphenylstannocene, Fig. 3; see below). Furthermore, with the Ti—Cp(centroid) distance [209 pm (97)] still much shorter than average Ge—Cp (225 pm) or Sn—Cp (240 pm) distances (106,120), the octaphenyl main group metallocenes most likely assume a bent structure. Of course, the decaphenylmetallocenes are insoluble in and stable toward tetrahydrofuran.

This last sentence brings us to the trade-off for the high stability of the decaphenylmetallocenes: The solubility decreases dramatically to the extent that these compounds can be considered "almost insoluble" in all common organic solvents. From the space-filling plot in Fig. 3 one can imagine that the tightly knit, spherelike particle minimizes van der Waals interactions with solvent molecules. Therefore, purification becomes a problem. A continuous extraction process with hot toluene had to be employed. In some cases it might be easier to remove the inorganic metal halide side products with water as can be done since the decaphenylmetallocenes are not wetted by water and hence are stable. Crystallization in the case of decaphenylstannocene succeeded only from hot 1-methylnaphthalene (bp 246°C) (*39,118*).

The low solubility induced by the Ph_5C_5 ligand, which is an obvious difficulty impeding the development of its complex chemistry, prompted the search for more soluble perphenylated systems. This idea was behind the synthesis of the octaphenylmetallocenes (*43*) and also the development of pentaphenyl-Cp systems where the phenyl rings carry solubility-improving methyl, ethyl, or *tert*-butyl groups, especially in the para position (*43,46*). These approaches can be considered successful, and substituents can be introduced on any of the phenyl rings. In both cases the metallocenes exhibit an improved solubility with respect to the normal decaphenyl systems. Furthermore, the transition metal octaphenylmetallocenes crystallize well because of the phenyl substituents (*96*).

b. Possible Charge Delocalization. The phenyl groups are not antiparallel (orthogonal) to the cyclopentadienyl ring (see the ball-and-stick drawing of Ph_5C_5 in Section I and Fig. 3) and can be thought of as helping in the charge delocalization from the Cp ring, and perhaps even from the (lone pair) electrons on the metal (*118*). Thus, electronic effects may play an additional role in the stabilization exerted by the pentaphenyl-Cp system. Although the authors of a Fenske–Hall calculation on decaphenylstannocene negate such a lone pair delocalization, they state that the HOMO is a "doubly occupied antibonding combination of the tin 5s and the Cp a_1 orbitals" (*121*). In terms of the total energy it is true that any mixing from filled orbital (Sn 5s)–filled orbital (Cp a_1) interactions can be eliminated by localizing the orbitals. However, if a frontier orbital perspective is taken, the character of the HOMO as a possible electron-donating orbital in an oxidation reaction becomes important, so the filled orbital–filled orbital interaction cannot be neglected. A Cp–phenyl interaction has not been commented on.

Experimentally, such a charge or electron delocalization ability is still subject to controversy. An increased potential of the redox couple Pd(II)/Pd(I) for $Ph_5C_5Pd(cod)^+$ [270 mV more positive than that of $H_5C_5Pd(cod)^+$; cod = cyclooctadiene] was originally explained by the electron-withdrawing ability of the phenyl substituents making the low oxidation states more accessible (*115*). This effect is opposite to that wrought by methyl substitution, which makes reduction more difficult by lowering the redox potential (*122*). A few years later, however, the thermodynamic stabilization, expressed in the $E°$ potentials, of a few hundred millivolts is viewed as mild (*116*). Redox potentials and electron paramagnetic measurements (EPR) for a series of octaphenyl-Cp complexes of first row transition metals are interpreted in the sense that the ligand Ph_4HC_5 is electronically similar to H_5C_5 (*96*). Tetraphenylcyclopentadienyl is suggested as a bulky substitute for normal Cp without altering the electronic properties of the metal complex.

Initially, one might have considered the electron-withdrawing character in octaphenylmetallocenes to be more pronounced than for pentaphenyl-Cp. Both phenyl groups adjacent to the hydrogen substituent in tetraphenyl-Cp are able to align their planes more parallel to the Cp ring plane, with angles as small as 20° to 30°, as is evident from the X-ray structures of tetraphenylcyclopentadiene (*123*) and the octaphenylmetallocenes of Fe (*48*), V, Cr, Co, and Ni (*96*) as well as Ti (*97*).

On the other hand, the ^{13}C-NMR spectra of the Ph_5C_5 anions were first interpreted as reflecting the inability of the phenyl rings to achieve a planar conformation, viewed necessary for an efficient resonance stabilization of the negative charge in the cyclopentadienyl ring (*124*). However, a more detailed comparative ^{13}C-NMR investigation encompassing the tetraphenyl- and (*tert*-butylphenyl)tetraphenylcyclopentadienyl anions with their neutral hydrocarbons and bromo derivatives revealed a high-field shift of the *para*-phenyl carbons in the anions by about 6 ppm (*43*). It seems to be accepted that the *para*-C atoms in phenyl substituents are a useful indicator for a charge delocalization in such substituents (*43,125,126*). Hence, the observed high-field shift was viewed as proving a delocalization of the negative charge of the C_5 anion onto the phenyl groups, which are neither coplanar nor orthogonal (*43*). In a number of Ph_5C_5- and Ph_4HC_5-transition metal complexes, though, the ^{13}C signal for the *para*-phenyl position deviates only very little (within 1 ppm) from the reference position in the hydrocarbon or bromide (*40,48,126*). This shows only a minor influence of the phenyl substituents on charge and electron density at the central metal (*126*), in agreement with the above interpretation of electrochemical and EPR data.

Moreover, $(Ph_5C_5)_2Sn$ exhibits conventional ^{119m}Sn Mössbauer parameters (isomer shift and quadrupole splitting) in comparison to other stannocenes (*120*) [the isomer shift is proportional to the $5s$ electron density at tin (*127*)]. One could, of course, argue that a much smaller charge in the Cp moiety of the metal complexes does not allow for as large a delocalization effect as in the Cp anions. We feel that further spectroscopic studies may be needed for a deeper understanding of the factors which influence the stabilization wrought by perphenylated Cp's, in particular the role of localization by the phenyl substituents.

2. Applications of Air-Stable Cyclopentadienyl–Metal Compounds

The improved air and water stability of bulky Cp–metal complexes makes it possible to investigate the potential uses of these compounds in areas which require such stability, thus excluding the use of the unstable, less bulky analogs.

a. Antitumor Agents. It was shown by experimental studies in the 1980s that organometallic compounds have the potential for being used as antitumor reagents (*128*). (With "organometallic" we adhere strictly to the definition of complexes having at least one metal–carbon bond.) Bis(cyclopentadienyl)metal complexes of early transition metals exhibit antiproliferative properties against various animal and human tumors, with the metallocene dichlorides, $(H_5C_5)_2MCl_2$, of titanium and vanadium being especially effective (*128,129*). Also, ferrocenyl complexes $[(H_5C_5)_2Fe]^+ X^-$ are examples of nonplatinum group metal antitumor agents (*130*).

Organometallic compounds containing the main group elements tin and germanium were also found to be effective against various experimental and, in some cases, human tumors. These group 14 complexes are The bis(carboxyethylgermanium) trioxide {"germanium sesquioxide," $[(GeCH_2CH_2COOH)_2O_3]_n$} (*131a*), 8,8-diethyl-2-[3-(*N*-dimethylamino)-propyl]-2-aza-8-germaspiro[4,5]decane ("spirogermanium") (*131b*), and diorganodihalotin(IV) derivatives $[R_2SnX_2(L_2); R =$ alkyl or phenyl, $L =$ unidentate ligand, e.g., pyridine, or $L_2 =$ bidentate chelating ligand, e.g., 2,2'-bipyridyl] (*131c,d*).

In view of these results the antitumor potential of Cp–Ge or –Sn combinations was explored. Before the advent of the stable, bulky Cp complexes, however, such an approach was destined to fail. Any attempt to deliver a normal Cp–Ge or –Sn compound (Cp being σ or π bonded) into a biological system would result in an almost instantaneous decomposition. Consequently, the stable decaphenyl- and decabenzylmetallocenes of germanium and tin were tested for their tumor-inhibiting properties against the Ehrlich ascites tumor in female mice. The complexes caused cure rates of 40 to 90% of the animals treated over rather broad dose ranges. Figure 4 illustrates the dose–activity and dose–lethality for decaphenylstannocene, which showed the best cure rates among the complexes tested (*132*).

With the germanocene complexes no strong dose–activity relationship was manifest. The toxicity of all four metallocenes was low, the LD_{10} values (lethal dose causing the death of 10% of the animals treated) of both stannocenes being 460 and 500 mg/kg, and those of both germanocenes higher than 700 mg/kg. For comparison purposes, the isolated hydrocar-

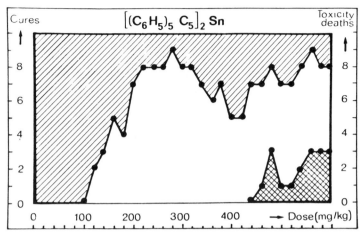

FIG. 4. Dose–activity (left) and dose–lethality (right) relationships of decaphenylstannocene against the fluid Ehrlich ascites tumor in mice. ▨ Tumor deaths, ▨ deaths due to substance toxicity, ☐ surviving, cured animals (*132*).

bon ligands pentaphenyl- and -benzylcyclopentadiene were also tested for their antiproliferative behavior. They also exhibit antitumor activity which was, however, less pronounced than that of the metal-containing sandwich complexes. It was concluded from these studies that decaphenyl- and -benzyl-substituted germanocene and stannocene represent a new type of nonplatinum antitumor agents (*132*).

b. Organometallic Liquid Crystals: Discoidal Metallocenes. The disklike structure of the pentaphenyl-Cp ligand makes it possible to envision discoidal metallocenes or more generally discoidal organometallic compounds with Cp units at their cores. Disklike mesogens (schematically illustrated in **11** with the column stacking in the mesophase) are known for

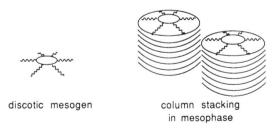

discotic mesogen   column stacking
                               in mesophase

11

organic compounds (*133*) and organic metal complexes (no direct M—C bond) (*134*). The only known liquid crystalline compounds with an organometallic core (albeit no discoidal mesogen) are ferrocene diesters, $(RO—C_6H_4—C_6H_4—O_2CH_4C_5)_2Fe$, with $R = n-C_5H_{11}$, $n-C_6H_{13}$, and $n-C_{11}H_{23}$ (*135*), aside from an example of a liquid crystal where a ferrocenyl group was attached to the end of a mesogenic unit (*136*).

Utilizing the pentaphenyl-Cp unit as a basis, one can substitute all five phenyl groups in the para position with high yield Friedel–Crafts acylation reactions (*137*) (Scheme 23). These modified pentaphenylcyclopenta-

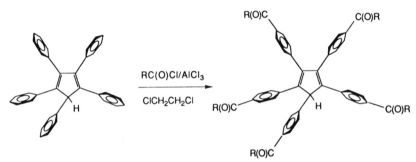

SCHEME 23.   Acylation of pentaphenylcyclopentadiene in dichloroethane (*137*). The yield is 60–85%. R can be CH_3, C_3H_7, C_5H_{11}, or C_7H_{15}.

dienes can readily be converted to the corresponding thallium(I) complexes on reaction with thallium ethoxide. Unfortunately, the thallium derivatives decompose before melting; therefore a liquid crystalline behavior or mesophase could not be observed. However, transformation of the R = methyl derivative to the pentaethyl or pentyl benzoates according to Scheme 24 and subsequent conversion to the thallium salts allow one to obtain the Tl–pentyl ester complex (Fig. 5) as the first real organometallic discoidal mesogene, forming a viscous, oily melt at 180°C, resolidifying at 176°C and undergoing another phase change at 164°C (*137*). In combination with the possible chirality of supra-Cp–metal complexes (see Section IV,D) one might also consider the formation of chiral, discoidal organometallic liquid crystals (*138*).

B. *Novel Molecular Structures*

It is conceivable that bulky ligands exert a much stronger influence on intra- or intermolecular forces than their normal, or less bulky, analogs. In the following we discuss the known examples where a supra-Cp system

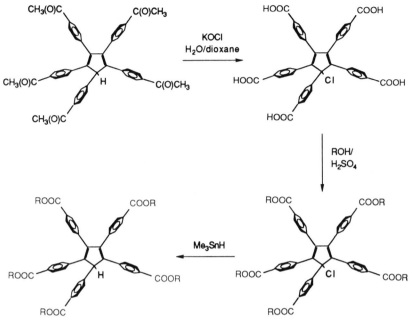

SCHEME 24. Conversion of the pentamethyl ketone into pentaalkyl benzoates (*137*). The yield is 90%. R can be C_2H_5 or C_5H_{11}.

leads to new molecular structures. Some of these structures have already been the subject of theoretical studies, which we refer to as well. The reader will also realize that most of the structures discussed below are Cp–main group metal complexes. Only a relatively few transition metal derivatives are mentioned. Special attention in this section is focused on the main group elements with normally stereochemically active lone pairs and the steric versus electronic interplay of the supra-Cp ligand with such a lone pair-carrying central metal. With very few exceptions so far, transition metal structures with supra-Cp's do not differ much from structures with the normal or pentamethyl-Cp groups.

In most cases bulky Cp–metal structures studied are unexceptional with respect to M—C(Cp) distances, as well as C—C values in the Cp moiety. It seems that the nature and length of the M—Cp bond are affected dramatically only by bulky substituents on the Cp ring if one starts from a rather ionic interaction between the metal and normal cyclopentadienyl. Such an ionic M—Cp interaction usually does not give localized M—Cp pairs but extended structures [e.g., $(Me_3Si)H_4C_5K$ (*139*), $(H_5C_5)_2Ca$ (*140,141*), $(Me_5C_5)_2Ba$ (*142*), $(H_5C_5)Tl$, and $(H_5C_5)In$ (*143*)]. Hence, a reduction of

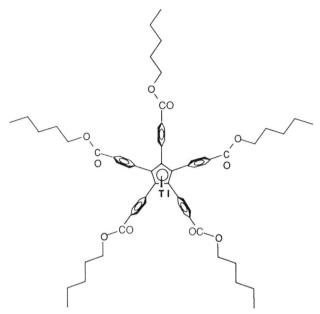

FIG. 5. Schematic drawing of the mesogenic pentakis(pentyl benzoate)cyclopentadien-ylthallium (mp 180°C) (137).

the "intermolecular" ionic interactions is often accompanied by an overall structural change [cf. to the vaporization or separation of, e.g., {H_5C_5Tl}$_\infty$ or In which is accompanied by a transition from a polymeric material with long, ionic C_5—M bonds to a monomeric species in the gas phase with a rather short, presumably covalent, C_5—M bond (144,145)]. Bulky Cp substituents have a similar "insulating" or separatory effect: An increased covalency by peralkylation (and -silylation) of the Cp ring is evident from a shortening of the M—Cp bond, a monomeric or oligomeric nature of the molecule (versus an often polymeric arrangement for the ionic species), and drastically improved solubility properties in nonpolar solvents. Effects like this can be observed in the pentabenzylcyclopentadienyl complexes of thallium (107,108,146) and indium (109), as well as in tris(trimethylsilyl)-Cp derivatives of lithium (147,148), magnesium (149), indium (150), and thallium (151). They are described below in more detail.

1. *Decaphenylstannocene*

The steric demand of large ligand systems can dominate the ligand arrangement around a central metal over its electronic requirements, to the extent of violating rules of chemical bonding. Such an extreme case

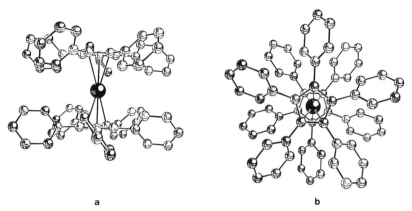

a   **b**

FIG. 6. Decaphenylstannocene, $(Ph_5C_5)_2Sn^{II}$, viewed parallel (a) and perpendicular (b) to the plane of the cyclopentadienyl rings (*118*). The hydrogen atoms have been omitted for clarity. See also the space-filling plot of decaphenylstannocene in Fig. 3.

is found in the structure of decaphenylstannocene (Fig. 6) (*118*). Bis(pentaphenylcyclopentadienyl)tin is the first highly symmetrical main group sandwich compound and one of the very few examples of a molecular main group species in which the lone pair electrons are stereochemically totally inert.[5] In this structure the tin atom sits on an inversion center between symmetry-related equidistant cyclopentadienyls which are planar, staggered, and exactly parallel (Fig. 6). The attached phenyl groups are canted to each cyclopentadienyl ring oppositely in a double-opposed paddle wheel fashion to give molecules of S_{10} symmetry. Normally, the lone pair electrons in subvalent, fourth main group compounds reveal their VSEPR- [valence shell electron pair repulsion (*154*)] expected stereochemical activity in distortions from high-symmetry geometries. The structures of these systems show voids in the central atom coordination sphere. In two-coordinated species the angles at these atoms are bent.

Comparative data for the solid state are available for $(R_5C_5)_2E$ derivatives in which for E = Sn, R = H (*155*) and Me (*111*), as well as for E = Ge, R = H (*112*) and E = Pb, R = H (*156*) and Me (*155*). In addition, gas phase data are available for E = Ge, R = Me (*157*), E = Sn and Pb, R = H (*158*), and for the 1,1′-methyl derivatives of E = Ge and Sn (*159*). All are severely bent. In direct comparison, the angles between the normals

---

[5] The only other two examples known to us are the planar structure of trisilylamine, $(H_3Si)_3N$, and related compounds in the gas phase, where the nitrogen lone pair is stereochemically inactive (*152*), and the solid-state structure of BrF_6^-, with an almost perfect octahedral geometry (*153*).

from Sn to the Cp ring planes for the R = H, Me, and Ph $(R_5C_5)_2Sn$ solids are 133°/134° (120), 143.6°/144.6° (111) (two independent molecules in each case), and 180° (Fig. 6). Other related structures also show severe bending at the tin(II) atom, for example, in $(H_5C_5)SnCl$ $(160a)$, $[(Me_5C_5)Sn \leftarrow NC_5H_5]^+$ $[CF_3SO_3]^-$ $(160b)$, $[(^iPr_2N)_2PH_4C_5]_2Sn$ $(160c)$, $[H_5C_5Co(C_2B_2C)]_2Sn$ $(160d)$, as well as in $\{[BF_4]^-(\mu-H_5C_5)_2Sn[\mu-H_5C_5Sn]^+$ (thf)$\}_n$ $(160e)$. Thus, the decaphenylstannocene structure represents quite a discontinuity from previous results.

The fact that the solid-state structure of decamethylsilicocene contains two structurally distinct molecules in the unit cell in a 2 : 1 ratio, one bent and one with the rings staggered and perfectly parallel (161), is no contradiction. The tendency to bend is expected to increase from the lighter to the heavier group cogeners in homologous complexes. Compare, for example, the series $H_2E{=}EH_2$, with E = C, Si, Ge, and Sn (162), where one goes from a planar ethene molecule to trans-bent digermylene and -stannylene **(12)** with $\theta = 148°$ for M = Ge and 138° for M = Sn in the structurally authenticated bis[bis(trimethylsilyl)methyl] derivatives, $\{[(Me_3Si)_2CH]_2M^{II}\}_2$ [M = Ge, Sn (163)].

12

Molecular orbital calculations of the extended-Hückel (111) and Fenske–Hall type (121) agree that the lone pair electrons in a linear subvalent group 14 stannocene reside in an antibonding combination of the tin $5s$ and the Cp a_1 orbitals. In point group C_{2v} (anticipating a possible bending) this $5s$ Cp level is of a_1 symmetry, and it decreases in energy with bending, mixing in an Sn $5p$ contribution from an empty Cp–Sn* orbital also of a_1 symmetry (in C_{2v}) a few electron volts above. This results in a stabilized filled orbital with Cp–Sn nonbonding character and an empty level which is destabilized and strongly Cp–Sn antibonding. The orbitals guiding these deformations are sketched in Fig. 7 [compare also the discussion to Fig. 12 and the analogy to the bending of AH_2 systems (164)].

In general the energy difference between two levels is one of the factors that influences the strength of its interactions: the smaller the gap, the more likely a strong mixing, that is, in the case at hand, the tendency to bend. However, in silicocene the initial energy difference between the two a_1

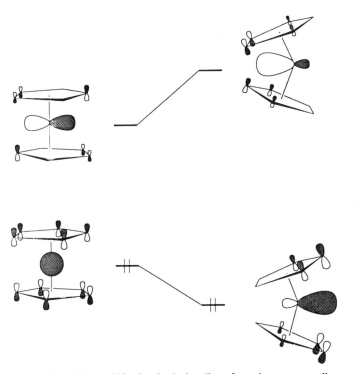

FIG. 7. Frontier orbital stabilization in the bending of a main group metallocene.

levels is larger than in stannocene owing to the higher $s-p$ separation in silicon versus tin, thus limiting the bending. [A hypothetical carbocene, "$(\eta^5\text{-}R_5C_5)_2C^{II}$," can be expected to prefer a linear geometry.] For decaphenylstannocene, however, the energy gain on bending is surpassed by the loss through steric repulsion if the phenyl rings on different Cp's are forced to approach each other past their van der Waals minimum. At a simple 1-electron level such a steric interaction of, say, two phenyl hydrogens can be expressed in a 4-electron two-orbital interaction, as sketched in **13**.

$$\text{C-H}_\sigma \quad \text{C-H}_\sigma$$

13

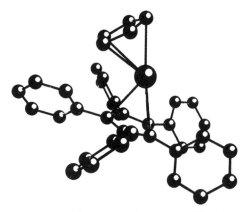

FIG. 8. Crystal structure of pentaphenylstannocene, $Ph_5C_5SnC_5H_5$ (120).

The space-filling plot of decaphenylstannocene in Fig. 3 clearly illustrates that there is no room for bending, which if it had to occur would require further separation of the cyclopentadienyls, that is, lengthening the Sn—Cp bonds from their preferred minimum position. The tin–ring centroid distance is within the same range as found for other stannocenes (112). A molecular mechanics study on the canting of the phenyl groups shows that the double-opposed paddle wheel arrangement serves to minimize steric interactions (121). The system is, however, able to assume the expected bent arrangement if one pentaphenyl-Cp ligand in decaphenylstannocene is replaced by a less sterically demanding, for example, normal, cyclopentadienyl moiety. Such is the case in pentaphenylstannocene (120), whose structure is given in Fig. 8 (hydrogens omitted for clarity).

2. Decaphenylgermanocene and Decaphenylplumbocene

It is not yet unequivocally proven if the analogous germanium and lead derivatives exhibit the same high symmetrical structure as decaphenylstannocene. Up to now, their low solubility has prevented the growth of suitable crystals for X-ray analysis. X-Ray powder diffraction patterns of decaphenylgermanocene and -stannocene each show prominent peaks at similar positions, but the patterns are not rich enough to establish isomorphism explicitly (the lead derivative failed to diffract) (120).

Another, perhaps clearer indication for isostructural character is given by solid-state, ^{13}C cross-polarization magic angle-spinning (CPMAS) NMR. A comparison of the high-field spectra revealed strong similarities for the decaphenylgermanium and tin compounds (the amorphous lead complex produced broader lines) (120).

3. Decaphenylferrocene

At first it may seem odd to expect a novel structural feature with iron as the central metal. The prototypical ferrocene structure (2) with its linear Cp–Fe–Cp D_{5d} or D_{5h} symmetrical arrangement (165) has so far showed no major changes (such as strong bending) on ligand substitution and iron oxidation.

Octaphenylferrocene also shows the normal structure with parallel, pentahapto-bonded cyclopentadienyl rings (48). It is, however, conceivable on examination of a space-filling model of a decaphenylmetallocene (see Fig. 3) that as the central metal gets smaller, the steric phenyl–phenyl repulsion will reach a point where no further shortening of the metal–Cp distance is possible. With tin being sandwiched the intraplane Cp–Cp distance is 480 pm (118), and with germanium it would still be about 450 pm. With iron, however, it decreases to about 330 pm assuming Ge—Cp and Fe—Cp distances as seen before (48,105,165). [In octaphenylferrocene, the voids from the two "missing" phenyl groups allow the remaining ones to tilt so as to enable the Cp planes an approach of 339 pm (48).]

The synthesis of what was assumed as "decaphenylferrocene" has been described (40) (Appendix), and no anomalies suggesting a deviation from normal structural behavior have been observed. There is, however, a recent report on the isolation of a linkage isomer of decaphenylferrocene (166). From ^{1}H- and ^{13}C-NMR studies it is deduced that one Ph_5C_5 ring shows the normal η^5-C_5 mode, whereas the other coordinates with a phenyl ring to the iron in an η^6-fashion giving rise to a zwitterionic structure as shown in Fig. 9. Personally, we have our doubts that a "decaphenylferrocene" can exist, as postulated in Ref. 40. All our attempts to prepare $(\eta^5$-$Ph_5C_5)_2Fe$ have failed, including efforts to follow the route given previously (40). From steric considerations, we feel strongly that the linkage isomer in Fig. 9 may be the only possible ligand arrangement for two pentaphenylcyclopentadienyls around the small iron center.

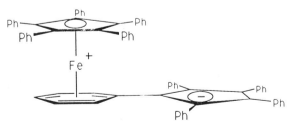

FIG. 9. Zwitterionic structure of a linkage isomer of decaphenylferrocene as deduced from NMR data (166).

4. *Pentabenzylcyclopentadienylthallium and Pentabenzylcyclopentadienylindium*

The indium and thallium derivatives of pentabenzylcyclopentadienyl crystallize in two modifications depending on the conditions. Fast, kinetically controlled crystallization affords needles, whereas slow, thermodynamically controlled crystallization yields parallelepipeds for both metals. X-Ray structural analyses of both thallium allotropes and the indium parallelepipeds show that both modifications represent novel structural features. Both the needles and the parallelepipeds consist of monomeric $(PhCH_2)_5C_5M$ species (M = Tl, In) with a covalent Cp—M bond (*107–109*). Most other structurally characterized Cp–Tl or Cp–In complexes exhibit a polymeric or at least oligomeric arrangement with rather ionic Cp–M interactions in the solid state. H_5C_5Tl and H_5C_5In (*143*), MeH_4C_5In (*145*), Me_5C_5Tl (*167*), $(Me_3Si)H_4C_5Tl$ (*168*) and (Me_3Si)-H_4C_5In (*169*), $(NC)_2C=C(CN)H_4C_5Tl$ (*170*), $[Ph(Me)_2Si]Me_4C_5Tl$, as well as $[PhCH_2(Me)_2Si]Me_4C_5Tl$ (*171*) adopt polymeric structures of zig-zag chains with equidistant Cp–M distances in the solid state, as depicted in **14**. $(Me_3Si)_2H_3C_5Tl$ forms a hexameric "doughnut" molecule (*169*), and Me_5C_5In features an octahedral "hexameric cluster" (*172*).

14

The needle modification of $(PhCH_2)_5C_5Tl$, on the other hand, consists of an almost linear chain (Cp ··· Tl—Cp 176°) of monomeric molecules with covalent thallium–cyclopentadienyl interactions (Tl—Cp 249 pm, see Fig. 10). The nonbonded Cp ··· Tl distance is 488 pm (*108*). There is a certain similarity to the structure of $(\eta^5\text{-}H_5C_5)Co(\mu\text{-}\eta^5\text{-}Me_4C_2B_2C)Tl$ (*173*), with an isolobal diborolyl ligand instead of Cp, which also exists as discrete molecules in a close to linear chain arrangement.

The packing of the monomeric species has to be attributed to ligand–ligand interaction [cf. geared stacking of benzene in its crystal structure (*174*)]. In a first approximation, the same kind of organic envelope interaction is also responsible for the dimeric arrangement of the monomeric molecules found in the parallelepiped modification of pentabenzylcyclopentadienylthallium (*107*) and -indium (*109*) as shown in Fig. 11. However, metal–metal interaction may be also present (shaping the bottom of the potential energy well, in line with the observation that the parallelepipeds are apparently the thermodynamically more stable allotrope). Both

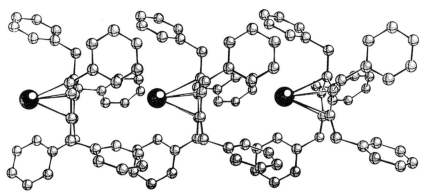

FIG. 10.   Arrangement of the molecules in the needle modification of pentabenzylcyclopentadienylthallium, (PhCH$_2$)$_5$C$_5$Tl, based on X-ray data (*108*). Hydrogens are omitted for
clarity.

centrosymmetric dimers feature a relatively covalent metal–cyclopentadienyl interaction, a metal–metal contact of 363 pm, and a Cp(centroid)–
M–M angle of 131.8° (Tl) or 136.5° (In).

The metal–metal contacts are the second structural highlight in these
pentabenzylcyclopentadienyl derivatives. Unambiguous Tl^I—Tl^I or
In^I—In^I bonds are not yet known in molecular complexes, as they are, for
example, for the related Ge^{II} or Sn^{II} in the solid state of {[(Me$_3$Si)$_2$CH]$_2$M}$_2$

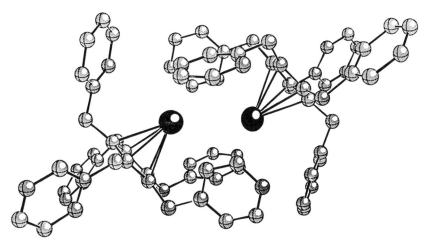

FIG. 11.   Crystal structure of the parallelepiped allotrope of pentabenzylcyclopentadienylthallium, (PhCH$_2$)$_5$C$_5$Tl, and its isostructural indium analog (*107,109*). Hydrogens are
omitted for clarity.

(12) (*163,175*). The question of possible metal–metal interactions in these dimeric forms was tackled theoretically (*176,177*). The influence of the ligand geometry on Tl^I—Tl^I or In^I—In^I bonding in $\{(PhCH_2)_5C_5M\}_2$ (M = Tl, In) has been studied within the extended Hückel framework and interpreted with the help of a HTlTlH model complex. For large metal–metal separations the ligand environment, especially the ligand–M–M angle, is found to be the dominant factor in determining the extent of the bonding interaction between the metals. It is shown that a mixing in of empty p levels into the filled s combinations is the basis for bonding in this formally s^2–s^2 closed subshell interaction. HTlTlH and CpTlTlCp want to bend at Tl (or at In in the respective indium compounds) as a result of a near resonance in the energy of the $2b_u$ (σ^*) and $3b_u$ (π) orbitals (Fig. 12). The orbital interactions in the trans bending of HTlTlH are related to the ones in the pyramidalization of AH_3 systems (e.g., NH_3, CH_3^-) and the bending of AH_2 molecules (e.g., H_2O, H_2S, but also Cp_2Sn; cf. Fig. 7) (*164*). The orbitals guiding these deformations are sketched in Fig. 12.

The controlling orbital (HOMO, for both AH_3 and AH_2) is a lone pair centered on the A atom. This level decreases in energy with bending, mixing in more s character and a hydrogen contribution from an empty A—H σ^* orbital above, to give a stabilized filled orbital with A–H non-bonding character and an empty level which is destabilized and strongly A–H antibonding. The lone pair begins as pure p in the planar AH_3 or linear AH_2 geometry. The closer the higher σ^* combination to the p orbital, the stronger the tendency to bend. Essentially the same thing

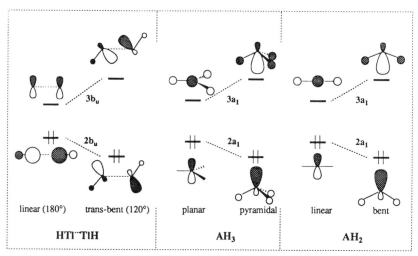

FIG. 12.   Analogy in the orbital diagram for the bending of HTl—TlH to AH_3 and AH_2 systems.

happens in HTlTlH. The only difference is that in the Tl case the upper orbital ($3b_u$) is of Tl p character, whereas in AH_3 and AH_2 it is the lower orbital. This can be traced to the long Tl–Tl separation. As soon as bending, and therefore mixing, begins, this distinction becomes unimportant.

The calculations on various systems show that a trans-bent geometry (as in Fig. 11) with a ligand–M–M angle close to 120° gives an optimum metal–metal overlap population at relatively long Tl^I–Tl^I or In^I–In^I separations. In this theoretical study it was concluded that molecular dimers where the trans-bent geometry has been observed, despite their long M–M contacts {such as 363 pm as in $[(PhCH_2)_5C_5Tl/In]_2$} can be assigned a definitive M^I—M^I bonding interaction (176,177).

In another theoretical study on metal–metal interactions in indium(I) and thallium(I) cyclopentadienyls (178), the authors interpret the short M—M contacts in the above dimers as weak donor–acceptor interactions, similar to the Sn—Sn bond in dimeric stannylenes (12) (163,175). However, they also attribute the general structural arrangement as being determined by "crystal packing forces."

5. Pentabenzylcyclopentadienylpotassium-tris(tetrahydrofuran)

The structure of $(PhCH_2)_5C_5K(thf)_3$ has been solved (179a). The potassium is η^5-bound to the planar cyclopentadienyl ring, and the three thf molecules complete the coordination polyhedron around potassium to the number eight. Two benzyl groups are directed to the side of the metal fragment. There are not enough comparative data, however, to discuss possible structural effects of the bulky ligand.

The only other Cp structure of potassium known is the one of the base-free $(Me_3Si)H_4C_5K$ (139) which consists of parallel, one-dimensional zig-zag strands (as in 14). Analogous to many CpTl and CpIn structures (see above) the potassium sits equidistantly between two Cp rings with an electrostatic interaction between the counterions. The tetrahydrofuran coordination surely plays a role in the monomerization of the structure of the title compound (although it is not clear to what extent). The K–Cp distances in both complexes, however, are essentially identical. The preparation and crystal structure of $Me_5C_5K(pyridine)_2$, which has a linear zig-zag chain structure, has been described recently (179b).

6. Decabenzylmetallocenes

There has been brief speculation in the literature on a possible metal lone pair–phenyl π interaction to explain in part the surrounding of the assumed lone pair position by three phenyl rings in the structures of the decabenzylmetallocenes of germanium, tin, and lead (see Fig. 2) (*105,106*). Also, in the discussion of the pentabenzylcyclopentadienylthallium and -indium structures (see Figs. 10 and 11) a comment was made on the protective shielding of the proposed lone pair space and the observed "shorter" phenyl carbon – metal distances implying the possibility of some kind of carbon π – metal interaction (*107–109, 146*).

These speculative suggestions were based on known and rather well-established Ge(II), Sn(II) (*180*), Ga(I). In(I), as well as Tl(I) arene interactions (*181*). The "lone pair – carbon π hypothesis" was probed by structural studies of model complexes such as [Ph(Me)$_2$Si]Me$_4$C$_5$Tl and [PhCH$_2$(Me)$_2$Si]Me$_4$C$_5$Tl (*171*) as well as (PhCH$_2$)$_5$C$_5$LuC$_8$H$_8$ (no lone pair electrons) (*98*). In the thallium complexes polymerization via the cyclopentadienyl rings of a neighboring molecule is preferred over the intra- or intermolecular interaction with a phenyl system. In [PhCH$_2$(Me)$_2$Si]Me$_4$C$_5$Tl (*171*) the phenyl ring is clearly turned away from the metal. Although the polymeric nature of these two thallium(I) complexes may not completely rule out a possible Tl(I) – arene interaction if they were monomeric, the lutetium complex (with a bent metallocene structure) also has one phenyl group oriented toward the lutetium atom, just in the area of maximum opening between the C$_5$ and C$_8$ ring planes. Since there are no lone pair electrons at the lutetium(III) center ($f^{14}d^0s^0p^0$ configuration), it must be concluded that the ordering of the five benzyl groups is determined only by steric effects (phenyl – phenyl repulsion, cf. **13**).

The half-sandwich transition metal complexes (PhCH$_2$)$_5$C$_5$Co(CO)$_2$ (*51*) and (PhCH$_2$)$_5$C$_5$Mn(CO)$_3$ (*182*) also have one phenyl ring in an orientation below the Cp plane approaching the metal carbonyl fragment, whereas four phenyls of the benzyl groups are situated above the cyclopentadienyl ring, or away from the metal moiety. In the pentabenzylcyclopentadienyl-potassium-tris(tetrahydrofuran) half-sandwich the ratio is 2 versus 3 (see above) (*179a*). Having all five benzyl groups lying on the same side of the cyclopentadienyl ring is apparently an unfavorable situation, as supported by model studies. However, the crystal structure of decabenzylferrocene, [(PhCH$_2$)$_5$C$_5$]$_2$Fe (given in Fig. 13), which was solved simultaneously and independently by two research groups (*98,182*), reveals just this orientation with all ten benzyl groups being directed away from the metal center. We would like to note the aesthetic aspects of supra-Cp structures, especially evident in the on-top views of decaphenylstannocene (Fig. 6b) and decabenzylferrocene (Fig. 13b).

The benzyl group orientation in the ferrocene derivative results from the small size of the iron center and consequently the closeness of the Cp rings. The ring planes are only 330 pm apart. Thus, repulsion and steric crowding from the methylene groups of the respective opposite pentabenzylcy-

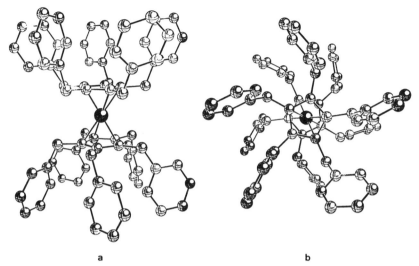

a   b

FIG. 13. Molecular structure of debenzylferrocene, [(PhCH₂)₅C₅]₂Fe, viewed parallel (a) and perpendicular (b) to the plane of the cyclopentadienyl rings (*98,182*). The hyⅾ ɔgen atoms have been omitted for clarity.

clopentadienyl ligand make it impossible for the benzyl groups to rotate around the $CH_2—C_5$ bond and to come to lie on the metal side of the Cp ring. Hence, the observed ordering of five benzyl groups on the same side of the ring may not be satisfying for the pentabenzyl moiety, but the apparent local minimum compromises steric constraints for the molecule as a whole (*98*). This structure supports our notion that a *decaphenyl*ferrocene in its "classic" form is not likely to exist (see Section IV,B,3).

The molecular structure of decabenzylferrocene also helped to explain the rather low solubility of this compound in toluene or benzene, in sharp contrast with the previously found excellent solubility properties of other decabenzylmetallocenes and the pentabenzyl-Cp thallium and indium complexes in these solvents. The steric crowding of the benzyl groups in the ferrocene and their limited mobility leave little space for the solvent arene molecules to interact and subsequently to dissolve the molecule. This is quite similar to the situation encountered in the pentaphenylcyclopentadienyl derivatives with their closely packed phenyl rings (see Fig. 3).

7. *Tris(trimethylsilyl)cyclopentadienyllithium Complexes*

The adducts of $(Me_3Si)_3H_2C_5Li$ with the nitrogen bases quinuclidine [$N(CH_2CH_2)_3CH$], tetramethylethylenediamine (tmeda, $Me_2NCH_2CH_2N-Me_2$), and pentamethyldiethylenetriamine [pmdeta, $Me_2NCH_2CH_2N-$

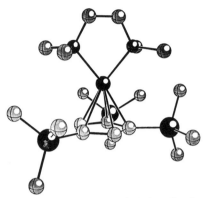

FIG. 14. Molecular structure of the tetramethylethylenediamine adduct of tris(trimethylsilyl)cyclopentadienyllithium, $(Me_3Si)_3H_2C_5Li(tmeda)$ (147).

(Me)CH$_2$CH$_2$NMe$_2$] were the first examples of Cp lithium complexes that were monomeric in solution, in the solid state, and in the gas phase (147). The same behavior was later confirmed as well for the analogous oxygen base adducts with thf, Et$_2$O, ^{n}Bu$_2$O, 1,4-dioxane, dme (dimethoxyethane, MeOCH$_2$CH$_2$OMe), diglyme (diethylene glycol dimethylether, MeOC$_2$H$_4$OC$_2$H$_4$OMe) (148). The lithium atom sits approximately above the centroid of the Cp ring in all structurally characterized complexes, as shown in Fig. 14 for the tmeda adduct as an example.

There are no structural data for H$_5$C$_5$Li or its adducts for comparison. Only nitrogen base adducts of indenyl- and fluorenyllithium have been described (183). It is noteworthy that $(Me_3Si)_3H_2C_5Li$ and its adducts are fairly soluble in hexane and other nonpolar solvents, despite their apparently still strong electrostatic Cp–Li interaction (147,148). For $(Me_3Si)_3H_2C_5Li$ this is interpreted as indicating a species of only oligomeric or even monomeric nature, depending on the concentration (9).

8. Hexakis(trimethylsilyl)magnesocene and Hexakis(trimethylsilyl)ferrocene

The two substituted cyclopentadienyl rings in hexakis(trimethylsilyl)-magnesocene and -ferrocene are nonparallel with a Cp–M–Cp angle of 171.5° [average for M = Mg (149)] and 173.9° [M = Fe (99)]. These bent metallocenic structures (Fig. 15a) are in contrast to parallel Cp arrangements observed otherwise for normal of even bulky Cp's [cf. octaphenylferrocene (48)]. The structure of normal magnesocene shows it to be isomorphic with ferrocene (184,185).

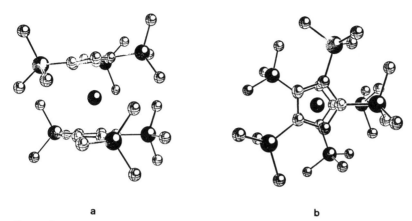

FIG. 15. Isomorphic hexakis(trimethylsilyl)magnesocene or -ferrocene, [(Me₃Si)₃-H₂C₅]₂Mg/Fe (*99,149*), viewed parallal (a) and perpendicular (b) to the plane of the cyclopentadienyl rings, with hydrogens omitted for clarity.

As in decaphenylstannocene, where the colinearity of the M – Cp vectors is a consequence of intraannular repulsion, so is the bending in hexakis(trimethylsilyl)magnesocene and -ferrocene. The source of the steric interaction in the latter two becomes clear from an on-top view of the molecule (Fig. 15b). The close to eclipsed conformation of the C_5 carbons [as in normal ferrocene (*165*)] allows for a staggered arrangement of four out of six Me₃Si groups. The remaining two, however, have to assume an eclipsed position that not only pushes the silicon atoms out of the Cp plane by 12° or more, but also bends the Cp – M vectors. Nevertheless, the conformation with two Me₃Si groups eclipsed, also observed in the germanium analog (*186*), seems to correspond to a global minimum. Other rotamers with staggered Cp rings and semieclipsed (or semistaggered) Me₃Si groups (see **18b**) are apparently of higher energy (*99,186*).

9. Tris(trimethylsilyl)cyclopentadienylthallium and Tris(trimethylsilyl)cyclopentadienylindium

The solid-state structure of the species (Me₃Si)₃H₂C₅M (M = Tl and In) is not yet known (*150,151*). However, solution studies [cryoscopic weight determinations (*151*), NMR] give strong indications for their monomeric nature with a dominant covalent Cp—Tl bonding interaction. It can be anticipated that the tendency to exist as a monomer is retained in the solid state, with results possibly similar to the pentabenzylcyclopentadienylthallium and -indium derivatives (see above).

C. Ring and Substituent Rotation Dynamics

In considering ring rotation dynamics in the following section, we refer primarily to the rotation of the Cp ring around the metal–ring (plane or centroid) vector (**15**). However, we also discuss rotations around Cp–substituent bonds (**16**).

15   **16**

1. C_5–M Rotations

The interest in the rotation of cyclopentadienyl rings (or aromatic organic rings in general) attached to metals (**15**) originated with the discovery of ferrocene itself (*3b*). Many methods have been employed to study the phenomenon, among them solid-state and solution NMR (*187*), dipole moment measurements (*188*), electron and X-ray diffraction techniques (*189,190*), mechanical spectroscopy (*117*), and last but not least molecular orbital calculations (*191*).

In normal ferrocene the barrier of ring rotation around the iron–Cp bond is only about 1–2 kcal/mol[6] (*187,192*). With ring substituents the rotational free activation enthalpy, $\Delta G\ddagger$,[7] is raised (in solution) to 13.1 kcal/mol in 1,1',3,3'-tetra(*tert*-butyl)ferrocene (*194*), to 11.0 ± 0.5 kcal/mol in the 1,1'3,3'-tetrakis(trimethylsilyl) analog (*195,196*), and to 12.6 and 10.7 ± 0.5 kcal/mol in the tetra(*tert*-butyl) and tetrakis(trimethylsilyl)cobalticinium ion, respectively (*197,198*). The single-crystal X-ray structure of tetrakis(trimethylsilyl)ferrocene apparently even shows a chiral C_2 symmetrical structure owing to a frozen ring rotation (*196*) (see below). Rigid rotational conformers were also found in 1,3-di(*tert*-butyl)-substituted cobalt half-sandwich complexes of the type $(^tBu_2H_3C_5)Co(PMe_3)_2$ and $[(^tBu_2H_3C_5)Co(PMe_3)_2X]^+$ (X = H, CH$_3$) in

---

[6] Energies are given here in the still more common unit kcal/mol instead of kJ/mol (1 cal = 4.184 joules).

[7] We note that in most cases the free activation enthalpy, $\Delta G\ddagger$ (estimated from the Eyring equation), is given as a measure for the rotational activation barrier. However, we also came across one example where the Arrhenius activation energy, E_a, had been calculated (*193*). Yet another manuscript listed both $\Delta G\ddagger$ and E_a and the activation enthalpy, $\Delta H\ddagger$, and activation entropy, $\Delta S\ddagger$ [see Mann *et al.* (*192*)].

the temperature range between 25° and 100°C (*199*). In a nickel half-sandwich, $(Me_3Si)_3H_2C_5Ni(PPh_3)Cl$, $\Delta G\ddagger$ was 10.5 kcal/mol (*200a*). A recent NMR bandshape analysis (*200b*) provided $\Delta G\ddagger$ values of 13.3 and 13.6 kcal/mol for $(1,3-R_2H_3C_5)_2Fe$ (R = *tert*-butyl, *tert*-pentyl) and 9.5 and 10.9 kcal/mol for the analogous ruthenocenes. In addition, ring rotation in 1,1′,3,3′-tetraphenylferrocene was found to be too rapid for measurement down to 173 K.

Unsubstituted bent titanocenes and zirconocenes, Cp_2ML_2, gave E_a values of approximately 2.0 kcal/mol in solution (*193*) and 4–5 kcal/mol in the solid (*201*), which increase to about 8–10 kcal/mol (depending on L_2) for *tert*-butyl-Cp (*202*) and 8.9 kcal/mol for bis(trimethylsilyl)-Cp (M = Ti, L = Cl) (*195*) (all in solution). In the solid, a *mono*methyl-substituted Cp already increases $\Delta G\ddagger$ to 16.9 kcal/mol (L = Cl) (*203a*).

The origin of the hindered rotation about the metal–ring bond is, of course, the interannular repulsion of the bulky ring substituents in the metallocenes, or a ring–substituent metal–ligand steric interaction in the half-sandwich complexes. Conversely, sterically demanding ligands on the metal (e.g., substituted olefins) can also raise the rotation barrier of an unsubstituted $\eta^5-H_5C_5$ group, as evidenced by a recent NMR study on cationic $[H_5C_5M(PPh_3)_2(\eta^2\text{-olefin})]^+$ complexes (M = Ru, Os) (*203b*). With the more highly substituted, even bulkier tris(trimethylsilyl)- or 4-*tert*-butyl-1,2-bis(trimethylsilyl)-substituted supra-Cp's one would expect the effect of a restricted ring rotation to be even more pronounced.

It is surprising at first that, from variable-temperature NMR studies, the free activation enthalpy was estimated to 11.1 ± 0.3 kcal/mol for hexakis(trimethylsilyl)ferrocene (*99*), 9.9 ± 0.5 kcal/mol for the cobaltocinium analog (*198*), 9.8 ± 0.5 kcal/mol for 4,4′-di(*tert*-butyl)-1,1′,2,2′-tetrakis(trimethylsilyl)ferrocene, and 8.8 ± 0.5 kcal/mol for the cobalticinium derivative (*204a*), values which are the same or even less than those for the tetrasubstituted analogs. We think this discontinuity can be explained by looking at the possibilities for the relative orientations of the two 1,3- or 1,2,4-substituted rings in a metallocene molecule. Although two disubstituted Cp's can assume a completely staggered conformation with respect to the ring substituents (**17**), this is not possible for trisubstituted Cp's in a metallocenic arrangement. There are a number of possibilities for the relative orientations for 1,2,4-substituted Cp rings in a molecule (*99,186*). Two lower energy configurations are sketched in **18a** and **b**. It becomes obvious that a fully staggered substituent conformation is no longer feasible. In **18a** we have four R groups staggered and two eclipsed with respect to each other, whereas, **b** shows all six groups semistaggered or semieclipsed to each other. However, **18a** is the minimum energy configuration judging from X-ray analyses: In the solid state, hexakis(trimethylsilyl)me-

17   18 a   b

tallocenes adopt structures which correspond very closely to **18a** (see Fig. 15b) *99,149,186*). Either way, because of the at least partly eclipsed steric interactions, the ground state energy in the hexakis metallocenes will be higher than in the related tetrakis complexes. Consequently, the activation barrier for the ring rotation, which involves an eclipsed orientation, will appear lower in the former.

In contrast to this situation encountered in silylated *linear* metallocenes, a recent study showed that *bent* hexakis(trimethylsilyl)zirconocene and hafnocene dichloride have substantially higher rotational, barriers than the analogous tetrakis(trimethylsilyl) complexes (*204b*): $\Delta G\ddagger = 11.0 \pm 0.2$ and 11.3 ± 0.2 kcal/mol for $[(Me_3Si)_3H_2C_5]_2MCl_2$ (M = Zr and Hf). No coalescence behavior in ^{1}H- and ^{13}C-NMR spectra was observed for $[(Me_3Si)_2H_3C_5]_2MCl_2$ (M = Zr, Hf) down to $-90°C$. Restricted $(Me_3Si)_3H_2C_5$ ring rotation was also observed at low temperature ($-80°C$) in the corresponding magnesocene, although no activation barrier has been determined (*149*). In the case of isopropyl-substituted Cp's, the free activation enthalpy has been estimated to 13.6 kcal/mol for the $^iPr_4HC_5$-ring rotation in octaisopropylferrocene (*35*).

Restricted ring rotations may also arise from nonbonded intermolecular interactions between the ring and a matrix. A mechanical spectroscopy study involving cyclopentadienyl platinum tripod complexes, $CpPtL_1L_2L_3$, with Cp = H_5C_5 and Ph_4HC_5, concluded that in the case of Ph_4HC_5 the ring is locked in place (at least in the temperature range from $-180°$ to $-160°C$) by interactions of the phenyl groups with the polystyrene matrix in which the samples were embedded (*117*).

In $(Ph_4HC_5)_2Fe$ and bent $(Ph_4HC_5)_2TiCl_2$ no evidence was found for a slowed rotation of the cyclopentadienyl rings down to $-95°C$ (*48,97*); a restricted rotation of the phenyl substituents was observed, however (see below). The other octaphenylmetallocenes with M = V, Cr, Co, and Ni did not lend themselves to NMR studies because of their paramagnetic nature. For the decaphenylmetallocenes, the inherently low solubility prevented solution studies; no variable-temperature solid-state NMR investigation has been done (120).

2. C_5-Substituent Rotations

There are several examples of substituted Cp's (including metal-free Cp systems) where hindered rotation about the substituent–C_5 bond (16) has been observed and studied in more detail. In the neutral pentaphenylcyclopentadienyl radical a small phenyl group oscillation about the planes orthogonal to the C_5 ring was deduced from EPR spectroscopy (205) and found to be in agreement with results from a gas phase electron-diffraction study of hexaphenylbenzene (206). However, in the solid state hexaphenylbenzene exhibits a propeller conformation with torsional angles of approximately 65° (207), corresponding well to the average value of 56° found for the phenyl canting (relative to the C_5 frame) in $Ph_5C_5Co(CO)_2$ (51). Detailed dynamic NMR studies of hexaarylbenzenes suggested an uncorrelated (one ring at a time) rotation of the aryl rings. Transition states involving more than one ring coplanar with the benzene ring are apparently energetically unfavorable (208).

A brief study on the 3,4 ring flipping in 3,4-di(ortho-tolyl)-2,5-diphenyl-cyclopentadienone yielded an activation barrier, $\Delta G‡$, of about 23 kcal/mol (209). The static and dynamic stereochemistry of tetra(ortho-tolyl)cyclopentadienone has been examined theoretically and by a variable-temperature ^{1}H- and ^{13}C-NMR study, with the results being interpreted in terms of the tetracyclone skeleton having C_{2v} symmetry with the four aryl rings perpendicular to the plane of the cyclopentadienone. Low and high temperature coalescence are observed and viewed as uncorrelated one-ring rotations of the aryl groups, with $\Delta G‡$ estimated at 20 kcal/mol (210).

In bent, $(Ph_4HC_5)_2TiCl_2$, and linear $(Ph_4HC_5)_2Fe$, octaphenylmetallocenes a restricted phenyl ring rotation for the inner (2,3) rings has been observed below $-20°C$. The free enthalpy of activation was estimated to be the same (~ 9 kcal/mol), within experimental error, for both cases, leading to the suggestion that the motions of the phenyl groups of each C_5 ring are independent of those on the other ring in a bent metallocene geometry (48,97).

For decabenzylferrocene (Fig. 13) it was reasoned, based on qualitative model studies, that a 180° rotation about the (benzyl)CH_2—C_5 bond should not be possible (Section IV,B,6) (98). No stronger evidence from molecular mechanics calculations or dynamic NMR studies is, as yet, available. Ambient temperature NMR measurements of other bent-metallocene or half-sandwich pentabenzyl-Cp complexes (cf. Figs. 2, 10, and 11) showed an equilibration of all five benzyl groups, indicating a free substituent as well as Cp rotation at 20°C (*51,98,105–109,182*). A variable-temperature NMR examination coupled with a theoretical molecular mechanics study might be well worth doing.

The barrier ($\Delta G\ddagger$) of an isopropyl group rotation in the penta(isopropyl)-Cp—Mo(CO)$_3$Me complex has been calculated as 13 ± 1 kcal/mol (*211*). However, the same barrier in [^{i}Pr$_5$C$_5$CoC$_5$H$_5$]$^+$ was determined as 17.1 ± 0.2 kcal/mol. The related penta(neopentyl)cobalticinium gave $\Delta G\ddagger = 19.4 \pm 0.2$ kcal/mol (*14*). For hexakis(isopropyl)*benzene*, the free activation enthalpy was estimated to be larger than 22 kcal/mol (*212*), whereas measurements on the tricarbonyl chromium complex of hexakis(dimethylsilyl)benzene yielded 14.2 kcal/mol (*213*). For the less crowded silicon analog a lower barrier is reasonable. Moreover, the π complexation of (Me$_2$SiH)$_6$C$_6$ will also result in a decrease of the activation energy by at least 4–5 kcal/mol (*213,214*) owing to an increase in ground state energy resulting from nonbonded interactions between the ML$_n$ fragment and the Me$_2$SiH groups. The large difference in rotational barriers for the isopropyl groups on a C$_5$ and C$_6$ ring can be attributed to the difference in steric bulk arising from a change in the internal angles from 72° for C$_5$ to 60° for C$_6$ (*14*). Nevertheless, the activation energies for the ^{i}Pr$_5$C$_5$ molybdenum and cobalt complexes might have been expected to be closer together.

There is little doubt that the isopropyl groups in ^{i}Pr$_6$C$_6$ are tightly interlocked so as to give a conformationally rigid and stable cyclic tongue-and-groove arrangement, which can be isolated as such (*212,213*). We just note briefly that the two faces of the ^{i}Pr$_6$C$_6$ ring (or related examples) would be rendered nonequivalent on complexation to a metal fragment. Although hexakis(isopropyl)benzene resists complexation, presumably because of its bulky substituents (*213*), metal coordination can be achieved successfully in other cases (see above), coinciding with the destruction of the plane and center of symmetry, and leading to a chiral structure (more about this phenomenon in Section IV,D).

In [^{i}Pr$_5$C$_5$CoC$_5$H$_5$]$^+$ the ^{i}Pr–C$_5$ rotational process is described as slow at 20°C. The coalescence temperature is 65°C in the NMR for the signal of the diastereotopic methyl groups (*14*). Even in hexaethylbenzene, the barrier to rotation for the ethyl groups is still substantial. For the tricarbonyl chromium and molybdenum complex, $\Delta G\ddagger$ was determined as 11.5 kcal/mol (*215*).

In all cases of hindered substituent–C$_5$ bond rotations discussed above, an uncorrelated, stepwise or one-ring rotation rather than a correlated, synchronous mechanism has been proposed, which appears to be the norm with alkyl groups attached to planar frameworks (*216*).

D. *Chirality*

There are (at least) four ways a Cp–metal fragment can assume a chiral character: (a) the Cp ring can carry an optically active substituent (R*, **19**); (b) with two (or more) different substituents a Cp–metal combination can exist in two enantiomeric forms (**20**); (c) if free rotation around a Cp–substituent bond is no longer possible (see also Section IV,C,2) a Cp–metal chirality arises as well (**21**); (d) if the rotation about the metal–ring vectors

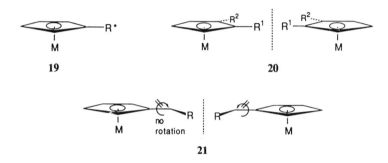

19   **20**

21

in a metallocene is frozen (cf. Section IV,C,1) because of bulky ring substituents, the metallocene can assume a chiral structure (**22**).

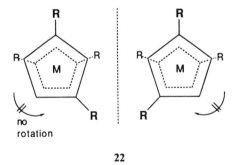

22

Option (a) is an interesting but clearcut and trivial case and will not concern us here (*217*). Possibility (b) and especially (c) and (d), on the other hand, specifically arise with certain examples of bulky Cp's. The substitution pattern in **20** (b), and the frozen substitutent or ring configurations in **21** (c) or (**22**) (d) effectively destroy the symmetry planes passing through the Cp ring and the metal center, thereby creating a chiral molecule. With both mirror images given in **20**, **21**, and **22** the reader can easily ascertain himself that they are nonsuperimposable. The drawing in **22** is rather schematic, though, and only meant to show the principle. We saw above that a frozen Cp–substituent rotation requires neighboring, rather bulky substituents.

We have to add a few comments on the concept of chirality in Cp–metal fragments according to **20**, **21**, and **22**. A very rapid site-exchange process involving metal–Cp bond dissociation would, of course, render an enantiomer distinction obsolete. Adding a second Cp ring of the same kind in **21** to form a metallocene destroys the chirality by creating a center of inversion at the metal (opposed R orientation, cf. decaphenylstannocene, Figs. 3 or 6) or a mirror plane passing through the metal between the Cp rings (same R directionalities). For **20** two enantiomers are retained in a metallocene, and the possibility of a meso form (achiral) is added. This is illustrated in **23**.

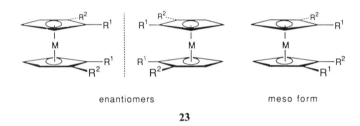

enantiomers   meso form

23

In addition, the reader may realize that axis of rotation can still be present in some chiral Cp–metal complexes (e.g., a C_2 axis in the enantiomeric forms in **22** and **23**, a C_5 axis in **24**). With rotation axes present the systems are not asymmetric, only dissymmetric (i.e., lacking mirror symmetry). This is, however, sufficient to induce the existence of enantiomeric forms (*218*). Moreover, it is known from numerous examples that chiral ligands with C_2 symmetry can provide for a higher stereoselectivity in (transition metal-catalyzed) reactions than comparable chiral ligands with a total lack of symmetry. The effect is explained by means of a reduced number of possible competing diastereomeric transition states (*218*). Hence, rotational symmetry elements may be advantageous for developing useful Cp–metal-based catalytic systems.

The separation of enantiomers in cases **20–22** is a separate problem, which we cannot discuss here. However, it can and has been done for type **20** racemic mixtures (*19,219*). Compared to the other three features attributable to supra-Cp's (stability, novel structures, and restricted rotations), the phenomenon of chirality, resulting specifically from bulky ligands (**21** and **22**), has received relatively little attention.

1. $R^1R^2R^3R^4R^5C_5$ – Metal Complexes (R^i Achiral)

In the following, we briefly present the type **20** chiral complexes with Cp ligands that also qualify as bulky according to our definition (see Section I). The condition for having a chiral $R^i_5C_5$ – M fragment ($i = 1,2,3,4,5$) is that out of five possible substituents R, at least three have to be different (R can also be hydrogen). In the special case of $R^1R^2_2R^3_2C_5$-M the 1-(R^1)-3,4-(R^2 or R^3) isomer is excluded. There are, of course, many examples for this kind of metallocenic chirality with Cp ligands containing just two large substituents (*220*), which, as pointed out earlier, could also be considered "bulky" but do not adhere to our strict definition. A review and introduction to the stereochemistry of metallocenes is given in Ref. *220*.

Examples for chiral type **20** supra-Cp metal complexes are $Ph_3RR'C_5$ – palladium compounds, $[(Ph_3RR'C_5)_2Pd]_2(\mu$-RC≡CR') (R/R' pair = 'Bu/ Me and Et/Me) (*80*); the 1,2,4 (*t,i,i*) isomer of 'Bu('Bu$_2$)H$_2$C$_5$TaCl$_4$ (*13*); the 1,2,3,4,5-('Bu,Me,Me,Et,Et) isomer of Et$_2$('Bu)Me$_2$C$_5$ – tungsten complexes (cf. Scheme 16) (*82*); and the 1,3,7,9-*tert*butyl substituted fulvalene derivatives (*221,222*). All are listed in the Appendix.

Furthermore, ethylene-bis(tetrahydroindenyl) complexes also exhibit a type **20** chirality (*19,219,223 – 225*) which was intentionally built in, unlike the previous cases. The bridge prevents rotation about the M—Cp bond. With Ti, Zr, and Hf as central metals the tetrahydroindenyl complexes are employed as catalysts in the polymerization of propylene, where the (ethylene) bridge-induced rigidity is necessary to enhance the isospecificity of the reaction, that is, the selectivity of the catalyst. The Cp$_2$M dichloride compounds listed in the Appendix are the immediate precursors to highly active, well-defined Ziegler – Natta catalysts. The active species is generated on addition of a trimethylaluminum/methylalumoxane mixture as a cocatalyst (*226 – 228*). During the last few years, these and other metallocene-based catalysts, primarily of titanium and zirconium, have become increasingly important industrially. They are regarded as a new generation of polyolefin catalysts, not necessarily replacing the conventional heterogeneous Ziegler – Natta catalysts but complementing them for specialty polymers. Metallocene catalysts, which can be employed homo- or heterogeneously, are apparently rapidly nearing their application in large-scale polyolefin production. Because of industrial interest, progress in this field is largely documented in the patent literature.

For stereospecific propylene polymerization, no enantiomer separation is necessary (*229*). The enantiomerically pure catalyst has been used in the enantioselective hydrooligomerization of α-olefins (*230*).

2. R_5C_5–Metal Complexes with Hindered R–C_5 Rotation (R Achiral)

In Section IV,C,2 octa- and pentaphenyl-, as well as penta(isopropyl)- and penta(neopentyl)-Cp, were cited as examples for hindered rotations about substituent–C_5 bonds. Half-sandwich metal complexes or *sym*-metallocenes with these ligands will be chiral when there is a barrrier for the interconversion of the enantiomers created by the phenyl canting or the isopropyl and neopentyl directionalities of the substituents (clockwise or counterclockwise) around the C_5 ring. For penta(isopropyl)- and penta(neopentyl)-Cp this difference in directionalities is commonly sketched (**24**) by showing the methine proton only and omitting the methyl or ethyl groups for clarity (*14*).

24

Although the barrier for phenyl flipping in Ph_4HC_5—M is probably too low (because of the void created by the missing phenyl group) to allow enantiomer separation, the barrier in Ph_5C_5–metal complexes has not been determined (*48,97*). Therefore, we simply note the chiral potential of this ligand and the fact that the molecular (solid-state) structure of pentaphenylstannocene in Fig. 8 depicts an enantiomer of this chiral *sym*-metallocene.

Penta(isopropyl)- and penta(neopentyl)-Cp cobalticinium salts, on the other hand, have a high enough barrier to interconversion ($\Delta G\ddagger$ 17.1 and 19.4 kcal/mol, respectively) with coalescence temperatures of 65° and 100°C, respectively, in the NMR (*14*). Thus, both enantiomers can coexist at 20°C with only a slow interconversion between them. In tricarbonylmethyl-penta(isopropyl)cyclopentadienyl molybdenum the activation energy has been estimated to 13 ± 1 kcal/mol (*211*) (cf. comment in Section IV,C,2).

One could imagine a complete freezing of substituent–C_5 rotations if vicinal substituents were tied into a ring system, somewhat similar to octahydrofluorene or related tricyclic systems (231) (see Appendix) or to the diborna-Cp sketched in **5** [except that **5** already contains chiral substituents (20)].

3. $(R^i{}_5C_5)_2M$ Complexes with Hindered C_5–M Rotation (R^i Achiral)

We saw in Section IV,C,1 that bulky substituents on a Cp ring in metallocenes (and also in half-sandwich compounds) raise the activation barrier for ring rotation about the M—Cp vectors. Although the barrier was still not high enough to prevent the rings from rotating completely [except maybe at low temperature and in a matrix; cf. the solid-state structure of tetrakis(trimethylsilyl)ferrocene (196)], the possibility exists of having a chiral metallocene with just two different achiral ring substituents (e.g., two R groups and three hydrogens, see **22**). If the sterically demanding R groups are placed in a 1,3 position on the ring, the energetically favored fully staggered R group rotamer in a metallocene represents a chiral array (see **22**).

Again, a method to prevent rotation completely is by linking the C_5 rings together to give a cyclophane- or metallocenophane-like species. This will, of course, automatically introduce a third ligand, along with the type **20** chirality based on three different ligands (see above). The ethylene-bridged bis(tetrahydroindenyl), so-called *ansa*-metallocenes (19,219,223–225) (cf. **4**, Appendix) provide an example of frozen rotation with three different substituents. Without the bridge, however, the barrier of rotation for bent bis(tetrahydroindenyl)metallocenes is likely to be too small to justify classification as a chiral system.

V

CONCLUSIONS

Research interest in supra-Cp's stems from a wide variety of thoughts with the major underlying concepts being stability, novel structures, rotation dynamics, and chirality. Cyclopentadienyl metal complexes with bulky substituents on the Cp ring show enhanced air and water stability compared to normal Cp analogs. In general, supra-Cp ligands increase the kinetic inertness for associative mechanisms by shielding the reactive metal and C_5 centers. The insulating or separatory effect of highly substituted Cp ligands with respect to intermolecular forces can lead to novel structural features. Furthermore, sterically demanding Cp ring substituents raise the activation barrier for rotations about the C_5-metal vector. Persubstituted Cp's can also exhibit hindered rotations about the substituent-C_5 bonds. Both modes of restricted rotations form the basis for chirality phenomena in the metal complexes, which are specifically attributable to the bulky ligand and its substituents.

APPENDIX:

TABULATION OF BULKY CYCLOPENTADIENYL METAL COMPLEXES

The tabulation compiles the metal complexes, method(s) of preparation, and characterization for the following bulky Cp's (in the given order):

Ph_5C_5 (pentaphenyl-Cp)
   Modifications of Ph_5C_5
Ph_4HC_5 (tetraphenyl-Cp)
   Modifications of Ph_4HC_5
$Ph_3H_2C_5$ (triphenyl-Cp): 1,2,4 isomers
                                     1,2,3 isomers
   Modifications of $Ph_3H_2C_5$
$(PhCH_2)_5C_5$ (pentabenzyl-Cp)
$(Me_3Si)_3H_2C_5$ [tris(trimethylsilyl)-Cp]
   Modifications of $(Me_3Si)_3H_2C_5$
$(Me_3Si)_2(^tBu)H_2C_5$ [bis(trimethylsilyl)($tert$-butyl)-Cp]
$^tBu_3H_2C_5$ [tri($tert$-butyl)-Cp]
$(^iBu)_2^tBuH_2C_5$ [di(isobutyl)($tert$-butyl)-Cp]
$(^iPentyl)_5C_5$ [penta(isopentyl)-Cp]
$(^tBuCh_2)HC_5$ [tetra(neopentyl)-Cp], modification of
$^iPr_5C_5$ [penta(isopropyl)-Cp]
$^iPr_4HC_5$ [tetra(isopropyl)-Cp]
$^iPr_3H_2C_5$ [tri(isopropyl)-Cp]
$^nPr_4HC_5$ [tetrapropyl-Cp], modification of
Et_5C_5 [pentaethyl-Cp]
   Modifications of Et_5C_5
Et_4HC_5 (tetraethyl-Cp)
   Modifications of Et_4HC_5
$Et_3H_2C_5$ (triethyl-Cp)
   Modifications of $Et_3H_2C_5$
$(MeS)_5C_5$ [pentakis(methylthio)-Cp]
$(MeS)_4HC_5$ [tetrakis(methylthio)-Cp], modification of
$(MeS)_3H_2C_5$ [tris(methylthio)-Cp], modification of
$(MeO_2C)_5C_5$ [penta(methoxycarbonyl)-Cp]
$(MeO_2C)_4HC_5$ [tetra(methoxycarbonyl)-Cp], modifications of
$(MeO_2C)_3H_2C_5$ [tri(methoxycarbonyl)-Cp]
Fulvalenes

Ethylene-bis(tetrahydroindenyl)-Cp
Octahydrofluorenyl and related tricyclic derivatives
Fulvenes, cyclopentadienylides
Mixed systems

The metals are divided into main group and transition metals and in each division are listed by increasing group number (on the periodic table). In the chemical formula of the complex, the bulky ligand is almost always abbreviated as **Cp**, with the respective supra-Cp given in the first column.

The method of preparation shows the principal reaction in an abridged form, with little consideration to stoichiometry. Reaction conditions (solvents, energy, Δ) are included only if they were considered crucial to the outcome. If a specific complex is modified further through, for example, ligand exchange, it is preceeded by a small or capital letter (in bold); in other words, a bold letter in the Methods of preparation column refers to a complex previously described. If a compound is further derivatized, a small letter is added to the letters of the respective starting material (xx**a**, xx**b**, xx**c**, . . ., depending if it is the first, second, or third product of this starting material which is part of a reaction sequence). For example, **gadaa** is derived from **gada**, in turn from **gad**, from **ga**, and from **g** as the initial educt; **gad** has been the fourth modification of **ga** (after **gaa**, **gab**, and **gac**), which will be again used as a starting material, somewhere below in the tabulation.

The listing of the spectroscopic methods of characterization may be useful if the reader searches for compounds to compare with his or her own results and may also serve as a measure of reliability for the assignment of the complex formula. The comment column indicates special features or points of interest for the complex or sections in the text where the respective compound is discussed in more detail.

We note again that the tabulation is not necessarily a (100%) complete list of all complexes for a respective ligand (see comment in Section I).

BULKY CYCLOPENTADIENYL METAL COMPLEXES[a]

Bulky Cp	Metal	Complex[b]	Method of preparation	Characterization[c]	Comments	Ref.
Ph5C5						
	Main group					
	Li	CpLi	Deprotonation of CpH with ^{n}BuLi, Li, Na, NaNH$_2$, K, or Cs	IR, UV, ^{1}H, ^{13}C	Greater stability in air than for corresponding unsubstituted Cp's noted	124
	Na	CpNa				124
	K	CpK(thf)				124
	Cs	CpCs				124
	Tl	CpTl	CpH + TlOEt	IR, MS	See section IV,B,2	146
	Ge	Cp$_2$Ge	CpLi + GeI$_2$ or GeCl$_2$(dioxane)	IR, Raman, ^{1}H, ^{13}C-CPMAS, MS, powder diffraction		39,120
	Sn	η^1-CpGeCl$_3$	Ox add. of CpCl to GeCl$_2$	X-ray, ^{1}H, ^{13}C, MS		232
		Cp$_2$Sn	CpLi/Na + SnCl$_2$	X-ray, IR, Raman, ^{1}H, ^{13}C, ^{119}Sn-CPMAS, ^{119}Sn-Mössbauer, MS, powder diffraction	S$_{10}$ symmetrical molecule with parallel Cp rings and stereochemically inert lone pair (cf. Figs. 3 and 6; Section IV,B,1) See Fig. 8	39,118,120
		CpSnC$_6$H$_5$	CpLi/Na + H$_5$C$_6$SnCl	X-ray, IR, ^{13}C-CPMAS, ^{119}Sn-Mössbauer, MS, powder diffraction		39,118,120
	Pb	η^1-CpSnCl$_3$	Ox add. of CpCl to SnCl$_2$	^{1}H, ^{13}C, MS		232
		Cp$_2$Pb	CpLi/Na + PbCl$_2$ or Pb(OAc)$_2$	IR, Raman, ^{13}C, ^{207}Pb-CPMAS, MS		39,120
	Transition metals					
	Ti	Cp$_2$TiCl$_2$	CpLi + TiCl, in xylenes, Δ		Synthesis in etheral solvents failed	40[d]
	Zr	Cp$_2$ZrCl$_2$	CpLi + ZrCl, in xylenes, Δ		Synthesis in etheral solvents failed	40
	Cr a	Li[CpCr(CO)$_3$]	Ox add. of CpLi to Cr(CO)$_6$	^{13}C		40
		[Et$_4$N][CpCr(CO)$_3$]	a + [Et$_4$N]Br			40
		Ph$_4$C$_5$H-(η^6-Ph)Cr(CO)$_3$	CpH + Cr(CO)$_6$	^{1}H	Aryl ring complexation	40
	Mo b	Cp$_2$Mo	PhC≡CPh + Mo(CO)$_6$ or diglyme-Mo(CO)$_3$	Magn. susc., density	Template synthesis, yield <5%	79
		[Cp$_2$Mo][Br$_3$]	b + Br$_2$	Magn. susc.	With Mg reduction back to b	79
		(CpMo(CO)$_3$)$_2$	CpH + K + Mo(CO)$_6$ + aq. Fe(NO$_3$)$_3$	IR	Extremely air sensitive, equilibrates with CpMo(CO)$_3$· in solution	233
	c	Li[CpMo(CO)$_3$]	Ox add. of CpLi to Mo(CO)$_6$	^{13}C		40
		[Et$_4$N][CpMo(CO)$_3$]	c + [Et$_4$N]B			40
		CpMo(CO)$_3$Me	c + MeI	^{13}C		40
		(CO)$_3$Mo(η^6-Ph-C$_5$Ph$_4$-η^5)FeCp	CpFeC$_5$H$_5$ (eba) + (OO)$_3$Mo(NCMe)$_3$		See also under Fe; Cp simultaneously η^5-Cp and η^6-Ph bonded	40

	Formula	Reaction	Characterization	Notes	Ref.
w d	Li[CpW(CO)$_3$]	Ox add. of CpLi to W(CO)$_6$	^{13}C		40
	[Et$_4$N][CpW(CO)$_3$]	d + [Et$_4$N]Br			40
	CpW(CO)$_3$Me	d + MeI	^{13}C		40
Fe e	(η^1-Cp)Fe(CO)$_3$	CpH + Fe(CO)$_5$	IR	Footnote in Ref. 79	79
	CpFe(CO)$_2$Br	Ox add. of CpBr to Fe(CO)$_5$	IR, ^{1}H, ^{13}C, X-ray	e utilized in catalytic study on hydroformylation of 1-octene (234)	40,95,46; 235
ea	CpFe(CO)$_2$H	e + NaBH$_4$	IR		95
	CpFe(CO)$_2$HgX (X = Cl, I)	ea + HgX$_2$	IR		95
	Hg[CpFe(CO)$_2$]$_2$	e + Na/Hg	IR		95
eb	CpFe(CO)$_2$(η^1-C$_5$H$_5$)	e + H$_5$C$_5$Na	IR		95
eba	CpFe$_5$R$_5$ (R = H, Me)	eb + Δ (R = H), CpLi + R$_5$C$_5$Fe(CO)$_2$ in xylenes/Δ, ox add. of CpH to [H$_5$C$_5$Fe(CO)$_2$]$_2$	^{1}H, ^{13}C, MS		40
				Trace amounts	95
ebaa	CpFe[C$_5$H$_4$C(O)Me]	eba + (MeCO)$_2$O/BF$_3$			40
ebaaa	CpFe[C$_5$H$_4$CH(OH)Me]	ebaa + NaBH$_4$/NaHCO$_3$			40
	CpFe[C$_5$H$_4$CH=CH$_2$]	ebaaa + Δ/alumina			40
	CpFeX(η^6-C$_6$H$_6$) (X = Br, I, [Cr(SCN)$_4$(NH$_3$)$_2$]]	e + C$_6$H$_6$ + AlCl$_3$, followed by KI or NH$_4$[Cr(SCN)$_4$(NH$_3$)$_2$]			95
	CpFe(CO)$_2$Me	Ox add. of CpLi + Fe(CO)$_5$ followed by MeI	^{13}C	Gives irreversible 1 electron ox	39,236
	[CpFe(CO)(μ-Co)]$_2$	e + (H$_5$C$_5$)$_2$Co	IR, MS	Doubtful assignment, cf. Section IV,B,3	236,95
	"Cp$_2$Fe"	e + CpLi in xylenes/Δ	^{1}H, MS		40
f	CpFe(η^6-Ph-C$_5$Ph$_4$)	Fe(CO)$_5$ + 2 Zn + 2 CpBr	IR, UV, ^{1}H, ^{13}C, MS, CV	Second Cp ring is η^6-Ph bonded, 50% yield, linkage isomer of "decaphenylferrocene" (see Fig. 9, Section IV,B,3)	166
	[CpFe(η^6-Ph-C$_5$Ph$_4$)][BF$_4$]	f + HBF$_4$	IR, ^{1}H, ^{13}C, MS		166
	CpFe(η^1-Ph$_4$C$_5$-Ph-η^6)Mo(CO)$_3$	eba (R = H) + (OC)$_3$Mo(NCMe)$_3$		See also under Mo	40
ec	CpFe(CO)$_2$K	e + K/ultrasound		Not isolated	237
eca	CpFe(CO)$_2$Et	ec + EtI/ultrasound			237
ecaa	[CpFe(CO)$_2$(C$_2$H$_4$)][PF$_6$]	eca + Ph$_3$CPF$_6$	^{1}H, ^{13}C	With NaBH$_4$, reverse reaction to eca	237
ecaaa	[CpFe(CO)$_3$][PF$_6$]	eca + CO			237
	[CpFe(CO)$_2$(NCCH$_3$)][PF$_6$]	eca + CH$_3$CN			237
	[CpFe(CO)$_2$(4-^{t}BuC$_5$H$_4$N)][PF$_6$]	ecaa + 4-^{t}Bu-pyridine			237
	[CpFe(CO)$_2$PR$_3$][PF$_6$] [R$_3$ = Me$_3$, Bu$_3$, Ph$_2$Me, (OMe)$_3$]	ecaaa + PR$_3$			237

(continued)

BULKY CYCLOPENTADIENYL METAL COMPLEXES (continued)

Bulky Cp	Metal	Complex[b]	Method of preparation	Characterization[c]	Comments	Ref.
Ph_4C_5 (cont.)		$[CpFe(CO)_2(CH_2CH_2PMe_3)][PF_6]$	ecaa + PMe$_3$			237
	Ru **g**	$CpRu(CO)_2Br$	Ox. add. of CpBr to Ru$_3$(CO)$_{12}$	X-Ray, IR, ^1H, ^{13}C, molecular weight, CV	**g** used in catalytic study on hydroformylation of 1-octene (234); kinetic study on CO displacement by P donor ligands (133)	40,236,238
		$CpRuC_5R_5$ (R = H, Me)	R$_5$C$_5$Ru(CO)$_2$I + CpLi in xylenes/Δ	^1H, ^{13}C, MS	Synthesis in etheral solvents failed	40
		"Cp$_2$Ru"	**g** + CpLi in xylenes/Δ	^1H, MS	Synthesis in etheral solvents failed	40
	ga	$CpRu(CO)LBr$ (L = PPh$_3$, PEt$_3$, P(OMe)$_3$, [P(OPh)$_3$])	**g** + L	IR, ^1H, ^{13}C, ^{31}P, MS, CV, EPR, X-ray (L = PPh$_3$)	**ga** allows reversible 1-electron oxidation[c]; kinetic study on CO displacement	236
	gaa	$[CpRu(CO)(L)(L')][PF_6]$ [L = PEt$_3$, P(OMe)$_3$], L' = CO, C$_2$H$_4$, MeC≡CMe]	**ga** + AgP$_6$/L'	IR, ^1H, ^{13}C, CV		113
		$[CpRu(CO)L_2][PF_6]$ [L = PEt$_3$, P(OMe)$_3$]	**gaa** (L' = CO) + L/ONMe$_3$(H$_2$O), Δ	IR, ^1H, ^{13}C	Allows reversible 1-electron oxidation but irreversible reduction	236
		$CpRu(CO)_2Me$	**g** + MeLi	IR, ^1H, ^{13}C, MS		236
	gab	$CpRu(CO)(PEt_3)Me$	**ga** (L = PEt$_3$) + MeLi	IR, ^1H, ^{13}C, ^{31}P, MS		236
	gaaa	$CpRu(CO)(PEt_3)R$ (R = COMe, CMe=CMe$_2$, or Et)	**gaa** (L = PEt$_3$, L' = CO or MeC≡CMe) + MeLi, or **gaa** (L = PEt$_3$, L' = C$_2$H$_4$) + [N^{n}Bu$_4$][BH$_4$]	IR, ^1H, ^{13}C, ^{31}P, MS		236
		$[CpRu(CO)(PEt_3)Br][SbCl_6]$	**gaa** (L = PEt$_3$) + [N(Ph'Br-p)$_4$][SbCl$_6$]	EPR		236
		$[CpRu(CO)(PEt_3)R][PF_6]$ (R = Me, Et, COMe, CMe=CMe$_2$)	**gab** or **gaaa** + AgPF$_6$	EPR		236
		$CpRu(CO)[P(OMe)_3]H$	Red of **gaa** [L = P(OMe)$_3$, L' = MeC≡CMe] with (H$_5$C$_5$)$_2$Co	^1H, MS		236
	gac	$[CpRu(CO)(\mu\text{-}CO)]_2$	**g** + (H$_5$C$_5$)$_2$Co	IR, MS		236
		$[CpRu(NO)(L)Br][PF_6]$ (L, see under **ga**)	**ga** + [NO][PF$_6$]	IR, ^1H, CV		239
	gaca	$[CpRu(CO)(NO)(PEt_3)][PF_6]_2$	**gaa** (L = PEt$_3$) + excess [NO][PF$_6$]	IR		239
		$CpRu(NO)(L)$ (L, see under **ga**)	**gac** + chemical red with (H$_5$C$_5$)$_2$Co or as a mixture with **gacb** on electrolytic red	IR, ^1H, ^{31}P, CV	With L = P(OPh)$_3$, reversible ox	239

Compound	Formula	Preparation	Spectroscopy	Notes	Ref.
gacb	$CpRu(NO)X_2$ (X = Br) (X = Br, I)	As a mixture with **gaca** on electrolytic red of **gac** or ox add. of X_2 to **gaca**	IR, CV		239
gad	$CpRu(CO)(CN^tBu)Br$	**ga** + CN^tBu	IR		239
gae, gada	$CpRu(CO)(L)Me$ [L = $P(OMe)_3$, $P(OPh)_3$, CN^tBu]	**ga** + Li[CuMe$_2$] or MeLi / **gad** + MeLi	IR		239 239
gaba, gaea, gadaa	$[CpRu(NO)(L)COMe][PF_6]$ [L = PEt_3, $P(OMe)$, $P(OPh)_3$, CN^tBu]	**gab**, **gae**, or **gada** + [NO][PF$_6$]	IR, ^{1}H, CV	Migratory insertion of CO in Ru—Me bond	239
	$CpRu(NO)(L)COMe$	1-electron red of **gaba**, **gaea**, and **gada**	EPR		239
Co h	$CpCo(CO)_2$	CpBr to K[Co(CO)$_4$], or CpBr + Co$_2$(CO)$_8$ + Zn/thf, or ox add. of CpH to Co$_2$(CO)$_8$	IR, MS / X-Ray, IR, ^{1}H, MS	Low yield	40,240,241 51
ha	$CpCo(CO)_2^-$	Electrochemical red of **h**	EPR	Radical anion	116
haa	$CpCo(CO)I_2$	**h** + I$_2$	IR		241
haaa	$CpCo(CO)[P(OR)_3]$ (R = Me, Ph)	**ha** + P(OR)$_3$ + (H$_5$C$_5$)$_2$Co	IR, CV		241
haaaa	$[CpCo(NO)(L)][PF_6]$ [L = $P(OMe)_3$, $P(OPh)_3$]	**haa** + [NO][PF$_6$]	IR, CV		242
	$[CpCo(\mu\text{-}NO)]_2$	Red of **haaa** with (H$_5$C$_5$)$_2$Co, or by potential electrolysis	IR, CV	Primary red product CpCo(NO)(L) detected by EPR	242
	$[(CpCo(\mu\text{-}NO))_2][PF_6]$	Ox of **haaaa** with [N$_2$PhF-p][PF$_6$]	IR		242
Rh i	$CpRh(CO)_2$	CpBr + [Rh(CO)$_2$Cl]$_2$ + Zn/thf, or CpNa + [Rh(CO)$_2$Cl]$_2$	IR, ^{1}H, ^{13}C, MS, X-Ray		51,241,243
j	$[CpRhBr(CO)_2][ZnCl_3]$	CpBr + [Rh(CO)$_2$Cl]$_2$ + Zn/benzene	—	Intermediate from synthesis of **i**	241
ia	$[CpRh(CO)_2]^-$	Electrochemical red of **i**	EPR	Radical anion	116
	$CpRh(CO)_2I$		IR		241
ib	$[CpRh(\mu\text{-}CO)]_2$	**ia** + (H$_5$C$_5$)$_2$Co	—		241
	$(CpRhBr_2)_2$	**i** + Br$_2$	—		241
iba	$CpRhBr_2[P(OMe)_3]$	**ib** + P(OMe)$_3$	—		241
ic, ibaa	$CpRh(CO)[P(OR)_3]$ (R = Me, Ph)	**i** + P(OPh$_3$)/cat., **iba** + CO + (H$_5$C$_5$)$_2$Co	IR, CV, IR, EPR of radical cationic complex (R = Ph)	cat = [(H$_5$C$_5$)$_2$Fe][PF$_6$].	241
	$CpRh(NO)[P(OR)_3]$	Red of **ica**/**ibaaa** with (H$_5$C$_5$)$_2$Co or by potential electrolysis	IR	Isolable	242
	$[CpRh(\mu\text{-}NO)]_2$	Red of **ib** with (H$_5$C$_5$)$_2$Co + NO, or **ibaaa** + [N^{n}Bu$_4$]I and red with (H$_5$C$_5$)$_2$Co	IR, CV	Gives reversible 1-electron ox and red	242

(continued)

BULKY CYCLOPENTADIENYL METAL COMPLEXES (*continued*)

Bulky Cp	Metal	Complex[b]	Method of preparation	Characterization[c]	Comments	Ref.
Ph_4C_5 (cont.)	Ni	$CpRh(CO)(EPh_3)(E = P, As)$	$j + EPh_3 + [N(PPh_3)_2][Mn(CO)_5]$	IR, CV	IR, EPR of radical cationic complex	241
		Cp_2Ni	Cp radical + (cdt)Ni or $(cod)_2Ni$	MS	High thermal stability, slightly paramagnetic	78
		$CpNi(\mu\text{-}\eta^3\text{-}C_3Ph_3)Ni(\eta^4\text{-}Ph_4C_4)$	$Ph_5C_4Al + Li + NiBr_2$	X-ray, IR, ^1H, ^{13}C	Substitution of Al in the C_4Al ring and formation of Cp, low yield; note: theoretical study of tilt position of $\mu\text{-}C_3R_3$ group (245)	244
	k	$[CpNi(\mu\text{-}X)]_2$ (X = Cl, Br)	Ox add. of CpX to $Ni(CO)_4$	IR		240
	ka	$CpNi(CO)X$ (X = Cl, Br)	k + CO	IR		240
		$CpNi(CO)I$	ka (X = Br) + NaI	IR		240
		$CpNiC_5H_5$	k (X = Br) + TlC_5H_5	MS, CV		240
	kaa	$[CpNi(NCMe)_2][BF_4]$	ka (X = Br) + $TlBF_4/MeCN$	IR, ^1H, ^{13}C, molecular weight		126
	kab	$CpNi(S_2CNMe_2)$	ka (X = Br) + NaS_2CNMe_2		Analogous n-Cp species cannot be isolated	126
	kaaa	$[CpNi(S_2CNMe_2)][BF_4]$	kaa + $[Me_2NC(S)]_2S$ or $[Me_2NC(S)]_2S_2$	X-Ray, IR, magn. susc., CV	Reversible chemical and electrochemical ox from kab to kaaa	126
	Pd l	$(CpPd)_2(\mu\text{-}PhC{\equiv}CPh)$	$PhC{\equiv}CPh + Pd(OAc)_2/MeOH$	X-ray, UV/VIS, MS, CV	Template synthesis (see Scheme 15), stable Pd(I) species	80,114,246, 115
	la	$[(CpPd)_2(\mu\text{-}PhC{\equiv}CPh)][PF_6]$	Electrochemical ox of l, l + $AgPF_6$	^1H, CV	Radical cation, source of diamagnetic Pd(II) complexes	115,247
	m	$CpPd(\eta^3\text{-}all)$ (all = allyl; 1- or 2-methylallyl; 1,1-dimethylallyl; 2-chloroallyl; 1-carbomethoxyallyl)	$CpNa + [(\eta^3\text{-}all)PdCl]_2$	^1H, MS	—	114
	lb/ma	$(CpPdX)_2$ (X = Cl, Br)	l + 2 HX, pyrolysis of m (all = chloroallyl, X = Cl)		—	80,114
		$(CpPd)_2(\mu\text{-}R^1C{\equiv}CR^2)$ ($R^1 = R^2 = H$, Et, p-tolyl, CO_2Me; $R^1 = Ph$, $R^2 = H$)	Red of lb/ma (X = Cl) with Zn in presence of $R^1C{\equiv}CR^2$		—	114
		$(CpPd)_2(\mu\text{-}CO)_2$	Red of lb/ma with Zn + CO	IR, UV/VIS, MS	Bridge displacement	114
		$CpPd(NO)$	l + NO	Molecular weight (Cl), UV/VIS		80
		$(CpPd)_2HX$ (X = Cl, Br)	l + HX	MS, ^1H (Cl)	Tentatively formulated as $CpPd(\mu\text{-}H)(\mu\text{-}Cl)PdCp$	80

	Complex	Preparation	Methods	Ref.	Comments
Iasa	CpPd(PMe₂Ph)Cl	Ib/ma (X = Cl) + PMe₂Ph	UV/VIS, MS, ¹H	80,114	Pd(II), reversible ox or red to
	[CpPd(L₂)][PF₆] (L = PPh₃; L₂ = dppe, cod, bipy, cot, dbcot, nbd)	Ia + L₂, or electrochemical ox of I in presence of L or L₂	EPR	115,247	Pd(III) or stable Pd(I) π radicals
	CpPd(L₂) (L₂ = cod, dbcot, nbd)	Electrochemical red of Ia	CV, EPR (L₂ = cod)	115,248	First isolable mononuclear Pd(I) species
Iasaa	CpPd(dbcot-OCOPh)	Iaa (L₂ = dbcot) + benzoylperoxide, (PhCO)₂O	IR, ¹H, CV	248	dbcot-OCOPh π,σ-bonded to Pd(I)
	CpPd(L₂-OH) (L₂ = nbd, dbcot, cod)	Column chromatography of Iaaa and Iaa	¹H, FD-MS, CV, ¹³C (L₂ = nbd)	248	Radical reaction with H₂O, chlorinated hydrocarbons, L₂-OH π, σ-bonded
	CpPdCl(π acid) (π acid = CO, C₂H₄, allene)	Ib (X = Cl) + π acid	IR, ¹H	114	Reversible reaction
	CpPdCl(L) [L = PMe₃, PMePh₂, PEtPh₂, PPh₃, P(OMe)₃, P(OEt)₃, PPh₃, P(OPh)₃]	m (all = chloroallyl) + L	¹H	114	Pd(II) complexes *not* synthetically accessible from CpNa + [PdCl₂(PR₃)]₂
Pt	CpPt(η³-allyl)	CpNa + [(η³-allyl)PtCl]₂	¹H, MS	114	
Lanthanides					
Lu	Cp₂LuCl(thf)	CpNa + LuCl₃ in a 2:1 or 1:1 ratio		249	
	CpLuCl₂			249	

Modifications of Ph₅C₅

Ph₄(RPh′)C₅ (Ph′ = C₆H₄)

R = p-ᵗBu

	Complex	Preparation	Methods	Ref.	Comments
Li	CpLi	Deprotonation of CpH with ᵗBuLi	¹H, ¹³C	43	
Na	CpNa	Deprotonation of CpH with NaNH₂	¹H, ¹³C	43	
Tl	CpTl	CpH + TlOEt	IR, MS	146	
Ge	Cp₂Ge	CpLi + GeI₂	IR, ¹H, MS, powder diffraction	43	
Sn	Cp₂Sn	CpLi + SnCl₂	IR, ¹H, MS, powder diffraction	43	
Pb	Cp₂Pb	CpLi + Pb(OAc)₂	IR, ¹H, MS, powder diffraction	43	
Fe	CpFe(CO)₂Br	Ox add. of CpBr to Fe(CO)₅	IR, ¹H, ¹³C	46	

R = p-Me

	Complex	Preparation	Methods	Ref.	Comments
Fe	CpFe(CO)₂Br	Ox add. of CpBr to Fe(CO)₅	IR, ¹H, ¹³C	46	
W	CpW(CO)(μ-CO)₂W(C₅H₅) (η-PhC≡CPh)	W(CO)(PhC≡CPh)₃ + W(≡CPh′Me)(CO)₂(C₅H₅)	X-ray, IR, ¹H, ¹³C	250	Template synthesis

R = p-Et

	Complex	Preparation	Methods	Ref.	Comments
Fe	CpFe(CO)₂Br	Ox add. of CpBr to Fe(CO)₅	IR, ¹H, ¹³C	45	

(continued)

359

BULKY CYCLOPENTADIENYL METAL COMPLEXES (*continued*)

Bulky Cp	Metal	Complex[b]	Method of preparation	Characterization[c]	Comments	Ref.
$Ph_3(RPh')_2C_5$ R = p-MeO, p-Me, or p-Br	Pd	$(CpPd)_2(\mu\text{-}RPh'C≡CPh'R$	"[endo-$Ph_4(MeOC_4)Pd(OAc)$" + $RPh'C≡CPh'R$	UV/VIS, ^{1}H, MS	Template synthesis	80
$(RPh')_3(R'Ph')_2C_5$ R = R' = p-Me[= penta(p-tolyl)-Cp]	Pd **n**	$(CpPd)_2(\mu\text{-}RPh'C≡CPh'R)$	$Pd(OAc)_2 + RPh'C≡CPh'R'$ or "[endo-(p-MePh')_4(MeOC_4]$ $Pd(OAc)$" + $RPh'C≡CPh'R'$	UV/VIS, ^{1}H	Low yield, <5% (see Scheme 15); high yield, 50%; noted to be considerably more soluble than complexes with unsubstituted Ph_5C_5	80
		$CpPd(NO)$	**n** + NO	IR, UV/VIS, MS	Bridge displacement	80
		$(CpPd)_2HCl$	**n** + HCl	UV/VIS, ^{1}H	Tentatively formulated as $CpPd(\mu\text{-}H)(\mu\text{-}Cl)PdCp$	80
		$(CpPdCl)_2$	**n** + 2 HCl	^{1}H		80
R = p-Me, R' = p-MeO	Pd	$(CpPd)_2(\mu\text{-}R'Ph'C≡CPh'R')$	"[endo-(p-MePh')_4(MeOC_4]$ $Pd(OAc)$" + $R'Ph'C≡CPh'R'$	UV/VIS, ^{1}H	Template synthesis	80
$[p\text{-}RC(O)Ph']_5C_5$ R = Me, ^nPr, $^nC_5H_{11}$, or $^nC_7H_{15}$	Na	CpNa	$CpH + NaNH_2$	^{1}H, ^{13}C	See Scheme 23 for CpH synthesis	251
	Tl	CpTl	$CpH + TlOEt$	^{1}H, ^{13}C, MS		137
R = EtO	Na	CpNa	$CpH + NaNH_2$	^{1}H, ^{13}C	See Scheme 24 for CpH synthesis	251
	Tl	CpTl	$CpH + TlOEt$	^{1}H, ^{13}C, MS		137
R = $^nC_5H_{11}O$	Tl	CpTl	$CpH + TlOEt$	^{1}H, ^{13}C, MS	Mesogene (cf. Scheme 24, Fig. 5)	137
Ph_4HC_5	Main group					
	Li	CpLi	Deprotonation of CpH with ^nBuLi, Na, or K in toluene; workup with thf; also deprotonation with NaH/Me_2SO or $NaNH_2$ in thf	^{1}H, ^{13}C		43,48
		$CpLi(thf)_3$		^{1}H, ^{13}C		43
	Na	$CpNa(thf)_3$		—		43
		CpNa				95
	K	$CpK(thf)_1$		^{1}H, ^{13}C		43
		$CpK(thf)_{0.5}$		—		96
	Tl	CpTl	$CpH + TlOEt$	IR, MS		146

360

	Compound	Synthesis	Characterization	Notes	Ref.
Ge	Cp$_2$Ge	CpLi + GeI$_2$	IR, Raman, ^{1}H, MS, powder diffraction	See Section IV,A,1	43
Sn	Cp$_2$Sn	CpLi + SnCl$_2$	IR, Raman, ^{1}H, MS, powder diffraction ^{119}Sn-CPMAS	See Section IV,A,1	43, 39
Pb	Cp$_2$Pb	CpLi + PbCl$_2$ or Pb(OAc)$_2$	IR, Raman, ^{1}H, MS	See Section IV,A,1	43
Transition metals					
Ti o	Cp$_2$TiCl(thf)$_{1.5}$	CpNa + TiCl$_2$ or TiCl$_3$	X-Ray, CV, EPR	No evidence for slowed Cp–Ti rotation above −95°C, $\Delta G\ddagger$ for Ph–C$_5$ rotation 9.6 ± 0.5 kcal/mol	97
	Cp$_2$TiCl$_2$(thf)$_{1.5}$	Ox of o with AgCl	^{1}H		97
	Cp$_2$Ti(CO)$_2$	Red of o with Na-naphthalide + CO	IR		97
V	Cp$_2$V	VCl$_3$(thf)$_3$ + Zn + CpK	X-Ray, CV, magn. susc., EPR		96
Cr	Cp$_2$Cr	CpK + CrCl$_2$	X-Ray, CV, magn. susc., EPR		96
	[Cp$_2$Cr][PF$_6$](CH$_2$Cl$_2$)$_{0.5}$	Ox of neutral complex with AgPF$_6$	EPR		96
p	Ph$_3$HCH-(η^6-Ph)Cr(CO)$_3$	CpH + (pyridine)Cr(CO)$_3$/BF$_3$(OEt$_2$)		Aryl ring complexation	252
	K[CpCr(CO)$_3$]	p + Me$_3$COK		Rearrangement from η^6-Ph to η^5-C$_5$ on deprotonation of p	252
Mo q	[CpMo(CO)$_3$]$_2$	CpK + Mo(CO)$_6$ + aq. Fe(NO$_3$)$_3$	IR		233
	CpMo(CO)$_2$(L$_2$-P,P') [L$_2$-P,P' = 2,3-bis(diphenylphosphino) maleic anhydride]	q + L$_2$, hv	X-Ray, IR, EPR	L$_2$-P,P' indicates P coordination of L$_2$ to Mo; $\Delta H\ddagger$ for Cp–Mo ring rotation 2.2 kcal/mol	233,253
r	(η^4-CpH)Mo(C$_5$H$_5$)(L)H [L = P(OMe)$_3$]	(PhC≡CPh)Mo(C$_5$H$_5$)(L)Me + PhC≡CPh	X-Ray, ^{1}H	CpH = 1,3,4,5-Ph$_4$C$_5$-1,3-diene	254
	CpMo(C$_5$H$_5$)H$_2$	r − L	^{1}H	H-transfer from CpH to Mo	254
s	Hg[CpMo(CO)$_3$]$_2$	CpLi + Mo(CO)$_6$ + Hg(CN)$_2$	IR, ^{95}Mo, ^{199}Hg-NMR	Mo–Hg bond	255
	CpMo(CO)$_3$HgX (X = Cl, Br, I)	s + HgX$_2$	IR, ^{95}Mo, ^{199}Hg-NMR		255
Fe	Cp$_2$Fe; [(d_5-Ph)$_2$Ph$_2$HC$_5$]$_2$Fe	CpNa + FeCl$_2$ or ox add. of CpH to Fe(CO)$_5$	X-Ray, UV, ^{1}H, ^{13}C	No evidence for restricted Cp–Fe ring rotation above −95°C, $\Delta G\ddagger$ 9 ± 1 kcal/mol for Ph–C$_5$ rotation, below −20°C ordering of phenyl–ring conformations.	48,95,95,245
	[Cp$_2$Fe][PF$_6$]	Ox of neutral complex with AgPF$_6$	UV		48
	CpFeC$_3$H$_5$	Ox add. of CpH to [C$_3$H$_5$Fe(CO)$_2$]$_2$ or PhC≡CPh + C$_3$H$_5$Fe(CO)$_2$Me	IR	Cp$_2$Fe obtained as side product	95, 257

(continued)

BULKY CYCLOPENTADIENYL METAL COMPLEXES (*continued*)

Bulky Cp	Metal	Complex[b]	Method of preparation	Characterization[c]	Comments	Ref.
Ph₄HC₅ (*cont.*)		CpFe(CO)₂Br	Ox. add. of CpBr to Fe(CO)₅	IR	Low yield	95
		[CpFe(CO)(μ-CO)]₂	Ph₄C₄C=N₂ + Fe₂(CO)₉ room temperature, hexane		Presumably hydride abstraction of the diazoalkane from the solvent to give Cp	258,259
		CpFe(CO)₃	Ph₄C₄C=N₂ + Fe₂(CO)₉ 124°C, octane	IR, ¹H, MS	Ph linked to Fe via acyl group (?)	259,260
	Co	Cp₂Co	CpK + CoBr₂	X-Ray, CV, magn. susc., EPR		96
		[Cp₂Co][PF₆](CH₂Cl₂)₀.₅	Ox of neutral complex with AgPF₆	EPR		96
	Rh	CpRh(η⁴-cod)	(η³-Ph₄C₄-4-Me)Rh(cod) + Δ	¹H, MS	Ring expansion of Ph₄-cyclobutadienyl	94
	Ni	Cp₂Ni	CpK + [Ni(NH₃)₆]Cl₂	X-Ray, CV, magn. susc., EPR		96
		[Cp₂Ni][PF₆](CH₂Cl₂)₀.₅	Ox of neutral complex with AgPF₆	EPR		96
	Pt	CpPtMe₃	—	—	*No* Cp–Pt ring rotation observed between −160° and −180°C	117
	Au	η¹-CpAuPPh₃	CpLi + ClAuPPh₃, or deprotonation of CpH with NaOH or K₂CO₃ + [(Ph₃PAu)₃O][BF₄]	X-Ray, IR, UV, ¹H		261,262
Modifications of Ph₄HC₅						
Ph₄(RO)C₅ R = H	Mn t	CpMn(CO)₃	Ph₄C₄C=O + Mn₂(CO)₁₀ + H₂O	IR		263
	Ru	CpRu(CO)₂X (X = Cl, O₂CCF₃, O₂CCH₃)(X = H)	(η⁴-Ph₄C₄C=O)Ru(CO)₂NMe₃ + HX	¹H, ¹³C		264
			(η⁴-Ph₄C₄C=O)Ru(CO)₃ + H₂O/− CO₂	IR		265
		[(Ph₄C₅O)Ru(CO)₂H]₂, also [(Ph₂(p-ClPh)C₅O)Ru(CO)₂H]₂	(η⁴-Ph₄C₄C=O)Ru(CO)₃ + Δ, propanol	IR, ¹H ¹³C	C₅—O—H—O—C₅ bridge and Ru—H—Ru bridge	266
		[(Ph₄(p-ClPh)C₅O)Ru(CO)₂)]²⁻ Ru(CO)₃ + Δ, propanol	[η⁴-Ph₄(p-ClPh)C₄C=O]	X-Ray, ¹H		266
	Co u	CpCo(Ph₄C₄C=O)	(Ph₄C₄C=O)₂Co₃ + HCl	IR	Proteolytic equilibrium, see **8** and **9**; also with perfluorinated Ph substituents	263
		CpCo(C₅H₅)	(Ph₄C₄C=O)Co(C₅H₅) + H⁺	IR, UV, ¹H, p*K*ₐ		267,268
	Rh	CpRh(C₅H₅)	(Ph₄C₄C=O)Rh(C₅H₅) + H⁺	IR UV, ¹H, p*K*ₐ		267,268
		(CpRhCl₂)₂	[(η⁴-Ph₄C₄C=O)RhCl₂] + HCl	¹H		264

362

	Complex	Reaction	Characterization	Comments	Ref.
R = Me					
Mn	$CpMn(CO)_3$	$t + CH_2N_2$	IR		263
Ru	$CpRu(CO)_2I$	$(\eta^4\text{-}Ph_4C_4C{=}O)Ru(CO)_2NMe_3 + MeI$	$^1H, {}^{13}C$		264
Co	$CpCo(Ph_4C_4C{=}O)$	$u + CH_2N_2$	IR		263
R = MeC(O)					
Ru	$CpRu(CO)_2Cl$	$(\eta^4\text{-}Ph_4C_4C{=}O)Ru(CO)_3 + MeC(O)Cl$	IR, $^1H, {}^{13}C$		264
Co	$CpCo(Ph_4C_4C{=}O)$	$u + (CH_3CO)_2O$	IR		263
Rh	$(CpRhCl_2)_2$	$[(\eta^4\text{-}Ph_4C_4C{=}O)RhCl_2]_2 + MeC(O)Cl$	IR, ^1H	Reverts to original complex within hours	264
R = PhC•H(OMeC(O))					
Ru v	$CpRu(CO)_2Cl$	$(\eta^4\text{-}Ph_4C_4C{=}O)Ru(CO)_3 + PhCH(OMeC(O))Cl$	X-Ray, IR, $^1H, {}^{13}C$	⎫ Use of this chiral ligand in an asymmetric hydroformylation and hydrogenation catalyst studied briefly	264
	$CpRu(CO)(PPh_3)Cl$	$v + PPh_3$	$^1H, {}^{13}C, {}^{31}P$		264
Rh w	$(CpRhCl_2)_2$	$[(\eta^4\text{-}Ph_4C_4C{=}O)RhCl_2]_2 + PhCH(OMeC(O))Cl$	IR, ^1H		264
	$CpRh(CO)_2$	$w + CO + Zn/MeOH$	IR, ^1H	⎭	264
R = Ph₃Sn					
Mn	$CpMn(CO)_3$	$Ph_4(HOC_5)Mn(CO)_3\ (t) + EtLi + Ph_3SnCl$	X-Ray, IR		269
Ph₄RC₅					
R = Me					
W	$CpW(CO)_2(\mu\text{-}PhC{\equiv}CPh)W(CO)_2(C_5H_5)$	$W(CO)(PhC{\equiv}CPh)_3 + W({\equiv}CMe)(CO)_2(C_5H_5)$	IR, $^1H, {}^{13}C$	Template synthesis	250
Pd x	$(CpPd)_2(\mu\text{-}PhC{\equiv}CMe)$	"[endo-$Ph_4(R'O_2C)Pd(OAc)$" + $PhC{\equiv}CR$ (R' = Me, Et)	UV/VIS, MS, ^1H	Template synthesis	80
	$(CpPd)_2(\mu\text{-}PhC{\equiv}CPh)$	$x + PhC{\equiv}CPh$, Δ, reduced pressure	UV/VIS, ^1H	Bridge displacement	80
R = naphthyl					
Pd	$(CpPd)_2[\mu\text{-}PhC{\equiv}C(naphthyl)]$	"[endo-$Ph_4(R'O_2C)Pd(OAc)$" + $PhC{\equiv}CR$ (R' = Me, Et)	UV/VIS, MS	Template synthesis	80
R = EtO₂C					
Co y	$(\eta^4\text{-}CpH)Co(C_5H_5)$	$\overline{PhC}{=}C(Ph)(Ph)C{=}(Ph)Co(C_5H_5)PPh_3 + N_2CH(CO_2Et)$	IR, $^1H, {}^{13}C$ (given in suppl. mat.)	Incorporation of a carbene in the metallacyclopentadiene	21,22
ya	$[CpCo(C_5H_5)][NO_3]$	$y + Fe(NO_3)_3$	IR, $^1H, {}^{13}C$, FAB-MS		21
	$CpCo(\eta^4\text{-}C_5H_6)$	$ya + NaBH_4$	IR, $^1H, {}^{13}C$, MS		21

(continued)

BULKY CYCLOPENTADIENYL METAL COMPLEXES (*continued*)

Bulky Cp	Metal	Complex[b]	Method of preparation	Characterization[c]	Comments	Ref.
R = Ph₃P⁺CH₂	Cr	CpCr(CO)₃	(Ph₄C₄C=CH₂)Cr(CO)₃ (see Fulvenes) + PPh₃			270
R = Cl, Br	Mn	CpMn(CO)₃ (R = Cl)	Ph₄C₄C=N₂ + Mn(CO)₅Cl	IR, ¹H	Insertion of the diazo-Cp into the metal–halogen bond to give the η⁵-halo substituted Cp complex; air stable	260
	Rh	CpRh(L₂) [L₂ = cod, (C₂H₄)₂, (CO)₂]	Ph₄C₄C=N₂ + (Rh(L₂)Cl/Br)₂	X-Ray [R = Cl, L₂ = (C₂H₄)₂], IR, ¹H		259,260
R = I	Fe	CpFe(CO)₂I	Ph₄C₄C=N₂ + Fe(CO)₄I₂	IR, ¹H	air stable	271
[p-RC(O)Ph']₄HC₅ R = Me, ⁿPr, ⁿC₅H₁₁, ⁿC₇H₁₅	Tl	CpTl	CpH + TlOEt	¹H, ¹³C, MS		251
[p-RC(O)Ph']ₘPh₄₋ₘHC₅ R = Me, ⁿPr; m = 1, 2, 3, 4	Fe	Cp₂Fe	(Ph₄HC₅)₂Fe (see above) + RC(O)Cl/AlCl₃	IR, ¹H, MS, X-Ray (R = ⁿPr, m = 1)	See Schemes 20 and 23	251
Ph₄C₅(= Ph₄HC₅ minus H) (=Cp-H)	Au z	(η¹-Cp-H)(AuPPh₃)₃ [(η¹-Cp-H)(AuPPh₃)₃][BF₄]	Ph₄HC₅H + 2[(Ph₃PAu)O][BF₄] z + [AuPPh₃][BF₄], or Ph₄HC₅Au(PPh₃)₃ (see above) + 2[AuPPh₃][BF₄]	IR X-Ray, IR, ³¹P		261 261,272
Ph₃H₂C₅ 1,2,4 isomer	Li	CpLi	Deprotonation with ⁿBuLi	—		50
	Fe	Cp₂Fe	CpLi + FeCl₃ (l)	IR, UV		50,273
		CpFeC₅H₅	Ox add. of CpH to [H₅C₅Fe(CO)₂]₂		Cp₂Fe is also obtained in similar amounts from the same reaction	95
1,2,3 isomer	Na	CpNa	Deprotonation with NaH/Me₂SO	—		95
	Fe	Cp₂Fe	CpNa + FeCl₃, or ox. add. of CpH to Fe(CO)₅ or [H₅C₅Fe(CO)₂]₂	—	No CpFeC₅H₅ observed	95
	Ru	[CpRu(η⁶-C₆H₆)][BF₄]	(η-Ph-allyl)Ru(C₆H₆)Cl + PhC≡CPh/AgBF₄	¹H	Template synthesis	274

364

	CpRu(C₅H₅)	Ru(C₅H₅)(cod)Cl + 1,2,3-Ph₃-3-vinyl-1-cyclopropene, or CpK, or CpH	^{1}H, ^{13}C, MS	Ring expansion of vinylcyclopropene, template synthesis	93
	CpRu(C₅Me₅)	[Ru(C₅Me₅)Cl]₄ + 1,2,3-Ph₃-3-vinyl-1-cyclopropene, or CpH	^{1}H, ^{13}C, MS	Ring expansion of vinylcyclopropene, template synthesis	93
	CpRu(indenyl)	Ru(indenyl)(cod)Cl + 1,2,3-Ph₃-3-vinyl-1-cyclopropene	^{1}H, ^{13}C, MS	Ring expansion of vinylcyclopropene, template synthesis	93
Rh A	(η⁴-CpH)Rh(indenyl)	Rh(indenyl)(η-C₂H₄)₂ + 1,2,3-Ph₃-3-vinyl-1-cyclopropene	^{1}H, ^{13}C	Ring expansion of vinylcyclopropene, template synthesis	92
	[CpRh(indenyl)][BF₄]	A + [Ph₃C][BF₄]	^{1}H, ^{13}C	Ring expansion of vinylcyclopropene, template synthesis	92 / 275
B	CpRh(PⁱPr₃)(H)Cl	"Rh(PⁱPr₃)₂Cl" + 1,2,3-Ph₃-3-vinyl-1-cyclopropene	^{1}H, ^{13}C, ^{31}P	Ring expansion of vinylcyclopropene, template synthesis	275
	CpRh(L)Cl₂ (L = PMe₃, PⁱPr₃)	"Rh(PMe₃)₂Cl" + 1,2,3-Ph₃-3-vinyl-1-cyclopropene, or B + CHCl₃	^{1}H, ^{13}C, ^{31}P	Ring expansion of vinylcyclopropene, template synthesis	275
Modification of 1,2,4-Ph₃H₂C₅ Ph₃(RPh)R'₂C₅ R = Me, R' = H Ru	CpRu(C₅H₅)	Ru(C₅H₅)(cod)Cl + 1,2,3-Ph₃-3-(β-Me-vinyl)-1-cyclopropene	^{1}H, ^{13}C, MS	Ring expansion of vinylcyclopropene, template synthesis	93
R = R' = Me W	CpW(CO)₂(μ-MeC≡CPh)W(CO)₂(C₅H₅)	W(CO)(MeC≡CPh)₃ W(≡CPhⁱMe)(CO)₂(C₅H₅)	IR, ^{1}H, ^{13}C	Template synthesis; 1,2,4-Cp as isomeric mixture	250
Modifications of 1,2,3-Ph₃H₂C₅ Ph₃RR'C₅ R = Ph, R' = Me Pd C	(CpPd)₂(μ-PhC≡CMe)	"[endo-Ph₄(R''O)C₄]Pd(OAc)" + RC≡CR' (R'' = Me, Et)	UV/VIS, ^{1}H, MS	Template synthesis	80
	(CpPd)₂(μ-PhC≡CPh)	C + PhC≡CPh, Δ, reduced pressure	UV/VIS, ^{1}H	Bridge displacement	80
R = ⁱBu, R' = Me Pd	(CpPd)₂(μ-ⁱBuC≡CMe)	"[endo-Ph₄(R''O)C₄]Pd(OAc)" + RC≡CR' (R'' = Me, Et)	UV/VIS	Not isolated; chiral compound (cf. Section IV,D,1)	80

BULKY CYCLOPENTADIENYL METAL COMPLEXES (*continued*)

Bulky Cp	Metal	Complex[b]	Method of preparation	Characterization[c]	Comments	Ref.
R = Et, R′ = Me	Pd	$(CpPd)_2(\mu\text{-}EtC{=}CMe)$	"[*endo*-Ph$_4$(R″OC$_4$]Pd(OAc)]" + RC≡CR′ (R″ = Me, Et)	UV/VIS	Not isolated; chiral compound (cf. Section IV,D,1)	80
R = R′ = *p*-MeOPh′, or *p*-MePh′, or *p*-BrPh′	Pd	$(CpPd)_2(\mu\text{-}RC{=}CR′)$	"[*endo*-Ph$_4$(R″OC$_4$]Pd(OAc)]" + RC≡CR′ (R″ = Me, Et)	UV/VIS, ¹H, MS		80
R = Ph, R′ = naphthyl	Pd	$(CpPd)_2[\mu\text{-}PhC{=}C(naphthyl)]$	"[*endo*-Ph$_4$(R″OC$_4$]Pd(OAc)]" + RC≡CR′ (R″ = Me, Et)	UV/VIS, MS		80
R = R′ = MeO$_2$C	Pd	$CpPd(C_4Ph_4OR)$ (R = Me, Et)	"[*endo*-Ph$_4$(R″OC$_4$]Pd(OAc)]" + RC≡CR′ (R″ = Me, Et)	UV/VIS, MS		80
R = R′ = Et	Pd **D**	$(CpPd)_2(\mu\text{-}EtC{=}CEt)$	"[*endo*-Ph$_4$(R″OC$_4$]Pd(OAc)]" + RC≡CR′ (R″ = Me, Et)	UV/VIS, ¹H, MS		80
		$(CpPd)_2(\mu\text{-}R″C{=}CR‴)$(R″ = R‴ = *p*-MeOPh′, *p*-FPh′, *p*-MePh′, Ph, *p*-ClPh′, *p*-BrPh′, *p*-O$_2$NPh′, R‴ = Ph, R‴ = Me)	D + R″C≡CR‴	UV/VIS, ¹H, MS	Bridge exchange	80
		$CpPd(NO)$	D + NO	IR, UV/VIS, MS		80
Ph$_2$(RPh)H$_2$C$_5$, R = F	Ru	$[CpRu(\eta^6\text{-}C_6H_6)][BF_4]$	$(\eta\text{-}FPh′\text{-allyl})Ru(C_6H_6)Cl$ + PhC≡CPh/AgBF$_4$	¹H	Template synthesis	274
(PhCH$_2$)$_5$C$_5$	Main group					
	Li	CpLi	Deprotonation with ″BuLi	¹H, ¹³C		51,106
	K	CpK	Deprotonation with K-metal	X-Ray		98
		CpK(thf)$_3$	Deprotonation with K-metal			179
	In	CpIn	CpLi + InCl	X-Ray, IR, ¹H, ¹³C, MS	Possible In–In interaction (177) (see Fig. 11 and Section IV,B,4)	109
	Tl	CpTl	CpH + TlOEt	X-Ray, IR, ¹H, ¹³C, MS	Two modifications: linear chain or dimeric pair with possible Tl–Tl interaction (176,177) (see Figs. 10, 11, and Section IV,B,4)	107,108, 146
	Ge	Cp$_2$Ge	CpLi + GeI$_2$	X-Ray, IR, Raman, ¹H, ¹³C, MS	air stable (see Fig. 2, Section IV,B,6)	105,106

Sn	Cp₂Sn	CpLi + SnCl₂	X-Ray, IR, Raman, ¹H, ¹³C, MS, ¹¹⁹Sn, ¹¹⁹Sn-CPMAS, ¹¹⁹Sn-Mössbauer	See Fig. 2, Section IV,B,6	106, 39, 120
Pb	Cp₂Pb	CpLi + Pb(OAc)₂	X-Ray, IR, Raman, ¹H, ¹³C, MS	See Fig. 2, Section IV,B,6	106
Transition metals					
Mn	CpMn(CO)₃	Ox add. of CpH to Mn₂(CO)₁₀	X-Ray, IR, ¹H, ¹³C	Low yield synthesis, air stable	182
Re	CpRe(CO)₃	Ox add. of CpH to Re₂(CO)₁₀	IR, ¹H, ¹³C		182
Fe	Cp₂Fe	CpLi + FeCl₂	X-Ray, IR, ¹H, ¹³C, MS powder diffraction	See Fig. 13, Section IV, B, 6	98,182
	CpFeC₅Me₅	CpLi + Me₅C₅Fe(acac)	¹H		182
	CpFeC₅H₅	CpLi + Fe(acac)₂ + TlC₅H₅	¹H, ¹³C		182
Co	CpCo(CO)₂	Ox add. of CpH to Co₂(CO)₈	X-Ray, IR, ¹H, MS		51
Rh	CpRh(CO)₂	CpLi + [Rh(CO)₂Cl]₂	IR, ¹H, MS		51
Au	(η¹-Cp)AuPPh₃	CpLi + (Ph₃P)AuCl	X-Ray, ¹H, ¹³C		276
Lanthanides					
Lu	CpLu(η⁴-cot)	CpK + Lu(η⁴-cot)Cl(thf)	X-Ray, ¹H, ¹³C	Not dissolved or attacked by 50% aqueous acetic acid for 30 minutes	98
(Me₃Si)₃H₂C₅ (1,2,4 isomer)					
Main group					
Li	CpLi	Deprotonation of 2,5,5-CpH with ⁿBuLi	¹H, ¹³C, ⁷Li, ²⁹Si	metallotropic rearrangement (1,2-shift of a Me₃Si group; cf. Section II,E)	53, 147
	CpLi(N-base) {N-base = quinuclidine, tmeda, pmdeta ([Me₂NCH₂CH₂]₂NMe), damp (Me₂N(CH₃)NMe₂), pyridine, bipy}	CpLi + N-base	X-Ray, ¹H, ¹³C, ⁷Li, ²⁹Si	Fig. 14 (cf. Section IV,7)	147
	(CpLi)₂(damp)		¹H, ¹³C, ⁷Li		148
	CpLi(O-base) (O-base = Et₂O, thf, ⁿBu₂O, 1,4-dioxane, dme, diglyme)	2 CpLi + damp / CpLi + O-base	¹H, ¹³C, ⁷Li, 2Si / X-Ray(thf), ¹H, ¹³C, ⁷Li, ²⁹Si	See Section IV,B,7, Fig. 14	148, 148
	CpLi(bmte) (bmte = MeSCH₂CH₂SMe)	CpLi + bmte	¹H, ¹³C, ⁷Li		148
K	CpK				
Mg	Cp₂Mg	Deprotonation of 2,5,5-CpH with Me₂Mg	X-Ray, IR, molecular weight ¹H, ¹³C, MS	Slightly bent metallocene, unlike ⁿCp₂Mg (Fig. 15, Section IV,B,8; Section IV,B,8); covalent character of metal–ring bond concluded for the CpMg species	149
	CpMg(tmeda)Me	CpH/tmeda + Me₂Mg	IR, molecular weight ¹H, ¹³C, MS		149
	CpMg(tmeda)Br	CpH/tmeda + MeMgBr	X-Ray, IR, molecular weight ¹H, ¹³C, MS		149
In	CpIn	CpLi + InCl	¹H, ¹³C, ²⁹Si, MS	Probably monomeric with covalent Cp—In/Tl bond (cf. Section IV,B,9)	150
Tl	CpTl	CpLi + TlCl	¹H, ¹³C, MS, molecular weight		151

(continued)

BULKY CYCLOPENTADIENYL METAL COMPLEXES (continued)

Bulky Cp	Metal	Complex[b]	Method of preparation	Characterization[c]	Comments	Ref.
$(Me_3Si)_3H_2C_5$ (cont.)	Ge	Cp_2Ge	$CpLi + GeCl_2$(dioxane)	X-Ray, ^1H, ^{13}C, MS		186
		$CpGeC_5Me_5$	$CpK + Me_5C_5GeCl$ or $[Me_5C_5Ge][BF_4]$	^1H, ^{13}C, MS		277
	Sn	Cp_2Sn	Sequential lithiation and silylation of $(H_5C_5)_2Sn$, or $CpLi + SnCl_2$	X-Ray	Template synthesis, see Scheme 18	85
	Pb	Cp_2Pb	$CpLi + PbCl_2$	^1H, ^{13}C, MS, molecular weight		278
	Transition metals					
	Ti E	$CpTiCl_3$	$CpLi + TiCl_3(thf)_3 + HCl$	IR, ^1H, ^{13}C, MS	Thermally stable up to 110°C	279
	Ea	$CpTiMe_3$	E + MeLi	IR, ^1H, ^{13}C, MS		279
	Eb	$(CpTiMe_2)_2(\mu\text{-}O)$	Ea + H_2O	IR, ^1H, ^{13}C, MS	Both Eb and Ec give a reversible	279
		$(CpTiCl)_2(\mu\text{-}O)$	E + ½ H_2O/NEt₃	IR, ^1H, ^{13}C, MS	reaction to E with Me_3SiCl/HCl;	279,280
	Ec	$(CpTiCl)_2(\mu\text{-}O)_2$	E + 1 H_2O/NEt₃ or Eb + ½ H_2O/NEt₃	IR, ^1H, ^{13}C, MS	Ec is noted to be thermally very stable, no decomposition if kept at 200°C for hours	280
		$CpTi(C_5H_3)Cl_2$	—	X-Ray		281
	Zr	Cp_2ZrCl_2	$CpLi + ZrCl_4$	IR, ^1H, ^{13}C	$\Delta G^\ddagger$ 11.0 kcal/mol for Cp–Zr rotation	204a
	Hf	Cp_2HfCl_2	$CpLi + HfCl_4$	IR, ^1H, ^{13}C	$\Delta G^\ddagger$ 11.3 kcal/mol for Cp–Hf rotation	204a
	Fe	Cp_2Fe	$CpLi$(tmeda) + $FeBr_2$, $CpLi + FeCl_2$ at low temperature (−95°C)	IR, ^1H	$\Delta G^\ddagger$ 11.0 kcal/mol for Cp–Fe rotation (see Fig. 15, Section IV,B,8)	282
				X-Ray, IR, ^1H, ^{13}C, MS		99
		$CpFeC_5R_5$ (R = H, Me)	$CpLi + FeCl_2 + H_5C_5Na$/ Me_5C_5Li, or $CpLi + Me_5C_5Fe$(acac)	IR, ^1H, ^{13}C, MS		99
		$CpFe(CO)_2X$ (X = Cl, Br, I)	$CpLi + FeX_2 + CO$ at low temperature	IR, ^1H, ^{13}C, MS	X = Cl derivative extremely air sensitive	100
	F	$[CpFeC_6H_5CH_3][IPF_6]$	$CpLi + FeCl_2 + C_6H_5CH_3$/ $AlCl_3 + KPF_6$	IR,^1H, ^{13}C		283
	G	$[CpFeP(OMe)_3][IPF_6]$	F + $P(OMe)_3$/$h\nu$	IR, ^1H, ^{13}C, ^{31}P		283
		$CpFe[P(OMe)_3]X$ (X = Cl, Br, I)	$CpLi + FeX_2 + 2$ $P(OMe)_3$	IR, ^1H, ^{13}C, ^{31}P, MS		284
		$CpFe[P(OMe)_3]H$	G (X = I) + ᵗBuMgCl	IR, ^1H, ^{13}C, ^{31}P, MS		284
		$CpFe[P(OMe)_3](CO)I$	G (X = I) + CO	IR, ^1H, ^{13}C, ^{31}P, MS		284
	Co	$[Cp_2Co][IPF_6]$	$CpLi + CoCl_2 + FeCl_3 + KPF_6$	IR, ^1H, ^{13}C	$\Delta G^\ddagger$ 9.9 kcal/mol for Cp–Co rotation	198

Ni H	(CpNiCl)₂	CpLi + NiCl₂	MS		200a
Ha	CpNi(COCl)	CpLi + NiCl₂ + CO, or H + CO	IR		200a
Hb	CpNi(PPh₃)X (X = Cl, Br)	CpLi + NiX₂ + PPh₃, or H + PPh₃	IR, ¹H, ¹³C, ³¹P, MS	ΔG‡ 10.5 kcal/mol for **Cp**–Ni rotation	200a
	CpNi(PPh₃)I	Hb(X = Cl) + NaI	IR, ¹³C, MS		200a
Actinides U	CpUCl₂(thf)(μ-Cl)₂Li(thf)₃	CpLi + UCl₄	X-Ray, ¹H, ⁷Li		285

Modifications of (Me₃Si)₂H₂C₅, 1,2,4-(Me₃Si)₃-2-[Me₂NC*H(Me)]HC₅

Fe	CpFeC₅H₃(SiMe₃)₂	(Me₃Si)₂[MeNCH(Me)] H₂C₅FeC₅H₄(SiMe₃) + ⁿBuLi/tmeda + Me₃SiCl	MS	See also Mixed systems	23

(Me₃Si)₂(ᵗBu)H₂C₅(1,2,5 isomer)

Li	CpLi	Deprotonation of 2,5,5 isomer with ⁿBuLi		1,2-Shift of Me₃Si group (Section II,E)	204a
Fe	Cp₂Fe	CpLi + FeCl₂ at low temperature	IR, ¹H, ¹³C, MS	ΔG‡ 9.7 kcal/mol for **Cp**–Fe rotation	204a
	CpFe(CO)₂Br	CpLi + FeBr₂(dme) + AlCl₃ + CO	IR, ¹H, ¹³C, MS		204a
Co	[Cp₂Co][PF₆]	CpLi + CoCl₂ + aq.FeCl₃ + KPF₆	IR, ¹H, ¹³C	ΔG‡ 8.8 kcal/mol for **Cp**–Co rotation	204a

ᵗBu₃H₂C₅(1,2,4 isomer)

Na	CpNa	CpH + NaNH₂	¹H, ¹³C, ²⁰⁷Pb		34,211
Pb	Cp₂Pb	CpNa + PbCl₂			36,286
Ta	CpTaCl₂(ᵗBuC≡CH)	[Cl₂(dme)Ta(μ-ᵗBu)]₂Zn(μ-Cl)₂ + ᵗBuC≡CH in toluene	¹H, ¹³C	See Scheme 17	13
	CpTaCl₄	[Cl₂(dme)Ta(μ-ᵗBu)]₂Zn(μ-Cl)₂ + ᵗBuC≡CH in CCl₄/CH₂Cl₂	¹H, ¹³C	See Scheme 17	13
Mo	CpMo(CO)₃Me	CpNa + Mo(CO)₆ + MeI	IR, ¹H, ¹³C, MS		211

ᵗBu₃H₂C₅(1,2,3 isomer)

Rh I	(η⁴-CpH/D)Rh(indenyl)	Ring closure to Cp of ᵗBuC=C(ᵗBu)—C(ᵗBu)—(η-CH=CH₂/D)Rh(indenyl)	¹H, ²H	Starting material is an intermediate in the vinylcyclopropene to Cp conversion (cf. Refs. 12, 93, and 275)	287
	[CpRh(indenyl)][PF₆]	I + [Ph₃C][BF₄], NBS or CDCl₃	X-Ray (footnote 12 in Ref. 287), ¹H, ²H	Cp can also be ᵗBu₃HDC₅	287

(continued)

369

BULKY CYCLOPENTADIENYL METAL COMPLEXES (*continued*)

Bulky Cp	Metal	Complex[b]	Method of preparation	Characterization[c]	Comments	Ref.
(^{t}Bu)$_2$^{n}BuH$_2$C$_5$						
	Ta	CpTaCl$_4$	[Cl$_2$(dme)Ta(μ-C^{t}Bu)]$_2$Zn(μ-Cl)$_2$ + EtC≡CEt	^{1}H, ^{13}C	**Cp** as 1,2,4- and 1,3,4-(^{t,t,n}Bu) isomeric mixture (Scheme 17), 1,2,4 isomer is chiral (Section IV,D,1)	13
^{t}Pentyl$_5$C$_5$[=(Et$_2$CH)$_5$C$_5$]						
	Co	[CpCoC$_5$H$_5$][PF$_6$]	Deprotonation of [Me$_5$CoC$_5$H$_5$][PF$_6$] with KOH; +EtI	^{1}H, ^{13}C, CV	Template synthesis (see Scheme 19); $\Delta G^{\ddagger}$ 19.4 kcal/mol for ^{t}Pentyl–C$_5$ rotation, chiral metallocene (see Sections IV,C,2 and IV,D,2)	14
(Neopentyl)$_x$HC$_5$						
Modification of (neopentyl)$_x$HC$_5$						
(neopentyl)$_4$^{t}BuC$_5$ (neopentyl = ^{t}BuCH$_2$)						
	Ta	(Cp-*H*)TaCl$_3$	[Cl$_2$(dme)Ta(μ-C^{t}Bu)]$_2$Zn(μ-Cl)$_2$ + ^{t}BuCH$_2$C≡CCH$_2$^{t}Bu in toluene/thf	^{1}H, ^{13}C, MS	Template synthesis (cf. Scheme 17), one neopentyl group is metallated	13
		(Cp-2*H*)TaCl$_2$	[Cl$_2$(dme)Ta(μ-C^{t}Bu)]$_2$Zn(μ-Cl)$_2$ + ^{t}BuCH$_2$C≡CCH$_2$^{t}Bu in CH$_2$Cl$_2$	X-Ray, ^{1}H, ^{13}C, MS	Template synthesis (cf. Scheme 17), two neopentyl groups metallated	13
^{i}Pr$_5$C$_5$						
	Main group					
	Na	CpNa	Deprotonation of CpH with NaNH$_2$	^{13}C		34
	Pb	Cp$_2$Pb	CpNa + PbCl$_2$	X-Ray, ^{1}H, ^{13}C, ^{207}Pb	Angle between Pb–Cp normals 170°	36,286
	Transition metals					
	Mo	CpMo(CO)$_3$Me	CpNa + Mo(CO)$_6$ + MeI	IR, ^{1}H, ^{13}C	$\Delta G^{\ddagger}$ 13 kcal/mol for ^{i}Pr–C$_5$ rotation, chiral compound (cf. Sections IV,C,2 and IV,D,2)	211
	Co **J**	[CpCoC$_5$H$_5$][PF$_6$]	Deprotonation of [Me$_5$CoC$_5$H$_5$][PF$_6$] with ^{t}BuOK; +MeI	^{1}H, ^{13}C, CV	Template synthesis (see Scheme 19), $\Delta G^{\ddagger}$ 17.1 kcal/mol for ^{i}Pr–C$_5$ rotation, chiral metallocene (cf. Sections IV,C,2 and IV,D,2)	14
		CpCoC$_5$H$_5$	Red of **J** with C$_5$H$_5$FeC$_6$Me$_6$	MS	Co(II) complex	14

iPr4HC5	Main group					
	Na	CpNa	Deprotonation of CpH with NaNH2	1H, 13C		35
	Ca	Cp2Ca	CpK + CaI2	X-Ray		288
	Pb	Cp2Pb	CpNa + PbCl2	1H, 13C, 207Pb		36,286
	Transition metals					
	Mo	CpMo(CO)3Me	CpNa + Mo(CO)6 + MeI	IR, 1H, 13C		35
	K	CpMo(CO)3H	Ox add. of CpH to Mo(CO)6	IR, 1H, 13C		211
	L	[CpMo(CO)2]2	Red of K with norbornadiene	IR, 1H, 13C	Mo—Mo quadruple bond in dimer	211
		Na[CpMo(CO)3]	CpNa + Mo(CO)6	1H, 13C		211
	La	CpMo(CO)3Cl	L + CuCl2	IR, 1H, 13C		211
		CpMo(CO)3SiHMe2	L + Me2SiHCl	IR, 1H, 13C		211
		Cp[CO2Mo≡Mo(CO)2]C5Me5	L + Me5C5Mo(CO)3Cl or La + Na[Me5C5Mo(CO)3]	IR, 1H, 13C, MS		211
	Fe	Cp2Fe	CpNa + FeCl2	1H, 13C, MS	ΔG‡ 13.6 kcal/mol for Cp–Fe rotation	35
iPr3H2C5	Na	CpNa	Deprotonation of CpH with NaNH2	13C		35
	Mo	CpMo(CO)3Me	CpNa + Mo(CO)6 + MeI	IR, 1H, 13C		35
	Fe	Cp2Fe	CpNa + FeCl2	1H, 13C, MS		35
nPr4HC5 modification of nPr4HC5 "Pr4BuC5	Ta	CpTaCl4	[Cl2(dme)Ta(μ-CnBu)]2Zn(μ-Cl)2 + "PrC≡C"Pr	1H, 13C	Template synthesis (see Scheme 17)	13
Et5C5	Main group					
	Li	CpLi	Deprotonation of CpH with nBuLi			36
	Na	CpNa	Deprotonation of CpH with NaNH2			36
	Transition metals					
	Mo	CpMo(CO)3Me	CpLi + Mo(CO)6 + MeI	IR, 1H, 13C		36
	W M	CpW(OtBu)O2	EtC≡CEt + W(CEt)(OtBu)3 or Et5C5W(OtBu)(OCMe2CMe2O)	1H, 13C, MS	See Scheme 16	81,82
	Ma	CpWCl4	M + PCl5	EPR		84
		CpWCl4(PMe3)	Ma + PMe3	1H, EPR		84

(continued)

371

BULKY CYCLOPENTADIENYL METAL COMPLEXES (continued)

Bulky Cp	Metal	Complex[b]	Method of preparation	Characterization[c]	Comments	Ref.
Et_5C_5 (cont.)		$CpWMe_4$	Ma + $MeMgCl$	^1H, EPR		84
		$(CpWH_4)_2$	Ma_2 + $LiAlH_4$	IR, ^1H, ^{13}C		37
	Mn	$CpMn(CO)_3$	Red of $Et_4(MeCO)C_5Mn(CO)_3$ (see below) with $LiAlH_4/AlCl_3$	IR	See Scheme 20, note study of basicity of metal carbonyls (289,290)	89
		$CpMn(CO)_2(PPh_3)$		^{13}C, ^{31}P	Study of basicity	290
	Fe	Cp_2Fe	$CpLi$ + $FeCl_2$, or $(EtH_4C_5)_2Fe$ + $MeC(O)Cl/AlCl_3$, + $LiAlH_4/AlCl_3$	^1H, ^{13}C, MS; IR, ^1H	Friedel–Crafts acylation, followed by reduction (cf. Scheme 20)	88,291
		$CpFeC_5H_5$	Fe + H_3C_5H + 2 $EtC{\equiv}CEt$		Metal atom synthesis, cocondensation, $(H_3C_5)_2Fe$ and $(Et_4HC_5)FeC_5H_5$ (below) are main products	292
modifications of Et_5C_5 $Et_5C_5C(Et)CH_2C(C_9H_6N)$	Pd	$[(\eta^2\text{-}Et_5C_5)\overline{C(Et)CH_2C(C_9H_6N)}Pd][BF_4]$	3 $EtC{\equiv}CEt$ + $[H_2\overline{C(C_9H_6N)}Pd][BF_4]$	X-Ray	Insertion of 3-hexyne in cyclopalladated 8-methylquinoline (C_9H_6N)	293
$Et_4C_5CH_2CH_2CH_2C_5Et_4$	W N	$Cp[W(O^tBu)O_2]_2$	$EtC{\equiv}CCH_2CH_2C{\equiv}CEt$ + $Et_3C_3W(O^tBu)(OCMe_2CMe_2O)$	IR, ^{13}C	Template synthesis (cf. Scheme 16)	37
		$Cp[W(O)Cl_3]_2$	N + HCl/Me_3SiCl	^{13}C		294
	Na	$Cp(WCl_4)_2$	N or Na + PCl_3	—		37,294
	Nb	$Cp[WCl_4(PMe_3)]_2$	Nb + PMe_3	X-Ray		294
	Nba	$Cp(WMe_4)_2$	Nba + Me_2Zn (excess)	EPR, MS		37
		$Cp(WH_4)_2$	Nb + $LiAlH_4$	X-Ray, IR, ^1H, ^{13}C, MS		37
	Nbb	$Cp_2W_2H_6(PMe_3)$	Nbb + PMe_3	^1H, ^{13}C, ^{31}P		37
$Et_4C_5CH_2CH_2CH_2C_5Me_4X$ $X = H$, or $M(CO)_2$ ($M = Co, Rh$)	W O	$CpW(O^tBu)O_2$	$EtC{\equiv}CCH_2CH_2C_5Me_4X$ + $Et_3C_3W(O^tBu)(OCMe_2CMe_2O)$, or O ($X = H$) + $Co_2(CO)_8/$ $^tBuCH{=}CH_2$	IR, ^1H, ^{13}C	Template synthesis (see Scheme 16)	84
$X = RhCl_2$:	W Oa	$CpW(ClO_2)$	O $[X = Rh(CO)_2]$ + Cl_2	IR, ^1H		84
	Oaa	$CpWCl_4$	Oa + PCl_5	IR		84

	Compound	Synthesis	Methods	Notes	Ref.
X = Rh(PMe₃)Cl₂ W Oaaa	CpWCl₄(PMe₃)	Oaa + PMe₃	^{1}H, EPR		84
X = Rh(PMe₃)Me₂ W Oaaaa	CpWMe₄	Oaaa = Me₂Zn	^{1}H, ^{13}C, EPR		84
	CpW(CO)₃Me	Oaaaa + CO	IR, ^{1}H, ^{13}C, ^{31}P		84
Et₄HC₅					
Li	CpLi	CpH + ^{n}BuLi		See Scheme 11	36
Mo	CpMo(CO)₃Me	CpLi + Mo(CO)₆ + MeI	IR, ^{1}H, ^{13}C		36
Mn	CpMn(CO)₃	Red of Et₃(MeCO)HC₅Mn(CO)₃ with LiAlH₄/AlCl₃	IR	See Scheme 20, note study of basicity of metal carbonyls (289,290) and study of isotopic C₅–H exchange as a function of alkyl substituent (295)	89
Fe	Cp₂Fe CpFeC₅H₅	CpLi + FeCl₂ Fe + H₅C₅H + 2 EtC≡CEt	^{1}H, ^{13}C, MS	Metal atom synthesis, cocondensation, (H₅C₅)₂Fe is main product	36 292
Modifications of Et₄HC₅					
Et₄ᵗBuC₅ Ta	CpTaCl₄	[Cl₂(dme)Ta(μ-C'Bu)]₂Zn(μ-Cl)₂ + EtC≡CEt	^{1}H, ^{13}C	See Scheme 17	13
W	CpW(EtC≡CEt)Cl₂ and (CpWCl₂)₂	W(CCMe₃)(dme)Cl₃ + EtC≡CEt	IR, ^{1}H, molecular weight		82
P	(CpWH₂)₂ CpW(O₂CCF₃)₄	P + LiAlH₄ W(CCMe₃)(dme)(O₂CCF₃)₃ + EtC≡CEt	IR, ^{1}H, ^{13}C EPR		37 83
Et₄(MePh')C₅ W	CpW(CO)₂(μ-EtC≡CEt)W(CO)₂(C₅H₅)	W(CO)(EtC≡CEt)₃ + W(CPh'Me)(CO)₂(C₅H₅)	X-Ray, IR, ^{1}H, ^{13}C	Template synthesis	250
Et₄(HO)C₅ Co	CpCo(C₅H₅)	(η⁴-Et₄C₄C=O)Co(C₅H₅) + HCl	IR, UV, pK_a	Proteolytic equilibrium (see **8/9**)	268
Rh	CpRh(C₅H₅)	(η⁴-Et₄C₄C=O)Rh(C₅H₅) + HCl	IR, UV, pK_a	Proteolytic equilibrium (see **8/9**)	268
Et₄(MeCO)C₅ Mn	CpMn(CO)₃	(Et₃H₂C₅)Mn(CO)₃ + MeC(O)Cl/AlCl₃	IR	Friedel–Crafts acylation (cf. Scheme 20)	89

The tricyclic derivatives of the form (H₂C)$_m$ (CH₂)$_n$, which could also be viewed as modifications [substituted on the Et moiety, (XEt')₄HC₅], are described toward the end of the Appendix

(continued)

BULKY CYCLOPENTADIENYL METAL COMPLEXES (continued)

Bulky Cp	Metal	Complex[b]	Method of preparation	Characterization[c]	Comments	Ref.
Et₃H₂C₅ (mixture of isomers)						
	Li	CpLi	CpH + "BuLi			36
	Mo	CpMo(CO)₃Me	CpLi + Mo(CO)₆ + MeI	IR, ¹H, ¹³C		36
	Mn	CpMn(CO)₃	Red of [Et₂(MeCO)HC₅]Mn(CO)₃ with LiAlH₄/AlCl₃	IR	Study of isotopic C₅–H exchange as a function of alkyl substituent (295)	89
	Fe	Cp₂Fe	CpLi + FeCl₂	¹H, ¹³C, MS		36
		CpFeC₅H₅	(Et₂H₂C₅)FeC₅H₅ + BrCH₂CH₂Br/AlCl₃ + NaBH₄		Template synthesis	296
(1,2,3 isomer)						
	Fe	CpFe(CO)(μ-EtCCHCH)Fe(CO)₃Fe(CO)₃	Fe₃(CO)₁₂ + EtC≡CH	X-Ray, IR, MS	Template synthesis, see structure drawing in ref. 297	297
Modifications of Et₃H₂C₅						
Et₃(MeCO)HC₅	Mn	CpMn(CO)₃	(Et₃H₂C₅)Mn(CO)₃ + MeC(O)Cl/AlCl₃	IR	Friedel–Crafts acylation, (cf. Scheme 20)	89
Et₃Me₂C₅	W Q	CpW(O'BuO)O₂	Et₃C₅W(O'Bu)(OCMe₂CMe₂O) + MeC≡CMe	¹H	See Scheme 16	82
	Qa	CpW(O)Cl₃	Q + HCl/Me₃SiCl		See Footnote 12a in Ref. 82	82
	Qaa	CpWCl₄(PMe₃)	Qa + PCl₅ + PMe₃		See footnote 12a in Ref. 82	82
		CpWMe₄	Qaa + 8 ZnMe₂	X-Ray	See footnote 12a in Ref. 82	82
Et₂('Bu)Me₂C₅	W	CpW(O'BuO)O₂	('Bu)Me₂C₅W(O'Bu)(OCMe₂CMe₂O) + EtC≡CEt	¹H	Cp as mixture of only two isomers (see Scheme 16)	82
	R	(CpWCl₂)₂	('Bu)Me₂C₅WCl₃ + EtC≡CEt			82
		CpW(C'Bu)Cl₂	R + Zn(CH₂'Bu)₂	¹H	Only two Cp isomers	82

The ethylene-bridged bis(tetrahydroindenyl) system (4), Et'(IndH₄)₂, could also be regarded as a modification [substituted on the Et moiety, 1,2,3-(XEt')₃H₂C₅], it is, however, listed in a separate section below

(MeS)₃C₅	Na	CpNa	H₅C₅H + NaH + Me₂S₂	X-Ray, IR, ¹H, ¹³C	See Scheme 13	61
	Mn S	CpMn(CO)₃	(MeS)₃ClC₅Mn(CO)₃ + "BuLi/Me₂S₂		See Scheme 21	90
		(OC)₃Mn(η⁵-Cp—[S]₂—μ)PdCl₂	S + Pd(NCPh₂)Cl₂	IR, ¹H	[S]₂-μ indicates chelation via two vicinal S on Cp	91
		(OC)₃Mn(η⁵-Cp—[S]₂—μ)M(CO)₄ (M = W, Mo)	S + W(CO)₆/hν, thf; or + Mo(CO)₄(C₇H₈)	X-Ray (M = Mo), IR, ¹H	Monochelate; [S]₂-μ indicates chelation via two vicinal S on Cp	298

$(OC)_3Mn(\eta^5\text{-}Cp\text{—}[S]_2\text{—}\mu)$ $[M(CO)_4]_2$ (M = W, Cr, Re$^+$ BF$_4^-$, Mo)	S + 2 W(CO)$_6$/hv, thf; or Cr(CO)$_5$(NCMe)$_3$; or Re(CO)$_4$(C$_6$H$_5$)/HBF$_4$; or Mo(CO)$_3$(p-xylene)	X-Ray (M = W), IR, ^{1}H	Dichelate; $[S]_2$-μ indicates chelation via two vicinal S on **Cp**	*298*

(MeS)$_4$HC$_5$
Modification of (MeS)$_4$HC$_5$
(MeS)$_4$ClC$_5$
Mn

CpMn(CO)$_3$	(MeS)$_3$Cl$_2$C$_5$Mn(CO)$_3$ + ^{n}BuLi/ Me$_2$S$_2$	IR, ^{1}H, ^{13}C	See Scheme 21	*90*

(MeS)$_3$H$_2$C$_5$
Modification of (MeS)$_3$H$_2$C$_5$
(MeS)$_3$Cl$_2$C$_5$
Mn T

CpMn(CO)$_3$	(MeS)$_2$Cl$_3$C$_5$Mn(CO)$_3$ + ^{n}BuLi/Me$_2$S$_2$	IR, ^{1}H, ^{13}C	See Scheme 21	*90*
$(OC)_3Mn(\eta^5\text{-}Cp\text{—}[S]_2\text{—}\mu)PdCl_2$	T + Pd(NCPh$_2$)Cl$_2$	IR, ^{1}H	$[S]_2$-μ indicates chelation via two vicinal S on **Cp**	*91*

Related (RS)$_3$Cl$_2$C$_5$ derivatives
R = ^{n}Bu or Ph
Mn U

CpMn(CO)$_3$	(RS)$_2$Cl$_3$C$_5$Mn(CO)$_3$ + ^{n}BuLi/ R$_2$S$_2$	IR, ^{13}C	See Scheme 21	*91*
$(OC)_3Mn(\eta^5\text{-}Cp\text{—}[S]_2\text{—}\mu)PdCl_2$	U + Pd(NCPh$_2$)Cl$_2$	IR	$[S]_2$-μ indicates chelation via two vicinal S on **Cp**	*91*

(MeO$_2$C)$_5$C$_5$
Complexes of (MeO$_2$C)$_5$C$_5$ where the metal is chelated by carboxylate groups with formation of a metal–oxygen ion pair, rather than a metal–C$_5$ hapto coordination, are listed summaringly

Li (*299,300*), Na, K, Rb, Cs (*300*);
Mg, Ca, Sr, Ba (*301*);
Tl (*300*);
Ge (*302*), Sn (*302,303*)
Cr (*304*)
Mn (*305*)
Fe (*64,305,306*); Ru (*305,306*); Os (*305*)
Co (*305,306*); Rh, Ir (*307*)
Ni (*305*)
Cu (*306,308*); Ag (*62,305,309,310*)
Zn; Cd (*301*)
Lanthanides (*311*)

η-C$_5$ bonded complexes
Mn V

CpMn(CO)$_3$	CpK + Mn(CO)$_5$Br	IR, ^{1}H, ^{13}C, ^{55}Mn, MS	**Cp** postulated as η^1 bonded	*312*
CpMn(CO)$_2$L (L = P^{n}Bu$_3$, PPh$_3$, PEt$_3$)	V + L, hv	IR, ^{1}H, ^{31}P, MS		*312*

(continued)

BULKY CYCLOPENTADIENYL METAL COMPLEXES (*continued*)

Bulky Cp	Metal	Complex[b]	Method of preparation	Characterization[c]	Comments	Ref.
$(MeO_2C)_4C_5$ (*cont.*)	Ru	$CpRuC_5H_5$	$CpTl + Ru(C_5H_5)(PPh_3)_2Cl$ in air, or $CpTl + Ru(CO)_7(C_5H_5)Cl$	X-Ray, IR, ^1H, ^{13}C	Both Cp rings η^5 bonded, **Cp** is easily displaced by PPh_3 in MeCN, yielding again the starting material	305,313
	Os	$CpOsC_5H_5$	$CpTl + Os(C_5H_5)(PPh_3)_2Cl$ in air	X-Ray, IR, ^1H, FAB-MS		305
	Rh	$CpRh(cod)$	$CpTl + [Rh(cod)(\mu\text{-}Cl)]_2$	IR, ^1H	Cp is η^5 bonded	308
	Ir	$CpIr(cod)$	$CpTl + [Ir(cod)(\mu\text{-}Cl)]_2$	X-Ray, IR, ^1H		308
	Au	$(\sigma\text{-}Cp)AuPPh_3$	$CpH + Au(OAc)PPh_3$			314
$(MeO_2C)_4HC_5$ Modifications of $(MeO_2C)_4HC_5$ $(MeO_2C)_4RC_5$ R = Me	Ru	$CpRuC_5H_5$	$CpTl + Ru(C_5H_3)(PPh_3)_2Cl$	^1H	In analogy to the above $(MeO_2C)_5C_5$ complexes, these compounds are expected to have the same bonding mode	17
	Rh	$CpRh(cod)$	$CpTl + [Rh(cod)(\mu\text{-}Cl)]_2$	^1H		17
	Au	$(\sigma\text{-}Cp)Au(PPh_3)$	$CpH + Au(OAc)PPh_3$	X-Ray, IR, ^1H		17
R = $^nBuNHC(O)$	Mn	$CpMn(CO)_3$	$(MeO_2C)_3C_5Mn(CO)_3 + {}^nBuNH_2$	IR, ^1H, MS		312
$(MeO_2C)_3H_2C_5$	Rh	$[1,2,3\text{-}Cp_3Rh][(MeO_2C)_5C_5]$	$(MeO_2C)_3C_5H + Rh_2(OAc)_4$	X-Ray, IR, UV, ^1H, ^{13}C	**Cp** rings η^5 bonded, two MeO_2C groups were replaced by H, net elimination of $C_2H_2O_2$ from two MeO_2C groups in each **Cp**	15,16
		$(1,2,3\text{-}Cp)Rh(\eta^4\text{-}1,4,5\text{-}CpH)$	$CpTl + (Rh(CO)_2Cl)_2, \Delta$	X-Ray, IR, ^1H, FAB-MS		16
Fulvalenes $[1,3\text{-}({}^tBu)_2H_2C_5\text{-}C_5H_2({}^tBu)\text{-}7,9 = Cp\text{-}Cp]$ Main group	In	InCp-CpIn				221
	Tl	TlCp-CpTl				221
	Ge	Ge(Cp-Cp)₂Ge			Chiral complex	221
	Sn	Sn(Cp-Cp)₂Sn			Chiral complex	221
Transition metals	Cr	$(Cp\text{-}Cp)[Cr(CO)_3]_2$	$HCp\text{-}CpH + Cr(NCMe)_3(CO)_3$	IR, ^1H, MS	Cr—Cr bond	222
	Mo	$(Cp\text{-}Cp)[Mo(CO)_6]_2$	$HCp\text{-}CpH + Mo(CO)_6$	IR, ^1H, ^{13}C, MS	Mo—Mo bond	222
		$(OC)_3MoCp\text{-}CpFe(CO)_2$	$HCp\text{-}CpH + Fe(CO)_3 + Mo(CO)_6$	IR, ^1H, MS		222

W	(Cp–Cp)[W(CO)$_3$]$_2$	HCp–CpH + W(NCMe)$_3$(CO)$_3$	IR, ^{1}H, ^{13}C, MS	W—W bond	222
Fe	(Cp–Cp)[Fe(CO)$_2$]$_2$	HCp–CpH + Fe(CO)$_5$	IR, ^{1}H, MS	Fe—Fe bond	222
	(OC)$_2$FeCp–CpMo(CO)$_3$	HCp–CpH + Fe(CO)$_5$ + Mo(CO)$_6$	IR, ^{1}H, Ms		222

Ethylene-bis(4,5,6,7-tetrahydro-1-indenyl)
[Et'(IndH$_4$)$_2$ = Cp$_2$]

Ti W	Cp$_2$TiCl$_2$ as (S)-W, (R)-W, and meso-W	Et'(Ind)$_2$Li$_2$ + TiCl$_4$ → Et'(Ind)$_2$TiCl$_2$; +H$_2$/Pd–C catalyst (hydrogenation) → **Et'(IndH$_4$)$_2$TiCl$_2$**	X-ray (for both the racemate and the separated meso form), ^{1}H; X-Ray (rac) redetermination, ^{1}H, ^{13}C	Racemic mixture and meso form, separation of meso form; racemate by chromatography, irradiation converts the meso form to the racemate	19 / 223
Wa	(S)-Cp$_2$Ti(binaphtholate)-(S)	W + (S)-1,1'-bi-2-naphthol/Na	X-Ray, ^{1}H, specific rotation	(R)-W enantiomer does not form a complex with (S)-binaphthol, cleavage of Wa with HCl gives the pure (S)-W enantiomer	19
	Cp$_2$Ti(L)$_2$ [L = N-acetyl-L-leucine, N-acetyl-L-methionine, (−)-camphonate, O-acetyl-R-mandelate]	W + 2 Et$_3$N + 2 HL	X-Ray, ^{1}H	Diastereomers separated by fractional crystallation, decomposition, with MeMgCl + HCl yields enantiomerically pure (S)- or (R)-W	219
Zr X	Cp$_2$ZrCl$_2$	Et'(Ind)$_2$Li$_2$ + ZrCl$_4$ → Et'(Ind)$_2$ZrCl$_2$; +H$_2$/PtO$_2$ catalyst (hydrogenation) → **Et'(IndH$_4$)$_2$ZrCl$_2$**	X-Ray, ^{1}H; X-Ray redetermination, ^{1}H, ^{13}C	Only racemic mixture obtained from reaction, no meso isomer	224 / 223
	Cp$_2$Zr(L)$_2$ (L = O-acetyl-R-mandelate)	X + 2 Et$_3$N + 2 HL	X-Ray, ^{1}H	Diastereomers, decomposition, with MeMgCl + HCl yields pure (S)- or (R)-X	219
Hf	Cp$_2$HfCl$_2$	Same as for Zr analog	X-ray	Racemic mixture	225

The above Cp$_2$MCl$_2$ species are immediate precursors for highly active Ziegler–Natta catalysts for the preparation of isotactic polypropylene; with methylalumoxane as a cocatalyst the racemic mixtures can be directly employed in stereospecific propylene polymerizations (see Section IV,D,2) (225,227,228); the enantiomers are employed in asymmetric hydrooligomerizations of α-olefins (230)

1,2,3,4,5,6,7,8-**Octahydrofluorenyl** (m = n = 4) and related tricyclic derivatives

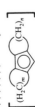

(continued)

BULKY CYCLOPENTADIENYL METAL COMPLEXES (*continued*)

Bulky Cp	Metal	Complex[b]	Method of preparation	Characterization[c]	Comments	Ref.
	Fe Y	$[CpFe(CO)(\mu\text{-}CO)]_2$ ($m = 4$, $n = 4$, 5, or 6)	$n = 4$: $Fe_3(CO)_{12}$ + 1,7-cyclotridecadiyne; $n = 5$: $Fe_3(CO)_{12}$ or $Fe(CO)_5$ + 1,8-cyclotetradecadiyne; $n = 6$: $Fe_3(CO)_{12}$ + 1,8-cyclopentadecadiyne	IR, ^1H		231
		$CpFe(CO)(SnPh_3)$ [$m = 4$; $n = 4$, 5, or 6 (or $m = n = 5$?)]	Y + Na/Hg + Ph_3SnCl	IR, ^1H		231
		$CpFe(CO)_2(C_6F_5)$ [$m = 4$; $n = 5$ or 6 (or $m = n = 5$?)]	Y + Na/Hg + C_6F_6	IR, ^1H, MS		231
		$CpFe(CO)_2I$ ($m = n = 5$)	Y + I_2	IR, ^1H		231

Diborna-cyclopentadienyl (5) (chiral ligand)

Bulky Cp	Metal	Complex[b]	Method of preparation	Characterization[c]	Comments	Ref.
	Li	$CpLi(OEt_2)$	CpH + $^nBuLi/Et_2O$	^1H, ^{13}C		20
	Zr	$CpZrCl_3$	CpLi + $ZrCl_4$	^1H, ^{13}C, specific rotation	Catalyst for asymmetric hydroxylation of 1-naphthol with pyruvate ester	20
	Hf	$CpHfCl_3$	CpLi + $HfCl_4$	^1H, ^{13}C, specific rotation		20

Fulvenes, cyclopentadienylides

$Ph_4C_5{=}CH_2 \rightleftharpoons Ph_4C_5^-\!{-}CH_2^+$

	Cr	$CpCr(CO)_3$	Cp + $(MeCN)_3Cr(CO)_3$, Δ	IR, ^1H		315

$Ph_4C_5{=}EPh_3 \rightleftharpoons Ph_4C_5^-\!{-}E^+Ph_3$ (E = P, As)

	Mo Z	$CpMo(CO)_3$ [$CpMo(CO)_2N_2Ph'OMe\text{-}p$] [$PF_6$] (E = P)	Cp + $(MeCN)_3Mo(CO)_3$; Z + $N_2Ph'OMe\text{-}p$	IR; IR		258; 258

$R_2H_2C_5{=}CHR \rightleftharpoons R_2H_2C_5^-\!{-}CHR^+$
R = Ph, Et, Pr

	Fe	$(OC)_3Fe(\eta^5\text{-}R_2H_2C_5{-}\!{-}CHR\text{-}\eta^1)Fe(CO)_4$			Mass spectral investigation	316

378

R = ^{t}Bu

Rh	CpRh(cod)	Rh(cod)Cl + AgPF$_6$ + ^{t}BuC≡CH	X-Ray, ^{1}H, ^{13}C	Template synthesis, structure and NMR show a preference for the ylidic resonance form	317
Mixed systems					
Ph$_2$Me(MeO$_2$C)(EtO$_2$C)C$_5$ Co	(η^4-CpH)CoC$_5$H$_5$	$\overline{\text{(MeO}_2\text{C)}}$C=C(Me)(Ph)C=C(Ph)Co (C$_5H_5$)PPh$_3$ + N$_2$CH(CO$_2$Et)	Data given as suppl. mat.	⎱ Incorporation of a carbene in the metallacyclopentadiene	21,22
Ph$_2$H(MeO$_2$C)$_2$C)(EtO$_2$C)C$_5$ Co	(η^4-CpH)CoC$_5$H$_5$	(MeO$_2$C)$_2$C=CH(Ph)C=C(Ph)Co (C$_5$H$_5$)PPh$_3$ + N$_2$CH(CO$_2$Et)	X-Ray, other data in suppl. mat.		22
Ph$_2$(MeO$_2$C)$_2$(HO)C$_5$ Ru	[(Ph$_2$(MeO$_2$C)$_2$C$_5$O(Ru(CO)$_2$H]$_2$	[η^4-Ph$_2$(MeO$_2$C)$_2$C$_5$)C$_4$C=O) Ru(CO)$_3$ + Δ, propanol	IR, ^{1}H	O—H—O and Ru—H—Ru bridge, see also under modifications of Ph$_4$HC$_5$	266
(Me$_3$Si)$_2$[Me$_2$NC*H(Me)]H$_2$C$_5$ ([2-(α-dimethylamino)ethyl]-1,3-bis(trimethylsilyl)-Cp)					
α ... Fe	CpFeC$_5$H$_5$	[Me$_2$NCH(Me)]H$_4$C$_5$FeC$_5$H$_5$ + 2 (^{n}BuLi + Me$_3$SiCl)	^{1}H, ^{29}Si, MS	Template synthesis	23
α	CpFeC$_5$H$_4$(SiMe$_3$)	[Me$_2$NCH(Me)]H$_4$C$_5$FeC$_5$H$_5$ + 2 (^{n}BuLi/tmeda + Me$_3$SiCl)	X-ray, ^{1}H, ^{29}Si, MS	Template synthesis	23
	CpFeC$_5$H$_3$(SiMe$_3$)$_2$	α + ^{n}BuLi/tmeda + Me$_3$SiCl	MS		23
	CpFeC$_5$H$_3$(SiMe$_3$)(PPh$_2$)	α + ^{n}BuLi/tmeda + Ph$_2$PCl	MS		23
1,2,4-(Me$_3$Si)$_3$-2-[Me$_2$NC*H(Me)]HC$_5$ Fe	CpFeC$_5$H$_3$(SiMe$_3$)$_2$	α + ^{n}BuLi/tmeda + Me$_3$SiCl	MS		23
1,3-(Me$_3$Si)$_2$-2-[Me$_2$NC*H(Me)]-4-(Ph$_2$P)HC$_5$ Fe	CpFeC$_5$H$_3$(SiMe$_3$)(PPh$_2$)	α + ^{n}BuLi/tmeda + Ph$_2$PCl	MS		23

^{a} Abbreviations: acac, acetylacetonate; bipy, 2,2'-bipyridine; $^{n/i/t}$Bu, normal-/iso-/tertiary-butyl; ^{13}C, carbon-13 NMR; CV, cyclovoltametry; cod, 1,5-cyclooctadiene; cot, 1,3,5,7-cyclooctatetraene; Cp, cyclopentadiene/-yl (cf. footnote 1 in text); CPMAS, cross-polarization magic angle-spinning (solid-state NMR); Δ, heat, thermal energy; dme, dimethoxyethane (MeOCH$_2$CH$_2$OMe); dppe, 1,2-diphenyl(phosphino)ethane (Ph$_2$PCH$_2$CH$_2$PPh$_2$); dbcot, dibenzo[a,e]cyclooctene; diglyme, diethylene glycol dimethylether (MeOC$_2$H$_4$OC$_2$H$_4$OMe); EPR, electron paramagnetic resonance; Et, ethyl; FAB-MS, fast atom bombardment MS; FD-MS, field desorption MS; ΔG‡, free activation enthalpy; ^{1}H, proton NMR; hν, light energy; IR, infrared spectrometry; Me, methyl; MS, mass spectrometry; nbd, norbornadiene; NMR, nuclear magnetic resonance; ox, oxidation/oxidative; Ph, phenyl; Ph', disubstituted phenyl (—C$_6$H$_4$—); $^{n/i}$Pr, normal-/iso-propyl; red, reduction/reductive; thf, tetrahydrofuran; tmeda, tetramethylethylenediamine (Me$_2$NCH$_2$CH$_2$NMe$_2$); UV/VIS, ultraviolet/visible spectroscopy.

^{b} A pentahapto, η^5, bonding mode can be assumed for Cp (R$_5$C$_5$) unless noted otherwise.

^{c} Elemental analyses and melting points are not listed separately as a mode of characterization.

^{d} Additional analytical data hinted in Ref. 40 for support of the formulations and structures of the compounds given therein.

^{e} Spectroscopic data for this and the following complexes (taken from Ref. 236) are not explicitly given in the reference. Only nonroutine data are discussed in the publication.

ACKNOWLEDGMENTS

The authors' work cited in this review could not have been carried out without the inspiring and engaging cooperation of D. Bernhardt, M. Dettlaff, Dr. A. Dietrich, L. Esser, T. Ghodsi, F. Görlitz, Dr. E. Hahn, U. Kieper, Dr. R. Köhn, Dr. A. Kucht, Dr. H. Kucht, A. Lentz, and J. Loebel. The crystallographic investigations supporting our own work were undertaken in collaboration with Prof. Dr. J. Pickardt, Technische Universität Berlin, and Prof. Dr. van der Helm and the crystallographic staff of the University of Oklahoma, Norman, Oklahoma. Their help is kindly acknowledged. We are also grateful to the Deutsche Forschungsgemeinschaft, the Fonds der Chemischen Industrie, and the Bundesministerium für Bildung und Wissenschaft (Graduiertenkolleg "Synthese und Strukturaufklärung niedermolekularer Verbindungen") for financial support of our work cited in this article. Furthermore, C.J. thanks the BASF AG, Ludwigshafen, for a postdoctoral fellowship and Dr. H. Sitzmann (University of Kaiserslautern, Germany) for making part of his work available to us prior to publication.

REFERENCES

1. T. J. Kealy and P. L. Pauson, *Nature (London)* **168**, 1039 (1951); S. A. MIller, J. A. Tebboth, and J. F. Tremaine, *J. Chem. Soc.,* 632 (1952).
2. E. O. Fischer and W. Pfab, *Z. Naturforsch.* **7B**, 377 (1952); E. Ruch and E. O. Fischer, *Z. Naturforsch.* **7B**, 632 (1952).
3a. G. Wilkinson, M. Rosenblum, M. C. Whiting, and R. B. Woodward, *J. Am. Chem. Soc.* **74**, 2125 (1952).
3b. R. B. Woodward, M. Rosenblum, and M. C. Whiting, *J. Am. Chem. Soc.* **74**, 3458 (1952).
4. The Nobel Prize in Chemistry in 1973 was awarded to E. O. Fischer and G. Wilkinson for their pioneering work, performed independently, on the chemistry of metallocenes, so-called sandwich compounds.
5. J. D. Dunitz and L. E. Orgel, *Nature (London)* **171**, 121 (1953).
6. G. Wilkinson, F. G. A. Stone, and W. Abel, eds., "Comprehensive Organometallic Chemistry," Vols. 3–7. Pergamon, Oxford, 1982; *Adv. Organomet. Chem.* Vols. 1 (1964)–29 (1989).
7. R. B. King and M. B. Bisnette, *J. Organomet. Chem.* **8**, 287 (1967).
8. See, for example, R. B. King, *Coord. Chem. Rev.* **20**, 155 (1976); P. M. Maitlis, *Acc. Chem. Res.* **11**, 301 (1978); P. T. Wolzanski and J. E. Bercaw, *Acc. Chem. Res.* **13**, 121 (1980); A. C. Campbell and D. L. Lichtenberger, *J. Am. Chem. Soc.* **103**, 6389 (1981); G. P. Pez and J. N. Armor, *Adv. Organomet. Chem.* **19**, 1 (1981); T. J. Marks, *Science* **217**, 989 (1982); E. J. Miller, S. J. Landon, and T. B. Brill, *Organometallics* **4**, 533 (1985).
9. P. Jutzi, *Pure Appl. Chem.* **61**, 1731 (1989).
10. M. A. El-Hinnawi and M. A. Kobeissi, *Inorg. Chim. Acta* **166**, 99 (1989); K. E. Du Ploy, C. F. Marais, L. Carlton, R. Hunter, J. C. A. Boeyens, and N. J. Coville, *Inorg. Chem.* **28**, 3855 (1989); P. C. Blake, M. F. Lappert, R. G. Taylor, J. L. Atwood, and H. Zhang, *Inorg. Chim. Acta* **139**, 13 (1987); M. F. Lappert, A. Singh, J. L. Atwood, and W. E. Hunter, *J. Chem. Soc., Chem. Commun.,* 1190, 1191 (1981); J. Jeffrey, M. F. Lappert, and P. I. Riley, *J. Organomet. Chem.* **181**, 25 (1979).
11. See, for example, I. A. Lobanova and V. I. Zdanovich, *Usp. Khim.* **57**, 1688 (1988); J. M. O'Connor and C. P. Casey, *Chem. Rev.* **87**, 307 (1987); B. Kanellakopulos and K. W. Bagnall, *in* "MTP (Med. Tech. Publ. Co.) International Reviews in Science: Inor-

ganic Chemistry, Series One" (K. W. Bagnall, ed.), Vol. 7, Issue Lanthanides Actinides, p. 299. Butterworth, London, 1972; *Chem. Abstr.* **76**, 85861y (1973).

12. For metallocenophanes, see, for example, W. E. Watts, *Organomet. Chem. Rev.* **2**, 231 (1967); U. T. Mueller-Westerhoff, *Angew. Chem.* **98**, 700 (1986); *Angew. Chem., Int. Ed. Engl.* **25**, 702 (1986). F. Vögtle, "Cyclophan-Chemie," p. 67, 455ff. Teubner, Stuttgart, Germany, 1990; M. Hisatome, J. Watanabe, Y. Kawajiri, K. Yamakawa, and Y. Iitaka, *Organometallics* **9**, 497 (1990); P. Jutzi, U. Siemeling, A. Müller, and H. Bögge, *Organometallics* **8**, 1744 (1989); M. Hisatome and K. Yamakawa, *Yuki Gosei Kagaku Kyokaishi* **48**, 319 (1990); *Chem. Abstr.* **113**, 78449z (1990), and references therein. For cyclophanes, see, in addition, P. M. Keehn and S. M. Rosenfield, eds., "Cyclophanes," Vols. 1 and 2. Academic Press, New York, 1983.

13. H. van der Heijden, A. W. Gal, P. Pasman, and A. G. Orpen, *Organometallics* **4**, 1847 (1985).

14. B. Gloaguen and D. Astruc, *J. Am. Chem. Soc.* **112**, 4607 (1990).

15. M. I. Bruce, J. R. Rodgers, and J. K. Walton, *J. Chem. Soc., Chem. Commun.*, 1253 (1981).

16. M. I. Bruce, P. A. Humphrey, J. K. Walton, B. W. Skelton, and A. H. White, *J. Organomet. Chem.* **333**, 393 (1987).

17. M. I. Bruce, P. A. Humphrey, M. L. Williams, B. W. Skelton, and A. H. White, *Aust. J. Chem.* **42**, 1847 (1989).

18. R. Brand, H.-P. Krimmer, H.-J. Lindner, V. Sturm, and K. Hafner, *Tetrahedron Lett.* **23**, 5131 (1982).

19. F. R. W. P. Wild, L. Zsolnai, G. Huttner, and H. H. Brintzinger, *J. Organomet. Chem.* **232**, 233 (1982).

20. G. Erker and A. A. H. van der Zeijden, *Angew. Chem.* **102**, 543 (1990); *Angew. Chem., Int. Ed. Engl.* **29**, 512 (1990); G. Erker, *J. Organomet. Chem.* **400**, 185 (1990).

21. J. M. O'Connor and J. A. Johnson, *Synlett.* **1**, 57 (1989).

22. J. M. O'Connor, L. Pu, R. Uhrhammer, J. A. Johnson, and A. L. Rheingold, *J. Am. Chem. Soc.* **111**, 1889 (1989).

23. I. R. Butler, W. R. Cullen, and S. J. Rettig, *Organometallics* **5**, 1320 (1986).

24. For $Ph_4C_4C=O$ complexes, see, for example, J. J. Eisch, J. E. Galle, A. A. Aradi, M. P. Boleslawski, and P. Marek, *J. Organomet. Chem.* **312**, 399 (1986); M. Abed, I. Goldberg, Z. Stein, and Y. Shvo, *Organometallics* **7**, 2054 (1988); M. J. Mays, M. J. Morris, P. R. Raithby, Y. Shvo, and D. Czarkie, *Organometallics* **8**, 1162 (1989); R. L. Beddoes, E. S. Cook, and M. J. Morris, *Polyhedron* **8**, 1810 (1989); Y. Shvo, M. Abed, Y. Blum, and R. M. Laine, *Isr. J. Chem.* **27**, 267 (1986), and references therein.

25. G. Maier, S. Pfriem, U. Schäfer, K.-D. Malsch, and R. Matusch, *Chem. Ber.* **114**, 3965 (1981).

26. S. S. Hirsch and W. J. Bailey, *J. Org. Chem.* **43**, 4090 (1978).

27. T. Niem and M. D. Rausch, *J. Org. Chem.* **42**, 275 (1977).

28. For a review on functionally substituted cyclopentadienyl metal compounds, see D. W. Macomber, W. P. Hart, and M. D. Rausch, *Adv. Organomet. Chem.* **21**, 1 (1982).

29. K. Hafner, K. H. Vöpel, G. Ploss, and C. König, *Justus Liebigs Ann. Chem.* **661**, 52 (1963); O. W. Webster, *J. Am. Chem. Soc.* **88**, 3046 (1966).

30. R. E. Christopher and L. M. Venanzi, *Inorg. Chim. Acta* **7**, 489 (1973).

31. J. Wislicenus and F. H. Newman, *Justus Liebigs Ann. Chem.* **302**, 236 (1898).

32. J. Wislicenus and H. Carpenter, *Justus Liebigs Ann. Chem.* **302**, 223 (1898).

33. K. Ziegler and B. Schnell, *Justus Liebigs Ann. Chem.* **445**, 266 (1925).

34. H. Sitzmann, *Z. Naturforsch.* **44B**, 1293 (1989).

35. H. Sitzmann, *J. Organomet. Chem.* **354**, 203 (1988).

36. D. Stein and H. Sitzmann, *J. Organomet. Chem.* **402**, 249 (1991).
37. J. Okuda, R. C. Murray, J. C. Dewan, and R. R. Schrock, *Organometallics* **5**, 1681 (1986).
38. The 1990–1991 Aldrich catalog lists tetraphenyl- and pentaphenylcyclopentadiene, as well as 5-bromopentaphenyl- and 2,5,5-tris(trimethylsilyl)cyclopentadiene.
39. C. Janiak, H. Schumann, C. Stader, B. Wrackmeyer, and J. J. Zuckerman, *Chem. Ber.* **121**, 1745 (1988).
40. D. W. Slocum, S. Duraj, M. Matusz, J. L. Cmarik, K. M. Simpson, and D. A. Owen, *in* "Metal Containing Polymeric Systems" (J. E. Sheats, C. E. Carraher, Jr., and C. U. Pittmann, Jr., eds.), p. 59. Plenum, New York, 1985; D. W. Slocum, S. Johnson, M. Matusz, S. Duraj, J. L. Cmarik, K. M. Simpson, and D. A. Owen, *Polym. Mater. Sci. Eng.* **49**,353 (1983).
41. M. A. Ogliaruso, M. G. Romanelli, and E. I. Becker, *Chem. Rev.* **65**, 261 (1965).
42. F. C. Leavitt, T. A. Manuel, F. Johnson, L. U. Matternas, and D. S. Lehman, *J. Am. Chem. Soc.* **82**, 5099 (1960); E. H. Braye, W. Hübel, and I. Caplier, *J. Am. Chem. Soc.* **83**, 4406 (1961).
43. H. Schumann, C. Janiak, and J. J. Zuckerman, *Chem. Ber.* **121**, 207 (1988); H. Schumann, C. Janiak, and J. J. Zuckerman, *Chem. Ber.* **121**, 1869 (1988).
44. H. Kainer, *Liebigs Ann. Chem.* **578**, 232 (1952).
45. L. Mehr, E. I. Becker, and P. E. Spoerri, *J. Am. Chem. Soc.* **77**, 984 (1955).
46. L. D. Field, K. M. Ho, C. M. Lindall, A. F. Masters, and A. G. Webb, *Aust. J. Chem.* **43**, 281 (1990).
47. M. P. Cava and K. Narasimhan, *J. Org. Chem.* **34**, 3641 (1969); H. M. N. Bandara, N. D. S. Rajasekera, and S. Sotheeswaran, *Tetrahedron* **30**, 2587 (1974).
48. M. P. Castellani, J. M. Wright, S. J. Geib, A. L. Rheingold, and W. C. Trogler, *Organometallics* **5**, 1116 (1986).
49. N. O. V. Sonntag, S. Linder, E. I. Becker, and P. E. Spoerri, *J. Am. Chem. Soc.* **75**, 2283 (1953); Y. N. Kreitsberga and O. Y. Neiland, *Zh. Org. Khim.* **14**, 1640 (1978).
50. P. L. Pauson, *J. Am. Chem. Soc.* **76**, 2187 (1954).
51. J. W. Chambers, A. J. Baskar, S. G. Bott, J. L. Atwood, and M. D. Rausch, *Organometallics* **5**, 1635 (1986).
52. A. Davison and P. E. Rakita, *J. Am. Chem. Soc.* **90**, 4479 (1968); A. Davison and P. E. Rakita, *Inorg. Chem.* **9**, 289 (1970); Y. A. Ustynyuk, A. V. Kisin, I. M. Pribytkova, A. A. Zenkin, and N. D. Antonova, *J. Organomet. Chem.* **42**, 47 (1972), and references therein.
53. P. Jutzi and R. Sauer, *J. Organomet. Chem.* **50**, C29 (1973).
54. Y. A. Ustynyuk, Y. N. Luzikov, V. I. Mstislavsky, A. A. Azizov, and I. M. Pribytkova, *J. Organomet. Chem.* **96**, 335 (1975).
55. Y. A. Ustynyuk, P. I. Zakharov, A. A. Azizov, V. K. Potapov, and I. M. Pribytkova, *J. Organomet. Chem.* **88**, 37 (1975).
56. P. Jutzi, *Chem. Rev.* **86**, 983 (1986), and references therein
57. R. Riemschneider, *Z. Naturforsch.* **18B**, 641 (1963).
58. C. G. Vernier and E. W. Casserly, *J. Am. Chem. Soc.* **112**, 2808 (1990).
59. K. Alder and H. J. Ache, *Chem. Ber.* **95**, 503 (1962).
60a. W. Best, B. Fell, and G. Schmitt, *Chem. Ber.* **109**, 2914 (1976).
60b. D. Stein and H. Sitzmann, *J. Organomet. Chem.* **402**, C1 (1991).
61. F. Wudl, D. Nalewajek, F. J. Rotella, and E. Gebert, *J. Am. Chem. Soc.* **103**, 5885 (1981).
62. E. Le Goff and R. B. LaCount, *J. Org. Chem.* **29**, 423 (1964).
63. O. Diels, *Chem. Ber.* **75**, 1452 (1942); O. Diels and U. Kock, *Liebigs Ann. Chem.* **556**, 38 (1944).

64. R. C. Cookson, J. B. Henstock, J. Hudec, and B. R. D. Whitear, *J. Chem. Soc. C,* 1986 (1967).
65. R. A. Brand and J. E. Mulvaney, *J. Org. Chem.* **45,** 633 (1980).
66. H. E. Zimmerman and S. M. Aasen, *J. Am. Chem. Soc.* **99,** 2342 (1977); H. E. Zimmerman and S. M. Aasen, *J. Org. Chem.* **43,** 1493 (1978); H. E. Zimmerman and D. J. Kreil, *J. Org. Chem.* **47,** 2060 (1982); K. H. Holm and L. Skattebøl, *Acta Chem. Scand.* **B38,** 783 (1984); I. N. Domin, V. N. Plotkin, and M. I. Komendantov, *Zh. Org. Khim.* **21,** 2223 (1985).
67. R. Breslow and H. W. Chang, *J. Am. Chem. Soc.* **87,** 2200 (1965).
68. W. Blum, H. Franke, W. J. Richter, and H. Schwarz, *Chem. Ber.* **116,** 2931 (1983).
69. F. G. Bordwell, J.-P. Cheng, and M. J. Bausch, *J. Am. Chem. Soc.* **110,** 2872 (1988).
70. B. J. Tabner and T. Walker, *J. Chem. Soc., Perkin Trans. 2,* 1304 (1975).
71. A. Sekiguchi, H. Tanikawa, and W. Ando, *Organometallics* **4,** 584 (1985); I. V. Borisova, Y. N. Luzikov, N. N. Zemlyanski, Y. A. Ustynyuk, and I. P. Beletskaya, *J. Organomet. Chem.* **268,** 11 (1984); G. S. Harris, D. Lloyd, W. A. MacDonald, and I. Gosney, *Tetrahedron* **39,** 297 (1983); G. Seitz, R. A. Olson, and T. Kämpchen, *Chem. Ber.* **115,** 3756 (1982); B. H. Freeman and D. Lloyd, *Tetrahedron* **30,** 2257 (1974), and references therein.
72. R. K. Haynes, J. M. Peters, and I. D. Wilmot, *Aust. J. Chem.* **33,** 2653 (1980); M. Mori, M. Nojima, and S. Kusabayashi, *J. Am. Chem. Soc.* **109,** 4407 (1987).
73. P. R. Kumar, *J. Chem. Soc., Chem. Commun.* 509 (1989); Y. D. Samuilov, R. L. Nurullina, and A. I. Konovalov, *Zh. Org. Khim.* **18,** 2253 (1982); Y. D. Samuilov, S. V. Bukharov, and A. I. Konovalov, *Zh. Org. Khim.* **17,** 2389 (1981).
74. B. H. Freeman, J. M. F. Gagan, and D. Lloyd, *Tetrahedron* **29,** 4307 (1973).
75. H. Dürr and L. Schrader, *Chem. Ber.* **102,** 2026 (1969).
76. C. Adachi, T. Tsutsui, and S. Saito, *Appl. Phys. Lett.* **56,** 799 (1990).
77. A. G. Lee, *Organomet. React.* **5,** 1 (1975); G. B. Deacon, A. J. Kopkick, and T. D. Tuong, *Aust. J. Chem.* **37,** 517 (1984).
78. A. Schott, H. Schott, G. Wilke, J. Brandt, H. Hoberg, and E. G. Hoffmann, *Justus Liebigs Ann. Chem.,* 508 (1973).
79. W. Hübel and R. Merényi, *J. Organomet. Chem.* **2,** 213 (1964).
80. T. R. Jack, C. J. May, and J. Powell, *J. Am. Chem. Soc.* **99,** 4707 (1977).
81. S. F. Pedersen, R. R. Schrock, M. R. Churchill, and H. J. Wasserman, *J. Am. Chem. Soc.* **104,** 6808 (1982).
82. R. R. Schrock, S. F. Pedersen, M. R. Churchill, and J. W. Ziller, *Organometallics* **3,** 1574 (1984).
83. R. R. Schrock, J. S. Murdzek, J. H. Freudenberger, M. R. Churchill, and J. W. Ziller, *Organometallics* **5,** 25 (1986).
84. J. F. Buzinkai and R. R. Schrock, *Organometallics* **6,** 1447 (1987).
85. A. H. Cowley, P. Jutzi, F. X. Kohl, J. G. Lasch, N. C. Norman, and E. Schlüter, *Angew. Chem.* **96,** 603 (1984); *Angew. Chem., Int. Ed. Engl.* **23,** 616 (1984).
86. See, for example: J. J. Bishop, A. Davison, M. L. Katcher, D. W. Lichtenberg, R. E. Merrill, and J. C. Smart, *J. Organomet. Chem.* **27,** 241 (1971); M. D. Rausch and D. J. Ciappenelli, *J. Organomet. Chem.* **10,** 127 (1967); S. I. Goldberg, D. W. Mayo, M. Vogel, H. Rosenberg, and M. Rausch, *J. Org. Chem.* **24,** 824 (1959). A. N. Nesmeyanov, E. G. Perevalova, E. G. Golovinya, and D. A. Nesmeyanov, *Dokl. Akad. Nauk. SSSR* **97,** 659 (1954).
87. F. Moulines and D. Astruc, *J. Chem. Soc., Chem. Commun.,* 614 (1989); F. Moulines and D. Astruc, *Angew. Chem.* **100,** 1394 (1988); *Angew. Chem., Int. Ed. Engl.* **27,** 1347 (1988); D. Astruc, *Acc. Chem. Res.* **19,** 377 (1986), and references therein.
88. K. Schlögl and M. Peterlik, *Monatsh. Chem.* **93,** 1328 (1962).

89. A. N. Nesmeyanov, K. N. Anisimov, N. E. Kolobova, and I. B. Zlotina, *Izv. Akad. Nauk SSSR, Ser. Khim.,* 1326 (1964).
90. K. Sünkel and D. Motz, *Angew. Chem.* **100,** 970 (1988); *Angew. Chem., Int. Ed. Engl.* **27,** 939 (1988).
91. K. Sünkel and D. Steiner, *Chem. Ber.* **122,** 609 (1989).
92. N. A. Grabowski, R. P. Hughes, B. S. Jaynes, and A. L. Rheingold, *J. Chem. Soc., Chem. Commun.,* 1694 (1986).
93. R. P. Hughes and D. J. Robinson, *Organometallics* **8,** 1015 (1989).
94. J. Müller, M. Tschampel, and C. Krüger, *Z. Naturforsch.* **43B,** 1519 (1988).
95. S. McVey and P. L. Pauson, *J. Chem. Soc.,* 4312 (1965).
96. M. P. Castellani, S. J. Geib, A. L. Rheingold, and W. C. Trogler, *Organometallics* **6,** 1703 (1987).
97. M. P. Castellani, S. J. Geib, A. L. Rheingold, and W. C. Trogler, *Organometallics* **6,** 2524 (1987).
98. H. Schumann, C. Janiak, R. D. Köhn, J. Loebel, and A. Dietrich, *J. Organomet. Chem.* **365,** 137 (1989).
99. J. Okuda and E. Herdtweck, *Chem. Ber.* **121,** 1899 (1988).
100. J. Okuda, *J. Organomet. Chem.* **333,** C41 (1987).
101. U. Kölle, B. Fuss, F. Khouzami, and J. Gersdorf, *J. Organomet. Chem.* **290,** 77 (1985).
102. J. E. Huheey, "Inorganic Chemistry," 3rd Ed., p. 547ff. Harper and Row, New York, 1983; F. A. Cotton and G. Wilkinson, "Advanced Inorganic Chemistry," 5th Ed., Chap. 29. Wiley (Interscience), New York, 1988.
103. R. Hoffmann, *Am. Sci.* **75,** 619 (1988).
104. S. G. Baxter, A. H. Cowley, J. G. Lasch, M. Lattman, W. P. Sharum, and C. A. Stewart, *J. Am. Chem. Soc.* **104,** 4064 (1982); M. Lattman and A. H. Cowley, *Inorg. Chem.* **23,** 241 (1984).
105. H. Schumann, C. Janiak, E. Hahn, J. Loebel, and J. J. Zuckerman, *Angew. Chem.* **97,** 765 (1985); *Angew. Chem., Int. Ed. Engl.* **24,** 773 (1985).
106. H. Schumann, C. Janiak, E. Hahn, C. Kolax, J. Loebel, M. D. Rausch, J. J. Zuckerman, and M. J. Heeg, *Chem. Ber.* **119,** 2656 (1986).
107. H. Schumann, C. Janiak, J. Pickardt, and U. Börner, *Angew. Chem.* **99,** 788 (1987); *Angew. Chem., Int. Ed. Engl.* **26,** 789 (1987).
108. H. Schumann, C. Janiak, M. A. Khan, and J. J. Zuckerman, *J. Organomet. Chem.* **354,** 7 (1988).
109. H. Schumann, C. Janiak, F. Görlitz, J. Loebel, and A. Dietrich, *J. Organomet. Chem.* **363,** 243 (1989).
110. E. O. Fischer and H. Grubert, *Z. Naturforsch,* **11B,** 423 (1956); L. D. Dave, D. F. Evans, and G. Wilkinson, *J. Chem. Soc.,* 3684 (1959); J. V. Scibelli and M. D. Curtis, *J. Am. Chem. Soc.* **95,** 924 (1973); P. Jutzi and F. Kohl, *J. Organomet. Chem.* **164,** 141 (1979).
111. P. Jutzi, F. Kohl, P. Hofmann, C. Krüger, and Y.-H. Tsay, *Chem. Ber.* **113,** 757 (1980).
112. M. Grenz, E. Hahn, W. W. du Mont, and J. Pickardt, *Angew. Chem.* **96,** 69 (1984); *Angew. Chem., Int. Ed. Engl.* **23,** 61 (1984).
113. H. Adams, N. A. Bailey, A. F. Browning, J. A. Ramsden, and C. White, *J. Organomet. Chem.* **387,** 305 (1990).
114. J. Powell and N. I. Dowling, *Organometallics* **2,** 1742 (1983).
115. K. Broadley, G. A. Lane, N. G. Connelly, and W. E. Geiger, *J. Am. Chem. Soc.* **105,** 2486 (1983).
116. N. G. Connelly, W. E. Geiger, G. A. Lane, S. J. Raven, and P. H. Rieger, *J. Am. Chem. Soc.* **108,** 6219 (1986).

117. A. Eisenberg, A. Shaver, and T. Tsutsui, *J. Am. Chem. Soc.* **102**, 1416 (1980); A. Shaver, A. Eisenberg, K. Yamada, A. J. F. Clark, and S. Farrokyzad, *Inorg. Chem.* **22**, 4154 (1983).

118. M. J. Heeg, C. Janiak, and J. J. Zuckerman, *J. Am. Chem. Soc.* **106**, 4259 (1984).

119. P. G. Harrison and J. J. Zuckerman, *J. Am. Chem. Soc.* **91**, 6885 (1969).

120. M. J. Heeg, R. H. Herber, C. Janiak, J. J. Zuckerman, H. Schumann, and W. F. Manders, *J. Organomet. Chem.* **346**, 321 (1988).

121. R. L. Williamson and M. B. Hall, *Organometallics* **5**, 2142 (1986).

122. U. Kölle and F. Khouzami, *Angew. Chem.* **92**, 658 (1980); *Angew. Chem., Int. Ed. Engl.* **19**, 640 (1980).

123. G. Evrard, P. Piret, G. Germain, and M. Van Meerssche, *Acta Crystallogr. Sect. B: Struct. Crystallogr. Cryst. Chem.* **B27**, 661 (1971).

124. R. Zhang, M. Tsutsui, and D. E. Bergbreiter, *J. Organomet. Chem.* **229**, 109 (1982).

125. L. Knothe, H. Prinzbach, and H. Fritz, *Liebigs Ann. Chem.*, 687 (1977).

126. M. Huhn, W. Kläui, L. Ramacher, R. Herbst-Irmer, and E. Egert, *J. Organomet. Chem.* **398**, 339 (1990).

127. J. J. Zuckerman, in "Chemical Mössbauer Spectroscopy" (R. H. Herber, ed.), p. 267. Plenum, New York, 1984; J. J. Zuckerman, *Adv. Organomet. Chem.* **9**, 21 (1970); N. W. G. Debye and J. J. Zuckerman, in "Determination of Organic Structure by Physical Methods" (F. C. Nachod and J. J. Zuckerman, Eds.), Vol. 5, p. 235. Academic Press, New York, 1973.

128. P. Köpf-Maier and H. Köpf, *Chem. Rev.* **87**, 1137 (1987); P. Köpf-Maier, *Prog. Clin. Biochem. Med.* **10**, 151 (1989).

129. H. Köpf and P. Köpf-Maier, *Angew. Chem.* **91**, 509 (1979); *Angew. Chem., Int. Ed. Engl.* **18**, 477 (1979); P. Köpf-Maier and H. Köpf, *Z. Naturforsch.* **34B**, 805 (1979); P. Köpf-Maier and H. Köpf, *Drugs Fut.* **11**, 297 (1986); J. H. Toney, L. N. Rao, M. S. Murthy, and T. J. Marks, *Breast Cancer Res. Treat.* **6**, 185 (1985).

130. P. Köpf-Maier, H. Köpf, and E. W. Neuse, *J. Cancer Res. Clin. Oncol.* **108**, 336 (1984).

131a. N. Kumano, Y. Nakai, T. Ishikawa, S. Koinumaru, S. Suzuki, and K. Konno, *Sci. Rep. Res. Inst. Tohoku Univ., Ser. A.* **25**, 89 (1978).

131b. M. G. Mulinos and P. Amin, *Fed. Am. Soc. Exp. Biol.* **39**, 747 (1980).

131c. A. J. Crowe, P. J. Smith, and G. Atassi, *Chem.–Biol. Interact.* **32**, 171 (1980).

131d. A. J. Crowe, P. J. Smith, and G. Atassi, *Inorg. Chim. Acta* **93**, 179 (1984).

132. P. Köpf-Maier, C. Janiak, and H. Schumann, *Inorg. Chim. Acta* **152**, 75 (1988); P. Köpf-Maier, C. Janiak, and H. Schumann, *J. Cancer Res. Clin. Oncol.* **114**, 502 (1988).

133. S. Chandrasekhar, B. K. Sadashiva, K. A. Suresh, N. V. Madhusudana, S. Kumar, R. Shashidhar, and G. Venkatesh, *J. Phys. Colloq.*, 120 (1979); R. Fugnitto, H. Strzelecka, A. Zann, J. C. Dubois, and J. Billard, *J. Chem. Soc., Chem. Commun.*, 271 (1980); P. Foucher, C. Destrade, N. H. Tinh, J. Malthete, and A. M. Levelut, *Mol. Cryst. Liq. Cryst.* **108**, 219 (1984), and references therein; B. Kohne and K. Praefcke, *Chimia* **41**, 196 (1987); T. Warmerdam, D. Frenkel, and R. J. J. Zijlstra, *Liq. Cryst.* **3**, 149 (1988).

134. A. M. Giroud-Godquin and J. Billard, *Mol. Cryst. Liq. Cryst.* **66**, 147 (1981); A. M. Giroud-Godquin and J. Billard, *Mol. Cryst. Liq. Cryst.* **97**, 287 (1983); K. Ohta, H. Muroki, A. Takagi, K. Hatada, Hiroshima, I. Yamamoto, and K. Matsuzaki, *Mol. Cryst. Liq. Cryst.* **140**, 131 (1986), and references therein; B. K. Sadashiva and S. Ramesha, *Mol. Cryst. Liq. Cryst.* **141**, 19 (1986); K. Ohta, H. Hasebe, H. Ema, T. Fujimoto, and I. Yamamoto, *J. Chem. Soc., Chem. Commun.*, 1610 (1989); D. Guillon, P. Weber, A. Skoulios, C. Piechocki, and J. Simon, *Mol. Cryst. Liq. Cryst.* **130**, 223 (1985), and references therein; M. J. Cook, M. F. Daniel, K. J. Harrison, N. B. McKeown, and A. J. Thomson, *J. Chem. Soc., Chem. Commun.*, 1086 (1987); J. F. Van

der Pol, E. Neeleman, J. W. Zwikker, R. J. M. Nolte, W. Drenth, J. Aerts, R. Visser, and S. J. Picken, *Liq. Cryst.* **6**, 577 (1989); A.-M. Giroud-Godquin and P. M. Maitlis, *Angew. Chem.* **103**, 370 (1991); *Agnew. Chem., Int. Ed. Engl.* **30**, 325 (1991).

135. J. Bhatt, B. M. Fung, K. M. Nicholas, and C.-D. Poon, *J. Chem. Soc., Chem. Commun.*, 1439 (1988); M. A. Khan, J. C. J. Bhatt, B. M. Fung, K. M. Nicholas, and E. Wachtel, *Liq. Cryst.* **5**, 285 (1989).

136. J. Malthete and J. Billard, *Mol. Cryst. Liq. Cryst.* **34**, 117 (1976).

137. H. Kucht, Ph.D. Thesis, Technische Universität Berlin (1990).

138. M. M. Green, H. Ringsdorf, J. Wagner, and R. Wüstefeld, *Angew. Chem.* **102**, 1525 (1990); *Angew. Chem., Int. Ed. Engl.* **29**, 1478 (1990), and references therein.

139. P. Jutzi, W. Leffers, B. Hampel, S. Pohl, and W. Saak, *Angew. Chem.* **99**, 563 (1987); *Angew. Chem., Int. Ed. Engl.* **26**, 583 (1987).

140. P. Jutzi, *Adv. Organomet. Chem.* **26**, 217 (1986).

141. R. Zerger and G. Stucky, *J. Organomet. Chem.* **80**, 7 (1974).

142. R. A. Williams, T. P. Hanusa, and J. C. Huffman, *J. Chem. Soc., Chem. Commun.*, 1045 (1988).

143. E. Frasson, F. Menegus, and C. Panattoni, *Nature (London)* **199**, 1087 (1963); J. F. Berar, G. Calvarin, C. Pommier, and D. Weigel, *J. Appl. Crystallogr.* **8**, 386 (1975).

144. S. Shibata, L. S. Bartell, and R. M. Gavin, Jr., *J. Chem. Phys.* **41**, 717 (1964); J. K. Tyler, A. P. Cox, and J. Sheridan, *Nature (London)* **183**, 1182 (1959); R. Blom, H. Werner, and J. Wolf. *J. Organomet. Chem.* **354**, 293 (1988); E. Canadell, O. Eisenstein, and J. Rubio, *Organometallics* **3**, 759 (1984).

145. O. T. Beachley, Jr., J. C. Pazik, T. E. Glassman, M. R. Churchill, J. C. Fettinger, and R. Blom, *Organometallics* **7**, 1051 (1988).

146. H. Schumann, C. Janiak, and H. Khani, *J. Organomet. Chem.* **330**, 347 (1987).

147. P. Jutzi, E. Schlüter, S. Pohl, and W. Saak, *Chem. Ber.* **118**, 1959 (1985); P. Jutzi, E. Schlüter, C. Krüger, and S. Pohl, *Angew. Chem.* **95**, 1015 (1983); *Angew. Chem., Int. Ed. Engl.* **22**, 994 (1983).

148. P. Jutzi, W. Leffers, S. Pohl, and W. Saak, *Chem. Ber.* **122**, 1449 (1989).

149. C. P. Morley, P. Jutzi, C. Krüger, and J. M. Wallis, *Organometallics* **6**, 1084 (1987).

150. P. Jutzi, W. Leffers, and G. Müller, *J. Organomet. Chem.* **334**, C24 (1987).

151. P. Jutzi and W. Leffers, *J. Chem. Soc., Chem. Commun.*, 1735 (1985).

152. K. Hedberg, *J. Am. Chem. Soc.* **77**, 6491 (1955); see also, B. Beagley and A. R. Conrad, *Trans. Faraday Soc.* **66**, 2740 (1970); C. Glidewell, D. W. H. Rankin, and A. G. Robiette, *J. Chem. Soc. A;*, 2935 (1970); Gmelin, "Handbook of Inorganic Chemistry," Si, Suppl. (F. Schröder, ed.), Vol. B4, p. 93ff. Springer, Berlin, 1989. H. Bock, I. Göbel, Z. Havlas, S. Liedle, and H. Oberhammer, *Angew. Chem.* **103**, 193 (1991); *Angew Chem. Int. Ed. Engl.* **30**, 187 (1991), and references therein.

153. A. R. Mahjoub, A. Hoser, J. Fuchs, and K. Seppelt, *Angew. Chem.* **101**, 1528 (1989); *Angew. Chem., Int. Ed. Engl.* **28**, 1526 (1989).

154. R. J. Gillespie, "Molecular Geometry." Van Nostrand-Reinhold, New York, 1970.

155. J. L. Atwood, W. E. Hunter, A. H. Cowley, R. A. Jones, and C. A. Stewart, *J. Chem. Soc., Chem. Commun.*, 925 (1981).

156. C. Panattoni, G. Bombieri, and U. Croatto, *Acta Crystallogr.* **21**, 823 (1966).

157. L. Fernholt, A. Haaland, P. Jutzi, F. X. Kohl, and R. Seip, *Acta Chem. Scand.* **A38**, 211 (1984).

158. A. Almenningen, A. Haaland, and T. Motzfeldt, *J. Organomet. Chem.* **7**, 97 (1967).

159. J. Almlöf, L. Fernholt, K. Faegri, A. Haaland, B. E. R. Schilling, R. Seip, and K. Taugbøl, *Acta Chem. Scand.* **A37**, 131 (1983).

160a. K. D. Bos, E. J. Bulten, J. G. Noltes, and A. L. Spek, *J. Organomet. Chem.* **99**, 71 (1975).

160b. P. Jutzi, F. Kohl, C. Krüger, G. Wolmershäuser, P. Hofmann, and P. Stauffert, *Angew. Chem.* **94**, 66 (1981); *Angew. Chem., Int. Ed. Engl.* **21**, 70 (1982).

160c. A. H. Cowley, J. G. Lasch, N. C. Norman, C. A. Stewart, and T. C. Wright, *Organometallics* **2**, 1691 (1983).

160d. H. Wadepohl, H. Pritzkow, and W. Siebert, *Organometallics* **2**, 1899 (1983).

160e. T. S. Dory, J. J. Zuckerman, and C. L. Barnes, *J. Organomet. Chem.* **281**, C1 (1985).

161. P. Jutzi, D. Kanne, and C. Krüger, *Angew. Chem.* **98**, 163 (1986); *Angew. Chem., Int. Ed. Engl.* **25**, 164 (1986).

162. T. Fjeldberg, A. Haaland, M. F. Lappert, B. E. R. Schilling, R. Seip, and A. J. Thorne, *J. Chem., Soc. Chem. Commun.*, 1407 (1982); J.-P. Malrieu and G. Trinquier, *J. Am. Chem. Soc.* **111**, 5916 (1989); G. Trinquier and J.-P. Malrieu, *J. Am. Chem. Soc.* **109**, 5303 (1987); C. Liang and L. C. Allen, *J. Am. Chem. Soc.* **112**, 1039 (1990).

163. D. E. Goldberg, D. H. Harris, M. F. Lappert, and K. M. Thomas, *J. Chem. Soc., Chem. Commun.*, 261 (1976); P. J. Davidson, D. H. Harris, and M. F. Lappert, *J. Chem. Soc., Dalton Trans.*, 2268 (1976); J. D. Cotton, P. J. Davidson, and M. F. Lappert, *J. Chem. Soc., Dalton Trans.*, 2275 (1976); J. D. Cotton, P. J. Davidson, M. F. Lappert, J. D. Donaldson, and J. Silver, *J. Chem. Soc., Dalton Trans.*, 2286 (1976); P. B. Hitchcock, M. F. Lappert, S. J. Miles, and A. J. Thorne, *J. Chem. Soc., Chem. Commun.*, 480 (1984).

164. T. A. Albright, J. K. Burdett, and M.-H. Whangbo, "Orbital Interactions in Chemistry." Wiley (Interscience), New York, 1985; B. M. Gimarc, "Molecular Structure and Bonding." Academic Press, New York, 1979.

165. P. Seiler and J. D. Dunitz, *Acta Crystallogr. Sect. B: Struct. Crystallogr. Cryst. Chem.* **B35**, 1068 (1979); F. Takusagawa and T. F. Koetzle, *Acta Crystallogr. Sect. B: Struct. Crystallogr. Cryst. Chem.* **B35**, 1074 (1979); Y. T. Struchkov, V. G. Andrianov, T. N. Sal'nikova, I. R. Lyatifov, and R. B. Materikova, *J. Organomet. Chem.* **145**, 213 (1978); A. Almenningen, A. Haaland, S. Samdal, J. Brunvoll, J. L. Robbins, and J. C. Smart, *J. Organomet Chem.* **173**, 293 (1979); D. F. Freyberg, J. L. Robbins, K. N. Raymond, and J. C. Smart, *J. Am. Chem. Soc.* **101**, 892 (1979).

166. K. N. Brown, L. D. Field, P. A. Lay, C. M. Lindall, and A. F. Masters, *J. Chem. Soc., Chem. Commun.*, 408 (1990).

167. H. Werner, H. Otto, and H. J. Kraus, *J. Organomet. Chem.* **315**, C57 (1986).

168. S. Harvey, C. L. Raston, B. W. Skelton, A. H. White, M. F. Lappert, and G. Srivastava, *J. Organomet. Chem.* **328**, C1 (1987).

169. O. T. Beachley, Jr., J. F. Lees, T. E. Glassman, M. R. Churchill, and L. A. Buttrey, *Organometallics* **9**, 2488 (1990).

170. M. B. Freeman, L. G. Sneddon, and J. C. Huffman, *J. Am. Chem. Soc.* **99**, 5194 (1977).

171. H. Schumann, H. Kucht, A. Dietrich, and L. Esser, *Chem. Ber.* **123**, 1811 (1990).

172. O. T. Beachley, Jr., R. Blom, M. R. Churchill, K. Faegri, Jr., J. C. Fettinger, J. C. Pazik, and L. Victoriano, *Organometallics* **8**, 346 (1989); O. T. Beachley, Jr., M. R. Churchill, J. C. Fettinger, J. C. Pazik, and L. Victoriano, *J. Am. Chem. Soc.* **108**, 4666 (1986).

173. K. Stumpf, H. Pritzkow, and W. Siebert, *Angew. Chem.* **97**, 64 (1985); *Angew. Chem., Int. Ed. Engl.* **24**, 71 (1985).

174. E. G. Cox, D. W. J. Cruickshank, and J. A. S. Smith, *Proc. R. Soc. London, Ser. A* **247**, 1 (1958); G. J. Piermarini, A. D. Mighell, C. E. Weir, and S. Block, *Science* **165**, 1250 (1969).

175. T. Fjeldberg, A. Haaland, B. E. R. Schilling, H. V. Volden, M. F. Lappert, and A. J.

Thorne, *J. Organomet. Chem.* **280,** C43 (1985); T. Fjeldberg, A. Haaland, B. E. R. Schilling, M. F. Lappert, and A. J. Thorne, *J. Chem. Soc., Dalton Trans.,* 1551 (1986); G. Trinquier, J.-P. Malrieu, and P. Rivière, *J. Am. Chem. Soc.* **104,** 4529 (1982); S. Nagase and T. Kudo, *J. Mol. Struct. (THEOCHEM)* **103,** 35 (1983); M. J. S. Dewar, G. L. Grady, D. R. Kuhn, and K. M. Merz, Jr., *J. Am. Chem. Soc.* **106,** 6773 (1984).

176. C. Janiak and R. Hoffmann, *Angew. Chem.* **101,** 1706 (1989); *Angew. Chem., Int. Ed. Engl.* **28,** 1688 (1989).

177. C. Janiak and R. Hoffmann, *J. Am. Chem. Soc.* **112,** 5924 (1990).

178. P. H. M. Budzelaar and J. Boersma, *Recl. Trav. Chim. Pays-Bas* **109,** 187 (1990).

179a. J. Lorberth, S.-H. Shin, S. Wocadlo, and W. Massa, *Angew. Chem.* **101,** 793 (1989); *Angew. Chem., Int. Ed. Engl.* **28,** 735 (1989).

179b. G. Rabe, H. W. Roesky, D. Stalke, F. Pauer, and G. M. Sheldrick, *J. Organomet. Chem.* **403,** 11 (1991).

180. P. F. Rodesiler, T. Auel, and E. L. Amma, *J. Am. Chem. Soc.* **97,** 7405 (1975); M. S. Weininger, P. F. Rodesiler, and E. L. Amma, *Inorg. Chem.* **18,** 751 (1979); J. L. Lefferts, K. C. Molloy, M. B. Hossain, D. van der Helm, and J. J. Zuckerman, *Inorg. Chem.* **21,** 1410 (1982); T. Probst, O. Steigelmann, J. Riede, and H. Schmidbaur, *Angew. Chem.* **102,** 1471 (1990); *Angew. Chem., Int. Ed. Engl.* **29,** 1397 (1990).

181. H. Schmidbaur, W. Bublak, B. Huber, J. Hofmann, and G. Müller, *Chem. Ber.* **122,** 265 (1989); H. Schmidbaur, R. Haager, B. Huber, and G. Müller, *Angew. Chem.* **99,** 354 (1987); *Angew. Chem., Int. Ed. Engl.* **26,** 338 (1987); M. D. Noirot, O. P. Anderson, and S. H. Strauss, *Inorg. Chem.* **26,** 2216 (1987); H. Schmidbaur, *Angew. Chem.* **97,** 893 (1985); *Angew. Chem., Int. Ed. Engl.* **24,** 893 (1985), and references therein.

182. M. D. Rausch, W.-M. Tsai, J. W. Chambers, R. D. Rogers, and H. G. Alt, *Organometallics* **8,** 816 (1989).

183. J. J. Brooks, W. Rhine, and G. D. Stucky, *J. Am. Chem. Soc.* **94,** 7339 (1972); W. E. Rhine and G. D. Stucky, *J. Am. Chem. Soc.* **97,** 737 (1975).

184. E. Weiss and E. O. Fischer, *Z. Anorg. Allg. Chem.* **278,** 219 (1955); W. Bünder and E. Weiss, *J. Organomet. Chem.* **92,** 1 (1975).

185. A. Haaland, J. Lusztyk, J. Brunvoll, and K. Starowieyski, *J. Organomet. Chem.* **85,** 279 (1975).

186. P. Jutzi, E. Schlüter, M. B. Hursthouse, A. M. Arif, and R. L. Short, *J. Organomet. Chem.* **299,** 285 (1986).

187. A. J. Campbell, C. A. Fyfe, D. Harold-Smith, and K. R. Jeffrey, *Mol. Cryst. Liq. Cryst.* **36,** 1 (1976); F. van Meurs, J. M. vander Toorn, and H. van Bekkum, *J. Organomet. Chem.* **113,** 341 (1976).

188. H. H. Richmond and H. Freiser, *J. Am. Chem. Soc.* **77,** 2020 (1955).

189. N.-S. Chiu, L. Schäfer, and R. Seip, *J. Organomet. Chem.* **101,** 331 (1975).

190. M. J. Bennett, W. L. Hutcheon, and B. M. Foxman, *Acta Crystallogr. Sect. A: Cryst. Phys. Diffr. Theor. Gen. Crystallogr.* **A31,** 488 (1975).

191. T. A. Albright, P. Hofmann, and R. Hoffmann, *J. Am. Chem. Soc.* **99,** 7546 (1977).

192. B. E. Mann, C. M. Spencer, B. F. Taylor, and P. Yavari, *J. Chem. Soc., Dalton Trans.,* 2027 (1984); P. Deslisle, G. Allegra, E. R. Mgnaschi, and A. Chierico, *J. Chem. Soc. Faraday Trans. 2* **71,** 207 (1975); A. Haaland, *Top. Curr. Chem.* **53,** 1 (1975), and references therein; F. Rocquet, L. Berreby, and J. P. Marsault, *Spectrochim. Acta* **A29,** 1101 (1973); L. N. Mulay and A. Attalla, *J. Am. Chem. Soc.* **85,** 702 (1963); C. H. Holm and J. I. Ibers, *J. Chem. Phys.* **30,** 885 (1969); M. Rosenblum, "Chemistry of the Iron Group Metallocenes," Part 1, p. 45. Wiley (Interscience), New York, 1965.

193. D. F. R. Gilson and G. Gomez, *J. Organomet. Chem.* **240,** 41 (1982).

194. W. D. Luke and A. Streitwieser, Jr., *J. Am. Chem. Soc.* **103,** 3241 (1981).

195. J. Okuda, *J. Organomet. Chem.* **356**, C43 (1988).
196. J. Okuda and E. Herdtweck, *J. Organomet. Chem.* **373**, 99 (1989).
197. J. Okuda, *J. Organomet. Chem.* **385**, C39 (1990).
198. J. Okuda, *J. Organomet. Chem.* **367**, C1 (1989).
199. W. Hofmann, W. Buchner, and H. Werner, *Angew. Chem.* **89**, 836 (1977); *Angew. Chem., Int. Ed. Engl.* **16**, 795 (1977); H. Werner and W. Hofmann, *Chem. Ber.* **114**, 2681 (1981).
200a. J. Okuda, *J. Organomet. Chem.* **353**, C1 (1988).
200b. E. W. Abel, N. J. Long, K. G. Orrell, A. G. Osborne, and V. Šik, *J. Organomet. Chem.* **403**, 195 (1991).
201. R. Benn, H. Grondey, G. Erker, R. Aul, and R. Nolte, *Organometallics* **9**, 2493 (1990).
202. G. Erker, T. Mühlenbernd, R. Benn, A. Rufinska, Y.-H. Tsay, and C. Krüger, *Angew. Chem.* **97**, 336 (1985); *Angew. Chem., Int. Ed. Engl.* **24**, 321 (1985); G. Erker, T. Mühlenbernd, A. Rufinska, and R. Benn, *Chem. Ber.* **120**, 507 (1987).
203a. R. Benn, H. Grondey, R. Nolte, and G. Erker, *Organometallics* **7**, 777 (1988).
203b. R. Mynott, H. Lehmkuhl, E.-M. Kreuzer, and E. Joussen, *Angew. Chem.* **102**, 314 (1990); *Angew. Chem., Int. Ed. Engl.* **29**, 289 (1990).
204a. J. Okuda, *Chem. Ber.* **122**, 1075 (1989).
204b. C. H. Winter, D. A. Dobbs, and X. X. Zhou, *J. Organomet. Chem.* **403**, 145 (1991).
205. K. Möbius, *Z. Naturforsch.* **20A**, 1117 (1965).
206. A. Almenningen, O. Bastiansen, and P. N. Skancke, *Acta Chem. Scand.* **12**, 1215 (1958).
207. J. C. J. Bart, *Acta Crystallogr. Sect. B: Struct. Crystallogr. Cryst. Chem.* **24B**, 1277 (1968).
208. D. Gust, *J. Am. Chem. Soc.* **99**, 6980 (1977); D. Gust and A. Patton, *J. Am. Chem. Soc.* **100**, 8175 (1978); A. Patton, J. W. Dirks, and D. Gust, *J. Org. Chem.* **44**, 4749 (1979).
209. J. Haywood-Farmer and M. A. Battiste, *Chem. Ind. (London),* 1232 (1971).
210. R. Willem, H. Pepermans, C. Hoogzand, K. Hallenga, and M. Gielen, *J. Am. Chem. Soc.* **103**, 2297 (1981); R. Willem, A. Jans, C. Hoogzand, M. Gielen, G. Van Binst, and H. Pepermans, *J. Am. Chem. Soc.* **107**, 28 (1985).
211. H. Sitzmann, *Chem. Ber.* **123**, 2311 (1990).
212. J. Siegel, A. Gutiérrez, W. B. Schweizer, O. Ermer, and K. Mislow, *J. Am. Chem. Soc.* **108**, 1569 (1986).
213. I. I. Schuster, W. Weissensteiner, and K. Mislow, *J. Am. Chem. Soc.* **108**, 6661 (1986); K. Mislow, *Chimia* **40**, 395 (1986).
214. D. J. Iverson and K. Mislow, *Organometallics* **1**, 3 (1982).
215. D. J. Iverson, G. Hunter, J. F. Blount, J. R. Damewood, Jr., and K. Mislow, *J. Am. Chem. Soc.* **103**, 6073 (1981).
216. U. Berg, T. Liljefors, C. Roussel, and J. Sandström, *Acc. Chem. Res.* **18**, 80 (1985).
217. See, for example, T. Hayashi and M. Kumada, *Acc. Chem. Res.* **15**, 395 (1982).
218. M. J. Burk and R. L. Harlow, *Angew. Chem.* **102**, 1511 (1990); *Angew. Chem., Int. Ed. Engl.* **29**, 1462 (1990); J. K. Whitesell, *Chem. Rev.* **89**, 1581 (1989).
219. A. Schäfer, E. Karl, L. Zsolnai, G. Huttner, and H.-H. Brintzinger, *J. Organomet. Chem.* **328**, 87 (1987).
220. K. Schlögl, *Top. Stereochem.* **1**, 39 (1967).
221. P. Jutzi, *J. Organomet. Chem.* **400**, 1 (1990); P. Jutzi, J. Schnittger, and M. B. Hursthouse, *Chem. Ber.* in press.
222. P. Jutzi and J. Schnittger, *Chem. Ber.* **122**, 624 (1989).
223. S. Collins, B. A. Kuntz, N. J. Taylor, and D. G. Ward, *J. Organomet. Chem.* **342**, 21 (1988).

224. F. R. W. P. Wild, M. Wasiucionek, G. Huttner, and H. H. Brintzinger, *J. Organomet. Chem.* **288**, 63 (1985).
225. J. A. Ewen, L. Haspeslagh, J. L. Atwood, and H. Zhang, *J. Am. Chem. Soc.* **109**, 6544 (1987).
226. J. A. Ewen, *J. Am. Chem. Soc.* **106**, 6355 (1984).
227. W. Kaminsky, K. Külper, H. H. Brintzinger, and F. R. W. P. Wild, *Angew. Chem.* **97**, 507 (1985); *Angew. Chem., Int. Ed. Engl.* **24**, 507 (1985).
228. W. Spaleck, M. Antberg, V. Dolle, R. Klein, J. Rohrmann, and A. Winter, *New J. Chem.* **14**, 499 (1990).
229. G. Wulff, *Angew. Chem.* **101**, 22 (1989); *Angew. Chem., Int. Ed. Engl.* **28**, 21 (1989).
230. P. Pino, P. Cioni, and J. Wei, *J. Am. Chem. Soc.* **109**, 6189 (1987); P. Pino, M. Galimberti, P. Prada, and G. Consiglio, *Makromol. Chem.* **191**, 1677 (1990).
231. R. B. King, I. Haiduc, and C. W. Eavenson, *J. Am. Chem. Soc.* **95**, 2508 (1973).
232. C. Janiak, M. Schwichtenberg, and F. E. Hahn, *J. Organomet. Chem.* **365**, 37 (1989).
233. F. Mao, C. E. Philbin, T. J. R. Weakley, and D. R. Tyler, *Organometallics* **9**, 1510 (1990).
234. A. Fusi, E. Cesarotti, and R. Ugo, *J. Mol. Catal.* **10**, 213 (1981).
235. L. D. Field, T. W. Hambley, C. M. Lindall, and A. F. Masters, *Polyhedron* **8**, 2425 (1989).
236. N. G. Connelly and I. Manners, *J. Chem. Soc., Dalton Trans.*, 283 (1989).
237. P. Brégaint, J.-R. Hamon, and C. Lapinte, *J. Organomet. Chem.* **398**, C25 (1990).
238. D. W. Slocum, M. Matusz, A. Clearfield, R. Peascoe, and S. A. Duraj, *J. Macromol. Sci. Chem.* **A27**, 1405 (1990).
239. N. G. Connelly, A. C. Lyons, I. Manners, D. L. Mercer, K. E. Richardson, and P. H. Rieger, *J. Chem. Soc., Dalton Trans.*, 2451 (1990).
240. W. Kläui and L. Ramacher, *Angew. Chem.* **98**, 107 (1986); *Angew. Chem., Int. Ed. Engl.* **25**, 97 (1986).
241. N. G. Connelly and S. J. Raven, *J. Chem. Soc., Dalton Trans.*, 1613 (1986).
242. N. G. Connelly, S. J. Raven, and W. E. Geiger, *J. Chem. Soc., Dalton Trans.*, 467 (1987).
243. U. Behrens and F. Edelmann, *Z. Naturforsch. B: Anorg. Chem., Org. Chem.* **41B**, 1426 (1986).
244. H. Hoberg, R. Krause-Göing, C. Krüger, and J. C. Sekutowski, *Angew. Chem.* **89**, 179 (1977); *Angew. Chem., Int. Ed. Engl.* **16**, 183 (1977).
245. E. D. Jemmis and B. V. Prasad, *J. Organomet. Chem.* **347**, 401 (1988).
246. E. Ban, P.-T. Cheng, T. Jack, S. C. Nyburg, and J. Powell, *J. Chem. Soc., Chem. Commun.*, 368 (1973).
247. K. Broadley, N. G. Connelly, G. A. Lane, and W. E. Geiger, *J. Chem. Soc., Dalton Trans.*, 373 (1986).
248. G. A. Lane, W. E. Geiger, and N. G. Connelly, *J. Am. Chem. Soc.* **109**, 402 (1987).
249. R. Zhang and M. Tsutsui, *Youji Huaxue*, 435 (1982); *Chem. Abstr.* **98**, 198374q (1983).
250. G. A. Carriedo, J. A. K. Howard, D. B. Lewis, G. E. Lewis, and F. G. A. Stone, *J. Chem. Soc., Dalton Trans.*, 905 (1985).
251. A. Kucht, Ph.D. Thesis, Technische Universität Berlin (1991).
252. O. I. Trifonova, Y. F. Oprunenko, and Y. A. Ustynyuk, *Metalloorg. Khim.* **3**, 465 (1990); *Chem. Abstr.* **113**, 59425x (1990).
253. F. Mao, S. K. Sur, and D. R. Tyler, *J. Am. Chem. Soc.* **111**, 7627 (1989).
254. F. J. Feher, M. Green, and A. G. Orpen, *J. Chem. Soc., Chem. Commun.*, 291 (1986).

255. M. M. Kubicki, J. Y. Le Gall, R. Pichon, J. Y. Salaun, M. Cano, and J. A. Campo, *J. Organomet. Chem.* **348**, 349 (1988).
256. V. Weinmayr *J. Am. Chem. Soc.* **77**, 3012 (1955).
257. A. Nakamura and N. Hagihara, *Nippon Kagaku Zasshi (J. Chem. Soc. Jpn Pure Chem. Sect.)* **84**, 344 (1963); *Chem. Abstr.* **59**, 14021f (1963); A. Nakamura, *Mem. Inst. Sci. Ind. Res., Osaka Univ.* **19**, 81 (1962); *Chem. Abstr.* **59**, 8786f (1963).
258. D. Cashman and F. J. Lalor, *J. Organomet. Chem.* **32**, 351 (1971).
259. V. W. Day, B. R. Stults, K. J. Reimer, and A. Shaver, *J. Am. Chem. Soc.* **96**, 1227 (1974).
260. K. J. Reimer and A. Shaver, *J. Organomet. Chem.* **93**, 239 (1975).
261. E. G. Perevalova, K. I. Grandberg, V. P. Dyadchenko, and T. V. Baukova, *J. Organomet. Chem.* **217**, 403 (1981).
262. T. V. Baukova, Y. L. Slovokhotov, and Y. T. Struchkov, *J. Organomet. Chem.* **220**, 125 (1981).
263. E. Weiss and W. Hübel, *J. Inorg. Nucl. Chem.* **11**, 42 (1959).
264. N. A. Bailey, V. S. Jassal, R. Vefghi, and C. White, *J. Chem. Soc., Dalton Trans.*, 2815 (1987).
265. Y. Shvo and D. Czarkie, *J. Organomet. Chem.* **368**, 375 (1989); Y. Shvo and D. Czarkie, *J. Organomet. Chem.* **315**, C25 (1986).
266. Y. Shvo, D. Czarkie, Y. Rahamim, and D. F. Chodosh, *J. Am. Chem. Soc.* **108**, 7400 (1986).
267. J. E. Sheats, W. Miller, M. D. Rausch, S. A. Gardner, P. S. Andrews, and F. A. Higbie, *J. Organomet. Chem.* **96**, 115 (1975).
268. J. E. Sheats, G. Hlatky, and R. S. Dickson, *J. Organomet. Chem.* **173**, 107 (1979).
269. R. D. Gorsich, *J. Organomet. Chem.* **5**, 105 (1966).
270. O. Koch, F. Edelmann, and U. Behrens, *Chem. Ber.* **115**, 1313 (1982).
271. W. A. Herrmann and M. Huber, *Chem. Ber.* **111**, 3124 (1978).
272. T. V. Baukova, Y. L. Slovokhotov, and Y. T. Struchkov, *J. Organomet. Chem.* **221**, 375 (1981).
273. R. T. Lundquist and M. Cais, *J. Org. Chem.* **27**, 1167 (1962).
274. Z. L. Lutsenko, G. G. Aleksandrov, P. V. Petrovskii, E. S. Shubina, V. G. Andrianov, Y. T. Struchkov, and A. Z. Rubezhov, *J. Organomet. Chem.* **281**, 349 (1985).
275. J. W. Egan, Jr., R. P. Hughes, and A. L. Rheingold, *Organometallics* **6**, 1578 (1987).
276. H. Schumann, F. H. Görlitz, and A. Dietrich, *Chem. Ber.* **122**, 1423 (1989).
277. P. Jutzi, B. Hampel, M. B. Hursthouse, and A. J. Howes, *Organometallics* **5**, 1944 (1986).
278. P. Jutzi and E. Schlüter, *J. Organomet. Chem.* **253**, 313 (1983).
279. J. Okuda, *Chem. Ber.* **123**, 87 (1990).
280. J. Okuda, *J. Organomet. Chem.* **397**, C37 (1990).
281. C. H. Winter, J. W. Kampf, and X.-X. Zhou, *Acta Crystallogr. Sect. C: Cryst. Struct. Commun.* **C46**, 1231 (1990).
282. M. S. Miftakhov and G. A. Tolstikov, *Zh. Obshch. Khim.* **46**, 930 (1976).
283. J. Okuda, *J. Organomet. Chem.* **375**, C13 (1989).
284. J. Okuda, *Chem. Ber.* **122**, 1259 (1989).
285. M. A. Edelman, M. F. Lappert, J. L. Atwood, and H. Zhang, *Inorg. Chim. Acta* **139**, 185 (1987).
286. H. Sitzmann, Abstracts of the 24th International Conference on Organometallic Chemistry, p. 71. Detroit, Michigan, 1990.
287. B. T. Donovan, R. P. Hughes, and H. A. Trujillo, *J. Am. Chem. Soc.* **112**, 7076 (1990).

288. T. P. Hanusa, *Polyhedron* **9**, 1345 (1990).
289. B. V. Lokshin, S. G. Kazaryan, and A. G. Ginzburg, *Izv. Akad. Nauk SSSR, Ser. Khim.*, 562 (1988).
290. A. G. Ginzburg, P. V. Petrovskii, V. N. Setkina, and D. N. Kursanov, *Izv. Akad. Nauk SSSR, Ser. Khim.*, 186 (1985); see also B. V. Lokshin, E. B. Nazarova, and A. G. Ginzburg, *J. Organomet. Chem.* **129**, 379 (1977).
291. K. Schlögl and M. Peterlik, *Tetrahedron Lett.* **13**, 573 (1962).
292. R. D. Cantrell and P. B. Shevlin, *J. Am. Chem. Soc.* **111**, 2348 (1989).
293. A. L. Rheingold, G. Wu, and R. F. Heck, *Inorg. Chim. Acta* **131**, 147 (1987); G. Wu, A. L. Rheingold, and R. F. Heck, *Organometallics* **5**, 1922 (1986).
294. S. A. MacLaughlin, R. C. Murray, J. C. Dewan, and R. R. Schrock, *Organometallics* **4**, 796 (1985).
295. A. N. Nesmeyanov, D. N. Kursanov, V. N. Setkina, N. V. Kislyakova, N. S. Kolobova, I. B. Zlotina, and K. N. Anisimov, *Izv. Akad. Nauk SSSR, Ser. Khim.*, 30 (1967).
296. V. P. Tverdokhlebov, B. V. Polyakov, I. V. Tselinskii, and L. I. Golubeva, *Zh. Obshch. Khim.* **52**, 2032 (1982).
297. E. Sappa, A. Tiripicchio, and A. M. M. Lanfredi, *J. Chem. Soc., Dalton Trans.*, 552 (1978).
298. K. Sünkel, A. Blum, K. Polborn, and E. Lippmann, *Chem. Ber.* **123**, 1227 (1990).
299. M. I. Bruce, J. K. Walton, M. L. Williams, B. W. Skelton, and A. H. White, *J. Organomet. Chem.* **212**, C35 (1981).
300. M. I. Bruce, J. K. Walton, M. L. Williams, S. R. Hall, B. W. Skelton, and A. H. White, *J. Chem. Soc., Dalton Trans.*, 2209 (1982).
301. M. I. Bruce, J. K. Walton, B. W. Skelton, and A. H. White, *J. Chem. Soc., Dalton Trans.*, 2221 (1982).
302. P. Jutzi, F.-X. Kohl, E. Schlüter, M. B. Hursthouse, and N. P. C. Walker, *J. Organomet. Chem.* **271**, 393 (1984).
303. F. X. Kohl, E. Schlüter, and P. Jutzi, *J. Organomet. Chem.* **243**, C37 (1983).
304. M. I. Bruce, J. K. Walton, B. W. Skelton, and A. H. White, *J. Chem. Soc., Dalton Trans.*, 2227 (1982).
305. M. I. Bruce, B. W. Skelton, R. C. Wallis, J. K. Walton, A. H. White, and M. L. Williams, *J. Chem. Soc., Chem. Commun.*, 428 (1981).
306. M. I. Bruce, J. K. Walton, M. L. Williams, J. M. Patrick, B. W. Skelton, and A. H. White, *J. Chem. Soc., Dalton Trans.*, 815 (1983).
307. M. I. Bruce, P. A. Humphrey, B. W. Skelton, and A. H. White, *J. Organomet. Chem.* **361**, 369 (1989).
308. M. I. Bruce, P. A. Humphrey, J. M. Patrick, B. W. Skelton, A. H. White, and M. L. Williams, *Aust. J. Chem.* **38**, 1441 (1985).
309. P. Schmidt, R. W. Hoffmann, and J. Backes, *Angew. Chem.* **84**, 534 (1972); *Angew. Chem., Int. Ed. Engl.* **11**, 513 (1972); R. W. Hoffmann, P. Schmidt, and J. Backes, *Chem. Ber.* **109**, 1918 (1976).
310. M. I. Bruce, M. L. Williams, B. W. Skelton, and A. H. White, *J. Chem. Soc., Dalton Trans.*, 799 (1983).
311. C. Qiu and Z. Zhou, *Huaxue Xuebao* **44**, 1058 (1986); *Chem. Abstr.* **107**, 59165p (1987).
312. C. Arsenault, P. Bougeard, B. G. Sayer, S. Yeroushalmi, and M. J. McGlinchey, *J. Organomet. Chem.* **265**, 283 (1984).
313. M. I. Bruce, R. C. Wallis, M. L. Williams, B. W. Skelton, and A. H. White, *J. Chem. Soc., Dalton Trans.*, 2183 (1983).

314. M. I. Bruce, J. K. Walton, B. W. Skelton, and A. H. White, *J. Chem. Soc., Dalton Trans.,* 809 (1983).
315. F. Edelmann and U. Behrens, *J. Organomet. Chem.* **134,** 31 (1977).
316. E. Sappa, M. L. Nanni-Marchino, and V. Raverdino, *Ann. Chim. (Rome)* **68,** 349 (1978).
317. G. Moran, M. Green, and A. G. Orpen, *J. Organomet. Chem.* **250,** C15 (1983).

Index

Cumulative List of Contributors